AGRONOMY

AGRONOMY

Editor

Elias Kerner

scitus
academics

Agronomy

Edited by **Elias Kerner**

This Book is Printed on Acid-Free Paper. ♾

ISBN: 9781681179070

Library of Congress Control Number: 2017948935

SCITUS Academics LLC,

616, Corporate Way, Suite 2, 4766,

Valley Cottage, NY 10989

www.scitusacademics.com

Notice

Printed in United States of America.

Preface

In recent times, agronomy has assumed newer dimensions and can be defined as a branch of agricultural science that deals with methods which provide a favorable environment to the crop for higher productivity. It considers the whole range of influences on crop production, including climate and adaptation, soil, water and water availability, crop genetics, the properties of the soil and how the soil interacts with the growing crop; what nutrients the crop needs; the ways that crops grow and develop; and how best to control weeds, insects, fungi, and other crop pests. Non-biological considerations such as economic requirements and consumer and farmer behavior also inform agronomic practices, as do environmental constraints. The emphasis of agronomy is now more towards the scientific study of the behavior of plant under the different environmental conditions like varing soils and climate, irrigation, fertilization etc. by conducting well laid out experiments in the fields, pots & laboratories. It is also involves application of research in the field or forming suitable packages of practices under a given set of conditions. There is a considerable amount of literature showing that an effective agro-ecosystem management, improving nutrient use efficiency, is a crucial challenge for a more sustainable production of horticultural, industrial and cereal crops. As a matter of fact, as agronomic efficiency (i.e., nutrients recovered within the soil-crop system) improves, economic (farm income is maximized from proper use of nutrient inputs) and environmental efficiency (reduced risk of nutrient losses) will benefit.

This volume 'Agronomy' collects current research findings dealing with different aspects of the understanding of agronomy focusing on crop production and management, as well as crop science and physiology. It also focuses on crop disease and protection, agroclimatology and soil science plant-soil relationships, crop quality and post-harvest physiology, cropping of sustainable development, farming and cropping systems, agro-ecosystems and the environment demonstrating new methods and technologies. This book shows a wide variety of functions that can be grouped according to their economic dimension, social dimension, and environmental dimension. The book is a valuable addition to the existing knowledge and is especially intended for university students and all professionals in the field of agriculture.

Table of Contents

CHAPTER 1

Climate Change and Sugarcane Production: Potential Impact and Mitigation Strategies

Duli Zhao[1] and Yang-Rui Li[2]

[1]USDA-ARS, Sugarcane Field Station, Canal Point, FL 33438, USA

[2]Sugarcane Research Center, Chinese Academy of Agricultural Science, Guangxi Academy of Agricultural Sciences, Nanning 530007, China

ABSTRACT

Sugarcane (Saccharum officinarum L.) is an important crop for sugar and bioenergy worldwide. The increasing greenhouse gas emission and global warming during climate change result in the increased frequency and intensity of extreme weather events. Climate change is expected to have important consequences for sugarcane production in the world, especially in the developing countries because of relatively low adaptive capacity, high vulnerability to natural hazards, and poor forecasting systems and mitigating strategies. Sugarcane production may have been negatively affected and will continue to be considerably affected by increases in the frequency and intensity of extreme environmental conditions due to climate change. The degree of climate change impact on sugarcane is associated with geographic location and adaptive capacity. In this paper, we briefly reviewed sugarcane response to climate change events, sugarcane production in several different countries, and challenges for sugarcane production in climate change in order for us to better understand effects of climate change on sugarcane production and to propose strategies for mitigating the negative impacts of climate change and improving sugarcane production sustainability and profitability.

1. INTRODUCTION

A combination of long-term change in the weather patterns worldwide (i.e., global climate change), caused by natural processes and anthropogenic factors, may result in major environmental issues that have affected and will continuously affect agriculture. Atmospheric CO_2 concentration ([CO_2]) has increased by about 30% since the mid-18th century due to increases in combustion of fossil fuels, industrial processes, and deforestation [1]. Projections indicate that atmospheric [CO_2] would increase to about 550 ppm in a low emission scenario or could double (800 ppm) from current levels in a high emission scenario by the end of the 21st century. Global warming is directly associated with increasing atmospheric [CO_2] and other greenhouse gases (GHG). Global surface mean temperatures had increased from 0.55 to 0.67°C in the last century and are project to rise from 1.1 to 2.9°C (low emission) or 2.0 to 5.4°C (high emission) by 2100 relative to 1980–1999, depending on GHG emission level, region, and geographic location [2].

Increases in atmospheric [CO_2] and air temperature can be beneficial for some crops (especially C_3 plants) in some places [3, 4]. Climate variability and climate change are projected to result in changes in sea levels, rainfall pattern, and the frequency of extreme high- and low-temperature events, floods, droughts, and other abiotic stresses [5, 6] as well as tornados and hurricanes [7]. High temperatures accompanied by drought stress have been two of the major issues influencing agricultural production and economic impacts in many regions of the world. The challenges, faced by the agricultural sector under the climate change scenarios, are to provide food security for an increasing world population while protecting the environment and the functioning of its ecosystems [8]. For most countries that are highly dependent on rainfall with limited or no proper irrigation conditions and/or that have poor mitigation systems, these challenges may be amplified [9].

Agriculture is vulnerable to climate change through the direct effects of changing climate conditions (e.g., changes in temperature and/or precipitation), as well as through the indirect effects arising from changes in the severity of pest pressures, availability of pollination services, and performance of other ecosystem services that affect agricultural productivity. Reduction of crop productivity is universally predicted in most status reports on effects of climate change [10]. Climate change poses unprecedented challenges to agriculture because of the sensitivity of agricultural productivity and costs of improving growth environmental conditions. Adaptive action offers the potential to manage the effects of climate change by altering patterns of agricultural activity to capitalize on emerging opportunities while minimizing the costs associated with negative effects.

2. SUGARCANE RESPONSE TO CLIMATE CHANGE EVENTS

Sugarcane is an important industrial crop used for sugar and bioenergy. It is one of the world's major C_4 crops that mainly grow in the tropic and subtropic regions. Weather and climate related events (i.e., growth environment of atmospheric [CO_2], temperature, precipitation, and other extreme weather) are the key factors for sugarcane production worldwide, especially in many developing countries. The potential negative impact of climate change, especially temperature and rainfall, on sugarcane production in Zimbabwe has been reviewed by Chandiposha [15]. The sugarcane and sugar yields have fluctuated with extreme climate events (drought and tropical cyclones) [7]. A record sugar production (516,529 tonnes) in Fiji in 1994 was recorded because of favorable weather, but sugar productions in 1997, 1998, and 2003 were 47, 50, and 43%, respectively, lower than that in 1994 due to drought environment [7].

By using crop simulation models, Marin at al. [16] reported that climate change improved sugarcane water use efficiency and cane yield in some areas of Brazil. They predicted that cane yield in 2050 could be 15–59% higher than that at the current average level. Studies have also indicated that elevated [CO_2] under controlled environment increased sugarcane photosynthesis, water use efficiency, biomass, and productivity [17–19]. Improved water use efficiency of sugarcane under elevated [CO_2] is mainly associated with the reduced stomatal conductance [17, 18]. Although these findings from the controlled environment are important for better understanding of physiological mechanisms of sugarcane plant response to elevated [CO_2], they may not completely reveal the interactions of [CO_2] and other climate factors under field conditions. The most significantly positive effect would be on reduced incidence of frost, which is a major limitation on production [20] in most regions, such as Louisiana of USA, where growing season is short. When realizing these benefits, however, we have to take serious consideration for long-term negative impact on nutrient levels, soil moisture, water availability, and other conditions. A negative effect of increased temperature may occur in the tropical regions where cool winters are required to slow plant growth and increase sucrose storage. Probably the most dramatic effect of climate change on sugarcane production in Australia would be from the increase in sea level [21]. A significant proportion of sugarcane is grown along coastal areas. Any increase in sea level would make these areas difficult to farm and a large increase in sea level would require large areas to be abandoned [20]. The same is true in South Florida, USA.

High temperature due to climate change in Northeastern Brazil will increase the evapotranspiration rates reducing the amount of water available in soils, making the planting of sugarcane increasingly difficult [22] and considerably increasing irrigation demand [23]. Knox et al. [24] assessed climate change impacts on sugarcane production in Swaziland using crop modeling and found a decreasing trend for future projections for cane yield unless irrigation was included in the model because of high demand of irrigation. In the South Caribbean, sugarcane yield may decrease by 20–40% under a doubled [CO_2] climate change scenario based on outputs of a crop model [25]. The decreased yield was mainly attributed to increased water deficit stress caused by the warmer climate. Although increases in atmospheric [CO_2] and air temperature may benefit sugarcane growth and biomass accumulation in some regions of the world based on studies in pots [17, 18] and under controlled conditions [19] or based on crop modeling prediction [16, 26], sugarcane production is highly vulnerable to climate change due to increased frequency and intensity of the extreme weather events, such as drought, heat, flooding, typhoon, and frost [7, 15, 24, 27–29].

Effects of drought due to climate change on sugarcane growth and development depend on plant growth stage, the degree of water deficit stress, and duration of the stress. In general, drought in early and mid growth stages mainly reduces cane yield leading to low sucrose yield. Moderate drought in late growth stage can improve sucrose content in stalks. Drought is the most important stress factor for sugarcane production in China, a country ranked the top third in sugarcane production in the world, because more than 80% of sugarcane grows under rainfed conditions [30]. Drought in 2003/2004 in Guangxi, China, resulted in an 18% decrease in cane yield [27]. Sugarcane production was recorded high in 2007/08 in Guangxi because of the ideal distribution of rainfall and other favorable growth environment conditions. Cane yield, cane production, and sugar production were 83.8t ha^{-1}, 77.1Mt, and 9.41Mt, respectively. However, a long duration of extremely low temperature and rainy weather in the region from January to February 2008 and the freeze temperatures caused severe damage of most sugarcane. In December 2009, the extreme freeze temperatures (−4 to −6°C) occurred again and drought in the 2010 early growing season (January–June) accompanied with severe freeze temperature in December resulted in considerable reduction in sugarcane production. Cane yield, cane production, and sugar production in 2010/11 dropped to 56.3t ha^{-1}, 55.7Mt, and 6.75Mt, respectively [30, 31]. Similarly, drought conditions in 1983, 1997, 1998, and 2003 and the 1997 tropic cyclones in Fuji led to big decline (16–46%) in sugarcane production, compared to production in record years [7, 13, 14]. Water logging is also a widespread phenomenon that drastically reduces the growth and survival of sugarcane and the water logging stress led to 18–64% reduction in cane yield [32], depending on duration of water logging, plant growth stage, and cultivars [32, 33].

A shift in temperature due to climate change will have an effect on some of diseases, insects, and weeds in sugarcane production [15]. For example, Matthieson [34] reported that the incidence of smut disease [caused by Sporisorium scitamineum (Syd.)] is likely to increase due to high temperatures. The prolific dry weather exacerbates the symptoms of ratoon stunting disease. It is difficult to predict the effect of climate change on sugarcane rust diseases, but severe storms and hurricane can spread leaf scald, caused by Xanthomonas albilineans [35]. The more extreme weather events due to climate change have caused more overwintering pests (weeds and insects), more disease pathogens, and more input costs for reducing these risks to maintain a certain level of sugarcane production. For instance, sugarcane leaf brown rust (caused by Puccinia melanocephala Syd. & P. Syd.) and orange rust [caused by P. kuehnii (W. Krüger) E. J. Butler] diseases, especially orange rust, are big challenges for sugarcane production in Florida, USA [36–38]. Severity of rusts is associated with winter temperature and relative humidity in the region. Sugarcane orange rust in 2012 and 2013 in South Florida was the most severe since it was first found in 2007 [39] due to favorable climate conditions of warmer winter and high humidity for the rust spores surviving and fast development [38]. Growers used fungicides to control the negative effects of rusts on yields, but the cost of three split applications of fungicides (at a hectarage level) during a growing season was equivalent to 3 tonnes (Mg) of cane yield lost per hectare. The economic impact just for controlling orange rust in South Florida was approximately $63 million in 2013 based on the estimate of the Florida sugarcane industry.

The adaptation of farming systems to climate change in sugarcane production requires taking advantage of the potential benefits and minimizing potential adverse impacts on crop production. Therefore, a better understanding of the functions of these climate/weather factors and their impacts on sugarcane production can help manipulate plants to meet human needs and formulate adaptation or mitigation strategies. In the following parts of this paper, we attempt to briefly review sugarcane production in several different countries, such as Brazil, India, China, Thailand, Pakistan, and USA (developing and developed countries), to better understand effects of climate change on sugarcane production and to propose strategies for mitigating the negative impacts of climate change and improving sugarcane production sustainability and profitability.

3. SUGARCANE PRODUCTION IN TOP 10 COUNTRIES

The top 10 sugarcane production countries in the world in 2013 were Brazil, India, China, Thailand, Pakistan, Mexico, Colombia, Indonesia, Philippines, and USA and their cane productions (million Mg of cane) accounted for 34.1, 15.8, 5.8, 4.6, 2.9, 2.8, 1.6, 1.6, 1.5, and 1.3% (a total of 72%) of the world total cane production, respectively (Table 1). Cane yields (Mg ha^{-1}) in these countries ranked 29th, 40th, 39th, 26th, 51st, 25th, 19th, 31st, 37th, and 27th, respectively, in the 103 sugarcane production countries [13, 14]. In last 41 years, sugarcane production was linearly increased with years from 1973 to 2013 in all the top seven sugarcane production countries. Both sugarcane area and cane yield contributed to the increases in cane production, but increased area was a dominant contributor compared with cane yield except for Pakistan where increases in sugarcane hectarage and cane yield had similar proportion. Hectarage in Brazil, India, China, Thailand, Pakistan, Mexico, and Colombia increased by 500, 94, 237, 286, 57, 52, and 61%, respectively, and cane yields increased by 60, 38, 59, 70, 58, 11, and 24%, respectively, in last 41 years (1973–2013) based on linear regression. In the same period of years, sugarcane hectarage in USA increased only 31% and yield had no big change or slightly decreased (7.0%) (Table 2).

Table 1: The world top 10 sugarcane production countries in 2013 for their cane production, hectarage, and cane yield as well as their ranks in 103 sugarcane production countries.

Country	Production		Area		Yield	
	Million Mg	Rank	×1000 ha	Rank	Mg ha^{-1}	Rank
Brazil	739.27	1	9835.2	1	75.17	29
India	341.20	2	5060.0	2	67.43	40
China	126.14	3	1827.3	3	69.03	39
Thailand	100.10	4	1321.6	4	75.74	26
Pakistan	63.75	5	1128.8	5	56.48	51
Mexico	61.18	6	782.8	6	78.16	25
Colombia	34.88	7	405.7	9	85.95	19
Indonesia	33.70	8	450.0	7	74.89	31
Philippines	32.00	9	435.4	8	73.49	37
USA	27.91	10	368.6	11	75.71	27
World total	2165.23		26522.7		81.64	

Table 2: Maximum, minimum, and mean cane yields and coefficient of variation (CV) across last 41 years (1973–2013) for the world top 10 sugarcane production countries. The slope and values of linear regression cane yield and year for each country are listed in the table.

Country	Maximum (Mg ha^{-1})	Minimum (Mg ha^{-1})	Mean (Mg ha^{-1})	CV (%)	Slope (Mg ha^{-1} yr^{-1})	r^2
Brazil	80.26	46.48	64.92	14.33	0.75	0.93
India	76.53	49.11	62.41	11.48	0.49	0.68
China	74.93	39.18	59.16	15.98	0.67	0.73
Thailand	76.20	30.14	53.93	20.42	0.70	0.58
Pakistan	57.23	31.57	44.19	15.48	0.50	0.76
Mexico	78.16	62.68	71.11	6.00	0.22	0.40
Colombia	101.81	57.23	84.87	10.68	0.45	0.36
Indonesia	149.02	55.17	84.08	26.57	−1.60	0.73
Philippines	96.52	58.59	74.27	11.94	0.25	0.12
USA	89.98	69.90	78.99	5.71	−0.14	0.14
World total	71.77	53.76	62.49	8.68	0.45	0.95

† Data source: FAO of the United Nations (FAOSTAT) and Factfish [13, 14].

Additionally, cane yield was lower and the yield variation (CV) across years was much greater in most developing countries than that in USA. Averaged across 41 years from 1973 to 2013, mean cane yields in Brazil, India, China, Thailand, and Pakistan were 17.8, 21.0, 25.1, 31.7, and 44.1% lower, respectively, than that in USA (Table 2). Coefficient of variation (CV) values for cane yields across years in these five countries ranged from 11.5 to 20.4% compared to a CV value of 5.7% in USA (Table 2). When plotting sugarcane hectarage and yield against year, neither hectarage nor yield in the top five sugarcane production countries leveled off and the slope (indicating cane yield increasing rate) of the linear regression in Table 2 ranged from 0.49 (India) to 0.75 (Brazil) Mg $ha^{-1} yr^{-1}$. Although impact of climate change on sugarcane production depends on geographic location and on degree of adaptation, cane yields in most developing countries still tend to increase by improved cultivars and management practices. Therefore, increases in both sugarcane area and cane yield are still feasible in these countries in current environment. To consider increasing population and land limitation, improving sugarcane yields in future is more important compared to hectarage for sugarcane production, especially in most developing countries.

4. CHALLENGES FOR SUGARCANE PRODUCTION

In general, great variation in sugarcane yields exists in most developing countries across years (Table 2) and regions with varying rainfall and temperature due to low adaptive capacity, high vulnerability to natural hazards, and poor forecasting system and mitigating strategies [7]. High inputs and high costs of the production and low cane price are also very common in these developing countries, which results in low profits for sugarcane growers. For instance, sugarcane growers in major production areas (Guangxi, Yunnan, Guangdong, and Hainan) in China have planted some more profitable crops because of the financial considerations [11]. Sugarcane hectarage in Guangxi, the largest cane producing province, is expected to drop 6% in 2014/15 as farmers grow the low-labor input and fast-growing tree species for industrial use according to the Provincial Sugar Industry Bureau. Cane hectarage in Hainan is estimated to decline 11% in 2014/15 due to low profits according to provincial statistics. In addition to low prices, high labor costs have also contributed to a major part of low profitability. More than half of sugarcane hectarage is located in hilly areas where mechanized operation is unavailable and the use of hand labor for planting, field management, and harvesting considerably increases the input of labors. As the cost of labor continues to rise ($20/Mg cane), which accounts for approximately 27% of cane price ($71/Mg) in 2013/14, grower's profit from sugarcane was impacted considerably (Figure 1).

Therefore, low prices of cane and high labor costs caused a great drop in net income for growers in 2013/14 [11]. Reducing production costs by introduction and development of creative technologies and expanding use of sugarcane products not only for sugar but also for ethanol, cellulosic biofuel, and other coproducts will improve profits under the current and future climate conditions.

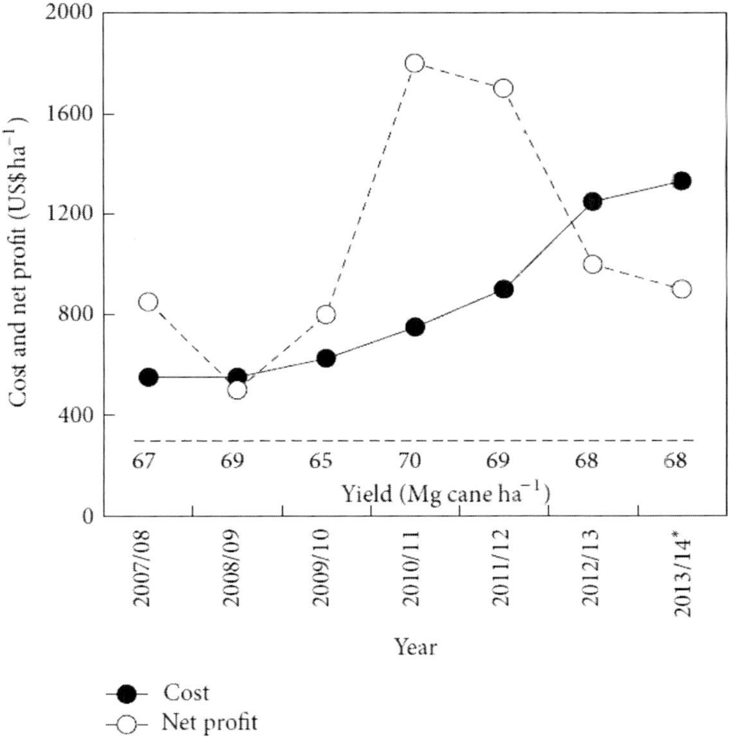

Figure 1: Average labor cost, net profit, and cane yield in major sugarcane production regions in China from 2007 to 2014. based on postestimate. Data are adopted from Anderson-Sprecher and Jiang [11].

When assessing agriculture and crop production systems as well as climate change and its negative impact on crop production, many economic, environmental, and social issues have to be thoroughly considered, such as how to (1) balance short-term and long-term goals; (2) increase productivity, profitability, and sustainability; (3) introduce new technologies and transfer them to growers; (4) meet environmental regulations; (5) deal with contradictions between climate change and crop production; and (6) balance competition of food and energy in resources. Certainly, sugarcane production systems are also challenged by these specific issues.

5. MITIGATING IMPACT OF STRESS ENVIRONMENT AND SUSTAINING SUGARCANE PRODUCTION

Although climate change increases the frequency and intensity of extreme weather events and uncertainty and vulnerability of adverse effects on agriculture [2, 6, 7, 9, 26, 29], the world sugarcane production was tripled in last 41 years [13, 14] because of increase in demand. The increased cane production was associated with increases in both hectarage and cane yield in most developing countries as described above. Much more efforts are needed to focus on increasing yield and improving profits under the current conditions and in the future climate change. Sugarcane yield relies on crop varieties (genotypes), biotic and abiotic growth environments (i.e., insects, diseases, weeds, and other climate related factors), and management practices (Figure 2).

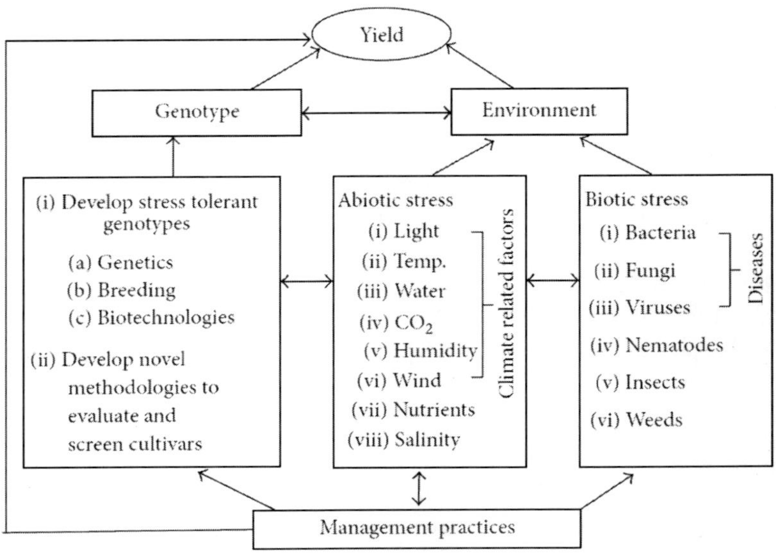

Figure 2: A flow chart to demonstrate major factors (genotype, environment, and management practices) influencing sugarcane yield as well as their interactions.

Some mitigation and adaptation strategies for climate change in sugarcane production in Zimbabwe have recently been proposed [15] and these mitigation strategies included planting drought tolerant varieties, investing irrigation infrastructure, improving irrigation efficiency and drainage systems, and improving cultural and management practices. Based on long-term data collected in South Africa, Deressa et al. [40] suggested that adaptation strategies should focus special attention on technologies and management regimes that will enhance sugarcane tolerance to warmer temperatures during winter and especially the harvesting phases. Thus, development of the stress tolerant and high-yielding sugarcane cultivars is one of the important strategies in adaptation of climate change (Figure 2). Sugarcane breeders and other scientists can develop computer data base to design hybridization (within or between species) for special requirement in the breeding programs, use growth and physiological traits to screen elite clones for resistance/tolerance to biotic and abiotic stresses [41], and use tissue culture, molecular biology, and gene transformation technologies to improve breeding and selection efficiencies. Studies have shown that some genotypes/cultivars are better than others in tolerance to water deficit [41–43] and low temperature [13] stresses, in radiation use efficiency [44], and in nutrient use efficiency [45, 46].

Using 33-year data of sugarcane yields in Florida to estimate the contribution of a breeding program to sugarcane production, Edmé et al. [47] found that sucrose content, cane tonnage, and sugar yield of the Florida commercial sugarcane cultivars linearly increased by 26.0, 15.5, and 47.0%, respectively, from 1968 to 2000. They found that the increases in yield components mainly occurred on the Florida organic soils. Underscoring the critical need for cultivar development for the Florida sugarcane industry, about 69% of the sugar yield gain came from genetic improvement attributable to the Canal Point (CP) cultivar development program. Recently, we planted 12 CP-sugarcane cultivars/genotypes that have a wide range of released years (from 1980 to 2013) on sand soils at two locations in Florida in 2011. The 3-year results of this study indicated that sucrose yield linearly and positively related to the cultivar-released year (). The increased sucrose yield on the Florida sand soils for the latest released cultivars was mainly associated with cane tonnage () rather than commercial recoverable sucrose () (unpublished data, Figure 3). Based on pot and field studies with intensive measurements of physiological, growth, and yield traits, we also found that some sugarcane genotypes are more tolerant to stress environment than others [43, 46, 48]. Therefore, development of new sugarcane cultivars that can contribute to adaptation to climate change (especially for elevated CO_2 and temperature) by discovering and introducing desirable genes for agronomic trait development [49] and using basic breeding [50], physiological screening [41, 43], and new technologies of molecular biology [51] can mitigate the negative effect of climate change and improve sugarcane yields, productivity, and sustainability.

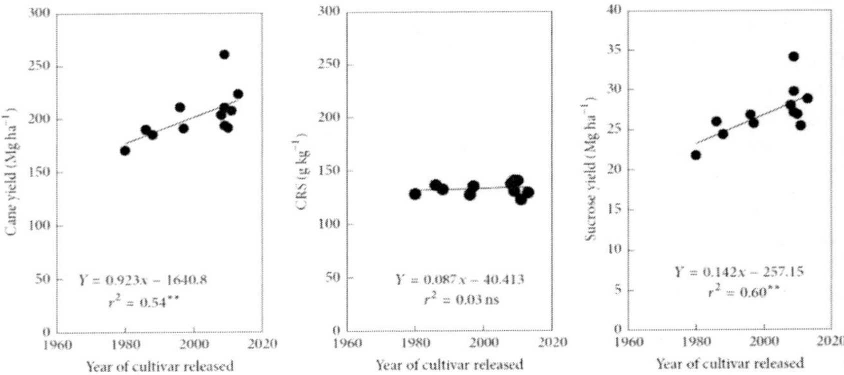

Figure 3: Trends of cane yield, commercial recoverable sucrose (CRS), and sucrose yield for 12 sugarcane cultivars or genotypes grown on sand soils at two locations in Florida. These cultivars had a wide range of released years from 1980 to 2013.

Using technologies of molecular biology and gene transformation to develop genetically modified (GM) sugarcane varieties [52–54], such as herbicide glyphosate resistance, drought tolerance, high sugar content, and disease resistance, may be one of the important ways to mitigate negative impacts of environmental stresses due to climate change. To address some of the potential concerns regarding safety of the GM sugarcane products, Joyce et al. [54] compared the GM sugarcane with non-GM control for product quality. They found that sugar crystallized from GM sugarcane plants did not contain residual DNA or proteins of introduced transgenes. The finding will improve the public perceptions surrounding GM sugar and its potential future incorporation within commercial sugarcane production.

Diversity of cropping systems, crops, and cultivars within a crop is also important for mitigating negative effect of climate change, biotic and abiotic stresses, or other uncertain extreme climate events because there are considerable differences among plant species, cultivars, and cropping systems in tolerance to stresses. Sugarcane cultivar Q124 in Queensland, Australia, in 2000 accounted for 45% of the crop, but a new race of orange rust pathogen devastated this high-performing cultivar and caused the industry Aus$ 150–210 million in yield losses [55, 56]. In a region, therefore, sugarcane variety diversity is also imperative for reducing risk of extreme climate factors, for mitigating negative effects of stress environment, and for improving sustainability of sugarcane production. Sugarcane cultivars with a wide range of maturity can buffer the harvest time and reduce the pressure of labor shortage and milling capacity, for instance, a total of 172,100 ha of sugarcane in the 2012-2013 harvest season in Florida with 12 major cultivars [12]. The fractions of these cultivars in Florida are listed in Figure 4. It is suggested that each of the leading sugarcane cultivars in a region may not be over 25% of total hectarage in order to mitigate negative effects of the extreme climate events on sugarcane production and to reduce risk of yield losses by some unexpected insects and diseases.

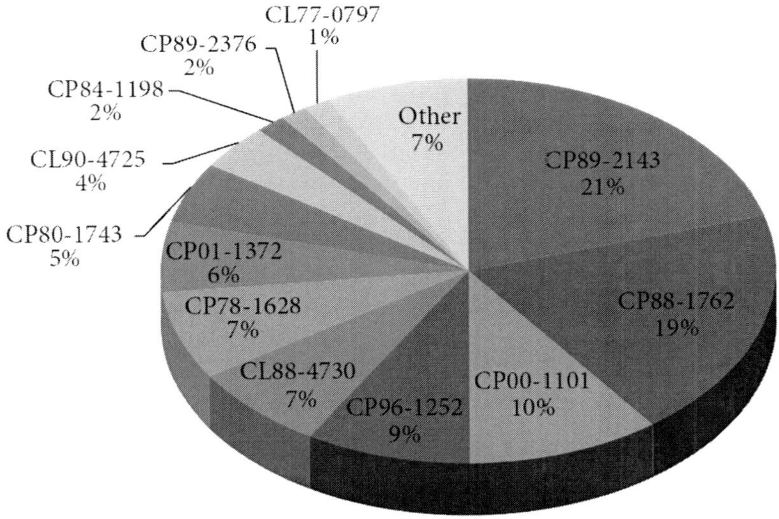

Figure 4: Commercial sugarcane cultivars and their % of total hectarage in the 2012-2013 harvest season in Florida. Total area = 172,118 ha. Data are adopted from Rice et al. [12].

Severity of most sugarcane diseases is associated with the climate related factors. Sugarcane orange rust disease in Florida was much severer in the 2012 and 2013 growing seasons than other years due to warmer winter and higher humidity [37]. Sugarcane smut disease was severer on sandy soils than on organic soils because of high temperature and relative dry conditions. In addition to development of disease resistant cultivars by breeding and variety development programs, integration of the best management practices (BMPs) for pest control and for increases in water and nutrient use efficiencies is also crucial for the climate change adaptation and sugarcane yield improvement.

These BMPs include carbon sequestration, soil tillage, irrigation methods and scheduling, drainage, nutrient monitoring, and fertilizer applications. All of these are associated with geographic locations and long-term climate change and have been reviewed in detail recently [15, 57]. Biggs et al. [58] found that the frequency of years with very high losses is predicted to increase under projected climate change and improved farming practices can more effectively limit losses than traditional practices. The use of ripeners and withholding irrigation prior to harvest can improve sucrose content of stalks. Additionally, integration of seasonal climate forecasting with proper management strategies has potential to benefit sugarcane production in many areas [59]. The details of cultivar development and management strategies for sugarcane production in China have been proposed [27, 30, 60]. Applications of plant growth regulators can be useful for improving plant tolerance to some specific stresses and for sugar yield enhancement. Studies have indicated that using low concentration of ethylene-producing substances such as ethephon to treat seed canes or yang plants at early growth stage can improve the drought tolerance and mitigate other abiotic stresses of sugarcane plants [61, 62]. Foliar application of low concentration of ethephon alleviated the injury of cell membrane caused by water deficit stress, maintained relatively lower osmotic rates of electrolytes and soluble sugar, increased the proline content and water potential in the leaf tissues, promoted the activities of cell protective enzymes (such as peroxidase, catalase, and polyphenol oxidase), and improved the gas exchange characteristics. The physiological and biochemical base of plant growth regulator applications and their beneficial effects on sugarcane growth under drought conditions have been reported in detail by Botha et al. [61] and Li [62]. The foliar application of ethephon in a low concentration may be used as a management practice to partly mitigate drought effect on sugarcane growth and yields.

6. SUGARCANE IMPACT ON LOCAL CLIMATE

In a region, changes in farming systems and crop combinations may directly or indirectly affect local climate factors. Residue burning before or after sugarcane harvest is a common management practice of sugarcane production in many countries. Greenhouse gas emission in sugarcane production is the major concern. A recent research [63] indicated that approximate 2.4 tonnes of CO_2 equivalent ha^{-1} was released to the atmosphere by sugarcane crop. The major contributors of the released CO_2 from sugarcane were residue burning (44%), the utilization of synthetic fertilizers (20%), and fossil fuel combustion (18%) [63]. Therefore, improving green harvest can increase soil organic carbon and reduce CO_2 emission from sugarcane production.

Georgescu et al. [64] investigated the direct climate effect of perennial bioenergy crops in the United States. Their results demonstrated that a thorough evaluation of costs and benefits of bioenergy-related land-use change must include potential impacts on the surface energy and water balance to comprehensively address important concerns for local, regional, and global climate change. Expansion of sugarcane can relatively increase carbon fixation and carbon sequestration because of its C_4 carbon fixation characteristics. A recent study [65] in Brazil, using maps and data from hundreds of satellite images, has revealed that, on a regional basis for clear-sky daytime conditions, conversion of natural vegetation to a crop/pasture warms the region by an average of 1.55 (1.45–1.65)°C, but subsequent conversion of the crop/pasture to sugarcane cools the region by an average of 0.93 (0.78–1.07)°C, resulting in a mean net increase of 0.6°C. They concluded that expanding sugarcane into existing crop and pasture land has a direct local cooling effect that reinforces the indirect climate benefits of this land-use option. Therefore, sugarcane may be better than other field crops for environmental protection in increasing atmospheric [CO_2] and surface temperature. Further research is required to reveal the mechanisms of the direct local cooling effect of sugarcane.

7. SUMMARY AND FUTURE PERSPECTIVES

Clearly, sugarcane production has been and will continue to be directly or indirectly affected by changes in climate conditions. The most significant challenges for sugarcane production are increases in frequency and intensity of extreme weather events, especially drought during climate change. Existing adaptation strategies can help offset many but not all effects in the future. The negative effects of climate change on sugarcane production are very likely to worsen after 2050, especially if greenhouse gas emissions still remain high. Therefore, agricultural scientists and decision makers need to work closely to mitigate the potential negative effects of climate change on agriculture and to improve sugarcane yields by multidisciplinary approaches, such as consistently developing new sugarcane cultivars using breeding and molecular biology, refining best management practices, improving new technology transfer, and increasing productivity and profitability. Improving the resilience of sugarcane production systems to climate change requires protection of the natural resource (especially water and soil) for sustainability. Expanding use of sugarcane products for sugar, ethanol, cellulosic biofuel, and other coproducts can further improve profits.

REFERENCES

1. R. A. Houghton, K. T. Lawrence, J. L. Hackler, and S. Brown, "The spatial distribution of forest biomass in the Brazilian Amazon: a comparison of estimates," Global Change Biology, vol. 7, no. 7, pp. 731–746, 2001.
2. IPCC, "Summary for policymakers," in Climate Change 2014: Impacts, Adaptation, and Vulnerability. Part A: Global and Sectoral Aspects. Contribution of Working Group II to the Fifth Assessment Report of the Intergovernmental Panel on Climate Change, C. B. Field, V. R. Barros, D. J. Dokken et al., Eds., pp. 1–32, Cambridge University Press, Cambridge, UK, 2014, http://www.ipcc.ch/pdf/assessment-report/ar5/wg2/ar5_wgII_spm_en.pdf.
3. B. A. Kimball, "Carbon dioxide and agricultural yield: an assemblage and analysis of 430 prior observations," Agronomy Journal, vol. 75, no. 5, pp. 779–788, 1983.
4. F. Tao, M. Yokozawa, Y. Xu, Y. Hayashi, and Z. Zhang, "Climate changes and trends in phenology and yields of field crops in China, 1981–2000," Agricultural and Forest Meteorology, vol. 138, no. 1–4, pp. 82–92, 2006.
5. K. E. Trenberth, P. D. Jones, P. Ambenje et al., "Observations: surface and atmospheric climate change: the physical science basis," in Contribution of WG 1 to the Fourth Assessment Report of the Intergovernmental Panel on Climate Change, S. Solomon, D. Qin, M. Manning et al., Eds., pp. 235–336, Cambridge University Press, Cambridge, UK, 2007.
6. R. S. Dhillon and G. von Wuehlisch, "Mitigation of global warming through renewable biomass," Biomass and Bioenergy, vol. 48, pp. 75–89, 2013.
7. J. Gawander, "Impact of climate change on sugar-cane production in Fiji," WMO Bulletin, vol. 56, pp. 34–39, 2007.
8. C. Rosenzweig, G. Casassa, D. J. Karoly et al., "Assessment of observed changes and responses in natural and managed systems," in Climate Change 2007: Impacts, Adaptation and Vulnerability. Contribution of Working Group II to the Fourth Assessment Report of the Intergovernmental Panel on Climate Change, M. L. Parry, O. F. Canziani, J. P. Palutikof, P. J. van der Linden, and C. E. Hanson, Eds., pp. 79–131, Cambridge University Press, Cambridge, UK, 2007.
9. P. K. Thornton, P. G. Jones, G. Alagarswamy, and J. Andresen, "Spatial variation of crop yield response to climate change in East Africa," Global Environmental Change, vol. 19, no. 1, pp. 54–65, 2009.
10. D. B. Lobell, M. B. Burke, C. Tebaldi, M. D. Mastrandrea, W. P. Falcon, and R. L. Naylor, "Prioritizing climate change adaptation needs for food security in 2030," Science, vol. 319, no. 5863, pp. 607–610, 2008.
11. Anderson-Sprecher and J. Jiang, "China: sugar annual," Report 14020, USDA Foreign Agricultural Service, Global Agricultural Information Network (GAIN), 2014, http://gain.fas.usda.gov/Recent%20GAIN%20Publications/Sugar%20Annual_Beijing_China%20-%20Peoples%20Republic%20of_4-30-2014.pdf.
12. R. Rice, L. Baucum, and B. Glaz, "Sugarcane variety census: Florida 2012," Sugar Journal, vol. 76, no. 2, pp. 11–19, 2013.
13. FAO, "Report, Food and Agricultural Organization, United Nations: Economic and Social Department," The Statistical Division, FAO 2013 FAOSTAT, 2014, http:// faostat3.fao.org/home/index.html#DOWNLOAD.
14. Factfish, "Sugar cane, production quantity (tons)—for all countries," 2015, http://www.factfish.com/statistic/sugar+cane,+production+quantity.

15. M. Chandiposha, "Potential impact of climate change in sugarcane and mitigation strategies in Zimbabwe," African Journal of Agricultural Research, vol. 8, pp. 2814–2818, 2013.
16. F. R. Marin, J. W. Jones, A. Singels et al., "Climate change impacts on sugarcane attainable yield in southern Brazil," Climatic Change, vol. 117, no. 1-2, pp. 227–239, 2013.
17. J. C. V. Vu and L. H. Allen Jr., "Growth at elevated CO_2 delays the adverse effects of drought stress on leaf photosynthesis of the C4 sugarcane," Journal of Plant Physiology, vol. 166, no. 2, pp. 107–116, 2009.
18. J. C. V. Vu and L. H. Allen Jr., "Stem juice production of the C_4 sugarcane (Saccharum officinarum) is enhanced by growth at double-ambient CO_2 and high temperature," Journal of Plant Physiology, vol. 166, no. 11, pp. 1141–1151, 2009.
19. P. de Souza, M. Gaspar, E. A. da Silva et al., "Elevated CO_2 increases photosynthesis, biomass and productivity, and modifies gene expression in sugarcane," Plant, Cell and Environment, vol. 31, no. 8, pp. 1116–1127, 2008.
20. S. Chakraborty, G. M. Murray, P. A. Magarey et al., "Potential impact of climate change on plant diseases of economic significance to Australia," Australasian Plant Pathology, vol. 27, no. 1, pp. 15–35, 1998.
21. G. Hayes, "Weather trends at Condong, NSW," Proceedings of the Australian Society of Sugar Cane Technologists, vol. 12, pp. 39–44, 1990.
22. L. de Carvalho, R. S. C. Menezes, R. S. Nóbrega et al., "Impact of climate changes on potential sugarcane yield in Pernambuco, northeastern region of Brazil," Renewable Energy, vol. 78, pp. 26–34, 2015.
23. D. L. Santos and P. C. Sentelhas, "Climate change scenarios and their impact on the water balance of sugarcane production areas in the State of São Paulo, Brazil," Ambiente e Agua, vol. 7, no. 2, pp. 7–17, 2012.
24. J. W. Knox, J. A. Rodríguez Díaz, D. J. Nixon, and M. Mkhwanazi, "A preliminary assessment of climate change impacts on sugarcane in Swaziland," Agricultural Systems, vol. 103, no. 2, pp. 63–72, 2010.
25. Singh and M. El Maayar, "Potential impacts of a greenhouse gas climate change scenarios on sugar cane yields in Trinidad, southern Caribbean," Tropical Agriculture, vol. 75, no. 3, pp. 348–355, 1998.
26. F. P. de Lucena, A. S. Szklo, R. Schaeffer et al., "The vulnerability of renewable energy to climate change in Brazil," Energy Policy, vol. 37, no. 3, pp. 879–889, 2009.
27. Y.-R. Li and Y.-A. Wei, "Sugar industry in China: R & D and policy initiatives to meet sugar and biofuel demand of future," Sugar Tech, vol. 8, no. 4, pp. 203–216, 2006.
28. R. A. Gilbert, C. R. Rainbolt, D. R. Morris, and A. C. Bennett, "Morphological responses of sugarcane to long-term flooding," Agronomy Journal, vol. 99, no. 6, pp. 1622–1628, 2007.
29. Challinor, T. Wheeler, C. Garforth, P. Craufurd, and A. Kassam, "Assessing the vulnerability of food crop systems in Africa to climate change," Climatic Change, vol. 83, no. 3, pp. 381–399, 2007.
30. Y. R. Li and L. T. Yang, "Chinese sugarcane and sugar industry in recent 10 years," in Proceedings of the 5th International Association of Professionals in Sugar and Integrated Technologies: Green Technologies for Sustanable Growth of Sugar & Integrated Industries in Developing Countries, Y.-R. Li, G. P. Rao, S. Solomon et al., Eds., pp. 1–10, 2014.
31. Y. R. Li, F. X. Fang, L. T. Yang et al., "Long duration cold stress and frost induced losses to sugarcane production in Guangxi (China) during 2010-11 and its control measures," in Proceedings of the 4th IAPSIT International Sugar Conference (IS '11) & Expo (Balancing Sugar and Energy Production in Developing Countries: Sustainable Technologies and Marketing Strategies), pp. 353–358, Army Printing Press, IBDC Publishers, Lucknow, India, 2011.
32. R. A. Gilbert, C. R. Rainbolt, D. R. Morris, and J. M. McCray, "Sugarcane growth and yield responses to a 3-month summer flood," Agricultural Water Management, vol. 95, no. 3, pp. 283–291, 2008.
33. Glaz and S. E. Lingle, "Flood duration and time of flood onset effects on recently planted sugarcane," Agronomy Journal, vol. 104, no. 3, pp. 575–583, 2012.

34. L. Matthieson, "Climate change and the Australian Sugarcane Industry: impacts, adaptation and R&D opportunities," SRDC Technical Report, Sugar Research and Development, Brisbane, Australia, 2007.
35. Y. L. Everingham, R. C. Muchow, R. C. Stone, N. G. Inman-Bamber, A. Singels, and C. N. Bezuidenhout, "Enhanced risk management and decision-making capability across the sugarcane industry value chain based on seasonal climate forecasts," Agricultural Systems, vol. 74, no. 3, pp. 459–477, 2002.
36. Zhao, N. C. Glynn, B. Glaz, J. C. Comstock, and S. Sood, "Orange rust effects on leaf photosynthesis and related characters of sugarcane," Plant Disease, vol. 95, no. 6, pp. 640–647, 2011.
37. Zhao, R. W. Davidson, M. Baltazar, and J. C. Comstock, "Field evaluation of sugarcane orange rust for first clonal stage of the CP cultivar development program," American Journal of Agricultural and Biological Sciences, vol. 10, no. 1, pp. 1–11, 2015.
38. R. N. Raid, J. C. Comstock, and S. Sood, "Sugarcane orange rust in Florida: its history, current status, and management," Journal of the American Society of Sugar Cane Technologists, vol. 33, p. 57, 2013.
39. J. C. Comstock, S. G. Sood, N. C. Glynn, J. M. Shine Jr., J. M. McKemy, and L. A. Castlebury, "First report of Puccinia kuehnii, causal agent of orange rust of sugarcane, in the United States and in the Western Hemisphere," Plant Disease, vol. 92, p. 175, 2008.
40. T. Deressa, R. Hassan, and D. Poonyth, "Measuring the impact of climate change on south african agriculture: the case of sugarcane growing regions," Agrekon, vol. 44, no. 4, pp. 524–542, 2005.
41. N. G. Inman-Bamber, P. Lakshmanan, and S. Park, "Sugarcane for water-limited environments: theoretical assessment of suitable traits," Field Crops Research, vol. 134, pp. 95–104, 2012.
42. P. P. da Silva, L. Soares, J. G. da Costa et al., "Path analysis for selection of drought tolerant sugarcane genotypes through physiological components," Industrial Crops and Products, vol. 37, no. 1, pp. 11–19, 2012.
43. D. Zhao, B. Glaz, and J. C. Comstock, "Sugarcane leaf photosynthesis and growth characters during development of water-deficit stress," Crop Science, vol. 53, no. 3, pp. 1066–1075, 2013.
44. L. C. De Silva and W. A. J. M. De Costa, "Growth and radiation use efficiency of sugarcane under irrigated and rain-fed conditions in Sri Lanka," Sugar Tech, vol. 14, no. 3, pp. 247–254, 2012.
45. N. Robinson, A. Fletcher, A. Whan et al., "Sugarcane genotypes differ in internal nitrogen use efficiency," Functional Plant Biology, vol. 34, no. 12, pp. 1122–1129, 2007.
46. D. Zhao, B. Glaz, and J. C. Comstock, "Physiological and growth responses of sugarcane genotypes to nitrogen rate on a sand soil," Journal of Agronomy and Crop Science, vol. 200, no. 4, pp. 290–301, 2014.
47. S. J. Edmé, J. D. Miller, B. Glaz, P. Y. P. Tai, and J. C. Comstock, "Genetic contribution to yield gains in the Florida sugarcane industry across 33 years," Crop Science, vol. 45, no. 1, pp. 92–97, 2005.
48. D. Zhao, B. Glaz, M. S. Irey, and C.-J. Hu, "Sugarcane genotype variation in leaf photosynthesis properties and yield as affected by mill mud application," Agronomy Journal, vol. 107, no. 2, pp. 506–514, 2015.
49. M. Menossi, M. C. Silva-Filho, M. Vincentz, M.-A. Van-Sluys, and G. M. Souza, "Sugarcane functional genomics: gene discovery for agronomic trait development," International Journal of Plant Genomics, vol. 2008, Article ID 458732, 11 pages, 2008.
50. J. Todd, B. Glaz, D. Burner, and C. Kimbeng, "Historical use of cultivars as parents in Florida and Louisiana sugarcane breeding programs," International Scholarly Research Notices, vol. 2015, Article ID 257417, 9 pages, 2015.
51. M. D. da Silva, R. L. D. O. Silva, J. R. Costa Ferreira Neto et al., "analysis of sugarcane aquaporin genes under water deficit," Journal of Nucleic Acids, vol. 2013, Article ID 763945, 14 pages, 2013.

52. J. Grice, M. K. Wegener, L. M. Romanach, S. Paton, P. Bonaventura, and S. Garrad, "Genetically modified sugarcane: a case for alternate products," AgBioForum, vol. 6, no. 4, pp. 162–168, 2003.
53. GRAIN, "Corporate candyland: the looming GM sugar cane invasion," Seedling, vol. 4, no. 1, pp. 2–12, 2009.
54. P. A. Joyce, S.-Q. Dinh, E. M. Burns, and M. G. O'shea, "Sugar from genetically modified sugarcane: tracking transgenes, transgene products and compositional analysis," International Sugar Journal, vol. 12, pp. 860–863, 2013.
55. R. C. Magarey, B. J. Croft, and T. G. Willcox, "An epidemic of orange rust on sugarcane in Australia," in Proceedings of the 24th Congress of the International Society of Sugar Cane Technologists, pp. 410–416, Brisbane, Australia, September 2001.
56. K. S. Braithwaite, B. J. Croft, R. C. Magarey, and T. Scharaschkin, "Phylogenetic placement of the sugarcane orange rust pathogen Puccinia kuehnii in a historical and regional context," Australasian Plant Pathology, vol. 38, no. 4, pp. 380–388, 2009.
57. Bhaskaran and N. V. Nair, "Challenges and opportunities in sugarcane cultivation under climate change scenario," Journal of Sugarcane Research, vol. 4, pp. 1–18, 2014.
58. J. S. Biggs, P. J. Thorburn, S. Crimp, B. Masters, and S. J. Attard, "Interactions between climate change and sugarcane management systems for improving water quality leaving farms in the Mackay Whitsunday region, Australia," Agriculture, Ecosystems and Environment, vol. 180, pp. 79–89, 2013.
59. R. C. Muchow, A. J. Higgins, N. G. Inman-Bamber, and P. J. Thorburn, "Towards sustainable sugarcane production using a whole systems analysis approach," in Proceedings of the 24th Congress of the International Society of Sugar Cane Technologists, pp. 91–94, Brisbane, Australia, September 2001.
60. Y.-R. Li and L.-T. Yang, "Research and development priorities for sugar industry of China: recent research highlights," Sugar Tech, vol. 17, no. 1, pp. 9–12, 2014.
61. C. Botha, P. Lakshmanan, A. O'Connell, and P. H. Moore, "Hormones and growth regulators," in Sugarcane: Physiology, Biochemistry, and Functional Biology, P. H. Moore and F. C. Botha, Eds., pp. 331–378, John Wiley & Sons, 2014.
62. Y.-R. Li, "Beneficial effects of ethephon application on sugarcane under sub-tropical climate of China," Sugar Tech, vol. 6, no. 4, pp. 235–240, 2004.
63. E. B. de Figueiredo, A. R. Panosso, R. Romão, and N. La Scala Jr., "Greenhouse gas emission associated with sugar production in southern Brazil," Carbon Balance and Management, vol. 5, article 3, 2010.
64. M. Georgescu, D. B. Lobell, and C. B. Field, "Direct climate effects of perennial bioenergy crops in the United States," Proceedings of the National Academy of Sciences of the United States of America, vol. 108, no. 11, pp. 4307–4312, 2011.
65. S. R. Loarie, D. B. Lobell, G. P. Asner, Q. Mu, and C. B. Field, "Direct impacts on local climate of sugar-cane expansion in Brazil," Nature Climate Change, vol. 1, no. 2, pp. 105–109, 2011.

CHAPTER 2

Bridging the Rice Yield Gaps under Drought: QTLs, Genes, and Their Use in Breeding Programs

Nitika Sandhu and Arvind Kumar *

International Rice Research Institute, DAPO BOX 7777, Metro Manila 1301, Philippines

AUTHOR CONTRIBUTIONS

N.S. was involved in conceptualizing and drafting the review article; A.K. was involved in the conceptualizing and critical revision of the review article.

ABSTRACT

Rice is the staple food for more than half of the world's population. Although rice production has doubled in the last 30 years as a result of the development of high-yield, widely adaptable, resource-responsive, semi-dwarf varieties, the threat of a food crisis remains as severe as it was 60 years ago due to the ever-increasing population, water scarcity, labor scarcity, shifting climatic conditions, pest/diseases, loss of productive land to housing, industries, rising sea levels, increasing incidences of drought, flood, urbanization, soil erosion, reduction in soil nutrient status, and environmental issues associated with high-input agriculture. Among these, drought is predicted to be the most severe stress that reduces rice yield. Systematic research on drought over the last 10 years has been conducted across institutes on physiology, breeding, molecular genetics, biotechnology, and cellular and molecular biology. This has provided a better understanding of plant drought mechanisms and has helped scientists to devise better strategies to reduce rice yield losses under drought stress. These include the identification of quantitative trait loci (QTLs) for grain yield under drought as well as many agronomically important traits related to drought tolerance, marker-assisted pyramiding of genetic regions that increase yield under drought, development of efficient techniques for genetic transformation, complete sequencing and annotation of rice genomes, and synteny studies of rice and other cereal genomes. Conventional and marker-assisted breeding rice lines containing useful introgressed genes or loci have been field tested and released as varieties. Still, there is a long way to go towards developing drought-tolerant rice varieties by exploiting existing genetic diversity, identifying superior alleles for drought tolerance, understanding interactions among alleles for drought tolerance and their interaction with genetic backgrounds, and pyramiding the best combination of alleles.

Keywords: drought; marker; pyramiding; QTLs; rice; genomics

1. INTRODUCTION

Rice feeds more than half of the global population. Global rice (paddy) production in 2015 trails 0.8 percent behind the 2014 outcome, 738.2 million tons (490.3 million tons, milled rice), obtained from an area of 160.6 million hectares, a decrease of 1.3% [1]. Asia, where 60% of the earth's population lives, is the major producer and consumer of the world's rice. Water, climate, season, rainfall, soil conditions, agriculture inputs, and genetic potential of germplasm are key determinants of crop productivity. Increasing population (Figure 1a), increasing demand for water (Figure 1b), water crisis (Figure 1c), drought (Figure 1d), failure to adapt to climate change, declining farm land, soil moisture, soil characteristics, deterioration in nutrient content, weed competitiveness, increasing intensity, and the frequency of biotic/abiotic stresses will amplify the challenges of achieving future food requirements. This will affect the economic growth and social stability of regions with food shortages. Farmers will earn a profit only if they successfully solve the algebraic puzzle of farming. Wheat, rice, maize, and other grains that are the staple food of the human population and the sources of feed for livestock account for more than 60% of the total crop evapotranspiration requirement, while soybeans and other oilseed crops account for 17%, and sugarcane 6% [2]. In such circumstances, the available water resources will not be sufficient to produce enough food for the increasing population. With changes in the climate and unpredictable rainfall, there is a possibility that nearly half of the world's population may face water scarcity by 2030 [3]. Water scarcity will worsen in the world's extremely dry regions and areas where water is already in short supply.

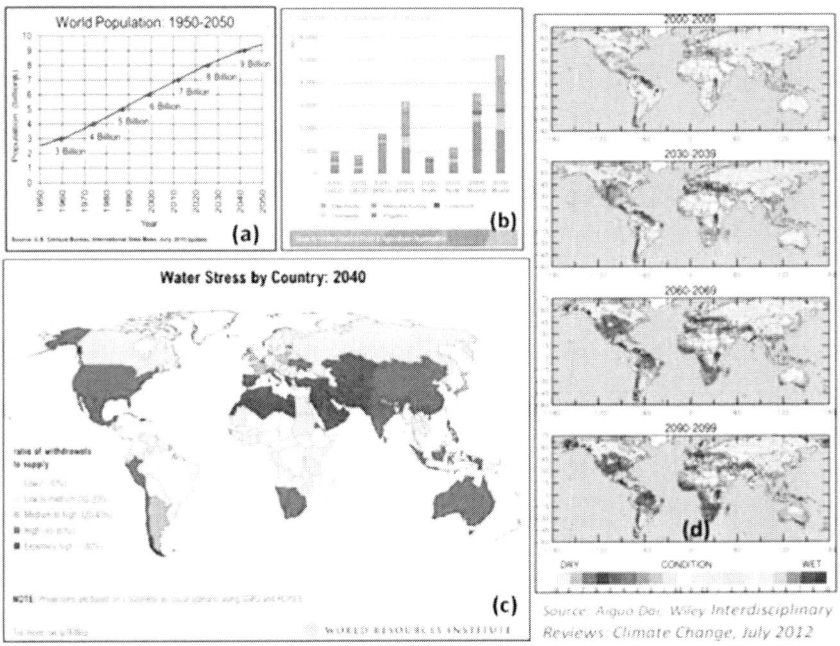

Figure 1. (**a**) Projected population curve (source: U.S. Census Bureau, International database 1950–2050, July 2015 update); (**b**) estimated global water demand (OECD: Organization for Economic Cooperation and Development; BRIC: Brazil, Russia, India and China; RoW: Rest of world; source: United Nation Food & Agriculture Organization); (**c**) severity pattern of water stress by country by 2040 (source: World Resource Institute); (**d**) estimated possibilities for future drought worldwide based on the Palmer Drought Severity Index (source: Aigup Dai, Wiley interdisciplinary Reviews: Climate Change, July 2012).

The contribution of plant breeding to improving commercially important crops, including major ones such as rice, maize, wheat, cotton, and pearl millet, at a global level is remarkable. Before the Green Revolution, traditional rice and wheat varieties were tall, photoperiod-sensitive, low-yielding and drought-tolerant, having a broad maturity duration and good grain quality. In the post-Green Revolution era, these traditional varieties were replaced by a few widely adapted varieties including inbreds and hybrids that are dwarf and photoperiod-insensitive, with early maturity, higher yield, poor grain quality, and low pest resistance. The dwarf rice varieties were bred by targeting irrigated ecosystems wherein ample water was thought to remain available for traditional practices of puddled transplanted system of rice cultivation. These varieties have high yield potential and good resistance to biotic stresses, but are highly susceptible to abiotic stresses such as drought. They are also prone to heavy yield losses even under mild drought stress [4]. In the course of post-Green Revolution breeding over the past 50 years, unknowingly, the drought tolerance contributing alleles of traditional cultivars have not been properly maintained in the modern cultivars. Recent understanding of molecular and physiological mechanisms for different abiotic stresses has opened up new opportunities to improve yield under adverse climatic conditions for many crops. There is still a need to bridge the large gap between yields in most favorable and stress conditions. Strategies involving bridging the yield gap and increasing yield stability and adaptability under variable environmental conditions are of importance in assuring food security and sustainability in the future. There is a need to move forward from the Green Revolution to a 'gene revolution,' which is more productive and more 'green' in terms of conserving natural resources and the environment [5].

2. DROUGHT: THE KEY CONCERN IN FOOD SECURITY

Drought has been the main catalyst of many large famines of the past and has a major destructive effect on rice production in rainfed areas across Asia and sub-Saharan Africa. The most vulnerable, drought-prone areas are shown in Table 1. The most devastating drought events around the world were the Deccan Famine and those in the Horn of Africa, the United States, Vietnam, Australia, China, Brazil, the Sahel, Malawi, East Africa, Ethiopia, India, and Bangladesh. From 2003 to 2013, at least one medium- to large-scale natural disaster caused $70 billion in crop and livestock production losses; drought alone accounted for 44%. Asia is the most affected region, with total crop and livestock production losses amounting to $28 billion (40% of total losses), followed by Africa with $25 billion (Table 2) [6]. The 1987 drought in India, the 2004 drought in Thailand, and the 1978–2003 drought in China were estimated to have affected 60% [7], 2 million ha [8], and 14 million ha of cropped area, respectively. Drought events between 1980 and 2014 in sub-Saharan Africa affected 203, 86, 74, 61, and 48 million people in eastern Africa, southern Africa, western Africa, Ethiopia, and Kenya, respectively [6].

Table 1. Most vulnerable drought-prone areas across the world.

Region	Areas Most Vulnerable to Drought	Drought Events
Asia/Pacific	India, Nepal, Bangladesh, China, Laos, Cambodia, Pakistan, Afghanistan, Sri Lanka, Bhutan, Indonesia, Thailand, Myanmar, Vietnam, Malaysia	1876, 1878, 1896, 1902, 1907, 1928, 1930, 1936, 1941, 1942, 1944, 1958, 1961, 1964, 1972, 1973, 1974, 1983, 1987, 1993, 1996, 2000, 2002, 2010
Middle East	Yemen, the United Arab Emirates, Saudi Arabia, Iraq, Iran, Syria	1940, 1998, 2000, 2007, 2010
Europe	France, Italy, Germany, northern Spain, Czech Republic	1955, 1957, 1962, 1968, 1971, 1974, 2005, 2009, 2012
United States	Arizona, Kansas, Arkansas, Georgia, Florida, Mississippi, Alabama, South, North Carolina, Texas, Oklahoma, California	1934, 1936, 1939, 1940, 1983, 2002, 2010, 2011
Africa	Ethiopia, Kenya, Eritrea, Somalia, Uganda, Djibouti, Mauritania, Angola, Zambia, Zimbabwe, Mozambique, Malawi, Lesotho, Swaziland	1888, 1972, 1973, 1983, 1985, 1991, 1992, 1999, 2002, 2002, 2003, 2010, 2011, 2012
Latin America	Peru, Chile, Argentina, Brazil, Mexico	1630, 1640, 1650, 1782, 1884, 1992, 1999, 2011, 2015
Australia	New south wales, Queensland, Victoria, Tasmania, Sydney, Northam, York area of Western Australia	1813, 1826, 1829, 1835, 1838, 1850, 1888, 1897, 1902, 1982, 1983, 2000

Source: Modified from Spring 2015 global attributes survey.

Table 2. Effect of drought on crops and livestock across the world.

Region	Crop Losses (Billion USD)	Livestock Losses (Billion USD)	Total (Billion USD)
Africa	21	4	25
Asia	27	1	28
Latin America and Caribbean	9	2	11
Near East	4	0	4
Central Asia	1	0	4
% share of total Global losses	42.4	35.3	78.2

Source: FAO based on data from FAOSTAT, 2003–2013.

Drought induces critical losses in crop yield. Yield integrates many of the physiological and biochemical responses at cellular and molecular levels, influenced by a number of predictable and unpredictable factors that are genetically difficult to understand and manipulate. Therefore, long-term and systematic attention should be given to the complex issues surrounding drought in order to develop a better understanding and devise sustainable solutions.

3. EFFECT OF DROUGHT ON DIFFERENT CROPS

Approximately 34% of rice is grown in rainfed lowland, 9% in rainfed upland, and 7% in flood-prone areas, while irrigated ecosystem covers 50% of total world rice area. Drought has been reported to produce devastating effects in rice at panicle initiation and flowering [4,9]; in maize at the tasseling and silking stages [10,11]; in sorghum and pearl millet at the booting and flowering stages [12]; in finger millet at the flowering stage; in sunflowers at head formation and the early grain-filling stage [13,14]; in groundnuts at the peg penetration and pod development stages; in soybean at the flowering and pod filling stages [15,16]; in black and green gram at the flowering and early pod development stages [17]; in cotton at the square formation and ball development stages [18,19]; and during the reproductive stage in rice [20,21]. Like in other crops, in rice drought has the most devastating effect at the reproductive stage. In rice, the damage to the crop is also significant at the seedling as well as vegetative stages. At the seedling stage, delay in monsoon rains, insufficient rain to puddle land, and preparation for transplanting force farmers to leave their land uncultivated. Severe drought at the vegetative stage reduces biomass production, causes the death of the plant and in severe cases, forces farmers to allow the grazing of the crops by cattle. Drought has a complex effect on plants [22,23,24,25,26,27,28,29,30,31,32,33,34, 35,36, 37,38, 39,40,41,42], and plants respond with many defensive adaptations (Figure 2). The major determinants of grain yield under drought are the variety [43], type of soil [44], length and timing of drought [45], severity of drought [46,47], season (early season, mid-season, or terminal stage, Table 3 [48,49,50,51,52,53,54,55,56,57, 58,59,60,61,62,63, 64,65,66,67, 68]), the age, period, and development stage of the plant [69], plant responses after stress elimination, and the interaction between the biotic/abiotic factors [70] and the region. Apart from this, drought stress also makes the rice crop more susceptible to biotic stresses (rice blast, brown spot, and bacterial blight), leading to a further decline in rice production. In many rice-growing areas in rainfed ecosystems, drought and submergence can occur in the same season at different growth stages of the plant or in different seasons, thus creating more complexity. Drought tolerance is a means for the rice plant to survive and produce a stable and satisfactory yield. There is urgent need for a strategy to get the highest yield out of every single drop of water on existing cropland to satisfy food needs in the future.

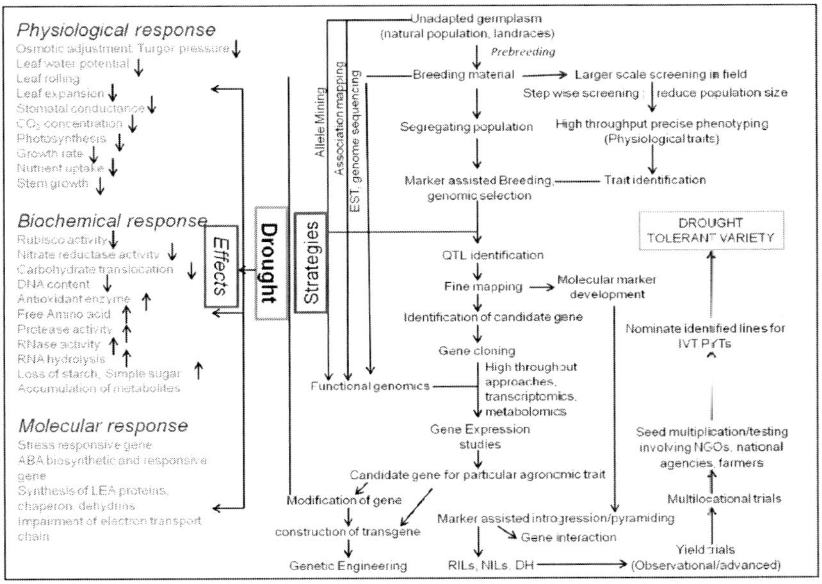

Figure 2. Effect of drought and approaches in developing drought-tolerant rice varieties. RILs: Recombinant inbred lines, NILs: Near-isogenic lines, DH: Double haploid, NGO: Non-Governmental Organization; IYT: Intermediate Yield Trial, PYTs: Preliminary yield trial, ↑ (increase/enhance), ↓ (decrease/reduce).

Water availability (drought and flood), soil problems (salinity, nutrient deficiencies, and toxicities), extreme temperatures (heat and cold) and biotic stresses (brown planthopper, gall midge, blast, tungro, bacterial blight) are the main constraints in South Asia, Southeast Asia, and Africa, where rice often suffers from extensive shock to sustain full yield potential. Surveys conducted by the Africa Rice Center in 12 sub-Saharan African countries reported a yield decline of 33% [71] when drought and flooding occurred together. Another study by the Africa Rice Center reported yield losses of 40% and 25% in Senegal and Uganda, respectively, due to salinity and iron toxicity [72]. Therefore, it is advantageous to select cultivars with multiple stress tolerance (drought, salinity, submergence, stagnant flooding, biotic stress, and high temperature) to allow the crop to survive if multiple stresses come at the same time.

Table 3. Yield losses in different crops as a result of drought.

Crop	Stress	Yield Reduction	Reference
Rice	Lowland moderate reproductive stage	45%–60%	[48–50]
Rice	Lowland severe reproductive stage	65%–91%	[48–51]
Rice	Upland mild reproductive stage	18%–39%	[48,52]
Rice	Upland moderate reproductive stage	70%–75%	[48,52]
Rice	Upland severe reproductive stage	80%–97%	[48,49,53]
Wheat	Moderate reproductive stage	10%–50%	[54–57]
Pearl Millet	Prior and beginning of flowering	65%	[58]
Pearl Millet	Early stress	62%	[59]
Pearl Millet	Late stress	28%	[59]
Maize	Mild-moderate-severe reproductive stage	1%–76%	[60–63]
Barley	Severe reproductive stage	73%–87%	[64]
Chickpea	Late terminal drought	49%–54%	[65]
Chickpea	Reproductive stage	45%–69%	[66]
Pigeon Pea	Reproductive stage	40%–55%	[67]
Canola	Reproductive stage	15%–35%	[68]

4. STRATEGIES TO MANAGE DROUGHT

Comprehensive information, early warning systems and cultivation of high-yielding, high-quality, drought- plus biotic stress-tolerant varieties in drought-prone areas could provide a solution to the problem of drought. Identification and introduction of suitable traits that narrow the gap between expected and actual yield; understanding realistic physio-morpho-molecular mechanisms of drought tolerance; and designing a standard screening method for a large population [73] could contribute to the development of drought-tolerant rice varieties. Adopting proper strategies such as larger scale standardized screening for grain yield under drought and understanding the components of yield based on morpho-physiological traits could contribute to breeders' efforts to develop better drought-tolerant varieties. Conventional and marker-assisted breeding strategies based on the use of drought-tolerant donors, pre-breeding to use the lines derived from crosses involving donors, and the development of suitable mapping populations to identify QTLs/genes affecting yield could result in yield improvement and stability under drought stress. Breaking undesirable linkages between drought tolerance and tall plant height, drought tolerance and earliness, and drought tolerance and low yield potential [74] could help to develop semi-dwarf drought-tolerant varieties without any yield penalty. Molecular, cellular, physiological, biochemical, and developmental responses to abiotic stress involve several genes and gene functions controlling drought tolerance. Several efforts have been made to better understand the expression of drought-tolerance-related traits and the complex network of drought-related genes. Exogenous application of hormones and osmoprotectants to seed or growing plants, engineering for drought resistance, and high-throughput novel technologies could be useful tools in identifying genes to improve yield under drought (Figure 2).

4.1. Screening Strategies

Although it is difficult to understand how plants build up, combine, and exhibit the changing processes over the entire growth and development cycle, efforts have been made to standardize screening protocols, understand the mechanisms related to drought tolerance, and develop varieties that are tolerant of drought. The assessment of the type, intensity, degree of drought, and appropriate selection/screening for drought tolerance is a very crucial step. Each method has some advantage and limitations. Identification of drought-tolerant and -susceptible cultivars based on a few physiological measures (such as canopy temperature, water potential, and osmotic adjustment) [75] and specific environmental factors (such as weather and soil water availability) may not be adequate for breeders to use such donors in the breeding program. Screening of donor lines for grain yield under drought, performance of such lines under both stress and non-stress conditions [76,77,78,79], and use of robust statistical methods to clearly differentiate drought-tolerant and drought-susceptible lines [80,81,82,83] could be considered an appropriate methodology for drought screening [84]. Simultaneous screening for resistance to multiple biotic and abiotic stresses could be more beneficial to improve yield under multiple stress-prone environments.

4.1.1. Secondary Traits

Secondary traits are distinct components of prime plant traits such as grain yield. Secondary traits are important indicators of different physiological, molecular, and developmental changes involved in drought resistance, tolerance, and adaptation mechanisms. The effectiveness of selection for secondary traits such as root thickness, penetration ability and depth, greater hydraulic conductance, xylem thickness and osmotic adjustment, leaf area [85,86], leaf water potential [87], fresh and dry root weight, root volume, relative water content [26], root length [25], photosynthesis [88], early flowering, and harvest index [89] in rice to improve yield under drought is yet to be successfully demonstrated. This also goes for the anthesis-silking interval in maize [90], greenness in sorghum [91], and water-use efficiency in wheat [92]. Improvement in yield potential and yield stability across variable environments has also been reported by considering stay-green [93,94], an essential trait in several crops (maize, rice, sorghum) that gives plants resistance to drought, premature senescence [95], and lodging.

Selection for effective mobilization of the reserves from source to sink [96], osmoregulation [97], cuticular resistance, surface roughness [98], and membrane composition [99] suggested the importance of these traits in reducing drought-dependent yield loss. Stomatal conductance, maximal rates of photosynthesis [100], and developmental plasticity [101] were reported to be positively correlated, whereas leaf temperatures were negatively correlated with yield increase under stress in semi-dwarf spring wheat cultivars [100]. Another example of a successful breeding program for drought stress using carbon isotope discrimination as a substitute for water-use efficiency in increasing yield in wheat was reported by Rebetzke et al. [102] and Cattivelli et al. [103]. The limitations associated with these techniques involved the screening of only a limited number of plants because of high cost and screening under controlled conditions that may not reflect field conditions.

A number of putative secondary traits such as root density, root thickness, root distribution pattern [104,105], rooting depth [106,107], root branching, root-to-shoot ratio, root penetration [108,109, 110,111,112], root length, root hydraulic conductance, transpiration demand [113], and water and nutrient uptake [111,114, 115] have been suggested to confer drought tolerance [116]. Traits such as transpiration rate, biomass accumulation, stomatal conductance, leaf area [117,118, 119], osmoregulation [93], relative water content, and leaf water potential [120] reported a positive association with grain yield under drought stress. Various reports suggested the role of genetic regions associated with secondary traits (Table 4, [121,122,123, 124,125,126, 127,128,129,130,131, 132,133,134,135,136]) in enhancing grain yield under drought stress.

4.1.2. Grain Yield as a Selection Criterion under Drought

Even though screening for physiological traits is more accurate than the screening of complex quantitative agronomic traits, drought is still a complex process involving multiple steps starting from moisture-nutrient uptake by roots to grain formation by the panicle. Each physiological trait in turn fulfills one or two of the multiple sequential components needed to produce higher yield. Moreover, the appropriate combinations of these components to achieve increased yield under drought are not well understood. Grain yield, being a complex quantitative trait, was not considered earlier as a suitable selection criterion in breeding [93,105,137]. On the contrary, exploitation of genetic variation using direct selection for the trait for grain yield under drought and combining high yield potential with this trait has now been suggested as an appropriate alternative [138,139,140,141,142]. Several studies on comparative phenotypic screening of breeding material for grain yield under reproductive-stage drought stress and under a controlled environment [138,139,140,141,142,143] showed moderate heritability of grain yield under drought stress. Several experiments to standardize the procedure of phenotypic screening involving direct selection for grain yield as selection criteria (Figure 3) reported grain yield advantage under reproductive-stage drought stress with comparable yield under irrigated situations in uplands [53] and lowlands [50,144], and in multiple locations [145]. This type of cyclical stress will allow development, phenotyping, and selection for drought resistance in populations consisting of genotypes with broad growth duration.

Table 4. Genetic regions reported to be associated with secondary traits enhancing drought tolerance.

Crop	Chr	Trait Improved	Reference
Rice	1	Root-shoot growth, deep root growth	[109,121]
	9	Root length, root thickness, straw yield	[122,123]
	12	Biomass, panicle number, lateral root, panicle branching	[124,125]
Wheat	2B, 4A, 5A, 7B	Carbon isotope ratio, osmotic potential, chlorophyll content, flag leaf, rolling index	[126]
	2A, 2B, 3A, 3B, 5A, 5B, 6B, 7A,	Osmotic adjustment	[126,127]
Barley	6HL	Relative water content, leaf osmotic potential, osmotic adjustment, carbon isotope discrimination	[128–130]
	2H, 3H, 6H, 7H	Carbon isotope discrimination	[131]
	2H, 4H, 6H, 7H	Chlorophyll, fluorescence	[132]
	1H, 2H, 3H, 5H, 6H,7H	Relative water content	[133,134]
	2H, 3H, 4H, 5H	Osmotic potential	[134]
Sorghum	1, 2, 3, 4	Leaf area, delayed leaf senescence, stay green	[91]
Cotton	06, 02, 25	Biomass production; panicle number, specific, leaf weight and chlorophyll, osmotic potential, stomatal density, stomatal conductance	[135,136]

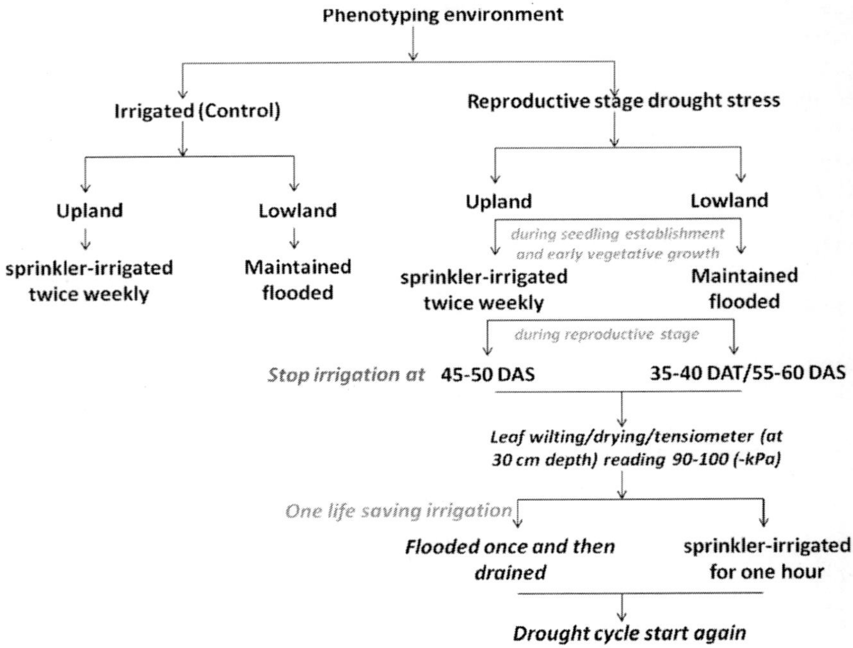

Figure 3. Standardized protocol for drought phenotyping screening at IRRI. DAS: days after seeding, DAT: days after transplanting.

4.1.3. High-Throughput Screening

The new tools of phenomics, such as carbon isotope discrimination (CID) [146], infrared thermography, canopy spectral reflectance [104,147], pulse amplitude-modulated fluorometry for chlorophyll fluorescence [148], normalized difference vegetation index (NDVI) [149] and photosynthetic reflective index (PRI) [150], positron emission tomography (PET), magnetic resonance imaging (MRI), and nuclear magnetic resonance [151,152] are now available to better understand the contribution of different morpho-physiological traits to grain yield. Planes, airborne instruments, and moving equipment with multispectral sensors can estimate the plant cover and nutrient needs of crops. The information collected from phenomics tools such as a high-density soil map to track porosity and mineral content, detectors to predict nutrient content and changes in response to inputs, contour mapping to observe water movements, and soil moisture detectors at multiple depths, when combined with GPS data, can give useful information about land productivity and will be useful for the following season's planting pattern. Well-developed analytical tools/packages are essential for analyzing and interpreting the large amount of data produced by these modern techniques in the future.

4.2. Breeding Strategies

Research work is needed in breeding rice varieties with high grain yield potential, good yield under drought, yield stability, resistance to existing biotic stresses, good grain and cooking quality, and good relative performance in multiple locations and environmental (managed under drought-stress and non-stress environments) conditions.

4.2.1. Donor Identification

The preliminary and important step of any breeding program involves the identification of suitable donors. Selection of a specific donor from a large germplasm collection is a crucial step. The use of a specific donor with special characteristics for a specific environment may lead to the success of any varietal and trait development program. Most of the traditional donors have several undesirable traits and therefore are not suitable for direct use in any breeding program. These landraces have undesirable traits such as little ground cover, tall plant height, low yield potential, and poor grain and eating quality, but they have a desirable drought tolerance trait. On the other hand, modern rice varieties have desirable traits such as high yield, improved plant type (early vigor, medium height, and lodging resistance), tolerance of biotic stress, and good grain type (medium to long slender). However, they are drought-susceptible. Breeding for any desired trait to get new gene combinations requires exploitation of genetic variation (intra-specific, inter-specific, or inter-generic) that exist in traditional landraces carrying desirable characteristics and modern improved varieties with high yield potential [153]. The genotype at par performance in the target environment [154] and the trait with high heritability [155] can account for further high-throughput screening. The identified drought-tolerant donors such as PSBRc68, PSBRc80, PSBRc82, Aday Sel, Dagaddeshi, Kali Aus, Aus276, Kalia, N22, Apo, Dular, and IR77298-14-1-2 have been used in conventional breeding and QTL mapping studies at IRRI. Among these, improved donors such as PSBRc68, PSBRc80, PSBRc82, and IR77298-14-1-2 have been directly used in conventional breeding programs, whereas improved drought-tolerant lines free from undesirable linkages were derived from the mapping populations that involve traditional donors such as Aday Sel, Dagaddeshi, Kali Aus, Aus 276, Kalia, N22, Apo, and Dular and used in conventional breeding programs. In marker-assisted breeding programs, lines possessing the identified QTLs for grain yield under drought, which come from mapping populations that involve traditional donors, were used to improve mega-varieties.

A model drought-resistant rice variety for drought-prone environments can be considered as having better yields than any other presently available cultivar, not only under drought stress but also under irrigated conditions across different seasons and environments, being less sensitive to variable conditions [83,156,157,158], and possessing good grain quality and resistance to biotic stresses.

4.2.2. Conventional Breeding

Over the last 10 years, conventional breeding at distinguished worldwide research centers has made significant progress in developing biotic and abiotic stress-tolerant lines/cultivars of some important food crops such as chickpea [159], soybean [160], wheat [161,162,163], barley [164,165], rice [89], and common bean [166] using different protocols and designs. The drought breeding program at IRRI has led to the development of several high-yielding, drought-tolerant lines with a release of varieties across South and Southeast Asia and Africa since 2009 (Table 5). However, it is time-consuming, costly, and labor-intensive, and there is a high probability of transferring undesirable genes. A modified conventional breeding approach (Figure 4) involving an integrative sequential phenotyping, genotyping, and selection strategy to screen a large number of plants will improve the assessment of plant response to drought stress. This efficient, precise, cost-effective breeding approach may expedite the development of drought-tolerant rice varieties with a high frequency of favorable genes.

4.2.3. Marker-Assisted Breeding: Identification, Introgression, and Pyramiding of QTLs

Marker-assisted breeding adopted at IRRI involves: the development of mapping populations involving traditional drought-tolerant donors and modern high-yielding varieties; precise phenotyping in multi-environment, controlled, and drought-stress conditions; repeated years; identification of polymorphic markers; genotyping with polymorphic markers; linkage map construction; and QTL mapping using genotypic and phenotypic data.

Large-scale systematic study with several mapping populations for identification of major quantitative trait loci (QTLs) using yield as a selection criterion [89] led to the identification of several QTLs for grain yield under drought, followed by introgression of identified QTLs to develop drought-tolerant rice cultivars.

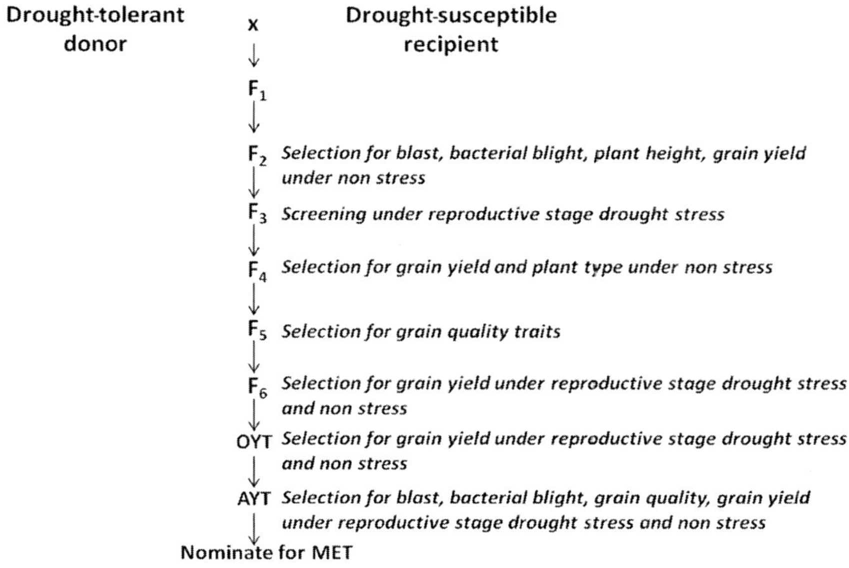

Figure 4. Modified conventional breeding approach. OYT: Observational yield trials, AYT: advanced yield trials, MET: multi environmental trials.

The success of screening strategies with careful assessment of size and structure of population has led to the development and release of several drought-tolerant lines with high yield under irrigated conditions [89]. Identification of genetic regions linked to drought tolerance using genotyping strategies such as selective genotyping (SG), whole-genome genotyping (WGG), bulk segregant analysis (BSA) [50,51,167,168], genome-wide association studies (GWAS, an improved version of marker-assisted selection) [169,170,171,172], and successful introgression in different genetic backgrounds using marker-assisted backcrossing [42,46,52,144,167,168,173,174], marker-assisted recurrent selection [175,176], and marker-assisted QTL pyramiding [89] has been reported. Mapping populations segregating for drought-tolerance-related traits led to the identification of 12 quantitative trait loci (QTLs) (Table 6) showing a large effect against high-yielding, drought-susceptible popular varieties: Swarna, IR64, MTU1010, TDK1, Sabitri, and Vandana [49,50,51,52,53,167,168,177,178,179,180] (Table 6). Gathering all data on the donors/recipients, factors, traits, genes, mechanisms, and technologies that sustain yield under drought and accumulating them into elite genotypes without negative effects on yield potential could be the best solution for rainfed environments.

Table 5. High-yielding drought-tolerant varieties released from IRRI's drought breeding program

Name	Designation	Country	Ecosystem [a]	Release Year	Days to Maturity	Plant Height (cm)
Katihan 1	IR 79913-B-176-B-4	Philippines	UP	2011	105	90
Sahod Ulan 3	IR 81412-B-B-82-1	Philippines	RL	2011	120	107
Sahod Ulan 5	IR 81023-B-116-1-2	Philippines	RL	2011	115	130
Sahod Ulan 6	IR 72667-16-1-B-B-3	Philippines	RL	2011	115	100
Sahod Ulan 8	IR 74963-262-5-1-3-3	Philippines	RL	2011	125	100
Inpago LIPI Go 1	IR 79971-B-191-B-B	Indonesia	UP	2011	110	115
Inpago LIPI Go 2	IR 79971-B-227-B-B	Indonesia	UP	2011	113	114
CR dhan 40	IR 55423-01	India	UP	2012	110	100
Sahod Ulan 12	IR 81047-B-106-2-4	Philippines	RL	2013	105	119
M'ZIVA	R77080-B-B-34-3	Mozambique	RL	2013	120	130
CR dhan 201	IR 83380-B-B-124-1	India	Aerobic	2014	118	100
CR dhan 202	IR 84899-B-154	India	Aerobic	2014	115	100
CR dhan 204	IR 83927-B-B-279	India	Aerobic	2014	110	100
Sukha dhan 5	IR 83388-B-B-108-3	Nepal	RL	2014	125	105
Sukha dhan 6	IR 83383-B-B-129-4	Nepal	RL	2014	125	105
BRRI dhan 66	IR 82635-B-B-75-2	Bangladesh	RL	2014	113	116
Katihan 3	IR 86857-101-2-1-3	Philippines	UP	2014	107	87
DRR dhan 43	IR 83876-B-RP	India	RL	2014	115	105
DRR dhan 44	IR 93376-B-B-130	India	RL	2014	115	105
Katihan 2	IR 82635-B-B-47-2	Philippines	UP	2014	107	84
BRRI dhan 71	IR 82589-B-B-84-3	Bangladesh	RL	2015	115	112
Swarna Shreya	IR 84899-B-179-16-1-1-1-1	India	RL	2015	112	121
Sahod Ulan 15	IR 83383-B-B-129-4	Philippines	RL	2015	115	110
Sahod Ulan 20	IR 86781-3-3-1-1	Philippines	RL	2015	115	112
MPTSA	IR 82077-B-B-71-1	Malawai	RL	2015	120	110
ATETE	IR 80411-B-49-1	Malawai	IR, RL	2015	118	112
CAR 14	IR80463-B-39-3	Cambodia	IR, RL	2015	115	110
Identified	IR 84878-B-60-4-1	Philippines	RL	2016	113	97

[a] UP: upland, RL: rainfed lowland, IR—irrigated ecology. Source: Modified from Kumar et al. [89].

Table 6. QTLs identified for grain yield under drought in different backgrounds.

QTLs	Donors	Backgrounds	Ecosystems	Reference
$qDTY_{1.1}$	N22, Dhagaddeshi, Apo, CT9993-10-1-M, Kali Aus, Basmati 334	Swarna, IR64, MTU1010	Lowland, Upland	[50–52,179]
$qDTY_{2.1}$	Apo, Aus 276	Swarna, MTU1010	Lowland	[52,144]
$qDTY_{2.2}$	Aday sel, Kali Aus	MTU1010, IR64, Samba Mahsuri	Lowland, Upland	[178,180]
$qDTY_{2.3}$	Kali Aus	IR64	Upland, Lowland	[52,180]
$qDTY_{3.1}$	Apo, IR55419-04	Swarna, TDK 1	Lowland	[49,144]
$qDTY_{3.2}$	N22, IR77298-5-6-18, Aday sel	Swarna, Sabitri	Lowland, Upland	[50,158]
$qDTY_{4.1}$	Aday Sel	IR64, Samba Mahsuri	Lowland	[178]
$qDTY_{6.1}$	Apo, Vandana, IR55419-04	IR72, TDK 1	Upland, Lowland	[49,177]
$qDTY_{6.2}$	IR55419-04	TDK 1	Lowland	[49]
$qDTY_{9.1}$	Aday sel	IR64	Lowland	[178]
$qDTY_{10.1}$	N22, Aday sel, Basmati 334	IR64, MTU1010, Swarna	Lowland	[50,178]
$qDTY_{12.1}$	Way Rarem, IR74371-46-1-1	Vandana, Sabitri	Upland, Lowland	[53,167]

Table 7. High-yielding drought-tolerant varieties released from IRRI's drought marker-assisted breeding program.

Name	Designation	Country	Ecosystem	Release Year	Days to Maturity	Plant Height (cm)
Sukha dhan 4	IR 87707-446-B-B-B	Nepal	RL	2014	125	102
DRR 44	IR 87707-445-B-B-B	India	RL	2014	115	110
Yaenelo 4	IR 87707-446-B-B-B	Myanmar	RL	2015	115	117
Yaenelo 5	IR 87705-44-4-B-B	Myanmar	RL	2016	115	117
Yaenelo 6	IR 87707-182-B-B-B	Myanmar	RL	2016	115	117
Yaenelo 7	IR 87705-83-12-B-B	Myanmar	RL	2016	115	117

Source: Modified from Kumar et al. [89].

The drought marker-assisted breeding program at IRRI has led to the development and release of high-yielding drought-tolerant lines (Table 7).

The major and consistent drought grain yield (GY) QTLs were reported to be collocated with QTLs for plant height and/or days to flowering [50,53,144,177]. The developed drought-tolerant lines possessed earliness, root plasticity traits, greater root length density, better water-use efficiency mechanism, better regulation of shoot growth [106,121,181], and a yield advantage of 0.8–1.0 t·ha^{-1} under severe drought. These short-duration varieties of 105–110 days without any yield decline possessed better adaptability to less water and variable environmental growing conditions. QTLs related to traits enhancing drought tolerance have been reported in cotton [136], pearl millet [182], maize [156], Sorghum [91], and barley [183]. Fine-mapping of QTLs to facilitate exact introgression devoid of undesirable linkages; identification of useful candidate genes; effectiveness in various genetic backgrounds and variable environment; and effective use, pyramiding, and interaction studies may now open new windows to the development of drought-tolerant rice cultivars. Fine-mapping of q*DTY*$_{12.1}$ resulted in the partitioning of the *qDTY*$_{12.1}$ into sub-QTLs and multiple intra-QTL genes (*OsNAM*$_{12.1}$transcription factor and co-localized target genes). This strengthened the view of more than a single gene underneath the functionality of one QTL and reiterate grain yield under drought, a complex trait [124]. Insertion mutants in the co-localized target genes in the q*DTY*$_{12.1}$ region lead to an increase in the lateral roots compared to the wild type [124]. Fine-mapping of *qDTY*$_{1.1}$ shows that *qDTY*$_{1.1}$ harbors the green revolution gene '*sd1*' [121].

Genetic linkages; complex gene network; QTL × QTL, QTL × background, QTL × environment interactions [175,184]; and pleiotropy are the most important aspects in breeding when studying the complexity of genetic regions related to drought biotic and abiotic stress traits. The linkage of *qDTY*$_{1.1}$ and *sd1* supports the fact that during the green revolution era the drought-tolerant alleles were not maintained properly during the development of dwarf varieties for the irrigated ecosystem. The debate continued on the pleiotropic effect of dominant allele of *sd1* on drought vs. linkage of dominant allele of *sd1* with drought tolerance. The possibility of a pleiotrophic effect indicated the separation of the drought-susceptible allele and dwarfness is impossible. Vikram et al. [121] have successfully demonstrated the linkage of *qDTY*$_{1.1}$with the *sd1* gene, nullifying the debate on the linkages or pleiotropic effects of the *sd1* gene. The development of new drought-tolerant dwarf lines is a successful example of breakage of linkages between *qDTY*$_{1.1}$ and *sd1* loci. Many studies reported the collocation of major and consistent drought grain yield (GY) QTLs such as *qDTY*$_{1.1}$, *qDTY*$_{2.3}$, *qDTY*$_{3.1}$, *qDTY*$_{3.2}$and *qDTY*$_{12.1}$, with QTLs for days to flowering and plant height [50,52,53,144]. The linkages of the drought QTLs were successfully broken and drought-tolerant lines in Swarna, IR64, and Vandana background were developed [74].

Pyramiding QTLs for a quantitative trait such as grain yield may be an effective approach to combine superior alleles and achieve the desirable phenotypic level of variation [185]. QTL pyramiding may be an appropriate approach to improve the efficiency of marker-assisted selection for desirable loci in rice breeding programs and to understand the interactions among genetic loci. Under severe reproductive-stage drought stress, grain yield advantage of 0.8–1.0 t·ha^{-1} was reported in QTL introgression programs involving popular high-yielding varieties IR64 and Swarna [144,178]. The QTL pyramiding program ongoing at IRRI in the background of popular rice varieties Swarna, IR64, Vandana, Sabitri, TDK1, Anjali, Samba Mahsuri, MRQ74, MR219, and some Korean lines (Jinmibyeo, Gayabyeo, Hanarumbyeo, and Sangnambatbyeo) uses the different marker-assisted breeding approaches shown in Table 8. It is evident from Table 8 that, even for the same QTL, researchers may have to find and use different sets of peak and flanking markers depending on the polymorphism of the donor and recipient and the identification of such polymorphic markers within the QTL region. Fine mapping, physiological and molecular characterization of the QTL interval to capture all the desirable genes with positive interactions contributing to drought tolerance is an important step before initiating a QTL introgression program.

4.3. Interactions between QTLs (Q × Q), QTLs and Genetic Background (Q × G), and QTLs and the Environment (Q × E)

Undesirable genetic linkages, *Q*TL × genetic background (*Q* × G), and *Q*TL × environmental interaction (*Q* × E) play an important role in restricting the use of *Q*TLs in marker-assisted breeding [109,186,187]. The combined effect of alleles at more than one locus on a trait of interest, which departs from simply adding up the effects of the alleles at each locus, represents the case of genetic interaction. Many examples of such interactions are known [188], but the relative contribution of interactions to trait variation is questionable. The large sample size population, effective screening strategy, screening under variable conditions and environment, accurate genotyping, and analytical approach increase the power to detect the *Q*TLs, *Q* × *Q*, and *Q* × E interactions. These interactions could be one of the possible reasons for the variable effect of *Q*TLs in different genetic backgrounds and environments. Identification and pyramiding of positively interacting large-effect *Q*TLs may provide a wider adaptability of *Q*TLs across genetic backgrounds and environments. The effect of the *Q*TLs varies with donors and recipients [50,51]. To achieve success in *Q*TL pyramiding, there is a need to identify *Q*TLs with large and consistent effect under variable environmental conditions; different intensities of stress; multiple genetic backgrounds; and positive interaction between *Q*TLs different genetic backgrounds, *Q*TLs × environment, and *Q*TLs × genotype × environment for appropriate yield increase under drought [50,125]. The selection of donor and recipient varieties in a breeding program requires the consideration of factors such as flowering synchronization, cross compatibility, maturity duration, resistance/susceptibility to biotic and abiotic stresses, and adaptability to environment, and grain quality traits. Stability of grain yield *Q*TLs under drought, different backgrounds, and environments have been reported by Bernier et al. [125] ($qDTY_{12.1}$; 21 experiments conducted at IRRI and in eastern India), Mishra et al. [167] ($qDTY_{12.1}$; at IRRI and Nepal) and Yadaw et al. [168] ($qDTY_{3.2}$ at IRRI, Nepal). Seven *DTY* QTLs—$qDTY_{1.1}$ [50,51,177], $qDTY_{2.2}$ [52,178], $qDTY_{3.1}$ [50,144], $qDTY_{3.2}$ [51,168], $qDTY_{4.1}$ [178], $qDTY_{6.1}$ [50,177], and $qDTY_{12.1}$ [54,167]—have shown consistent effect across two or more genetic backgrounds and ecosystems. Four of the identified *qDTY* QTLs—$qDTY_{1.1}$, $qDTY_{2.2}$, $qDTY_{6.1}$and $qDTY_{12.1}$ [49,52,173] are also known to be associated with increased yield under dry direct-seeded/aerobic situation. Dixit et al. [189] reported positive interaction of $qDTY_{2.3}$ and $qDTY_{3.2}$ with $qDTY_{12.1}$ and Shamsudin et al. [190] reported the positive interaction of $qDTY_{2.2}$ and $qDTY_{3.1}$ with $qDTY_{12.1}$, significantly increasing the yield of $qDTY_{12.1}$ positive lines. Identification of major QTLs for grain yield under drought with a larger and more consistent effect across genetic backgrounds and ecosystems has opened new opportunities of developing new rice varieties with better adaptations to predicted future scenarios.

Besides the contribution of a single genetic region, linkage, pleiotropy [191], and epistasis were reported to be key factors of quantitative traits [192] in wheat, soybean, and rice [109,193-197]. However, few studies have been conducted on the existing positive and negative interactions among different rice yield-related traits/*Q*TLs under drought stress. Unfavorable linkages between desirable and undesirable traits such as high yield under drought, tall plant height, and very early flowering were successfully broken through breeding to develop high-yielding, medium-duration, drought-tolerant rice varieties [121,178]. $qDTY_{3.2}$ was reported to interact with $qDTY_{1.1}$ and $qDTY_{12.1}$ for reduction in flowering duration [74]. Strong interactions between QTL-affecting quantitative traits have also been observed in maize, soybean, and other cereal crops [198-201].

Table 8. QTL pyramiding program ongoing at IRRI in the background of popular rice varieties through marker-assisted breeding.

Breeding Approach [a]	QTLs	Marker	Target Variety	Target Ecosystem
MAS	$qDTY_{3.1}$, $qDTY_{12.1}$	$qDTY_{3.1}$: RM416, RM16030, RM520 $qDTY_{12.1}$: RM28048, RM28130, RM28099, CG29430, indel8	Anjali	Rainfed upland
	$qDTY_{12.1}$	$qDTY_{12.1}$: RM28048, RM28130, RM28099, CG29430, indel8	Kalinga	Rainfed upland
MAB	$qDTY_{2.2}$, $qDTY_{4.1}$	$qDTY_{2.2}$: RM236, RM279, RM555 $qDTY_{4.1}$: RM518, RM335, RM16368	IR64	Rainfed lowland
	$qDTY_{1.1}$, $qDTY_{1.2}$, $qDTY_{2.2}$, $DTY_{2.3}$, $qDTY_{3.2}$, $qDTY_{4.1}$, $qDTY_{12.1}$	$qDTY_{1.1}$:RM11943, RM12023, RM12233 $qDTY_{1.2}$:RM212, RM3825, RM315 $qDTY_{2.2}$: RM236, RM279, RM555 $qDTY_{12.1}$: RM28048, RM28130, RM28099, CG29430, indel8 $qDTY_{2.3}$: RM3212, RM573, RM1367 $qDTY_{3.2}$:RM523, RM22, RM545 $qDTY_{4.1}$: RM518, RM335, RM16368	IR64	Rainfed lowland
	$qDTY_{1.1}$, $qDTY_{2.1}$, $qDTY_{3.1}$	$qDTY_{1.1}$:RM11943, RM12023, RM12091, RM12233 $qDTY_{2.1}$: RM5791, RM521, RM3549, RM324, RM6374 $qDTY_{3.1}$: RM416, RM16030, RM520	Swarna	Rainfed lowland
	$qDTY_{12.1}$	$qDTY_{12.1}$: RM28048, RM28130, RM28099, CG29430, indel8	Vandana	Rainfed upland
	$qDTY_{2.2}$, $qDTY_{4.1}$	$qDTY_{2.2}$: RM236, RM279, RM555 $qDTY_{4.1}$: RM518, RM335, RM16368	Samba Mahsuri	Rainfed lowland
	$qDTY_{3.1}$, $qDTY_{6.1}$, $qDTY_{6.2}$	$qDTY_{3.1}$: RM55, RM168, RM186, RM293, RM468 $qDTY_{6.1}$:RM204, RM217, RM508, RM586, RM587 $qDTY_{6.2}$: RM3, RM541	TDK1	Rainfed lowland
	$qDTY_{3.2}$, $qDTY_{12.1}$	$qDTY_{3.2}$: RM231, RM517 $qDTY_{12.1}$: RM28048, RM511, RM28199, RM28166	Sabitri	Rainfed lowland
	$qDTY_{2.2}$, $qDTY_{3.1}$, $qDTY_{12.1}$	$qDTY_{2.2}$: RM236, RM279, RM12460 $qDTY_{3.1}$: RM416, RM16030, RM520 $qDTY_{12.1}$: RM28048, RM511, RM28099, RM28166, CG29430, indel8, RM28130	MR219	Rainfed lowland
	$qDTY_{2.2}$, $qDTY_{3.1}$, $qDTY_{12.1}$	$qDTY_{2.2}$: RM154, OSR17, RM12460 $qDTY_{3.1}$: RM416, RM15935, RM520 $qDTY_{12.1}$: RM28048, RM511, RM28099, RM28166, CG29430, indel8, RM28130	MRQ74	Rainfed lowland
	$qDTY_{1.1}$, $qDTY_{2.2}$	$qDTY_{1.1}$:RM431, RM11943, RM12023, RM12091, RM12233 $qDTY_{2.2}$: RM236, RM279	Jinmibyeo	Rainfed lowland
	$qDTY_{1.1}$, $qDTY_{2.2}$	$qDTY_{1.1}$: RM12023, RM12146 $qDTY_{2.2}$: RM236, RM279	Gayabyeo	Rainfed lowland
MAB	$qDTY_{1.1}$, $qDTY_{2.2}$	$qDTY_{1.1}$: RM11943, RM12233 $qDTY_{2.2}$: RM236, RM279	Hanarumbyeo	Rainfed lowland
	$qDTY_{1.1}$, $qDTY_{2.2}$	$qDTY_{1.1}$: RM11943, RM12233 $qDTY_{2.2}$: RM109, RM279	Sangnambatbyeo	Rainfed lowland
MARS	$qDTY_{1.1}$, $qDTY_{2.1}$, $qDTY_{3.1}$, $qDTY_{11.1}$	$qDTY_{1.1}$: RM212, RM486 $qDTY_{2.1}$: RM525, RM221 $qDTY_{3.1}$: RM16, RM520 $qDTY_{11.1}$: RM287	Samba Mahsuri	Rainfed lowland

[a] MAS: marker-assisted selection, MAB: marker-assisted backcrossing, MARS: marker-assisted recurrent selection.

A multi-disciplinary approach involving understanding physiological and molecular mechanisms associated with QTLs/genes across variable environments, identification and validation of genomic coordinates for correlated traits, differential expression of genes involved in metabolic processes, signal transductions, and response of identified genes can be used to explain drought tolerance in detail and to select/identify genotypes with stable and improved yield under multiple stresses.

4.4. Transgenic Approaches

Transgenic approaches involve the incorporation of specifically cloned genes by limiting the transfer of unwanted genes from the donor organism. Transgenic approach is being practiced throughout the world to improve resistance to biotic stresses and tolerance of abiotic stresses in a number of crops. Rapid progress in recombinant-DNA technology and development of accurate and efficient gene-transfer protocols have resulted in efficient engineering of genes encoding compatible organic solutes [202], and biosynthesis of glycine betaine in tobacco and maize [203,204]; trehalose-6-phosphate synthase or phosphatase (TPSP) in rice [205], and tobacco [206,207]; choline dehydrogenase in maize [204]; and pyrroline-5-carboxylate synthetase (P5CS) in wheat [208], tobacco [209], soybeans [210], and petunias [211].

Although the transgenic approach is expected to be faster and more precise, there are still constraints associated with it, including gene silencing, undesirable genetic alterations resulting from the transformation process, ethical issues, public acceptability, and the assurance time in biosafety regulations and release. Sometimes the transgenic lines that had shown remarkable performance under controlled laboratory or glasshouse conditions would not be able to survive under natural field conditions where they encounter a myriad of environmental factors. The growth and development stages of plants play a significant role in defining tolerance as the tolerance seen in transgenic lines at one particular stage may not be the same at other growth stages.

4.5. Novel Strategies

Besides conventional and marker-assisted selection, heterosis breeding, recurrent selection, bi-parental mating, disruptive mating, candidate gene identification, gene cloning, plant tissue culture, and foreign gene transfer, novel opportunities of exploiting the full potential of genomics-assisted breeding are on the way and will require an integrated knowledge of high-throughput phenotyping and molecular, physiological, and developmental processes that influence drought tolerance. Genomic selection allows breeders to consider the effect of a huge number of markers to calculate the Genomic Estimated Breeding Value (GEBV), and select a few desired individual plants for phenotypic selection in the field. On the other hand, traditional breeding involves many cycles of selections based on plant phenotypic evaluation or taking the result of a few trait-linked markers into account for quality, disease, and pest resistance. Breeders no longer need to select for individual traits; instead, they can select the combination of traits based on breeding value. This allows for easy selection; breeding cycles are shortened and several breeding programs can run at the same time by planting even a few good progenies within a limited budget.

The supplementation of old with modern breeding techniques and innovative technologies based on the science of genomics may greatly help in increasing crop productivity under drought. With the rapid progress in structural and functional genomics, proteomics will certainly be beneficial to polish existing approaches to achieve significant progress in future crop improvement. The development of genome-wide analytical tools may constitute a turning point towards the easier transfer of beneficial traits to locally adapted varieties. Genome-wide association studies (GWAS) have been widely used as a popular method to identify genetic regions related to drought tolerance traits in plants [169,170,171,172]. GWAS provides a better platform in screening a large number of accessions for genetic variation underlying diverse complex traits. Recent studies reported the combined approach of GWAS and candidate-gene sequencing as a more powerful approach than separate individual approaches [212].

The available rice genome sequence information will make it feasible to produce comprehensive datasets on all existing information on genes; gene function; biochemical and molecular pathways; protein profiles; metabolites and gene expression; comparison of the genome, genes, and intergenic regions between cereal species; and allele mining in the large collection of rice germplasm and wild species. A compilation of all this information will be a boon for the scientific community as it tries to develop new varieties with high yield and stabilize this trait along with resistance to pests and disease; tolerance to drought, salinity, flood, and cold; and improved nutritional quality. The involvement of similar transcription factors, various common stress-inducible genes, and similar physiological and molecular responses in both dicotyledonous and monocotyledonous plants under abiotic stress was reported in *Arabidopsis*, wheat, and rice [213,214,215,216]. The syntenic relationships between different cereal crops and grasses allow developmental biologists, biochemists, and physiologists to inspect the gene complements in related species to see which pathways are common and which are unique, and how these pathways may have been modified. The vast reservoir of available genetic resources (introgression lines, mapping populations, wild species, mutants, NILs (near-isogenic lines), RILs (recombinant inbred lines), improved breeding populations, and double haploids) and the huge amount of genomic, transcriptomic, proteomic, and metabolomic information in rice would be valuable materials in the structural and functional genomics of designing novel rice varieties for a particular ecosystem. High-throughput approaches such as DNA sequencing, SNP chips, microarray, serial analysis of gene expression (SAGE), site directed mutagenesis (T-DNA insertion, transposon tagging and homologous recombination), RNA-mediated interference, yeast two-hybrid screening, and metabolite quantification will help in identifying the conditions under which various genes are expressed and the phenotype that results when they are knocked out or when their expression is altered. This will assist with the identification of alleles conferring a superior phenotype. Bioinformatics will be useful to inter-link the phenotypic data gathered from different locations under different conditions for diverse germplasm with sequence information, which will ultimately provide information on candidate gene, gene function, and phenotypic and genotypic expression of specific genotypes, thereby helping with breeders' development of elite cultivars [217]. Crop models involving the interaction of breeding, genomics, physiology, and system and functional biology will enable us to fill the gap between genotype and complex phenotype [218].

5. CONCLUSIONS

Agriculture has undergone dramatic shifts starting from the introduction of new semi-dwarf rice varieties in 1966. This shift has been less evident in rainfed areas due to the susceptibility of modern semi-dwarf varieties to most of the abiotic stresses prevalent in rainfed ecosystems. Under ongoing climate change, which is predicted to increase the frequency of moderate to severe drought, there is an immediate need to improve existing technologies and compile all the information we have for developing better rice varieties for drought-prone areas. This challenge can only be met with long-term systematic research on drought to generate a better understanding of rice plants that can survive with less water like other cereals.

ACKNOWLEDGMENTS

The authors thank the Generation Challenge Program (GCP), Mexico; the Bill & Melinda Gates Foundation (BMGF); and BMZ, Germany, for financial support for QTL identification and introgression work. We thank the All India Coordinated Rice Improvement Program (AICRIP), NARES (National agricultural research and extension systems) partners and associated scientists for the evaluation of lines in different locations.

REFERENCES

1. Food Agriculture Organization (FAO). Rice Marker Monitor. Available online: http://www.fao.org/economic/RMM (accessed on 28 April 2016).
2. Lorey, D.E. Global Environmental Challenges of the Twenty-First Century: Resources, Consumption, and Sustainable Solutions (No. 3); Rowman and Littlefield Publishers: Lanham, MD, USA, 2002.
3. United Nations Convention to Combat Desertification (UNCCD). Desertification: The Invisible Frontline. Available online: http://www.unccd.int/Lists/ SiteDocument Library/Publications/Desertification_The%20invisible_frontline.pdf
4. Kumar, A.; Bernier, J.; Verulkar, S.; Lafitte, H.R.; Atlin, G.N. Breeding for drought tolerance: Direct selection for yield, response to selection and use of drought-tolerant donors in upland and lowland-adapted populations. Field Crops Res. **2008**, 107, 221–231.
5. Pingali, P.; Raney, T. From the Green Revolution to the Gene Revolution: How will the poor fare? Mansholt Publ. Ser. **2005**, 4, 407.
6. Food Agriculture Organization. The Impact of Natural Hazards and Disasters on Agriculture and Food and Nutrition Security: A Call for Action to Build Resilient Livelihoods; FAO: Rome, Italy, 2015.
7. Sinha, A. Natural Disaster Management in India; A Country Report from Member Countries; Asian Disaster Reduction Center (ADRC): Kobe, Japan, 1999.
8. Bank of Thailand. The Inflation Report; The Bank of Thailand: Bangkok, Thailand, 2005.
9. Sabetfar, S.; Ashouri, M.; Amiri, E.; Babazadeh, S. Effect of drought stress at different growth stages on yield and yield component of rice plant. Persian Gulf Crop Prot. **2013**, 2, 14–18.
10. Boonjung, H.; Fukai, S. Effects of soil water deficit at different growth stages on rice growth and yield under upland conditions. 2. Phenology, biomass production and yield. Field Crops Res. **1996**, 48, 47–55.
11. Song, Y.; Qu, C.; Birch, S.; Doherty, A.; Hanan, J. Analysis and modelling of the effects of water stress on maize growth and yield in dryland conditions. Plant Prod. Sci. **2010**, 13, 199–208.
12. Mostafavi, K.; Shoahosseini, M.; Sadeghi, G.H. Multivariate analysis of variation among traits of corn hybrids traits under drought stress. Int. J. Agric. Sci. **2011**, 1, 416–422.
13. Nezami, A.; Hamid, R.K.; Rezazadeh, Z.B. Effects of Drought Stress and Defoliation on Sunflower. Desert **2008**, 12, 99–104.
14. Chimenti, C.A.; Pearson, J.; Hall, A.J. Osmotic adjustment and yield maintenance under drought in sunflower. Field Crops Res. **2002**, 75, 235–246.
15. Nazariyana, G.; Mehrpooyanb, M.; Khiyavic, M. Study of effects of drought stress on yield and yield components of four sunflower cultivars in Zanjan, Iran. Plant Ecophysiol. **2009**, 3, 135–139.
16. Samarah, N.H.; Mullen, R.E.; Cianzio, S.R.; Scott, P. Dehydrin-like proteins in soybean seeds in response to drought stress during seed filling. Crop Sci. **2006**, 46, 2141–2150.

17. Kobraei, S.; Etminan, A.; Mohammadi, R.; Kobraee, S. Effect of drought stress on yield and components of yield of soyabean. Ann. Biol. Res. **2011**, 2, 504–509.
18. Baroowa, B.; Gogoi, N. Morpho-physiological and yield responses of black gram (Vigna mungo L.) and green gram (Vigna radiate L.) genotypes under drought at different Growth stages. Res. J. Rec. Sci. **2016**, 5, 43–50.
19. Ma, F.Y.; Li, M.C.; Yang, J.R.; Ji, X.J.; Shentu, X.D.; Tao, H.J. A study of effect of water deficit of three periods during cotton anthesis on canopy apparent photosynthesis and WUE. Sci. Agric. Sin. **2002**, 12, 1467–1472.
20. Loka, D.A.; Oosterhuis, D.M. Water stress and reproductive development in cotton. In Flowering and Fruiting in Cotton; Oosterhuis, D.M., Cothren, J.T., Eds.; The Cotton Foundation: Candova, TN, USA, 2012; pp. 51–58.
21. Hsiao, T.C. The soil plant atmosphere continuum in relation to drought and crop production. In Drought Resistance in Crops with Emphasis on Rice; IRRI: Los Banos, Philippines, 1982; pp. 39–52.
22. O'Toole, J.C. Adaptation of rice to drought prone environments. In Drought Resistance in Crops with Emphasis on Rice; IRRI: Los Baños, Philippines, 1982; pp. 195–213.
23. Harris, D.; Tripathi, R.S.; Joshi, A. On-farm seed priming to improve crop establishment and yield in dry direct-seeded rice. In Direct Seeding: Research Strategies and Opportunities; Pandey, S., Mortimer, M., Wade, L., Tuong, T.P., Lopes, K., Hardy, B., Eds.; International Research Institute: Manila, Philippines, 2002; pp. 231–240.
24. Kaya, M.D.; Okçub, G.; Ataka, M.; Çıkılıc, Y.; Kolsarıcıa, Ö. Seed treatments to overcome salt and drought stress during germination in sunflower (Helianthus annuus L.). Eur. J. Agron. **2006**, 24, 291–295.
25. Asch, F.; Dingkuhn, M.; Sow, A.; Audebert, A. Drought-induced changes in rooting patterns and assimilate partitioning between root and shoot in upland rice. Field Crops Res. **2005**, 93, 223–236.
26. Sandhu, N.; Jain, S.; Battan, K.R.; Jain, R.K. Aerobic rice genotypes displayed greater adaptation to water-limited cultivation and tolerance to polyethyleneglycol-6000 induced stress. Physiol. Mol. Biol. Plants **2011**, 18, 33–43.
27. Sinclair, T.R.; Ludlow, M.M. Who taught plants thermodynamics? The unfulfilled potential of plant water potential. Funct. Plant Biol. **1985**, 12, 213–217.
28. Anjum, F.; Yaseen, M.; Rasul, E.; Wahid, A.; Anjum, S. Water stress in barley (Hordeum vulgare L.). I. Effect on chemical composition and chlorophyll contents. Pak. J. Agric. Sci. **2003**, 40, 45–49.
29. Lei, D.; Ying, L.I.; Yong, L.I.; Qi rong, S.; Shiwei, G. Effects of Drought Stress on Photosynthesis and water status of rice leaves. Chin. J. Rice Sci. **2014**, 28, 65–70.
30. Hussain, M.; Malik, M.A.; Farooq, M.; Ashraf, M.Y.; Cheema, M.A. Improving Drought tolerance by exogenous application of glycine betaine and salicylic acid in sunflower. J. Agron. Crop Sci. **2008**, 194, 193–199.
31. Wahid, A.; Rasul, E. Photosynthesis in leaf, stem, flower and fruit. In Handbook of Photosynthesis, 2nd ed.; Pessarakli, M., Ed.; CRC Press: Boca Raton, FL, USA, 2005; pp. 479–497.
32. Farooq, M.; Wahid, A.; Kobayashi, D.N.; Fujita, S.; Basra, M.A. Plant drought stress: Effects, mechanisms and management. In Agronomy for Sustainable Development; Springer: Berlin, Germany, 2009; Volume 29, pp. 185–212.
33. Turner, N.C.; Wright, G.C.; Siddique, K.H.M. Adaptation of grain legumes (Pulses) to water-limited environments. Adv. Agron. **2000**, 71, 193–231.
34. Grossman, A.; Takahashi, H. Macronutrient utilization by photosynthetic eukaryotes and the fabric of interactions. Annu. Rev. Plant Phys. **2001**, 52, 163–210.
35. McWilliams, D. Drought Strategies for Cotton, Cooperative Extension Service Circular 582; College of Agriculture and Home Economics, New Mexico State University: Las Cruces, NM, USA, 2003.
36. Ekanayake, I.J.; Steponkus, P.L.; De Datta, S.K. Spikelet sterility and flowering response of rice to water stress at anthesis. Ann. Bot. **1989**, 63, 257–264.
37. Ekanayake, I.J.; Steponkus, P.L.; De Datta, S.K. Sensitivity of pollination to water deficits at anthesis in upland rice. Crop Sci. **1990**, 30, 310–315.

38. Liu, J.X.; Liao, D.Q.; Oane, R.; Estenor, L.; Yang, X.E.; Li, Z.C.; Bennett, J. Genetic variation in the sensitivity of anther dehiscence to drought stress in rice. Field Crops Res. **2006**, 97, 87–100.
39. Fu, J.; Huang, B. Involvement of antioxidants and lipid peroxidation in the adaptation of two cool-season grasses to localized drought stress. Environ. Exp. Bot. **2001**, 45, 105–114.
40. Reddy, A.R.; Chaitanya, K.V.; Vivekanandan, M. Drought-induced responses of photosynthesis and antioxidant metabolism in higher plants. J. Plant Physiol. **2004**, 161, 1189–1202.
41. Kasuga, M.; Liu, Q.; Miura, S.; Yamaguchi-Shinozaki, K.; Shinozaki, K. Improving plant drought, salt, and freezing tolerance by gene transfer of a single stress-inducible transcription factor. Nat. Biotechnol. **1999**, 17, 287–291.
42. Rontein, D.; Basset, G.; Hanson, A.D. Metabolic engineering of osmoprotectant accumulation in plants. Metab. Eng. **2002**, 4, 49–56.
43. Rampino, P.; Pataleo, S.; Gerardi, C.; Mita, G.; Perrotta, C. Drought stress response in wheat: Physiological and molecular analysis of resistant and sensitive genotypes. Plant Cell Environ. **2006**, 29, 2143–2152.
44. Cairns, J.E.; Impa, S.M.; O'Toole, J.C.; Jagadish, S.V.K.; Price, A.H. Influence of the soil physical environment on rice (Oryza sativa L.) response to drought stress and its implications for drought research. Field Crops Res. **2011**, 121, 303–310.
45. Fischer, K.S.; Lafitte, R.; Fukai, S.; Atlin, G.; Hardy, B. Breeding Rice for Drought-Prone Environments; International Rice Research Institute: Los Baños, Philippines, 2003.
46. Araus, J.L.; Slafer, G.A.; Reynolds, M.P.; Royo, C. Plant breeding and water relations in C3 cereals: What should we breed for? Ann. Bot. **2002**, 89, 925–940.
47. Bartels, D.; Sour, E. Molecular responses of higher plants to dehydration. In Plant Responses to Abiotic Stress; Hirt, H., Shinozaki, K., Eds.; Springer: Berlin, Germany, 2004; pp. 13–37.
48. Dixit, S.; Swamy, B.P.M.; Vikram, P.; Ahmed, H.U.; Sta Cruz, M.T.; Amante, M.; Atri, D.; Leung, H.; Kumar, A. Fine mapping of QTLs for rice grain yield under drought reveals sub-QTLs conferring a response to variable drought severities. Theor. Appl. Genet. **2012**, 125, 155–169.
49. Dixit, S.; Singh, A.; Sta Cruz, M.T.; Maturan, P.T.; Amante, M.; Kumar, A. Multiple major QTL lead to stable yield performance of rice cultivars across varying drought intensities. BMC Genet. **2014**, 15, 16.
50. Vikram, P.; Swamy, B.P.M.; Dixit, S.; Ahmed, H.U.; Sta Cruz, M.T.; Singh, A.K.; Kumar, A. qDTY1.1, a major QTL for rice grain yield under reproductive-stage drought stress with a consistent effect in multiple elite genetic backgrounds. BMC Genet. **2011**, 12, 89.
51. Ghimire, K.H.; Quiatchon, L.A.; Vikram, P.; Swamy, B.P.M.; Dixit, S.; Ahmed, H.; Hernandez, J.E.; Borromeo, T.H.; Kumar, A. Identification and mapping of a QTL (qDTY1.1) with a consistent effect on grain yield under drought. Field Crops Res. **2012**, 131, 88–96.
52. Sandhu, N.; Singh, A.; Dixit, S.; Sta Cruz, M.T.; Maturan, P.C.; Jain, R.K.; Kumar, A. Identification and mapping of stable QTL with main and epistasis effect on rice grain yield under upland drought stress. BMC Genet. **2014**, 15, 63.
53. Bernier, J.; Kumar, A.; Venuprasad, R.; Spaner, D.; Atlin, G.N. A large-effect QTL for grain yield under reproductive-stage drought stress in upland rice. Crop Sci. **2007**, 47, 507–516.
54. Bauder, J. Irrigating with Limited Water Supplies; Montana State Univ. Comm. Ser. Montana Hall: Bozeman, MT, USA, 2011.
55. Amiri, M.; Naseri, R.; Soleimani, R. Response of different growth stages of wheat to moisture tension in a semiarid land. World Appl. Sci. J. **2011**, 12, 83–89.
56. Wang,W.; Vinocur, B.; Altman, A. Plant responses to drought, salinity and extreme temperatures: Towards genetic engineering for stress tolerance. Planta **2003**, 218, 1–14.
57. Praba, M.L.; Cairns, J.E.; Babu, R.C.; Lafitte, H.R. Identification of physiological traits underlying cultivar differences in drought tolerance in rice and wheat. J. Agron. Crop Sci. **2009**, 195, 30–46.

58. Winkel, T.; Renno, J.F.; Payne,W.A. Effect of the timing of water deficit on growth, phenology and yield of pearl millet (Pennisetum glaucum (L.) R. Br.) grown in Sahelian conditions. J. Exp. Bot. **1997**, 48, 1001–1009.
59. Yadav, R.S.; Hash, C.T.; Bidinger, F.R.; Cavan, G.P.; Howarth, C.J. Quantitative trait loci associated with traits determining grain and stover yield in pearl millet under terminal drought-stress conditions. Theor. Appl. Genet. **2002**, 104, 67–83.
60. Adebayo, M.A.; Menkir, A. Assessment of hybrids of drought tolerant maize (Zea mays L.) inbred lines for grain yield and other traits under stress managed conditions. Nig. J. Genet. **2014**, 28, 19–23.
61. Adee, E.; Roozeboom, K.; Balboa, G.R.; Schlegel, A.; Ciampitti, I.A. Drought-Tolerant Corn Hybrids Yield More in Drought-Stressed Environments with No Penalty in Non-stressed Environments. Front. Plant Sci. **2016**, 7, 1534–1542.
62. Zarabi, M.; Alahdadi, I.; Akbari, G.A.; Akbari, G.A. A study on the effects of different biofertilizer combinations on yield, its components and growth indices of corn (Zea mays L.) under drought stress condition. Afr. J. Agric. Res. **2011**, 6, 681–685.
63. Daryanto, S.;Wang, L.; Jacinthe, P.A. Global synthesis of drought effects on maize and wheat production. PLoS ONE **2016**, 11, e0156362.
64. Samarah, N.H.; Alquda, A.M.; Amayreh, J.A.; McAndrews, G.M. The Effect of late-terminal drought stress on yield components of four barley cultivars. J. Agron. Crop Sci. **2009**, 195, 427–441.
65. Samarah, N.H.; Haddad, N.; Alqudah, A.M. Yield potential evaluation in Chickpea genotypes under late terminal drought in relation to the length of reproductive stage. Ital. J. Agron. Riv. Agron. **2009**, 3, 111–117.
66. Nayyar, H.; Kaur, S.; Singh, S.; Upadhyaya, H.D. Differential sensitivity of Desi (small-seeded) and Kabuli (large seeded) chickpea genotypes to water stress during seed filling: Effects on accumulation of seed reserves and yield. J. Sci. Food Agric. **2006**, 86, 2076–2082.
67. Nam, N.H.; Chauhany, S.; Johansen, C. Effect of timing of drought stress on growth and grain yield of extra-short-duration pigeon pea lines. J. Agric. Sci. **2001**, 136, 179–189.
68. Tanveer ul, H.; Anser, A.; Sajid, M.N.; Muhammad, M.M.; Muhammad, I. Performance of canola cultivars under drought stress induced by withholding irrigation at different growth stages. Soil Environ. **2014**, 33, 43–50.
69. Zhu, X.; Gong, H.; Chen, G.;Wang, S.; Zhang, C. Different solute levels in two spring wheat cultivars induced by progressive field water stress at different developmental stages. J. Arid Environ. **2005**, 62, 1–14.
70. Plaut, Z.; Butow, B.J.; Blumenthal, C.S.; Wrigley, C.W. Transport of dry matter into developing wheat kernels and its contribution to grain yield under post anthesis water deficit and elevated temperature. Field Crops Res. **2004**, 86, 185–198.
71. Africa Rice. Boosting Africa's Rice Sector: A Research for Development Strategy 2011–2020; Africa Rice Center: Cotonou, Benin, 2011.
72. Africa Rice Center. Overcoming abiotic stresses to increase production. In Africa Rice Center Annual Report 2010: Building African Capacity on Policy Analysis and Impact Assessment; Africa Rice Center: Cotonou, Benin, 2011; pp. 10–14.
73. Kamoshita, A.; Babu, R.C.; Boopathi, N.; Fukai, S. Phenotypic and genotypic analysis of drought-resistance traits for development of rice cultivars adapted to rainfed environments. Field Crops Res. **2008**, 109, 1–23.
74. Vikram, P.; Swamy, B.P.M.; Dixit, S.; Trinidad, J.; Sta Cruz, M.T.; Maturan, P.C.; Amante, M.; Kumar, A. Linkages and interactions analysis of major effect drought grain yield QTLs in rice. PLoS ONE **2016**, 11, e0151532.
75. Turner, N.C.; Abbo, S.; Berger, J.D.; Chaturvedi, S.K.; French, R.J.; Ludwig, C.; Mannur, D.M.; Singh, S.J.; Yadava, H.S. Osmotic adjustment in chickpea (Cicer arietinum L.) results in no yield benefit under terminal drought. J. Exp. Bot. **2007**, 58, 187–194.
76. Finlay, K.W.; Wilkinson, G.N. The analysis of adaptation in a plant breeding programme. Aust. J. Agric. Res. **1963**, 14, 742–754.
77. Eberhart, S.A.; Russell,W.A. Stability parameters for comparing varieties. Crop Sci. **1966**, 6, 36–40.

78. Fischer, R.A.; Maurer, R. Drought resistance in spring wheat cultivars. I. Grain yield response. Aust. J. Agric. Res. **1978**, 29, 897–912.
79. Yadav, O.P.; Bhatnagar, S.K. Evaluation of indices for identification of pearl millet cultivars adapted to stress and non-stress conditions. Field Crops Res. **2001**, 70, 201–208.
80. Karamanos, A.J.; Papatheohari, A.Y. Assessment of drought resistance of crop genotypes by means of the water potential Index. Crop Sci. **1999**, 39, 1792–1797.
81. Idso, S.B.; Reginato, R.; Reicosky, D.; Hatfield, J. Determining soil induced plant water potential depression in alfalfa by means of infrared thermometer. Agron. J. **1981**, 73, 826–830.
82. Motzo, R.; Giunta, F.; Deidda, M. Factors affecting the genotype x environment interaction in spring triticale grown in a Mediterranean environment. Euphytica **2001**, 121, 317–324.
83. Rizza, F.; Badeck, F.W.; Cattivelli, L.; Li Destri, O.; Di Fonzo, N.; Stanca, A.M. Use of a water stress index to identify barley genotypes adapted to rainfed and irrigated conditions. Crop Sci. **2004**, 44, 2127–2137.
84. Voltas, J.; Lopez-Corcoles, H.; Borras, G. Use of biplot analysis and factorial regression for the investigation of superior genotypes in multienvironment trials. Eur. J. Agron. **2005**, 22, 309–324.
85. Babu, C.R.; Nguyen, B.D.; Chamarerk, V.; Shanmugasundaram, P.; Chezhian, P.; Juyaprakash, P.; Ganesh, S.K.; Palchamy, A.; Sadasivam, S.; Sarkarung, S.; et al. Genetic analysis of drought resistance in rice by molecular markers: Association between secondary traits and field performance. Crop Sci. **2003**, 43, 1457–1469.
86. Lanceras, J.C.; Pantuwan, G.; Jongdee, B.; Toojinda, T. Quantitative trait loci associated with drought tolerance at reproductive stage in rice. Plant Physiol. **2004**, 135, 384–399.
87. Jongdee, B.; Pantuwan, G.; Fukai, S.; Fischer, K. Improving drought tolerance in rainfed lowland rice: An example from Thailand. Agric. Water Manag. **2006**, 80, 225–240.
88. Long, S.P.; Zhu, X.G.; Naidu, S.L.; Ort, D.R. Can improvement in photosynthesis increase crop yields? Plant Cell Environ. **2006**, 29, 315–330.
89. Kumar, A.; Dixit, S.; Ram, T.; Yadaw, R.B.; Mishra, K.K.; Mandal, N.P. Breeding high-yielding drought-tolerant rice: Genetic variations and conventional and molecular approaches. J. Exp. Bot. **2014**, 65, 6265–6278.
90. Liu, Y.; Li, S.; Chen, F.; Yang, S.; Chen, X. Soil water dynamics and water use efficiency in spring maize (Zea mays L.) fields subjected to different water management practices on the Loess Plateau, China. Agric. Water Manag. **2010**, 97, 769–775.
91. Harris, K.; Subudhi, P.K.; Borrel, A.; Jordan, D.; Rosenow, D.; Nguyen, H.; Klein, P.; Klein, R.; Mullet, J. Sorghum stay-green QTL individually reduce post-flowering drought-induced leaf senescence. J. Exp. Bot. **2007**, 58, 327–338.
92. Condon, A.G.; Richards, R.A.; Rebetzke, G.J.; Farquhar, G.D. Breeding for high water-use efficiency. J. Exp. Bot. **2004**, 55, 2447–2460.
93. Campos, H.; Cooper, M.; Habben, J.E.; Edmeades, G.O.; Schussler, J.R. Improving drought tolerance in maize: A view from industry. Field Crops Res. **2004**, 90, 19–34.
94. Tollenar, M.; Wu, J. Yield in temperate maize is attributable to greater stress tolerance. Crop Sci. **1999**, 39, 1604–1897.
95. Rajcan, I.; Tollenaar, M. Source-sink ratio and leaf senescence in maize. I. Dry matter accumulation and partitioning during the grain-filling period. Field Crops Res. **1999**, 90, 245–253.
96. Blum, A. Improving wheat grain filling under stress by stem reserve mobilisation. Euphytica **1988**, 100, 77–83.
97. Moinuddin; Fischer, R.A.; Sayre, K.D.; Reynolds, M.P. Osmotic Adjustment in wheat in relation to grain yield under water deficit. Environ. Agron. J. **2005**, 97, 1062–1071.
98. Kerstiens, G. Cuticular water permeability and its physiological significance. J. Exp. Bot. **1996**, 47, 1813–1832.
99. Tyerman, S.D.; Niemietz, C.M.; Bramley, H. Plant aquaporins: Multifunctional water and solute channels with expanding roles. Plant Cell Environ. **2002**, 25, 173–194.

100. Fischer, R.A.; Rees, D.; Sayre, K.D.; Lu, Z.M.; Condon, A.G.; Larque Saavedra, A. Wheat yield progress associated with higher stomatal conductance and photosynthetic rate and cooler canopies. Crop Sci. **1998**, 38, 1467–1475.
101. Siddique, K.H.M.; Tennant, D.; Perry, M.W.; Belford, R.K. Water use and water use efficiency of old and modern wheat cultivars in a Mediterranean type environment. Aust. J. Agric. Res. **1990**, 41, 431–447.
102. Rebetzke, G.J.; Condon, A.G.; Richards, R.A.; Farquhar, G.D. Selection for reduced carbon isotope discrimination increases aerial biomass and grain yield of rainfed bread wheat. Crop Sci. **2002**, 42, 739–745.
103. Cattivelli, L.; Rizza, F.; Badeck, F.W.; Mazzucotelli, E.; Mastrangelo, A.M.; Francia, E.; Mare, C.; Tondelli, A.; Stanca, A.M. Drought tolerance improvement in crop plants: An integrated view from breeding to genomics. Field Crops Res. **2008**, 105, 1–14.
104. Sirault, X.R.R.; James, R.A.; Furbank, R.T. A new screening method for osmotic component of salinity tolerance in cereals using infrared thermography. Funct. Plant Biol. **2009**, 36, 970–977.
105. Fukai, S.; Cooper, M. Development of drought-resistant cultivars using physiomorphological traits in rice. Field Crops Res. **1995**, 40, 67–86.
106. Wade, L.; Bartolome, V.; Mauleon, R. Environmental response and genomic regions correlated with rice root growth and yield under drought in the OryzaSNP panel across multiple study systems. PLoS ONE **2015**, 10, e0124127.
107. Pantuwan, G.; Ingram, K.T.; Sharma, P.K. Rice root system development under rainfed conditions. In Proceedings of the Thematic Conference on Stress Physiology, Rainfed Lowland Rice Research Consortium, Lucknow, India, 23 February–5 March 1994; International Rice Research Centre: Manila, Philippines, 1996; pp. 198–206.
108. Pantuwan, G.; Fukai, S.; Cooper, M.; Rajatasereekul, S.; O'Toole, J.C. Yield response of rice (Oryza sativa L.) genotypes to different types of drought under rainfed lowlands. Part1. Grain yield and yield components. Field Crops Res. **2002**, 73, 153–168.
109. Price, A.H.; Cairns, J.E.; Horton, P.; Jones, R.G.W.; Griffiths, H. Linking drought-resistance mechanisms to drought avoidance in upland rice during a QTL approach: Progress and new opportunities to integrate stomatal and mesophyll responses. J. Exp. Bot. **2002**, 53, 989–1004.
110. Yadav, R.; Courtois, B.; Huang, N.; McLaren, G. Mapping genes controlling root morphology and root distribution in a double haploid population of rice. Theor. Appl. Genet. **1997**, 619–632.
111. Lafitte, H.R.; Champoux, M.C.; McLaren, G.; O'Toole, J.C. Rice root morphological traits are related to isozyme group and adaptation. Field Crops Res. **2001**, 71, 57–70.
112. Nhan, D.Q.; Thaw, S.; Matsuo, N.; Xuan, T.D.; Hong, N.H.; Mochizuki, T. Evaluation of root penetration ability in rice using the wax-layers and the soil cake methods. J. Fac. Agric. Kyushu Univ. **2006**, 51, 251–256.
113. Nobel, P.S. Physiochemical and Environmental Plant Physiology, 3rd ed.; Elsevier: Amsterdam, The Netherlands, 2005.
114. Nguyen, H.T.; Babu, R.C.; Blum, A. Breeding for drought resistance in rice: Physiological and molecular genetics considerations. Crop Sci. **1997**, 37, 1426–1434.
115. Kato, Y.; Abe, J.; Kamoshita, A.; Yamagishi, J. Genotypic variation in root growth angle in rice (Oryza sativa L.) and its association with deep root development in upland fields with different water regimes. Plant Soil **2006**, 287, 117–129.
116. Deivanai, S.; Devi, S.S.; Sharrmila Rengeswari, P. Physiochemical traits as potential indicators for determining drought tolerance during active tillering stage in rice (Oryza sativa L.). Pert. J. Trop. Agric. Sci. **2010**, 33, 61–70.
117. Cabuslay, G.S.; Ito, O.; Alejar, A.A. Physiological evaluation of responses of rice (Oryza sativa L.) to water deficit. Plant Sci. **2002**, 163, 815–827.
118. Richards, R.A. Selectable traits to increase crop photosynthesis and yield of grain crops. J. Exp Bot. **2000**, 51, 447–458.
119. Tardieu, F.; Tuberosa, R. Dissection and modeling of abiotic tolerance plants. Curr. Opin. Plant Biol. **2010**, 13, 206–212.

120. Kumar, R.; Malaiya, S.; Srivastava, M.N. Evaluation of morpho-physiological traits associated with drought tolerance in rice. Indian J. Plant Physiol. **2004**, 9, 305–307.
121. Vikram, P.; Swamy, M.; Dixit, S.; Singh, R.; Singh, B.P.; Miro, B.; Kohli, A.; Henry, A.; Singh, N.K.; Kumar, A. Drought susceptibility of modern rice varieties: An effect of linkage of drought tolerance with undesirable traits. Sci. Rep. **2015**, 5, 14799122. Steele, K.A.; Price, A.H.; Shashidar, H.E.;Witcombe, J.R. Marker-assisted selection to introgress rice QTLs controlling root traits into an Indian upland rice variety. Theor. Appl. Genet. **2006**, 112, 208–221.
122. Steele, K.A.; Virk, D.S.; Kumar, R.; Prasad, S.C.; Witcombe, J.R. Field evaluation of upland rice lines selected for QTLs controlling root traits. Field Crops Res. **2007**, 101, 180–186.
123. Dixit, S.; Biswal, A.K.; Min, A.; Henry, A.; Oane, R.H.; Raorane, M.L.; Longkumer, T.; Pabuayon, I.M.; Mutte, S.K.; Vardarajan, A.R.; et al. Action of multiple intra-QTL genes concerted around a co-localized transcription factor underpins a large effect QTL. Sci. Rep. **2015**, 5, 15183.
124. Bernier, J.; Kumar, A.; Venuprasad, R. Characterization of the effect of a QTL for drought resistance in rice, qtl12.1, over a range of environments in the Philippines and eastern India. Euphytica **2009**, 166, 207–217.
125. Peleg, Z.; Fahima, T.; Krugman, T.; Abbo, S.; Yakir, D.; Korol, A.B.; Saranga, Y. Genomic dissection of drought resistance in durum wheat wild emmer wheat recombinant inbreed line population. Plant Cell Environ. **2009**, 32, 758–779.
126. Morgan, J.M.; Tan, M.K. Chromosomal location of a wheat osmoregulation gene using RFLP analysis. Aust. J. Plant Physiol. **1996**, 23, 803–806.
127. Teulat, B.; This, D.; Khairallah, M.; Borries, C.; Ragot, C.; Sourdille, P.; Leroy, P.; Monneveux, P.; Charrier, A. Several QTLs involved in osmotic adjustment trait variation in barley (Hordeum vulgare L.). Theor. Appl. Genet. **1998**, 96, 688–698.
128. Teulat, B.; Borries, C.; This, D. New QTLs identified for plant water status, water-soluble carbohydrate and osmotic adjustment in a barley population grown in a growth-chamber under two water regimes. Theor. Appl. Genet. **2001**, 103, 161–170.
129. Teulat, B.; Zoumarou-Wallis, N.; Rotter, B.; Ben Salem, M.; Bahri, H.; This, D. QTL for relative water content in field-grown barley and their stability across Mediterranean environments. Theor. Appl. Genet. **2003**, 108, 181–188.
130. Teulat, B.; Merah, O.; Sirault, X.; Borries, C.;Waugh, R.; This, D. QTLs for grain carbon isotope discrimination in field-grown barley. Theor. Appl. Genet. **2002**, 106, 118–126.
131. Guo, P.G.; Baum, M.; Varshney, R.K.; Graner, A.; Grando, S.; Ceccarelli, S. QTLs for chlorophyll and chlorophyll fluorescence parameters in barley under post-flowering drought. Euphytica **2008**, 163, 203–214.
132. Chen, G.X.; Krugman, T.; Fahima, T.; Chen, K.G.; Hu, Y.G.; Roder, M.; Nevo, E.; Korol, A. Chromosomal regions controlling seedling drought resistance in Israeli wild barley, Hordeum spontaneum C. Koch. Genet. Resour. Crop Evol. **2010**, 57, 85–99.
133. Diab, A.A.; Teulat-Merah, B.; This, D.; Ozturk, N.Z.; Benscher, D.; Sorrells, M.E. Identification of drought-inducible genes and differentially expressed sequence tags in barley. Theor. Appl. Genet. **2004**, 109, 1417–1425.
134. Levi, A.; Paterson, A.H.; Barak, V.; Yakir, D.;Wang, B.; Chee, P.W. Field evaluation of cotton near-isogenic lines introgressed with QTLs for productivity and drought related traits. Mol. Breed. **2009**, 23, 179–195.
135. Levi, A.; Ovnat, L.; Paterson, A.H.; Saranga, Y. Photosynthesis of cotton near-isogenic lines introgressed with QTLs for productivity and drought related traits. Plant Sci. **2009**, 177, 88–96.
136. Fukai, S.; Basnayake, J.; Cooper, M. Modelling water availability, crop growth, and yield of rainfed lowland rice genotypes in northeast Thailand. In Proceedings of the International Workshop on Characterizing and Understanding Rainfed Environments, Bali, Indonesia, 5–9 December 1999; Tuong, T.P., Kam, S.P., Wade, L., Pandey, S., Bouman, B.A.M., Hardy, B., Eds.; International Rice Research Institute: Los Banos, Philippines, 2001; pp. 111–130.
137. Venuprasad, R.; Lafitte, H.R.; Atlin, G.N. Response to direct selection for grain yield under drought stress in rice. Crop Sci. **2007**, 47, 285–293.

138. Venuprasad, R.; Sta Cruz, M.T.; Amante, M.; Magbanua, R.; Kumar, A.; Atlin, G.N. Response to two cycles of divergent selection for grain yield under drought stress in four rice breeding populations. Field Crops Res. **2008**, 107, 232–244.
139. Dixit, S.; Huang, B.E.; Cruz, M.T.S.; Maturan, P.T.; Ontoy, J.C.E.; Kumar, A. QTLs for tolerance of drought and breeding for tolerance of abiotic and biotic stress: An integrated approach. PloS ONE **2014**, 9, e109574.
140. Kumar, A.; Verulkar, S.B.; Dixit, S.; Chauhan, B.; Bernier, J.; Venuprasad, R.; Zhao, D.; Shrivastava, M.N. Yield and yield-attributing traits of rice (Oryza sativa L.) under lowland drought and suitability of early vigor as a selection criterion. Field Crops Res. **2009**, 114, 99–107.
141. Verulkar, S.B.; Mandal, N.P.; Dwivedi, J.L.; Singh, B.N.; Sinha, P.K.; Mahato, R.N.; Dongre, P.; Singh, O.N.; Bose, L.K.; Swain, P.; et al. Breeding resilient and productive genotypes adapted to drought-prone rainfed ecosystem of India. Field Crops Res. **2010**, 117, 197–208.
142. Lafitte, H.R.; Price, A.H.; Courtois, B. Yield response to water deficit in an upland rice mapping population: Associations among traits and genetic markers. Theor Appl Genet. **2004**, 109, 1237–1246.
143. Venuprasad, R.; Dalid, C.O.; Del Valle, M.; Zhao, D.; Espiritu, M.; Sta Cruz, M.T.; Amante, M.; Kumar, A.; Atlin, G.N. Identification and characterization of large-effect quantitative trait loci for grain yield under lowland drought stress in rice using bulk-segregant analysis. Theor. App. Genet. **2009**, 120, 177–190.
144. Ouk, M.; Basnayake, J.; Tsubo, M.; Fukai, S.; Fischer, K.S.; Cooper, M.; Nesbitt, H. Use of drought response index for identification of drought tolerant genotypes in rainfed lowland rice. Field Crops Res. **2006**, 99, 48–58.
145. Richards, R.A.; Rebetzke, G.J.; Watt, M.; Condon, A.G.; Spielmeyer, W.; Dolferus, R. Breeding for improved water productivity in temperate cereals: Phenotyping, quantitative trait loci, markers and the selection environment. Funct. Plant Biol. **2010**, 37, 85–97.
146. Gutierrez, M.; Reynolds, M.P.; Klatt, A.R. Association of water spectral indices with plant and soil water relations in contrasting wheat genotypes. J. Exp. Bot. **2010**, 61, 3291–3303.
147. Baker, N.R. Chlorophyll fluorescence: A probe of photosynthesis in vivo. Annu. Rev. Plant Biol. **2008**, 59, 89–113.
148. Jones, H.G.; Vaughan, R.A. Remote Sensing of Vegetation: Principles, Techniques and Applications; Oxford University Press: Oxford, UK, 2010.
149. Gamon, J.A.; Penuelas, J.; Field, C.B. A narrow waveband spectral index that tracks diurnal changes in photosynthetic efficiency. Remote Sens. Environ. **1992**, 41, 35–44.
150. Sardans, J.; Peñuelas, J.; Lope-Piedrafita, S. Changes in water content and distribution in Quercus ilex leaves during progressive drought assessed by in vivo 1H magnetic resonance imaging. BMC Plant Biol. **2010**, 10, 188.
151. Melkus, G.; Rolletschek, H.; Fuchs, J.; Radchuk, V.; Grafahrend-Belau, E.; Sreenivasulu, N.; Rutten, T.; Weier, D.; Heinzel, N.; Schreiber, F.; et al. Dynamic 13C/1H NMR imaging uncovers sugar allocation in the living seed. Plant Biotechnol. J. **2011**, 9, 1022–1037.
152. Ashraf, M.; Akram, N.A. Improving salinity tolerance of plants through conventional breeding and genetic engineering: An analytical comparison. Biotechnol. Adv. **2009**, 27, 744–752.
153. Bänziger, M.; Setimela, P.S.; Hodson, D.; Vivek, B. Breeding for improved drought tolerance in maize adapted to southern Africa. In Proceedings of the 4th International Crop Science Congress, Brisbane, Australia, 26 September–1 October 2004.
154. Falconer, D.S. Introduction to Quantitative Genetics; Longman: London, UK; New York, NY, USA, 1989.
155. Dixit, S.; Singh, A.; Kumar, A. Rice breeding for high grain yield under drought: a strategic solution to a complex problem. Int. J. Agron. **2014**, 15.
156. Ober, E.S.; Clark, C.J.A.; Le Bloa, M.; Royal, A.; Jaggard, K.W.; Pidgeon, J.D. Assessing the genetic resources to improve drought tolerance in sugar beet: Agronomic traits of diverse genotypes under drought and irrigated conditions. Field Crops Res. **2004**, 90, 213–234.

157. Pidgeon, J.D.; Ober, E.S.; Qi, A.; Clark, C.J.A.; Royal, A.; Jaggard, K.W. Using multi-environment sugar beet variety trails to screen for drought tolerance. Field Crops Res. **2006**, 95, 268–279.
158. Singh, K.B.; Omar, M.; Saxena, M.C.; Johansen, C. Registration of FLIP 87-59C, a drought tolerant chickpea germplasm line. Crop Sci. **1996**, 36, 1–2.
159. Chen, M.; Wang, Q.Y.; Cheng, X.G.; Xu, Z.S.; Li, L.C.; Ye, X.G. GmDREB2, a soybean DRE binding transcription factor, conferred drought and high-salt tolerance in transgenic plants. Biochem. Biophys. Res. Commun. **2007**, 353, 299–305.
160. Xinglai, P.; Sangang, X.; Qiannying, P.; Yinhong, S. Registration of 'Jinmai 50' wheat. Crop Sci. **2006**, 46, 983–995.
161. Baenziger, P.S.; Beecher, B.; Graybosch, R.A.; Ibrahim, A.M.H.; Baltensperger, D.D.; Nelson, L.A. Registration of 'NE01643' wheat. J. Plant Registr. **2008**, 2, 36–42.
162. Haley, S.D.; Johnson, J.J.; Peairs, F.B.; Quick, J.S.; Stromberger, J.A.; Clayshulte, S.R. Registration of 'Ripper' wheat. J. Plant Registr. **2007**, 1, 1–6.
163. Obert, D.E.; Evans, C.P.; Wesenberg, D.M.; Windes, J.M.; Erickson, C.A.; Jackson, E.W. Registration of 'Lenetah' spring barley. J. Plant Registr. **2008**, 2, 85–97.
164. Noaman, M.M.; El Sayad, A.A.; Asaad, F.A.; El Sherbini, A.M.; El Bawab, A.O.; El Moselhi, M.A. Registration of 'Giza 126' barley. Crop Sci. **1995**, 35, 1710.
165. Brick, M.A.; Ogg, J.B.; Singh, S.P.; Schwartz, H.F.; Johnson, J.J.; Pastor-Corrales, MA. Registration of drought-tolerant, rust-resistant, high-yielding pinto bean germplasm line CO46348. J. Plant Registr. **2008**, 2, 120–134.
166. Mishra, K.K.; Vikram, P.; Yadaw, R.B.; Swamy, B.P.M.; Dixit, S.; Sta Cruz, M.T.; Maturan, P.; Marker, S.; Kumar, A. qDTY 12.1: A locus with a consistent effect on grain yield under drought in rice. BMC Genet. **2013**, 14, 12.
167. Yadaw, R.B.; Dixit, S.; Raman, A.; Mishra, K.K.; Vikram, P.; Swamy, B.P.M.; Sta Cruz, M.T.; Maturan, P.T.; Pandey, M.; Kumar, A. A QTL for high grain yield under lowland drought in the background of popular rice variety Sabitri from Nepal. Field Crops Res. **2013**, 144, 281–287.
168. Brachi, B.; Aimé, C.; Glorieux, C.; Cuguen, J.; Roux, F. Adaptive value of phenological traits in stressful environments: Predictions based on seed production and laboratory natural selection. PLoS ONE **2012**, 7, e32069.
169. Elshire, R.J.; Glaubitz, J.C.; Sun, Q.; Poland, J.A.; Kawamoto, K.; Buckler, E.S. A robust, simple genotyping-by-sequencing (GBS) approach for high diversity species. PLoS ONE **2011**, 6, e19379.
170. Huang, X.; Wei, X.; Sang, T.; Zhao, Q.; Feng, Q.; Zha, Y.; Li, C.; Zhu, C.; Lu, T.; Zhang, Z.; et al. Genome-wide association studies of 14 agronomic traits in rice landraces. Nat. Genet. **2010**, 42, 961–966.
171. Begum, H.; Spindel, J.E.; Lalusin, A.; Borromeo, T.; Gregorio, G.; Hernandez, J. Genome-wide association mapping for yield and other agronomic traits in an elite breeding population of tropical rice (Oryza sativa). PLoS ONE **2015**, 10, e0119873.
172. Sandhu, N.; Jain, S.; Kumar, A.; Mehla, B.S.; Jain, R. Genetic variation, linkage mapping of QTL and correlation studies for yield, root, and agronomic traits for aerobic adaptation. BMC Genet. **2013**, 14, 104–119.
173. Sandhu, N.; Torres, R.; Sta Cruz, M.T.; Maturan, P.C.; Jain, R.; Kumar, A.; Henry, A. Traits and QTLs for development of dry direct seeded rainfed rice varieties. J. Exp. Bot. **2015**, 66, 225–244.
174. Xu, Y.; Crouch, J.H. Marker-assisted selection in plant breeding: From publications to practice. Crop Sci. **2008**, 48, 391–407.
175. Ribaut, J.M.; Ragot, M. Marker-assisted selection to improve drought adaptation in maize: The backcross approach, perspectives, limitations, and alternatives. J. Exp. Bot. **2006**, 58, 351–360.
176. Venuprasad, R.; Bool, M.E.; Quiatchon, L.; Sta Cruz, M.T.; Amante, M.; Atlin, G.N. A large effect QTL for rice grain yield under upland drought stress on chromosome 1. Mol. Breed. **2012**, 30, 535–547.

177. Swamy, B.P.M.; Ahmed, H.U.; Henry, A. Genetic, physiological, and gene expression analyses reveal that multiple QTL enhance yield of rice mega-variety IR64 under drought. PLoS ONE **2013**, 8, e62795.
178. Kumar, R.; Venuprasad, R.; Atlin, G.N. Genetic analysis of rainfed lowland rice drought tolerance under naturally occurring stress in eastern India: Heritability and QTL effects. Field Crops Res. **2007**, 103, 42–52.
179. Palanog, A.D.; Mallikarjuna Swamy, B.P.; Shamsudin, N.A.A.; Dixit, S.; Hernandez, J.E.; Boromeo, T.H.; Cruz, P.C.S.; Kumar, A. Grain yield QTLs with consistent-effect under reproductive-stage drought stress in rice. Field Crops Res. **2014**, 161, 46–54.
180. Henry, A.; Gowda, V.R.P.; Torres, R.O.; McNally, K.L.; Serraj, R. Variation in root system architecture and drought response in rice (Oryza sativa): Phenotyping of the OryzaSNP panel in rainfed lowland fields. Field Crops Res. **2011**, 120, 205–214.
181. Howarth, C.J.; Yadav, R.S. Successful marker assisted selection for drought tolerance and disease resistance in pearl millet. IGER Innov. **2002**, 6, 18–21.
182. Tuberosa, R.; Salvi, S. Genomics approaches to improve drought tolerance in crops. Trends Plant Sci. **2006**, 11, 405–412.
183. Bovill, W.D.; Horne, M.; Herde, D.; Davis, M.; Wildermuth, G.B. and Sutherland, M.W.; Pyramiding QTL increases seedling resistance to crown rot (Fusarium pseudograminearum) of wheat (Triticum aestivum). Theor. Appl. Genet. **2010**, 121, 127–136.
184. Nagai, K.; Kuroha, T.; Ayano, M.; Kurokawa, Y.; Angeles-Shim, R.B.; Shim, J.; Yasui, H.; Yoshimura, A.; Ashikari, M. Two novel QTLs regulate internode elongation in deepwater rice during the early vegetative stage. Breed. Sci. **2012**, 62, 178–185.
185. Courtois, B.; Shen, L.; Petalcorin, W.; Carandang, S.; Mauleon, R.; Li, Z.K. Locating QTLs controlling constitutive root traits in the rice population IAC 165 _ Co39. Euphytica **2003**, 134, 335–345.
186. Bernier, J.; Atlin, G.N.; Serraj, R.; Kumar, A.; Spaner, D. Breeding upland rice for drought resistance. J. Sci. Food Agric. **2008**, 88, 927–939.
187. Mäki-Tanila, A.; Hill, W.G. Influence of gene interaction on complex trait variation with multilocus models. Genetics **2014**, 198, 355–367.
188. Dixit, S.; Swamy, B.P.M.; Vikram, P.; Bernier, J.; Sta Cruz, M.T.; Amante, M.; Atri, D.; Kumar, A. Increased drought tolerance and wider adaptability of qDTY12.1 conferred by its interaction with qDTY2.3 and qDTY3.2. Mol. Breed. **2012**, 30, 1767–1779.
189. Shamsudin, N.A.A.; Swamy, B.P.M.; Ratnam, W.; Cruz, M.T.S.; Sandhu, N.; Raman, A.K.; Kumar, A. Pyramiding of drought yield QTLs into a high quality Malaysian rice cultivar MRQ74 improves yield under reproductive stage drought. Rice **2016**, 9, 21.
190. Lebreton, C.; Lazic-Jancic, V.; Steel, A.; Pekic, S.; Quarrie, S.A. Identification of QTL for drought responses in maize and their use in testing causal relationships between traits. J. Exp. Bot. **1995**, 46, 853–865.
191. Phillips, P.C. The language of gene interaction. Genetics **1998**, 149, 1167–1171.
192. Li, Z.K.; Yu, S.B.; Lafitte, H.R.; Huang, N.; Courtois, B.; Hittalmani, S.; Vijayakumar, C.H.M.; Liu, G.F.; Wang, G.C.; Shashidhar, H.E.; et al. QTL _ environment interactions in rice. I. Heading date and plant height. Theor. Appl. Genet. **2003**, 108, 141–153.
193. Veronica, C.; Roncallo, P.F.; Beaufort, V.; Cervigni, G.L.; Miranda, R.; Jensen, C.A.; Echenique, V.C. Mapping of main and epistatic effect QTLs associated to grain protein and gluten strength using a RIL population of durum wheat. J. App. Genet **2011**, 52, 287–298.
194. Zhao, Z.G.; Zhu, S.S.; Zhang, Y.H.; Bian, X.F.;Wang, Y. Molecular analysis of an additional case of hybrid sterility in rice (Oryza sativa L.). Planta **2010**, 233, 485–494.
195. Han, Y.; Xie, D.; Teng, W.; Zhang, S.; Chang, W.; Li, W. Dynamic QTL analysis of linolenic acid content in different developmental stages of soybean seed. Theor. Appl. Genet. **2011**, 122, 1481–1488.
196. Niu, Y.; Xu, Y.; Liu, X.F.; Yang, S.X.;Wei, S.P.; Xie, F.T.; Zhang, Y.M. Association mapping for seed size and shape traits in soybean cultivars. Mol. Breed. **2013**, 31, 785–794.
197. Lukens, L.N.; Doebley, J. Epistatic and environmental interactions for quantitative trait loci involved in maize evolution. Genet. Res. **1999**, 74, 291–302.

198. Lark, K.G.; Chase, K.; Adler, F.; Mansur, L.M.; Orf, J.H. Interactions between quantitative trait loci in soybean in which trait variation at one locus is conditional upon a specific allele at another. Proc. Natl. Acad. Sci. USA **1995**, 92, 4656–4660.
199. Wang, B.; Wu, Y.; Guo, W.; Zhu, X.; Huang, N.; Zhang, T. QTL analysis and epistasis effects dissection of fiber qualities in an elite cotton hybrid grown in second generation. Crop Sci. **2007**, 47, 1384–1392.
200. Zhang, K.; Tian, J.; Zhao, L.; Wang, S. Mapping QTLs with epistatic effects and QTL X environment interactions for plant height using a doubled haploid population in cultivated wheat. J. Genet. Genom. **2008**, 35, 119–127.
201. Ashraf, M.; Foolad, M.R. Roles of glycinebetaine and proline in improving plant abiotic stress resistance. Environ. Exp. Bot. **2007**, 59, 206–216.
202. Zhang, G.H.; Su, Q.; An, L.J.; Wu, S. Characterization and expression of a vacuolar Na+/H+ antiporter gene from the monocot halophyte Aeluropus littoralis. Plant Physiol. Biochem. **2008**, 46, 117–126.
203. Quan, R.; Shang, M.; Zhang, H.; Zhao, Y.; Zhang, J. Engineering of enhanced glycinebetaine synthesis improves drought tolerance in maize. Plant Biotechnol. J. **2004**, 2, 477–486.
204. Wu, R.; Garg, A. Engineering Rice Plants with Trehalose-Producing Genes Improves Tolerance to Drought, Salt, and Low Temperature. ISB News Report 2003; pp. 3–7. Available online: http://www.isb.vt.edu/news/2003/mar03.pdf (accessed on 20 May 2016).
205. Romero, C.; Belles, J.M.; Vaya, J.L.; Serrano, R.; Culianez-Macia, F.A. Expression of the yeast trehalose-6-phosphate synthase gene in transgenic tobacco plants: Pleiotropic phenotypes include drought tolerance. Planta **1997**, 201, 293–297.
206. Karim, S.; Aronsson, H.; Ericson, H.; Pirhonen, M.; Leyman, B.; Welin, B. Improved drought tolerance without undesired side effects in transgenic plants producing trehalose. Plant Mol. Biol. **2007**, 64, 371–386.
207. Vendruscolo, E.C.G.; Schuster, I.; Pileggi, M.; Scapim, C.A.; Molinari, H.B.C.; Marur, C.J. Stress induced synthesis of proline confers tolerance to water deficit in transgenic wheat. J. Plant Physiol. **2007**, 164, 1367–1376.
208. Gubis, J.; Vanˇková, R.; Cˇ ervená, V.; Dragúnˇová, M.; Hudcovicová, M.; Lichtnerová, H. Transformed tobacco plants with increased tolerance to drought. S. Afr. J. Bot. **2007**, 73, 505–511.
209. Ronde, J.A.D.; Cress, W.A.; Krugerd, G.H.J.; Strasserd, R.J.; Van Staden, J. Photosynthetic response of transgenic soybean plants, containing an Arabidopsis P5CR gene, during heat and drought stress. J. Plant Physiol. **2004**, 161, 1211–1224.
210. Yamada, M.; Morishita, H.; Urano, K.; Shiozaki, N.; Yamaguchi-Shinozaki, K.; Shinozaki, K. Effects of free proline accumulation in petunias under drought stress. J. Exp. Bot. **2005**, 56, 1975–1981.
211. Wilkening, S.; Chen, B.; Bermejo, J.L.; Canzian, F. Is there still a need for candidate gene approaches in the era of genome-wide association studies? Genomics **2009**, 93, 415–419.
212. Xue, G.P.; Intyre, C.L.; Chapman, S.; Bower, N.I.; Way, H.; Reverter, A.; Clarke, B.; Shorter, R. Differential gene expression of wheat progeny with contrasting levels of transpiration efficiency. Plant Mol. Biol. **2006**, 61, 863–881.
213. Mohammadi, M.; Kav, N.N.; Deyholos, M.K. Transcript expression profile of water-limited roots of hexaploid wheat (Triticum aestivum 'Opata'). Genome **2008**, 51, 357–367.
214. Aprile, A.; Mastrangelo, A.M.; De-Leonardis, A.M.; Galiba, G.; Roncaglia, E.; Ferrari, F.; De-Bellis, L.; Turchi, L.; Giuliano, G.; Cattivelli, L. Transcriptional profiling in response to terminal drought stress reveals differential responses along the wheat genome. BMC Genom. **2009**, 10, 279.
215. Ergen, N.Z.; Thimmapuram, J.; Bohnert, H.J.; Budak, H. Transcriptome pathways unique to dehydration tolerant relatives of modern wheat. Funct. Integr. Genom. **2009**, 9, 377–396.
216. Delseny, M.; Sales, J.; Richard, C.; Sallaud, C.; Regad, F. Rice genomics: Present and future. Plant Physiol. Biochem. **2001**, 39, 323–334.
217. Yin, X.; Struik, P.C.; Kropff, M.J. Role of crop physiology in predicting gene-to-phenotype relationships. Trends Plant Sci. **2004**, 9, 426–432.

CHAPTER 3

Management of Vegetation by Alternative Practices in Fields and Roadsides

Allen V. Barker and Randall G. Prostak

Department of Plant, Soil, and Insect Sciences, University of Massachusetts, Amherst, MA 01003, USA

ABSTRACT

In attempts to reduce the amounts of conventional herbicides used, alternative practices are sought in the management of roadside vegetation. In this investigation, alternative herbicides (citric-acetic acids, clove oil, corn gluten meal, limonene, and pelargonic acid), flaming, and mulching were assessed in management of annual and perennial, herbaceous vegetation in field and roadside plots. Several formulations of alternative herbicides applied singly or repeatedly during the growing season were evaluated and compared with conventional herbicides (glyphosate and glufosinate ammonium) or with flaming or mulching. Citric-acetic acid formulations, clove oil, limonene, or pelargonic acid applied as foliar sprays immediately desiccated foliage, but the efficacy lasted for no longer than five weeks. Repeated applications were better than single applications of these herbicides in suppressing plant vegetative growth. Corn gluten meal imparted little or no early control and stimulated late-season growth of vegetation. A single flaming of vegetation gave no better control than the alternative herbicides, but repeated flaming strongly restricted growth. Mulching with wood chips or bark gave season-long suppression of vegetation. Glyphosate gave season-long inhibition of vegetation, but the efficacy of glufosinate ammonium waned as the growing season progressed.

For season-long suppression of vegetation with alternative herbicides or flaming repeated applications will be required.

1. INTRODUCTION

Management of vegetation is an important element for safety and aesthetics in healthy roadside environments. Methods of management to produce this environment include cultural, biological, mechanical, and chemical procedures for weed control. Mechanical methods, primarily mowing, are the most widely used methods [1]. Chemical control involving the use of herbicides is used substantially in management of roadside vegetation [1].

Use of herbicides provides flexibility and low costs in control of vegetation, considering the equipment and the wide spectrum of materials that are available for use [1]. However, some of the public does not believe that commonly used herbicides are safe for use along roadsides [2]. Consequently, state departments of transportation are investigating alternatives to conventional herbicides in weed management. Alternatives include chemical treatments with plant-derived herbicides and mechanical treatments. Several alternative herbicides that are marketed as products to manage growth of vegetation have an active ingredient that is of plant origin or include various by-products of food and feed processing [3, 4].

Corn gluten meal (CGM) is the proteinaceous by-product from the milling of corn grain to make cornstarch and corn syrup and is about 60% protein or about 10% nitrogen by weight [5]. Peptides in the material impart preemergence herbicidal activity by inhibiting root growth from germinating seeds in soil [6, 7]. Corn gluten meal is nonselective, and banded application for row-crop production was recommended over broadcast application to avoid yield suppression of crops [8]. Corn gluten meal also can fertilize crops or weeds and may thereby improve crop growth or increase weed competition [9, 10].

Acetic acid and citric acid [11–19] are alterative materials that work as nonselective, postemergence, contact herbicides causing rapid desiccation of plant foliage. Their use as herbicides involves application as a foliar spray at concentrations ranging from 10 to 20% (v:v) acetic acid, citric acid, or blends of the two on weeds that are about 6 to 9 inches tall or less [16]. As growth stage advances, control and biomass suppression decreases and survival increases [12, 18]. Generally, 80 to 100% kill rates can be expected for small annual and perennial weeds, but perennial species with persistent root systems will regrow within several weeks. Repeated applications are needed to control large annual and perennial weeds and plants that grow from newly emerged seedlings [1, 19]. Reports noted that acetic acid or citric was nonselective and that crop damage and yield suppression would occur if these herbicides came in contact with the crops [13, 14].

Pelargonic acid (nonanoic acid) is a naturally occurring fatty acid in the oil of geranium (Pelargonium spp. L'Her.) [20, 21]. The commercial product is manufactured from other fatty acids and is a nonselective, contact herbicide, which effectively controls annual broadleaf and grass weeds that are less than 6 inches high [22]. Repeated applications may be required with large annual plants and perennials. Pelargonic acid has been investigated also as a preharvest desiccant [23]. Research on use along highways is not evident, but uses in the other applications indicate that this material could be used for weed control on roadsides.

Clove oil, derived from the clove plant (Syzygium aromaticum Merr. and Perry), is active against various organisms including nematodes [24], insects [25], microorganisms [26–28], and plants [4, 29, 30] and may have potential as a postemergence desiccant in roadside environments. Brainard et al. [11] reported that a concentration of at least 7.5% (v/v) of clove oil was needed for control of annual broadleaf weeds but that poor control of annual grasses was achieved.

Limonene is a cyclic terpene derived from oil of the rind of citrus fruits [21]. It has common uses as a cleaner or degreaser in household and industrial uses and as an adjuvant to increase the penetration of agricultural chemical into plant leaves. Limonene also has insecticidal and antimicrobial activities [31–35]. The phytotoxic response of some plants to limonene suggests that it has herbicidal properties [36]. Limonene has not been evaluated for herbicidal activity in roadside environments.

Burning of fields to reduce weed populations has been a historic practice but has many disadvantages including air pollution, reduced visibility caused by smoke, and the chance of uncontrolled fire. Burner-type weeders deliver heat energy to the plants directly by close contact to a flame. Ascard [37, 38] and Rifai et al. [39] reported that weeds with thin leaves and unprotected growing points were more susceptible to flaming than grasses and other species with protected growing points. All species were most susceptible to flaming when small (0–4 true leaves) compared to larger plants (6–12 true leaves), and significantly more heat energy was needed to control the larger plants.

Covering the soil with a mulch of a layer of organic matter or with a barrier, such as plastic or landscape fabric, has potential use in roadside weed control in small areas, such as around sign posts, along fence rows, and under guardrails. The weed-control function of mulches is based on the blocking of sunlight from the plants or on the presentation of a barrier through which the weeds cannot emerge [40, 41]. Some preparation of land, such as shallow cultivation, is necessary to reduce early weed competition for organic mulches such as wood chips to be effective in weed control [42].

In field and roadside plots, this research investigated the use of corn gluten meal, acetic-citric acid formulations, pelargonic acid, clove oil, limonene, flaming, and mulches as alternative practices for management of vegetation. These practices were compared with use of conventional herbicides of glyphosate and glufosinate ammonium. Mowing of vegetation also was investigated for its effects on the efficacy of alternative herbicides.

2. MATERIALS AND METHODS

Two studies were conducted. This work was part of an investigation in which several alternative methods of weed control on roadsides were evaluated [1]. One study involved experiments in field plots at a University of Massachusetts research farm in South Deerfield, and the other involved experiments on curbsides of a driveway and parking area at a weigh station along Interstate 91 in Deerfield, Massachusetts.

2.1. Field Plot Research

This research was conducted in 2005 to gain information on the efficacy of alternative herbicides and mechanical treatments. Field plot experimentation allowed for research on land with more uniform stands of vegetation than those that existed along roadsides.

2.1.1. Evaluation of Conventional and Alternative Herbicides and Mechanical Treatments for Control of Vegetation in Field Plots

Two experiments were conducted. Experiment 1 investigated the efficacy of various alternative herbicides applied at different amounts, mechanical treatments, and conventional herbicides (Table 1). Experiment 2 investigated the effects of repeated applications of alternative practices (Table 2). These experiments were conducted in plots on two side-by-side areas of land at the South Deerfield farm of the University of Massachusetts. One area was infested with annual grasses crabgrass, Digitaria spp. Haller; foxtail (Setaria glauca Roem. and Schult); and annual broadleaf species (lambsquarters, Chenopodium album L.; marestail, Conyza canadensis Cronquist; redroot pigweed, Amaranthus retroflexus L.), a situation that was created by tillage of the land in the spring of the year of research. The other area was infested mainly with perennial grasses, almost entirely quackgrass (Agropyron repens Beauv.) in untilled land. Each area received all of the treatments described in the next paragraphs. Plot size was 1.2 m by 4.6 m, and the experimental design was a randomized, complete block with three replications in each experiment.

Table 1: Treatments in experiments involving alternative herbicides applied in differing amounts, conventional herbicides, and mechanical treatments in field plot research (Field Experiment 1).

Alternative practice	Application
Herbicide†	Active ingredient, L ha^{-1}
Untreated	0
Clove oil	34
"	68
"	102
Pelargonic acid	13
"	26
"	39
Citric-acetic acid	34
Citric acid	45
Glyphosate	2.8
Glufosinate ammonium	1.8
Corn gluten meal	975 kg ha^{-1}
"	2,925 kg ha^{-1}
"	4,875 kg ha^{-1}
Mechanical treatments‡	
Flaming	Vegetation ashed
Bark mulch	760 m^3 ha^{-1}
Woodchips mulch	760 m^3 ha^{-1}

† Clove oil, Matran EC, 500 mL/L a.i., EcoSMART Technologies, Franklin, TN; pelargonic acid, Scythe, 570 mL a.i./L., Mycogen Corporation, San Diego; citric-acetic acid, Ground Force, 50 mL citric acid/L, acetic acid unspecified, Abby Laboratories, Ramsey, MN; citric acid, Brush and Blackberry Block, 200 mL a.i./L, Greenergy, Brookings, OR; glyphosate, Roundup Pro, 410 mL a.i./L, Monsanto Company, St. Louis, MO; glufosinate ammonium, Finale, 200 mL a.i./L, Bayer Environmental Science, Montcalm, NJ; and corn gluten meal, about 600 g crude protein/kg, applied to surface after flaming vegetation by torch; tabulated application is total product.

‡ Flaming with propane-fired torch to ash vegetation to soil surface; bark mulch, mixed tree-bark mulch, 7.5 cm thick; woodchips mulch, mixed tree woodchips, 7.5 cm thick following flaming.

Table 2: Treatments involved in repeated applications of alternative herbicides or flaming in field plot experiments (Field Experiment 2).

Treatment	Application, active ingredient, L ha^{-1}†
Clove oil	34
Pelargonic acid	13
Citric acid-acetic acid	34
Citric acid	45
Flaming	‡Ashing of vegetation

† Applications occurred in July, July and August, and July, August, and September (see Table 1 for identity of herbicides).

‡ Vegetation was ashed to ground surface on the three dates noted above for applications.

Field Experiment 1. Alternative herbicides (clove oil, pelargonic acid, citric-acetic acid, and corn gluten meal) were applied in different amounts (Table 1). These applications were developed from the label recommendations and from preliminary research in a greenhouse [1]. Also, included in the experiment was the evaluation of flaming with a handheld torch (500,000 BTU, Red Dragon, Flame Engineering, Inc., LaCrosse, KS) and mulching with 3-inch layers of hardwood bark or woodchips. Flaming to ash vegetation was used to prepare plots for mulching and for application of corn gluten meal. Treatments with conventional herbicides, glyphosate and glufosinate ammonium (Table 1), were included in the trial. Plots were treated on July 20, 2005, with alternative or conventional herbicides, mulching, or flaming and were evaluated on August 1, September 9, and September 23, 2005, by visual indexing (Table 3) and in the following year without indexing. In the second year, the flamed and herbicide-treated plots were overgrown with the original week mass, and the mulched plots had sparse emergence of dandelion (Taraxacum officinale F. H. Wigg) and milkweed (Asclepias syriaca L.). In the first year, mass of shoot growth was determined by weighing 0.25 m^2 of the vegetation cut to ground level after the final visual rating.

Field Experiment 2. Another experiment was conducted at the same time and in the same areas as described above. In the second experiment, several alternative herbicides were evaluated with different times of repeated applications (Table 2). Treatments of alternative herbicides or flaming were applied on July 21, August 15, and September 9, 2005. Some plots received the treatments only in July; some received the treatments in July and August, and some received the treatments in July, August, and September.

Mulching and conventional herbicides were not included in this experiment. Treatments were evaluated on August 1, September 9, and September 23, 2005, by visual indexing (Table 3), and mass of shoot growth was determined by weighing 0.25 m^2 of the vegetation cut at ground level on September 28. The plots also were evaluated in the spring of 2006.

Table 3: Visual indexing for scoring efficacy of treatments.

Index score	Description of score
0	No control of vegetation
1	Less than 10% of vegetation controlled
2	Between 11 and 20% of vegetation controlled
3	Between 21 and 30% of vegetation controlled
4	Between 31 and 40% of vegetation controlled
5	Between 41 and 50% of vegetation controlled
6	Between 51 and 60% of vegetation controlled
7	Between 61 and 70% of vegetation controlled
8	Between 71 and 80% of vegetation controlled
9	Between 81 and 90% of vegetation controlled
9.5	A rating used when more than 90% of the vegetation was controlled with only one or two live plants remaining in a plot.
10	This rating was applied when all of the vegetation was killed or if no regrowth occurred at the time of rating.

2.2. Roadside Research

This research involved two experiments in 2006 run simultaneously with alternative herbicides at curbsides along driveways at a weighing station along Interstate 91 in South Deerfield, Massachusetts. In one experiment, alternative herbicides of clove oil, pelargonic acid, and limonene were applied in two amounts (Table 4). In the second experiment, the lower amounts of these herbicides were used, and a citric acid-acetic acid based material and a conventional herbicide (glyphosate) were additional treatments (Table 4). The principal species was tall fescue (Festuca arundinacea Schreb.). The experimental design was split plot with half of a block being mowed to 10 cm heights with a string mower and with half of

Table 4: Treatments employed in two experiments in roadside research with alternative herbicides (Roadside Experiments).

Experiment 1		Experiment 2	
Material†	Application, active ingredient, L ha^{-1}	Material†	Application, active ingredient, L ha^{-1}
Clove oil	34	Clove oil	34
"	68		
Pelargonic acid	13	Pelargonic acid	13
"	26		
Limonene	18	Limonene	18
"	36		
Untreated	0	Clove oil	34
		and limonene	18
		Citric acid-	25
		acetic acid	19
		Glyphosate	2.8
		Untreated	0

† Clove oil, Matran EC, 500 mL a.i./L., EcoSMART Technologies, Franklin, TN; limonene, Natures Avenger, 175 mL a.i./L, Cutting Edge Formulations, Buford, GA; pelargonic acid, Scythe, 570 mL a.i./L, Mycogen Corporation, San Diego; citric acid-acetic acid, AllDown, 80 mL citric acid/L and 60 mL acetic acid/L, Summerset Products, Chaska, MN; and glyphosate, Roundup Pro, 410 mL a.i./L, Monsanto Company, St. Louis, MO.

the block remaining unmowed. The vegetation was about 20 cm high at mowing. Mowing was on July 3, 2006, and treatments were applied on July 6, 2006, in a randomized, complete block design within the split plots and with three replications. Plot size was 1.2 m by 4.6 m along on the shoulders. Control of growth of vegetation (Table 3) was assessed at weekly or biweekly intervals after application of the treatments. On September 28, the mass of shoot growth was determined by cutting plants at ground level from 0.25 m^2 of plot.

2.3. Statistical Procedures

Data were processed by analysis of variance employing statistical software (SAS Institute, Cary, NC), and means were separated by -test or by least significant difference (LSD, *P*) [43].

3. RESULTS

3.1. Field Plot Research

3.1.1. Effects of Applications of Conventional Herbicides and Alternative Herbicides and Effects of Mechanical Treatments on Control of Annual and Perennial Vegetation

Field Experiment 1 (effects of different amounts of alternative herbicides and mechanical treatments on management of vegetation). No differences in efficacy in weed management among amounts of application of alternative herbicides occurred at the first two ratings of the field plots (August 1 and September 9); hence, means of the various rates are reported (Table 5). The effects of the two citric acid-based herbicides (Ground Force or Brush and Blackberry Block, Table 1) or the two mulches (bark or woodchip) were not different and are included as means in the table. The effects of treatments at the end of the experiment (September 23) are reported in Table 6, which gives the visual indices and the mass of the plants. Averages of treatments in Table 6 are reported as described in Table 5.

Table 5: Assessments of conventional and alternative herbicides and mechanical treatments on early season control of vegetation in field plots with annual and perennial vegetation (Field Experiment 1).

Treatment	Rating and type of vegetation and date[†]			
	August 1		September 9	
	Annual	Perennial	Annual	Perennial
	Visual index[‡]			
None	0	0	0	0
Clove oil[§]	2.6	4.1	0.7	0.7
Pelargonic acid[§]	4.8	4.1	0.7	0.7
Citric acid[¶]	0.5	2.4	0	0.4
Corn gluten meal[§]	9.7	8.5	4.7	3.2
Flaming	10.0	8.0	7.8	5.0
Mulches[¶]	10.0	10.0	9.0	9.2
Glyphosate	10.0	10.0	8.8	7.0
Glufosinate ammonium	10.0	10.0	8.5	1.0
LSD ($P = 0.05$)	1.0	1.6	1.3	1.0

[†] Treatments applied on July 20, 2005. See Table 1.

[‡] Visual rating index of efficacy of materials; see Table 3.

[§] Means of 3 rates of alternative herbicides; effects of rates of application were nonsignificant.

[¶] Including means of two citric acid based herbicides, the effects of which were not significantly different. Bark mulch or woodchips mulch did not differ in effects, and means of the two mulches are tabulated.

Table 6: End-of-season weed mass and visual indexing of treatments with alternative herbicides, mulches, burning, or conventional herbicides in field plot research with annual and perennial vegetation (Field Experiment 1).

Treatment[†]	Weed mass, g fresh wt m^{-2}[‡]		Visual index rating[§]	
	Annual	Perennial	Annual	Perennial
Untreated	2,910	1,574	0	0
Clove oil	2,332	954	0.2	0.2
Pelargonic acid	2,360	774	0.3	0.3
Citric acid	2,340	808	0.5	0.4
Corn gluten meal	1,743	1,350	3.2	3.1
Mulch	32	208	9.3	9.3
Flaming	1,486	384	5.0	5.0
Glyphosate	638	741	7.0	7.0
Glufosinate ammonium	854	644	1.0	1.0
LSD ($P = 0.05$)	1,164	472	1.4	1.1

[†] All treatments were applied on July 20 and July 21, 2005. Amounts of alternative herbicides, sources of citric acid, or kind of mulch had no significant effect on efficacy; hence, means of the applications are reported. See Table 1 for presentation of treatments.

[‡] Weed fresh mass was measured on September 28, 2005.

[§] Visual rating index on September 23, 2005. See Table 3.

Annual Weeds. For plots infested with annual vegetation, the visual indexing indicated that about 10 days (August 1) after application of treatments, the citric acid-based treatments gave little suppression of annual vegetation and that the clove oil and pelargonic acid gave weak to moderate control (Table 5). Flaming or application of the two conventional herbicides (glyphosate or glufosinate ammonium) gave full control of annual vegetation at this time. As the season progressed to about 6 weeks after treatment (September 9), suppression of annual vegetation from burning or application of glyphosate or glufosinate ammonium persisted, but the alternative herbicides had few suppressing effects on vegetation. At 6 weeks, plant growth from seeds or from surviving crowns of plants in the soil gave return of vegetation to the plots that were flamed. The conventional herbicides (glyphosate and glufosinate ammonium) fully controlled the annual vegetation at the September 9 date of assessment.

In August, the benefits of the corn gluten meal in the suppression of annual weeds may be due to flaming before application of the meal. The meal stimulated growth of vegetation relative to the growth with flaming alone as is indicated by the lesser suppression with the meal than with flaming alone in the September 9 rating. Mulch applied after flaming suppressed weeds through the September 9 date of evaluation. No visual differences were noted between bark mulch and woodchip mulch; hence, the ratings of these plots are pooled into one entry. No weeds were emerging through either of the mulching materials, whereas, with the flaming alone, weeds were reappearing in the plots.

The end-of-season visual indexing (September 23) indicated that the efficacy of all alternative, sprayed herbicides and glufosinate ammonium had waned to 10% or less control of vegetation (Table 3, Table 6). The different amounts of alternative herbicides applied had no effect on growth or indexing, and results are combined for tabulation. Glyphosate continued to provide strong efficacy for weed control until the end of the experiment. Increased amounts of application of corn gluten meal did not improve its efficacy; hence, means of all rates of the meal are reported. Based on visual indexing, corn gluten meal following flaming gave somewhat better control than the alternative herbicides. Also, based on visual indexing or shoot mass, corn gluten meal, which was applied to flamed plots, had lower efficacy than flaming alone, indicating that the corn gluten meal had a stimulatory effect on weed growth, likely through its action as a nitrogen fertilizer. The shoot mass of weeds in the flamed plots was about half of that in the untreated plots, whereas the average mass of vegetation from the plots treated with corn gluten meal following burning was about 60% of the untreated plot biomass. At the end of the growing season, the treatments with corn gluten meal kept the weeds green and sustained growth, whereas the plants in the other treatments were senescing.

The pelargonic acid herbicide had high efficacy in initial killing of annual vegetation, but this efficacy dissipated during the growing season as the surviving roots and crowns of the plants recovered from the initial dieback and emerging seedlings grew to produce shoot biomass that was not different from that of the untreated plots or of the clove oil, citric acid, or corn gluten meal treatments. Glyphosate or glufosinate ammonium restricted weed mass more than the alternative treatments except flaming or mulching. The mulches gave the best season-long control of vegetation of all of the treatments employed in this experiment.

Perennial Weeds. In August (10 days after treatment), the citric acid-acetic acid treatments gave only a slight growth suppression of perennial weeds, which were almost all quackgrass (Table 3, Table 5). No suppression by these treatments was evident on September 9, about 6 weeks after application. The clove oil or pelargonic acid treatment showed some suppression in August, but this control also had abated by the early September date. After giving growth suppression at the August rating, the efficacy of glufosinate ammonium was lost by September 9. Glyphosate was effective in control of perennial vegetation on both dates. In the plots cleared of vegetation by flaming, annual and perennial weeds were emerging in the plots initially dominated by the perennial vegetation, but the control remained moderately effective through early September. Few weeds were emerging through the mulches on any date. Corn gluten meal gave only weak suppression of vegetation at the early September date and was not as effective as flaming alone even though the plots with corn gluten meal had been preburned.

Generally, the plots treated with alternative, sprayed herbicides had smaller end-of-season (September 28) (Table 6) biomasses than untreated plots, but visual indexing did not make this distinction, suggesting that the aesthetics of the plots treated with alternative herbicides was not better than no treatment. The mass of weeds in the plots treated with corn gluten meal did not differ from that of the untreated plots, although visual indexing suggested some control from the meal. The visual rating may have assessed a sparser population of larger plants in the plots treated with corn gluten meal than in the untreated plots.

Glyphosate gave strong suppression of weed growth at the end of the season. Based on visual indexing, glufosinate ammonium had little efficacy in season-long control of weeds, indicating that the early season control had dissipated. Also, the mass of vegetation did not differ from many of the treatments with alternative herbicides, although the mass of weeds in plots treated with glufosinate ammonium was less than that from the untreated plots.

Plots flamed with a weed torch had weed masses that were about 20% of the end-of-season growth of the untreated perennial weed plots (Table 6).

Flamed plots also had higher weed suppression indexes than the untreated plots or plots treated with alternative herbicides.

Mulching with wood chips or bark chips had strong suppressive effects on perennial vegetation at the end of the season, continuing the early season suppression that was noted. Suppression of weed mass was greater with the mulches than with any other treatment.

With either the annual weeds or the perennial weeds, visual observations in the spring of the following year showed that only the mulched plots had a residual effect on weed control. In these plots, milkweed and dandelion were growing sparse populations. All of the other treatments showed no residual control and were not distinguishable from the treatment of no herbicide applied in the previous season.

Field Experiment 2 (effects of repeated applications of alternative treatments on control of vegetation). This experiment was conducted at the same time and site as Experiment 1. Treatments of alternative herbicides or flaming were applied in July, August, and September. One-third of the plots received the treatments only in July; one-third received the treatments in July and August, and one-third received the treatments in July, August, and September (Table 2). Mulching and conventional herbicides were not included in this experiment.

Annual and Perennial Weeds. Immediately after application of the treatments, vegetation treated with clove oil, pelargonic acid, or either of citric-acid based materials showed wilting and burning of foliage. At 10 days (August 1) and at 6 weeks (September 9) after treatment applications, the efficacies of the one-time applications were assessed (Table 7). The effects of the citric-acetic acid herbicide and the citric acid herbicide differed, and the results of these materials are reported separately. The alternative herbicides had only weakly suppressive effects on growth of vegetation, and the effects diminished as time after treatment advanced from 10 days to 6 weeks. These materials are contact herbicides, which kill the foliage that is exposed to contact but which do not damage the entire plant so that perennials resumed growth. Also, some of the returning growth was from newly emerged plants, especially in the plots with annual weeds. Flaming had much better short-term effects on suppression of vegetation than the alternative herbicides, but the effects of flaming also diminished with time, as a result of emergence of new plants from the soil and regrowth from roots and crowns.

Table 7: Efficacy of alternative herbicides and burning of vegetation at ten days and six weeks after one application of treatments in field plot experiment with repeated applications on annual and perennial vegetation (Field Experiment 2).

Treatment[†]	Rating by date and type of vegetation			
	August 1		September 9	
	Annual	Perennial	Annual	Perennial
	Visual index[‡]			
Clove oil	1.7	0.9	0	1.0
Pelargonic acid	3.3	7.3	0.7	1.4
Citric-acetic acid[§]	2.7	4.7	0.7	1.2
Citric acid[§]	0.3	0.7	0	0.5
Flaming	9.5	8.7	4.7	2.7
LSD ($P = 0.05$)	1.0	1.0	1.3	1.3

[†] One application of the herbicides or burning was made on July 21, 2005.

[‡] Visual index. See Table 3 for indexing scale.

[§] Citric-acetic acid was Ground Force; citric acid was Brush and Black Berry Block. See Table 1.

Table 8: End-of-season mass of shoots of weeds and ratings of efficacy of treatments with repeated applications of alternative practices in field plot research with annual and perennial vegetation (Field Experiment 2).

Treatment	Times repeated†	Weed mass, g fresh wt m^{-2}‡		Visual index rating§	
	1	2,174	860	0.3	1.0
Clove oil	2	1,226	708	0.3	1.0
	3	496	415	0.3	1.0
	1	1,241	546	0.7	1.7
Pelargonic acid	2	1,128	353	1.3	1.3
	3	474	241	1.3	1.3
	1	2,037	875	0.7	1.0
Citric acid-acetic acid	2	924	776	1.3	1.3
	3	827	394	1.0	1.3
	1	1,960	686	0	0
Citric acid	2	1,779	753	0	1.0
	3	1,954	717	0	0.7
	1	1,046	381	4.7	4.0
Flaming	2	453	336	9.5	4.3
	3	150	90	9.5	4.7
LSD ($P = 0.05$)		611	234	1.0	1.0

† 1, first application, July 21; 2, second application, August 15; and 3, third application, September 9, 2005.
‡ Fresh mass on September 28, 2005.
§ Visual rating index on September 23, 2005. See Table 3 for indexing scale.

Suppression in weed mass indicates that increasing the number of applications of alternative herbicides increased the control of weeds (Table 8). However, visual indexing, based on observations of the appearance of the plot of vegetative coverage of the plots, suggested that the multiple treatments of the alternative herbicides were not effective on the aesthetics of the plots at the end of the season. Each flaming eliminated all aboveground vegetation so that the multiple treatments with fire gave good end of the season control of weeds.

Based on weed masses, pelargonic acid or flaming gave the best control of annual or perennial weeds (Table 9). Based on weed mass, the citric acid or clove oil treatments were better than no application of herbicide with annual or perennial vegetation. Judging of the plants by visual indexing suggested that only the flaming treatment was better than no treatment with the annual species and that all treatments were better than no treatment with the perennial species.

Table 9: Main effects of herbicide treatments on end-of-season weed mass and visual index rating of annual and perennial vegetation (Field Experiment 2).

Treatment	Weed mass, g m^{-2} [†]		Visual index rating [‡]	
	Annual	Perennial	Annual	Perennial
Clove oil	1,299	661	0.3	1.0
Pelargonic acid	948	380	1.1	1.4
Citric-acetic acid	1,263	682	1.0	1.2
Citric acid	1,898	722	0	0.6
Flaming	550	269	0	0
Untreated	2,910	1,174	0	0
LSD (P = 0.05)	270	270	1.0	1.0

† Fresh mass on September 28, 2005.
‡ Visual rating index on September 23, 2005. See Table 3 for indexing scale.

3.2. Roadside Research

Roadside Experiment 1 (clove oil, pelargonic acid, and limonene applied at two concentrations to mowed or unmowed plots). Mowing of the plots did not increase the efficacy of the herbicides, as no differences occurred between mowing treatments or between the interaction of mowing with herbicides or with date of observation (Table 3, Table 10). In fact, in the short term (1 week), the efficacy on the unmowed plots (mean index rating

5.6 on the 10-point scale) was slightly larger than with the mowed plots (4.6 rating), and the mean of all season-long ratings showed a similar relationship with the unmowed mean being 3.3 and the mowed mean being 3.0 on the visual index for the herbicide-treated plots. The mass of harvested top growth did not differ between mowing treatments (Table 11).

Pelargonic acid applied to mowed or unmowed plots gave strong control of vegetation (mean rating of 7.5) at one week after application (Table 10). The efficacy of this material declined with time after treatment so that, after 5 weeks, control of vegetation was weak (mean rating of 1.8 for both amounts applied). Clove oil (mean rating of 4.0) or limonene (mean rating of 3.8) was less effective than pelargonic acid in management of vegetation at one week following treatment. At 5 weeks after application, the efficacy of clove oil and limonene declined to ratings of 1.0 or less, thereby imparting essentially no suppression of weeds. Application of higher amounts of clove oil or pelargonic acid appeared to increase the short-term (1 week) efficacy of treatments, but this effect was not evident after 5 weeks. The higher amount of limonene was less effective than the lower amount at some dates (Table 10). The higher concentration was too viscous for successful application by spraying. This preparation clogged the nozzle of the sprayer, thereby giving uneven and perhaps inadequate coverage of the plots.

Roadside Experiment 2 (experiment with clove oil, pelargonic acid, limonene, citric-acetic acid, and glyphosate applied to mowed or unmowed plots). Mowing of the plots did not affect the efficacy of the herbicides, as the mean index rating for mowed or unmowed plots was 3.2 for each mowing regime (Table 12). The interaction of mowing with herbicide treatment or with date of observation also was nonsignificant.

In this experiment, the short-term (1 week) efficacy of pelargonic acid by index rating (mean 7.3) was the highest of the alternative herbicides (Table 12). At this time, limonene also gave good control of vegetation with a mean rating of 5.8, but control with clove oil was weak with a mean rating of 2.6. A mixture of clove oil and limonene (rating of 1.7 at 1 week) was less effective than the herbicides used individually. Perhaps, this effect resulted from uneven application of the viscous mixture of clove oil and limonene. The citric acid formulation at 1 week was rated as 5.0 on the indexing scale. The efficacy of each of the alternative herbicides declined with time over a 5-week period to impart little or no observable control at the end of the growing season. On the other hand, glyphosate gave strong initial control (rating of 7.8 at 1 week) and increased in efficacy with time. The mass of weeds harvested at the end of the season (September 26, 2006) indicated that the treatments with alternative herbicides or with mowing did not differ in season-long control of vegetation (Table 13).

Table 10: Assessment of control of vegetation by alternative herbicides applied at two concentrations to mowed or unmowed plots in roadside research (Roadside Experiment 1).

Herbicide	Application[†] a.i. L ha^{-1}	Management of vegetation and time after treatment					
		Mowed[‡]			Unmowed		
		1 wk	3 wk	5 wk	1 wk	3 wk	5 wk
		Visual index[§]					
Clove oil	34	1.3	0.7	0.3	2.3	1.7	1.0
	68	5.0	4.0	1.0	7.3	4.0	1.0
Pelargonic acid	13	6.3	5.0	0.7	7.7	5.0	1.3
	26	7.3	6.3	1.7	8.7	7.7	3.3
Limonene	18	5.0	3.3	1.0	3.7	4.3	1.0
	36	2.7	1.7	0.3	3.7	2.3	0.3
None		0	0	0	0	0	0
Mean			3.0 NS			3.3	

[†] a.i.: active ingredient.
[‡] Mowed to 10 cm height above ground.
[§] Visual rating index. See Table 3 for rating scale.
LSD (P = 0.05) is 1.1 for comparing herbicides in columns and is 0.3 for comparing dates in rows under mowed or unmowed treatments. NS: no significant difference between mowed and unmowed plots by F-test.

Table 11: Mass of vegetation harvested from roadside plots treated with alternative herbicides or with mowing (Roadside Experiment 1).

Herbicide	Application[†] a.i. L ha^{-1}	Mass of vegetation, g fresh wt m^{-2}	
		Mowed[‡]	Unmowed
Clove oil	34	1,163	1,210
	68	1,156	1,120
Pelargonic acid	13	1,116	1,340
	26	1,273	1,180
Limonene	18	1,172	1,035
	36	1,150	1,250
None		1,487	1,126
Mean		1,217	1,180

[†] a.i.: active ingredient.
[‡] Mowed to 10 cm height above ground.
Neither herbicide treatment, mowing, nor their interaction had a significant effect on mass by *F*-test ($P > 0.05$).

Table 12: Assessment of control of vegetation by alternative herbicides and a conventional herbicide applied to mowed or unmowed roadside plots (Roadside Experiment 2).

Herbicide	Application[†] a.i. L ha^{-1}	Management of vegetation and time after treatment					
		Mowed[‡]			Unmowed		
		1 wk	3 wk	5 wk	1 wk	3 wk	5 wk
		Visual index[§]					
Clove oil	34	2.3	0.7	0	3.0	0.7	0
Pelargonic acid	13	6.3	3.3	1.3	8.3	4.7	1.3
Limonene	18	6.0	3.3	1.3	5.7	3.7	1.0
Clove oil and	34	1.7	0.3	0	1.7	2.3	1.0
limonene	18						
Citric acid-	25	5.7	1.7	0.3	4.3	1.3	0.3
acetic acid	19						
Glyphosate	2.8	8.3	9.8	10.0	7.3	10.0	10.0
None		0	0	0	0	0	0

[†] a.i.: active ingredient.
[‡] Mowed to 10 cm height above ground.
[§] Visual rating index. See Table 3 for indexing scale.
LSD ($P = 0.05$) is 1.4 for comparing herbicides in columns and is 0.7 for comparing dates in rows under mowed or unmowed treatments.

Table 13: Mass of vegetation harvested from plots treated with alternative herbicides or a conventional herbicide applied to mowed or unmowed roadside plots (Roadside Experiment 2).

Herbicide	Application[†] a.i. L ha^{-1}	Mass of vegetation, g fresh wt m^{-2}	
		Mowed[‡]	Unmowed
Clove oil	34	950	773
Pelargonic acid	13	773	926
Limonene	18	980	900
Clove oil and	34	1,011	814
limonene	18		
Citric acid-	25	1,074	1,059
acetic acid	19		
Glyphosate	2.8	0	0
None		971	968
Mean		823	786

[†] a.i.: active ingredient.
[‡] Mowed to 10 cm height above ground.
Except for the effect of glyphosate, which differed from all other treatments, neither herbicide treatment or mowing nor their interaction had a significant effect on mass ($P > 0.05$). LSD ($P = 0.05$) = 350 for the interaction. Mowing and not mowing did not differ by F-test, $P > 0.05$.

4. DISCUSSION

An urgent need exists to develop and implement effective organic-compliant herbicides to meet consumer demand for organic products for farm and highway use [1, 18, 19]. Natural products represent a vast repository of materials and compounds with biological activity, including phytotoxicity [44]. Selected organic acids, such as pelargonic acid, citric acid, and acetic acid, are effective nonselective herbicides for a wide spectrum of annual weed species [20, 44]. The present study evaluated many organic products derived from plants. Research has shown that concentration used in this study with at least 15% to 20% acetic acid provided acceptable weed control of broadleaf annual weeds but gave poor control of annual grasses [11, 14]. Application and timing are critical to maximizing weed control with alternative herbicides [12]. As growth stage advances, control and biomass suppression is decreased, and survival is increased. In the current study and in prior work, a single application was not as effective as multiple applications over the growing season. For example, a single application of pelargonic acid demonstrated immediate or short-term suppression of growth of vegetation; however, the efficacy lasted for no more than 6 weeks, after which regrowth was not distinguishable from untreated vegetation [1]. Also, formulations of citric-acetic acid gave no control or only weak suppression of vegetative growth soon after application, and no suppression was evident after 6 weeks [1]. Repeated applications of alternative herbicides were necessary for season-long efficacy. Two seasons of application were needed for control of stiltgrass (Microstegium vimineum A. Camus) [45].

The effects of flaming lasted for about 6 weeks, similar to the effects of alternative herbicides in the current study and in prior work [1]. Flaming, however, appeared to improve the efficacy of applications of corn gluten meal and mulches. Other research showed that corn gluten meal or flaming used individually was not an effective herbicide in strawberry (Fragaria × ananassa Duchesne) production [46]. No suppression of growth of roadside vegetation occurred with the use of corn gluten meal, which acted as a nitrogen fertilizer to promote growth [1]. Strawberry yield in plots receiving corn gluten hydrolysate as a fertilizer increased in one year but did not show an increase in another two years of research [9]. Even when used in combination with cultivation, corn gluten meal or clove oil was ineffective and offered little potential in a weed management system for organic peanut (Arachis hypogaea L.) production [47].

In this study, mulches of bark or woodchips were equally effective in imparting strongly suppressive against emerging vegetation for 2 years but were more effective in the first year than in the second year after application. Law et al. [42] reported that no treatment with mulches or corn gluten mean provided acceptable season-long weed control and resulted in low yields of pepper (Capsicum spp. L.). If treatment applications were delayed for 6 weeks, with three shallow cultivations which were used to reduce early weed competition before mulching, average weed control rating provided by the mulches ranged from 45% to 86% and resulted in greatly improved yields [42]. Previous reports noted that the costs of materials and labor for the alternative practices, especially mulching, were substantially more than for the conventional herbicides used in this study [1, 19]. Costs of materials alone for natural products were two to ten times the cost of glufosinate ammonium and six to thirty times the cost of glyphosate. Pelargonic acid was the most economical alternative herbicide with materials costing about three times that of glufosinate ammonium and six times the cost of glyphosate. Cost of mulch was about 200 times the cost of glyphosate. Young [48] reported that, for control of annual vegetation compared to glyphosate, a single application of acetic acid was 79% as effective and pine oil was 40% or less as effective. Essential oils from plants were about 80% as effective as glyphosate. Generally, multiple applications of natural products were needed to attain relative efficacy of 80% compared to glyphosate. The cost of one or more applications of natural products was more than ten times the cost of one or two applications of glyphosate [48]. Economics may not always be the limiting factor in the use of natural products for pest control in agriculture.

Concerns over the potential impact of pesticides on the environment and stringent pesticide registration procedures are growing. New regulations have reduced the number of synthetic pesticides available in agriculture, and reliance on these pesticides needs to be considered [49]. Dayan et al. [49, 50] reviewed the use of pesticides derived from natural products and noted their current applications and potential for impacts in organic and conventional agriculture. These researchers emphasized the needs for development of pesticides from natural products to meet the requirements for new products especially in weed management. The current and previous work demonstrates that alternative methods can be effective means of weed management at roadside sites and for crops in fields but at high economic costs. Multiple applications of the materials will be necessary for effective season-long weed control. End of season results suggested that no second-season control except for mulching would be imparted.

ACKNOWLEDGMENT

This research was supported in part by the Massachusetts Department of Transportation—Highway Division, Boston, Massachusetts.

REFERENCES

1. V. Barker and R. G. Prostak, "Herbicides alternatives research," Final Report, Massachusetts Highway Department, Boston, Mass, USA, 2008.
2. K. Owens, "The right way to vegetation management," Pesticides and You, vol. 19, no. 1, pp. 9–17, 1999.
3. S. O. Duke, "Natural pesticides from plants," in Advances in New Crops, J. Janick and J. E. Simon, Eds., pp. 511–517, Timber Press, Portland, Ore, USA, 1990.
4. T. Tworkoski, "Herbicide effects of essential oils," Weed Science, vol. 50, no. 4, pp. 425–431, 2002.
5. N. E. Christians, "Use of corn gluten meal as a natural pre-emergent weed control," Turfgrass Trends, vol. 11, no. 1, pp. t14–t16, 2002.
6. D. L. Y. Liu and N. E. Christians, "Isolation and identification of root-inhibiting compounds from corn gluten hydrolysate," Journal of Plant Growth Regulation, vol. 13, no. 4, pp. 227–230, 1994.
7. D. L. Liu and N. E. Christians, "Bioactivity of a pentapeptide isolated from corn gluten hydrolysate on Lolium perenne L.," Journal of Plant Growth Regulation, vol. 15, no. 1, pp. 13–17, 1996.
8. L. Webber, J. W. Shrefler, and M. J. Taylor, "Influence of corn gluten meal on squash plant survival and yields," HortTechnology, vol. 19, no. 2, pp. 346–352, 2010.
9. A. Dilley, G. R. Nonnecke, and N. E. Christians, "Corn-based extracts to manage weeds and provide nitrogen in matted-row strawberry culture," HortScience, vol. 37, no. 7, pp. 1053–1056, 2002.
10. F. Forcella, T. James, and A. Rahman, "Post-emergence weed control through abrasion with an approved organic fertilizer," Renewable Agriculture and Food Systems, vol. 26, no. 1, pp. 31–37, 2011.
11. C. Brainard, W. S. Curran, R. R. Bellinder et al., "Temperature and relative humidity affect weed response to vinegar and clove oil," Weed Technology, vol. 27, no. 1, pp. 156–164, 2013.
12. G. J. Evans, R. R. Bellinder, and M. C. Goffinet, "Herbicidal effects of vinegar and a clove oil product on redroot pigweed (amaranthus retroflexus) and velvetleaf (abutilon theophrasti)," Weed Technology, vol. 23, no. 2, pp. 292–299, 2009.
13. G. J. Evans, R. R. Bellinder, and R. R. Hahn, "Integration of vinegar for in-row weed control in transplanted bell pepper and broccoli," Weed Technology, vol. 25, no. 3, pp. 459–465, 2011.
14. J. A. Ivany, "Acetic acid for weed control in potato (Solanum tuberosum L.)," Canadian Journal of Plant Science, vol. 90, no. 4, pp. 537–542, 2010.
15. Johnson, T. Wolf, B. Caldwell, R. Barbour, R. Holm, and K. Sapsford, "Efficacy of vinegar (acetic acid) as an organic herbicide. ADF Project no. 20020202 and AAFC Project A03637," Final Report, University of Saskatchewan, Saskatoon, Canada, 2004.
16. J. Radhakrishnan, J. Teasdale, and C. Coffman, "Agricultural applications of vinegar," Proceedings of Northeastern Weed Science Society, vol. 57, pp. 63–64, 2003.
17. C. L. Webber III and J. W. Shrefler, "Organic weed control with vinegar: application volumes and adjuvants," Proceedings of Horticultural Industry Show, vol. 26, pp. 149–151, 2007.
18. H. F. H. Abouziena, A. A. M. Omar, S. D. Sharma, and M. Singh, "Efficacy comparison of some new natural-product herbicides for weed control at two growth stages," Weed Technology, vol. 23, no. 3, pp. 431–437, 2009.
19. V. Barker and R. G. Prostak, "Alternative management of roadside vegetation," HortTechnology, vol. 19, no. 2, pp. 346–352, 2009.
20. R. Coleman and D. Penner, "Desiccant activity of short chain fatty acids," Weed Technology, vol. 20, no. 2, pp. 410–415, 2006.
21. M. Windholz, Ed., The Merck Index, Merck & Co, Rahway, NJ, USA, 1983.

22. O. Ayeni, B. A. Majek, J. R. Johnson, and R. G. Obal, "Container nursery weed control: Bittercress , groundsel, and oxalis," Rutgers Coop. Ext. Fact Sheet 939.
23. J. E. Arboleya, J. G. Masabni, M. G. Particka, and B. H. Zandstra, "Identification of preharvesr desiccants for use in onion production," HortTechnology, vol. 15, no. 4, pp. 808–811, 2005.
24. S. L. F. Meyer, D. K. Lakshman, I. A. Zasada, B. T. Vinyard, and D. J. Chitwood, "Dose-response effects of clove oil from Syzygium aromaticum on the root-knot nematode Meloidogyne incognita," Pest Management Science, vol. 64, no. 3, pp. 223–229, 2008.
25. Y. C. Yang, H. S. Lee, J. M. Clark, and U. J. Ahn, "Insecticidal activity of plant essential oils against Pediculus humanus capitis (Anoplura: Pediculidae)," Journal of Medical Entomology, vol. 41, no. 4, pp. 699–704, 2004.
26. K. Kishore, S. Pande, and S. Harish, "Evaluation of essential oils and their components for broad-spectrum antifungal activity and control of late leaf spot and crown rot diseases in peanut," Plant Disease Journal, vol. 91, no. 4, pp. 375–379, 2007.
27. C. Lucas, E. Alves, R. B. Pereira, A. B. Zacaroni, F. J. Perina, and R. M. de Souza, "Indan clove essential oil in the control to tomato bacterial spot," Journal of Plant Pathology, vol. 94, no. 1, pp. 45–51, 2012.
28. J. M. van der Wolf, Y. Birnbaum, P. S. van der Zouwen, and S. P. C. Groot, "Disinfection of vegetable seed by treatment with essential oils, organic acids and plant extracts," Seed Science and Technology, vol. 36, no. 1, pp. 76–88, 2008.
29. N. S. Boyd and E. B. Brennan, "Burning nettle, common purslane, and rye response to a clove oil herbicide," Weed Technology, vol. 20, no. 3, pp. 646–650, 2006.
30. Stokłosa, R. Matraszek, M. B. Isman, and M. K. Upadhyaya, "Phytotoxic activity of clove oil, its constituents, and its modification by light intensity in broccoli and common lambsquarters (Chenopodium album)," Weed Science, vol. 60, no. 4, pp. 607–611, 2012.
31. R. Di Pasqua, G. Betts, N. Hoskins, M. Edwards, D. Ercolini, and G. Mauriello, "Membrane toxicity of antimicrobial compounds from essential oils," Journal of Agricultural and Food Chemistry, vol. 55, no. 12, pp. 4863–4870, 2007.
32. Gupta and V. A. Krischik, "Professional and consumer insecticides for management of adult Japanese beetle on hybrid tea rose," Journal of Economic Entomology, vol. 100, no. 3, pp. 830–837, 2007.
33. R. G. Hollingsworth, "Limonene, a citrus extract, for control of mealybugs and scale insects," Journal of Economic Entomology, vol. 98, no. 3, pp. 772–779, 2005.
34. Park, L. Kim, I. Choi, Y. Lee, and S. Shin, "Fumigant activity of plant essential oils and components from Schizonepeta tenuifolia against Lycoriella ingenua (Diptera: Sciaridae)," Journal of Economic Entomology, vol. 99, no. 5, pp. 1717–1721, 2006.
35. Rasooli and M. B. Rezaei, "Bioactivity and chemical properties of essential oils from Zataria multiflora Boiss and Mentha longifolia (L.) Huds.," Journal of Essential Oil Research, vol. 14, no. 2, pp. 141–146, 2002.
36. M. A. Ibrahim, E. J. Oksanen, and J. K. Holopainen, "Effects of limonene on the growth and physiology of cabbage (Brassica oleracea L) and carrot (Daucus carota L) plants," Journal of the Science of Food and Agriculture, vol. 84, no. 11, pp. 1319–1326, 2004.
37. Ascard, "Dose-response models for flame weeding in relation to plant size and density," Weed Research, vol. 34, no. 5, pp. 377–385, 1994.
38. J. Ascard, "Effects of flame weeding on weed species at different developmental stages," Weed Research, vol. 35, no. 5, pp. 397–411, 1995.
39. N. Rifai, T. Astatkie, M. Lacko-Bartosova, and J. Gadus, "Effect of two different thermal units and three types of mulch on weeds in apple orchards," Journal of Environmental Engineering and Science, vol. 1, no. 5, pp. 331–338, 2002.
40. V. Barker and P. C. Bhowmik, "Weed control with crop residues in vegetable cropping systems," Journal of Crop Production, vol. 4, no. 2, pp. 163–183, 2001.
41. V. Barker and T. A. O'Brien, "Weed control in establishment of wildflower sods and meadows," Proceedings of the NorthEastern Weed Science Society, vol. 49, pp. 56–60, 1995.

42. D. M. Law, A. B. Rowell, J. C. Snyder, and M. A. Williams, "Weed control efficacy of organic mulches in two organically managed bell pepper production systems," HortTechnology, vol. 16, no. 2, pp. 225–232, 2006.
43. R. G. D. Steel and J. A. Torrie, Principles and Procedures of Statistics, McGraw Hill, New York, NY, USA, 2nd edition, 1980.
44. S. O. Duke, F. E. Dayan, A. M. Rimando et al., "Chemicals from nature for weed management," Weed Science, vol. 50, no. 2, pp. 138–151, 2002. View at Publisher · View at Google Scholar · View at Scopus
45. J. S. Ward and T. L. Mervosh, "Nonchemical and herbicide treatments for management of Japanese stiltgrass (Microstegium vimineum)," Invasive Plant Science and Management, vol. 5, no. 1, pp. 9–19, 2012.
46. T. W. Miller, C. R. Libbey, and B. G. Maupin, "Evaluation of organic amendments and flaming for weed control in matted-row strawberry," HortScience, vol. 48, no. 3, pp. 304–310, 2013.
47. W. C. Johnson III, M. A. Boudreau, and J. W. Davis, "Combinations of corn gluten meal, clove oil, and sweep cultivation are ineffective for weed control in organic peanut production," Weed Technology, vol. 27, no. 2, pp. 417–421, 2013.
48. S. L. Young, "Natural product herbicides for control of annual vegetation along roadsides," Weed Technology, vol. 18, no. 3, pp. 580–587, 2004.
49. F. E. Dayan, C. L. Cantrell, and S. O. Duke, "Natural products in crop protection," Bioorganic and Medicinal Chemistry, vol. 17, no. 12, pp. 4022–4034, 2009.
50. F. E. Dayan, D. K. Owens, and S. O. Duke, "Rationale for a natural products approach to herbicide discovery," Pest Management Science, vol. 68, no. 4, pp. 519–528, 2012.

CHAPTER 4

Plant Responses to Salt Stress: Adaptive Mechanisms

Jose Ramón Acosta-Motos [1], Maria Fernanda Ortuño [2], Agustina Bernal-Vicente [1], Pedro Diaz-Vivancos [1], Maria Jesus Sanchez-Blanco [2] and Jose Antonio Hernandez [1,*]

[1] *Fruit Tree Biotechnology Group, Department of Plant Breeding, CEBAS-CSIC, Campus Universitario de Espinardo, P.O. Box 164, E-30100 Murcia*

[2] *Irrigation Department, CEBAS-CSIC, Campus Universitario de Espinardo, P.O. Box 164, E-30100 Murcia, Spai*

ABSTRACT

This review deals with the adaptive mechanisms that plants can implement to cope with the challenge of salt stress. Plants tolerant to NaCl implement a series of adaptations to acclimate to salinity, including morphological, physiological and biochemical changes. These changes include increases in the root/canopy ratio and in the chlorophyll content in addition to changes in the leaf anatomy that ultimately lead to preventing leaf ion toxicity, thus maintaining the water status in order to limit water loss and protect the photosynthesis process. Furthermore, we deal with the effect of salt stress on photosynthesis and chlorophyll fluorescence and some of the mechanisms thought to protect the photosynthetic machinery, including the xanthophyll cycle, photorespiration pathway, and water-water cycle. Finally, we also provide an updated discussion on salt-induced oxidative stress at the subcellular level and its effect on the antioxidant machinery in both salt-tolerant and salt-sensitive plants. The aim is to extend our understanding of how salinity may affect the physiological characteristics of plants.

Keywords: adaptive mechanisms; antioxidative metabolism; chloroplast; osmotic regulation; oxidative stress; photosynthesis; salinity; water relations

AUTHOR CONTRIBUTIONS

JAHC and JRAM projected the design of the review. JAHC, PDV and ABV wrote all the information related with Photosynthesis and Chlorophyll Fluorescence, Biochemical Adaptations: JRAM wrote all the information related with Morphological Adaptations of Plants to Salinity and Leaf Anatomy and Ultrastructure Changes in Leaves under Salinity. MJSB and MFO wrote all the information related with Water Relations, Gas Exchange and Osmotic Regulation and Related Salt Tolerance Mechanisms. JAHC and JRAM made all the Figures and Tables. JRAM ordered all the references.

1. INTRODUCTION

Salinity is one of the most significant environmental challenges limiting plant productivity, particularly in arid and semi-arid climates [1,2]. Salinity in irrigation water and in soils is one of the major abiotic constraints on agriculture worldwide, and the situation has worsened over the last 20 years due to the increase in irrigation requirements in arid and semi-arid regions such as those found in the Mediterranean area [3,4,5,6,7,8]. Soil salinity affects about 800 million hectares of arable lands worldwide [9]. A soil is considered to be saline when the electric conductivity (EC) of the soil solution reaches 4 dS m^{-1} (equivalent to 40 mM NaCl), generating an osmotic pressure of about 0.2 MPa and significantly reducing the yields of most crops [9]. As a consequence, ion toxicity, lead to chlorosis and necrosis, mainly due to Na^+ accumulation that interferes with many physiological processes in plants [10].

The harmful effect of salinity can vary depending on climatic conditions, light intensity, plant species or soil conditions [11]. Depending on the ability of plants to grow in saline environments, they are classified as either glycophytes or euhalophytes, and their response to salt stress differs in terms of toxic ion uptake, ion compartmentation and/or exclusion, osmotic regulation, CO_2 assimilation, photosynthetic electron transport, chlorophyll content and fluorescence, reactive oxygen species (ROS) generation, and antioxidant defences [11,12,13,14]. Most salinity adaptive mechanisms in plants are accompanied by certain morphological and anatomical changes [15]. Glycophytes, which includes most crop plants, cannot grow in the presence of high salt levels; their growth is inhibited or even completely prevented by NaCl concentrations of 100–200 mM, resulting in plant death [16]. Such growth inhibition can even occur in the short term [17]. In contrast, halophytes can survive in the presence of high NaCl concentrations (300–500 mM) because they have developed better salt resistance mechanisms, described above, characteristic of these plants [18,19].

Euhalophythes (plants growing in saline habitats) can cope with the effects of salt stress by developing different resistance mechanisms. These plants can regulate their salt content in the following ways (for more information, please consult [15]):

Salt exclusion: prevents the entry of salts into the vascular system.

Salt elimination: Salt-secreting glands and hairs actively eliminate salts, thus keeping the salt concentration in the leaves beneath a certain threshold.

Salt succulence: If the storage volume of the cells increases progressively with the uptake of salt (as the cells steadily take up water), the salt concentration can be kept reasonably constant for extended periods. Salt redistribution: $Na^{+}$and Cl^{-} can be readily translocated in the phloem so that the high concentrations arising in actively transpiring leaves can be redistributed throughout the plant.

Euhalophytes can also accumulate salt in their cell sap up to a level at which their osmotic potentials are lower than in the soil solution. In addition to salts, the accumulation of soluble carbohydrates plays an important role in maintaining a low osmotic cell sap potential. The ability of the protoplasm to tolerate high concentrations of salt also depends on the selective compartmentalisation of ions entering the cell. The majority of the salt ions are accumulated in the vacuoles (includer mechanisms).

Salt stress is first perceived by the root system and impairs plant growth both in the short term, by inducing osmotic stress caused by reduced water availability, and in the long term, by salt-induced ion toxicity due to nutrient imbalance in the cytosol [12]. Therefore, the two main threats imposed by salinity are induced by osmotic stress and ionic toxicity associated with excessive Cl^{-} and Na^{+} uptake, leading to Ca^{2+} and K^{+} deficiency and to other nutrient imbalances [20]. In addition, salt stress is also manifested as oxidative stress mediated by ROS. All these responses to salinity contribute to the deleterious effects on plants [21,22,23, 24,25,26].

Under saline conditions, plants have to activate different physiological and biochemical mechanisms in order to cope with the resulting stress.

Such mechanisms include changes in morphology, anatomy, water relations, photosynthesis, the hormonal profile, toxic ion distribution and biochemical adaptation (such as the antioxidative metabolism response) [18,23,27,28,29].

2. MORPHOLOGICAL ADAPTATIONS OF PLANTS TO SALINITY

Irrigation with saline water has different effects on plant growth according to different effects and mechanisms in (1) glycophyte vs. halophyte plants, as well as between different species of the same family and genus, and even between cultivars; (2) different salt levels in the irrigation water and (3) the time of exposure to such stress (short-term assays vs. long-term assays).

Munns [30] concluded that the salts absorbed by plants do not control growth directly, but that they do influence turgor, photosynthesis and/or the activity of specific enzymes. Demonstrating the complexity of salt stress, this author developed a model showing a two-phase effect of salinity on plant growth. Growth is first reduced by a decrease in the soil water potential (osmotic phase) and, later, a specific effect appears as salt injury in leaves, which die because of a rapid increase in salt in the cell walls or cytoplasm when the vacuoles can no longer sequester incoming salts (ionic phase). Munns [30] found that this salt accumulation in the old leaves accelerates their death and thus decreases the supply of carbohydrates and/or growth hormones to the meristematic regions, thereby inhibiting growth. The fact that plant growth is limited by a reduction in the photosynthesis rate and by an excessive uptake of salts affects the production of specific metabolites that directly inhibit growth [31].

2.1. Roots and Aerial Part Morphology

The anatomy of the root system (length, root diameter, etc.) determines root performance, enabling plants to acquire water and nutrients and thereby increase the replacement rate of plant water lost [32]. Optimum root systems can support shoot growth and improve plant yields, since roots serve as an interface between plants and the soil [33]. A proliferated root system would therefore appear to be better for plants, for it allows them to penetrate deeper layers of soil to acquire water and nutrients [34]. Recent studies, however, have shown that species with other root features, including small roots, can be more advantageous for shoot development [35]. For example, just a few roots in moist soil can provide amounts of water independent of the root number.

Other root characteristics, such as the number and diameter of xylem vessels, width of the root cortex, number of root hairs, and the suberin deposition in both the root exodermis and endodermis, also determine the permeability of roots to water [36,37]. Furthermore, environmental factors in the soil (changes in temperature, lack of O_2, mechanical impedance, salinity) can also produce marked impacts on root anatomy. The cell walls of root cells of salinized plants are often unevenly thickened and convoluted [38]. Salts often promote the suberisation of the hypodermis and endodermis in woody tree roots, resulting in the formation of a well-developed casparian strip closer to the root apex, different to that found in non-salinized roots [39]. Furthermore, the morphology of some plants shows their sensitivity to salinity. For example, in avocado trees, the root system is quite superficial and presents low ramification, thus reducing the water and nutrient absorption capacity [40], resulting in a higher sensitivity to soil salinity [41]. These morphological features limit the distribution of this crop to areas where irrigation water is of good quality. Saline water irrigation has also been found to alter the root system morphology of *Callistemon citrinus* plants [42]. Furthermore, similar responses were found by Álvarez et al. [43] and Álvarez and Sánchez-Blanco [44] in *C. citrinus* under deficit irrigation, which means that the effect of water stress on *C. citrinus* root system morphology is very similar to that produced by irrigation with saline water. The same behaviour was also observed in *Euonymus japonica* plants irrigated with an NaCl solution and reclaimed water containing different salt concentrations, which decreased the total root length of the plants, more specifically in thin ($Ø \leq 0.5$ mm) and medium thickness ($0.5 < Ø \leq 2.0$ mm) roots [45]. In other plant species (*Picea* sp., *Pinus banksiana*, *Portulaca oleracea*), Franco et al. [34] and Croser et al. [46] also found an increase in root diameter (hypertrophy) in response to salinity. The greater root density observed in these plants suggests greater robustness and, presumably, a higher accumulation of reserves [34,45,47], which would improve plant resistance to saline situations and speed up the establishment of plants, especially ornamental plants for gardening and landscaping purposes [48,49].

Salinity reduces plant growth through osmotic and toxic effects, and high sodium uptake ratio values cause sodicity, which increases soil resistance, reduces root growth, and reduces water movement through the root with a decrease in hydraulic conductivity [50]. Root hydraulic conductivity in field crops may vary in response to the salt content of the irrigation water applied [36]. In general, the root hydraulic conductivity in plants irrigated with poor quality water tends to decrease. In fact, the effect on root hydraulic conductivity is one of the main factors determining the sustainability of a reclaimed water irrigation system. Normally, root hydraulic conductance is expressed in terms of the whole root dry weight, without taking into account the role of root architecture in the water uptake capacity. Nonetheless, for any given root dry weight value, the amount of fine roots, which determine the root length and surface area, may vary greatly, thus, affecting the water absorption level [51,52].

Moreover, in assays where the grafted technique has been applied, the rootstock properties can affect the plant response to salinity. Salt-tolerant rootstocks alleviate the negative effects of abiotic stress to a greater extent than salt-sensitive rootstocks. Navarro et al. [53] described worse fruit yield and quality in Clemenules mandarin trees grafted on Carrizo (salt-sensitive rootstock) compared with Cleopatra (salt-tolerant rootstock), both irrigated with an NaCl solution (30 mM). In experiments carried out by Penella et al. [54,55], salt-tolerant rootstocks were found to increase the yield of a commercial pepper cultivar under salinity irrigations. Different nutritional and physiological responses were observed, such as ionic regulation, appropriate photosynthetic performance or maintenance of sink strength.

Another possible way to relieve or protect plants from saline stress is to inoculate the roots with arbuscular mycorrhizal fungi (AMF), which are considered to be important bio-ameliorators for saline soils. This role of AMF was studied by Navaro et al. [56] in citrus seedlings inoculated with a mixture of AMF (*Rhizofagus irregularis* and *Funneliformis mosseae*), which alleviated the negative effect of salt stress. This protective effect or synergic interaction is even better when it is established with salt-tolerant rootstocks. Other protective effects produced by other AMF (*Glomus iranicum* var. *tenuihypharum)* have also been observed on soil and *Viburnum tinus* irrigated with reclaimed water. Good AMF-soil-plant interactions could make it possible to reuse reclaimed water, particularly when the roots are growing in a saline soil [57].

A general decrease in fresh weight (FW) or dry weight (DW) has been observed in all plant tissues subjected to salt stress, but it is especially noticeable in the aerial part. Different authors have associated the decrease in FW or DW with a reduction in the number of leaves or in leaf abscissions (Table 1). To offer an explanation, some experiments have shown that a specific Cl^- build-up in the leaves of salt-stressed plants triggers 1-aminocyclopropane-1-carboxilic acid (ACC) synthesis and its conversion to ethylene with high efficiency, releasing enough hormone to trigger leaf abscission as occurred in citrus leaves and other plants [58,59,60,61]. Salt and osmotic stress also promoted the conversion of ACC to ethylene in the halophyte *Allenrolfea occidentalis* [62]. Other experiments in tomato plants irrigated with high salt concentrations describe that early events during the osmotic phase of salt stress promote leaf senescence prior to the massive accumulation of toxic ions. An indirect effect related to an initial abscisic acid (ABA) accumulation and a decrease in indol-3-acetic acid (IAA) and cytokinin (CK) contents certainly favours the progression of senescence in salinized leaves. However, the ethylene precursor ACC is the major hormonal signal that is temporally correlated with the onset of the oxidative damage, decreases in chlorophyll fluorescence and massive Na^+ accumulation [63,64].

Another typical response to salt stress described in different papers is a reduction in total leaf area. Indeed, decreased leaf growth is the earliest response of glycophytes exposed to salt stress [16]. The observed reduction in the canopy area may be considered as an avoidance mechanism, which minimises water loss by transpiration when the stomata are closed [65,66]. This effect can favour the retention of toxic ions in roots, limiting the accumulation of these ions in the aerial part of the plant [9,67]. Under saline conditions cell wall properties change and leaf turgor and photosynthesis rates (P_N) decrease, leading to a reduction in total leaf area [68,69]. Furthermore, the stem growth (a component of the aerial part) is also normally reduced by high salt concentrations. Decreases in leaf and stem provoke a reduction in all aerial part sizes and in the plant height. All these responses have been described by many different authors (Table 1).

The increase in root to shoot ratio or decrease in shoot to root ratio is a common response to salt stress, related to factors associated with water stress (osmotic effect) rather than a salt-specific effect [70]. A greater root proportion under salt stress can favour the retention of toxic ions in this organ, controlling their translocation to the aerial parts. This response can constitute a typical mechanism of plant resistance/survival under saline conditions [71,72].

High salt concentrations in the irrigation water result in reduced plant growth [9], limiting leaf expansion [73] and changing the relationship between the aerial and root parts [74]. Under high salinity, different plant species have shown a higher dry root mass than shoot dry mass, resulting in an increased root to shoot ratio (Table 1), which is thought to improve the source/sink ratio for water and nutrients under such conditions [75]. Also, a decrease in the shoot to root ratio has also been described by other authors (Table 1).

Table 1. Effect of salt stress on growth parameters in different plant species.

Plant Species	Salt Treatment	Type of Water [a]	Time	Parameter	References
			(a)		
Pisum sativum cv. Challis *Pisum sativum* cv. Granada	30–210 mM	NaCl	15 days	Leaf area, leaf FW, leaf DW	[22]
Carrizo citrange	200 mM	NaCl	30 days	Leaf abscission	[60]
Argyranthemum coronopifolium	70 mM	NaCl	15 days	Leaf area, RGR, NAR, LAR, SLW	[76]
Limonium latifolium cv. Avignon *Limonium latifolium* cv. Beltlaard	140 mM	NaCl	20 days	Total DW, shoot DW, Leaf area	[77]
Pisum sativum cv. Puget	70–160 mM	NaCl	15 days	Shoot FW, shoot DW	[78]
(*Pisum sativum* cv. Lincoln	50–90 mM	NaCl	15 days	Shoot FW, shoot DW	[23]
Pisum sativum cv. Puget					
7 *Olea europea* L. cvs.	25–200 mM	NaCl	150 days	Total DW (shoot DW + root DW), Leaf area, shoot length, root:shoot DW ratio	[79]
Carrizo citrange (citrus genotype)	30, 60, 90 mM	NaCl	28 days	Damage in aerial part, leaf abscission	[80]
Cistus monspeliensis L. *Cistus albidus* L.	70, 140 mM	NaCl	150 days	Leaf area, leaf damage Shoot DW, root DW, RGR, NAR, LAR, LWR, SLA	[81]
Cistus monspeliensis L. *Cistus albidus* L.	Anion surfactant, Sea-aerosol, AN + SA		120 days	Total DW, leaf area, leaf number, stem DW, root DW	[82]
Asteriscus maritimus	70, 140 mM	NaCl	126 days	Leaf DW, leaf area, stem DW, shoot DW, flower DW, root DW, RGR, NAR, LAR, LWR, SLA	[69]
Rosmarinus officinalis	35, 70 mM	NaCl	180 days	Leaf DW, total leaf area, stem DW, plant height, root DW, shoot:root DW ratio	[83]
5 perennial herbaceous	0.8, 2, 4 dS·m $^{-1}$	RW	90 days (Summer) 90 days (Fall)	Visual quality, shoot DW, plant height	[84]
Arbutus unedo	52 mM (5.45 dS·m $^{-1}$) 105 mM (9.45 dS·m $^{-1}$)	NaCl	112 days	Total DW, leaf area, plant height	[85]
Solanum lycopersicum L. cv. Moneymaker	100 mM NaCl	NaCl	21 days	Leaf number, leaf FW, leaf area, shoot FW, root FW, root:shoot FW ratio	[63]
5 Citrus genotypes	60 mM	NaCl	86 days	% of damaged leaves, leaf abscission	[86]
12 ornamental shrubs	10, 40, 70 mM	NaCl	120 days	Necrotic leaf percentage, RGR	[72]
Lantana camara *Polygala myrtifolia*	2.46 dS·m $^{-1}$ 5.63 dS·m $^{-1}$ 4.32 dS·m $^{-1}$	NaCl RW NaCl + RW	160 days	Plant dry weight, leaf area, shoot:root ratio, visual quality index	[87]

Plant Species	Salt Treatment	Type of Water [a]	Time	Parameter	References
			(b)		
5 ornamental species	0.6, 4, 6, 8, 12 dS·m^{-1}	NaCl + $CaCl_2$	180 days (S-S) 180 days (F-W)	Visual quality rating, leaf DW, stem DW, root DW, shoot:root DW ratio	[88]
Phlomis purpurea	4 dS·m^{-1}	NaCl	182 days	Leaf number, total leaf area, stem diameter, shoot DW, plant height, root DW, shoot:root DW ratio	[89]
Viburnum tinus *Metrosideros excelsa*	NaCl (2, 6 dS·m^{-1}) B (1, 6 mg/L)	NaCl+B	170 days	Leaf fall, plant dry weight, shoot:root ratio	[90]
Echinacea purpurea *Echinacea pallida* *Echinacea angustifolia*	50 mM (5.50 dS·m^{-1}), 75 mM (7.89 dS·m^{-1}) 100 mM (9.96 dS·m^{-1})	NaCl	14 days	Salt leaf injury index, leaf area, shoot DW, shoot WC (%), root DW, root WC (%)	[91]
Euonymus japonica	1.7 dS·m^{-1}, 4 dS·m^{-1}, 4 dS·m^{-1} RW	NaCl	140 days	Leaf DW, leaf number, total leaf area, root DW, total root length, root:shoot DW ratio, visual characteristic	[45]
Evonymus japonica *Viburnum tinus*	4 dS·m^{-1} 4 dS·m^{-1}	NaCl RW	140 days	Total biomass DW, leaf number, total leaf area, plant height	[92]
Myrtus communis L.	1.7, 4, 8 dS·m^{-1}	RW	154 days	Leaf area, Shoot DW, root DW, plant height, plant compactness (leaf area/height)	[93]
Callistemon citrinus	4 dS·m^{-1}	NaCl	56 days	Leaf DW, stem DW, stem diameter, root DW, total DW, root:shoot DW ratio, total leaf area, plant height, root morphology parameters, RGR	[42]
Lawsonia inermis L.	75, 150 mM	NaCl	53 days	Total FW, shoot:root DW ratio, RGR	[94]
Myrtus communis L.	4, 8, 12 dS·m^{-1}	NaCl	30 days	Leaf DW, stem DW, root DW, root:shoot DW ratio	[28]
Eugenia myrtifolia	4, 8, 12 dS·m^{-1}	NaCl	30 days	Leaf FW, leaf DW, total leaf area, leaf water content, stem DW, root DW, root:shoot DW ratio	[29]
Callistemon laevis	4 dS·m^{-1}	NaCl	300 days	Leaf DW, leaf area, stem DW, root DW, plant height, root:shoot DW ratio, plant compactness (leaf area/ height)	[95]
Myrtus communis L.	2.0, 5 dS·m^{-1} 10.0 dS·m^{-1}	RW NaCl	90 days	Leaf DW, stem DW, root DW, total DW, root:shoot DW ratio	[6]
Viburnum lucidum *Callistemon citrinus*	200 mM	NaCl	103 days	Leaf number, leaf area, shoot DW, apical shoot length, lateral shoot length, lateral shoots, plant height, root DW, root:shoot DW ratio	[8]
Verbana bonariensis L.	200 mM	NaCl	2013–2014	Total plant height, FW of aboveground part, visual score	[96]

[a] NaCl solutions (NaCl) or reclaimed water (RW).

Gómez-Bellot et al. [45] reported increases in the root to shoot ratio in *E. japonica* plants irrigated with an NaCl solution with an electric conductivity (EC) of 4 dS m^{-1}. The use of saline water in *C. citrinus* plants also slightly decreased aerial growth, increased the root to shoot ratio and improved the root system (increase in root diameter and root density) [42]. Acosta-Motos et al. [28] observed increases in the root to shoot ratio in *Myrtus communis* plants irrigated for one month with an NaCl solution with an EC of 8 dS m^{-1}. Over the same stress period, it was necessary to irrigate with an NaCl solution with a higher EC (12 dS m^{-1}) in order to observe any increases in the root to shoot ratio in *Eugenia mirtyfolia* plants [29]. Moreover, Acosta-Motos et al. [6] did not observe in the root to shoot ratio in *Myrtus communis* plants during three months of salt stress using reclaimed waters with ECs of 2 and 5 dS·m^{-1}. However, a significant increase in the root to shoot ratio was observed when an NaCl solution with an EC of 10 dS m^{-1} was employed.

2.2. Visual and Aesthetic Quality

Another interesting growth parameter linking leaf area with plant height is plant compactness. From an aesthetic and commercial point of view, high values in foliage size in relation to plant height give the plant architecture and compactness that are highly appreciated by nurseries and consumers [44,93]. Also, these changes, induced by salinity, have the potential to improve crop quality by reducing excessive vigour and promoting a more compact habit [49]. However, the level of soil salinity is important, and irrigation with high salt concentrations provokes a decrease in this ratio (leaf area/plant height) due to a reduction in both components which could negatively affect the commercial quality of plants [93,95].

Visual measurements provide a rapid and non-destructive way to diagnose leaf damage from salts. The different classification criteria that have been used for such measurements are described in the following paragraphs. For example, the number of damaged leaves in different *Citrus* genotype and pea plants was expressed as an injury percentage (%). When such percentage is equal o above 50%, leaves were considered affected by salinity (chlorosis and/or necrotic lesions) [80,86,97].

Visual quality of five herbaceous perennials was assessed on a scale from 1 to 5 points, where 1 indicated severely stunted growth with more than 50% of foliage either showing salt damage (leaf necrosis, browning) or dead; 2 indicated somewhat stunted growth with moderate (25% to 50%) foliar salt damage; 3 reflected average quality with slight (>25%) foliar salt damage; 4 meant good quality with acceptable growth reduction and little foliar damage (acceptable in terms of landscape performance); and 5 indicated excellent plants with vigorous growth and no foliar damage [84].

Bañón et al. [87] and Salachna and Piechocki [96], on the other hand, used a visual quality index to evaluate the aesthetic quality of plants at the end of their experiment. The index ranged from 1 (non-commercial) to 5 (very good quality), is based on the amount of leaf discolouration, leaf necrosis, defoliation and flowering.

In other experiment carried out by Valdez-Aguilar et al. [88], five ornamental species were visually rated to determine their ornamental attributes using a scale from 1 (poor quality, leaf bronzing higher than 75% or dead plants) to 5 (best quality, no leaf bronzing). Moreover, the same authors determined the visual quality rating using the piece-wise linear response function as proposed by Van Genuchten [98]. This function contains the following three independent parameters: (1) the non-saline control quality rate; (2) the salinity threshold (the maximum salinity without a quality rate reduction as compared with non-saline conditions); and (3) the slope (the fractional quality rate decline per unit increase in salinity beyond the threshold).

Sabra et al. [91] studied the salt injury index in three different *Echinacea* species. To measure this parameter, they assessed the visual appearance of the leaves using a five-point scale according to the severity of necrotic tissues and the number of injured plants, as previously described [99]. The scale was as follows: level 1 (no necrotic areas); level 2 (1%–25% necrotic areas); level 3 (26%–50% necrotic areas); level 4 (51%–75% necrotic areas); and level 5 (76%–100% necrotic areas or death of the plant).

Finally, Gómez-Bellot et al. [45] studied the percentage of plants affected according to visible characteristics linked to quality in *E. japonica* plants irrigated with water from different sources. These plants were classified into the following four groups: percentage of plants in acceptable conditions (PAC); percentage of plants with dry branches (PDB); percentage of plants in ideal conditions (PIC) and; percentage of dry plants (DP).

2.3. Growth Indexes Used with Salinity Studies

To observe the effects of salinity on plant morphology, many authors have studied different growth indexes such as relative growth rate (RGR), net assimilation rate (NAR), leaf area ratio (LAR), specific leaf area (SLA) and leaf weight ratio (LWR), according to Beadle [100], as follows:

$$RGR = (\ln W2/W1)/(t2-t1)$$

$$NAR = [(W2-W1)/(A2-A1)]\ [(\ln A2/A1)/(t2-t1)]$$

$$LAR = [(A2-A1)/(W2-W1)]\ [(\ln W2/W1)/(\ln A2/A1)]$$

$$SLA = [(A2-A1)/(WL2\text{-}WL1)]\ [(\ln WL2/WL1)/(\ln A2/A1)]$$

$$LWR = [(WL2-WL1)/(W2-W1)]\ [(\ln W2/W1)/(\ln WL2/WL1)],$$

where W is the total dry weight, t is the time, A is the total leaf area, WL is the total dry weight of leaves, and 1 and 2 are the beginning and end of a period (stress or recovery), respectively. Generally, high salt concentrations considerably reduce the RGR values (Table 1).

To give an example, Cassaniti et al. [71] studied the effect of NaCl on 12 widely cultivated ornamental shrubs in order to classify the plants. The following salt concentrations were used for irrigation over a period of 120 days: 10, 40 and 70 mM. The effect the salt had on growth depended on the species and salt concentration employed. Based on the results, the plants were distributed into the following four groups, using the decrease in RGR as criterion: (1) salt tolerant species, RGR decrease of less than 25%; (2) moderately salt tolerant species, RGR decrease between 25%–50%; (3) moderately salt sensitive species, RGR decrease between 50%–75%; and (4) salt sensitive species, RGR decrease greater than 75%.

Other papers have observed good correlation between RGR and LAR, but not with NAR, showing that leaf expansion is the growth limiting factor [101]. Other authors, however, have described a correlation between RGR and NAR, but not with LAR, indicating that photosynthesis could be the limiting factor and that a reduction in RGR may be due to a direct effect of salt stress on stomatal conductance and/or the photosynthetic apparatus [69,81,102].

Finally, high salt concentrations can produce a significant reduction in NAR and LAR values, indicating that both the photosynthetic rate and leaf expansion act as limiting factors in plant growth [76]. Decreases in leaf area linked to salinity directly affect the specific leaf area (SLA). However, salinity can increase the biomass fraction in leaves (LWR), associated with lower values in dry stem weight. Plants showing this behaviour are able to maintain their leaf characteristics and bloom well in spite of their total decrease in biomass [69].

3. LEAF ANATOMY AND ULTRASTRUCTURE CHANGES IN LEAVES UNDER SALINITY

Studying the changes produced in leaf anatomy is an appropriate way to study abiotic stress situations, including salt stress [28,29,94,103]. A pioneer assay in this field was carried out by Longstreth and Nobel [104]. These authors studied the effect of salinity on three plants with different responses to salt stress (*Phaseolus vulgaris*, salt-sensitive; *Gossypium hirsutum*, moderately salt-tolerant; and *Atriplex patula*, salt-tolerant). To study changes in leaf anatomy, these authors measured different parameters, such as epidermal, mesophyll and leaf thickness; the length and diameter of palisade cells; the diameter of spongy cells; the surface area of mesophylls per unit of leaf surface area; the ratio of mesophyll cell surface area to leaf surface area; and leaf succulence. The salt-tolerant specie, *Atriplex patula*, irrigated with different NaCl solutions (0.05, 0.1, 0.2, 0.3 and 0.4 M), showed more leaf thickness due to an increase in epidermal and mesophyll thickness. Finally, major increases were observed in the succulence values. Opposite effects were observed in the other two species that were less tolerant to salt irrigation.

Romero-Aranda et al. [105] studied anatomical disturbances produced by chloride salts (KCl, $ClCa_2$, NaCl) in both sensitive (*Carrizo citrange*) and tolerant (*Cleopatra mandarin*) citrus varieties. The salt-induced declines in P_N were linked to changes in leaf anatomical properties, such as the increase in leaf thickness and the lower area/volume ratio of mesophyll cells. Salinity also increased the succulence of leaves and reduced the intercellular air spaces, the surface/volume ratio of cells and tissue density. These results suggest that in salinized citrus, the increase in leaf thickness, in combination with several metabolic components such as chloride overloading, low Mg^{2+}, stomatal closure and chlorophyll loss, may contribute, among other factors, to the decline in photosynthesis. In addition, the contribution of leaf anatomical disturbance to the decline in net CO_2 assimilation is also dependent on the degree of genotype tolerance. On the other hand, a recent work carried out by Franco-Navarro et al. [106] proposes that chloride (Cl^-) plays a beneficial role as a nutrient in tobacco plants. These plants improve their leaf water balance with a decrease in transpiration associated with a reduction in stomatal conductance, which results in a decrease in water loss, an increase in the photosynthetic rate and improved water use efficiency. In addition, Cl^- in particular produces anatomical changes in the leaves, increasing their size and succulence.

Navarro et al. [85] observed the anatomical changes associated with salt stress in *Arbutus unedo* leaves by optical microscopy in semi-thin sections.

A comparison of control plants with saline-treated plants showed that the cell size of the first layer of palisade cells was not significantly altered. However, the cell size of the second layer of palisade cells significantly increased in parallel with the levels of salinity (0 mM, 52 mM and 105 mM NaCl). These authors also noticed an important decrease in the intercellular spaces in the spongy mesophyll in both saline treatments compared with control leaves, and this decrease affected the stomatal and mesophyll conductance to CO_2.

Fernández-García et al. [94] studied the ratio of leaf mass to leaf area (LSM) in henna plants irrigated with high salt concentrations. LSM is a key trait in plant growth and an important indicator of plant strategies to salt stress. Salt-treated henna plants respond to low and high salinity by increasing LSM. The same authors also noticed that high salt levels increased leaf thickness and suggested that the increase in leaf thickness and therefore in LSM in salt-stressed henna plants may imply a greater investment in assimilatory tissue as a strategy to maximise the photosynthesis potential.

Acosta-Motos et al. [28,29] studied leaf anatomical changes in *Myrtus communis* and *Eugenia myrtifolia* plants subjected to an NaCl solution of 8 $ds \cdot m^{-1}$ for the same exposure time (30 days). In myrtle, no significant changes were observed in the palisade parenchyma, but a decrease in spongy parenchyma and an increase in intercellular space were found. In eugenia leaves, the same anatomical changes were observed. In addition, eugenia leaves displayed a remarkable increase in palisade parenchyma. These traits in eugenia plants improve CO_2 diffusion, making it easier for CO_2 to reach the chloroplasts, which have a greater presence in palisade parenchyma. These alterations can protect and improve the $P_N$performance in eugenia plants, especially in a situation of reduced stomatal aperture (Figure 1). Similar results were observed by Gómez-Bellot et al. [107] in *Viburnum tinus* L. plants irrigated with treated wastewater. The leaf thickness of these plants increased due to an increase in the palisade parenchyma, thus maximising the photosynthesis potential in a saline situation.

Similar anatomical changes were also described by Rouphael et al. [108] in zucchini squash grown under saline conditions. In this experiment salt stress was accompanied by an increase in mesophyll thickness due to increased intercellular spaces. Such anatomical modifications in the leaves of salt stressed zucchini squash plants agree with previous findings in the ornamental shrubs *Myrtus communis* and *Eugenia myrtifolia* [28,29]. However, despite these modifications, zucchini squash plants supplied with NaCl had sharply reduced leaf P_N. Ultimately, salt stress induced decreases in P_N in zucchini squash as well as in eugenia and myrtle plants.

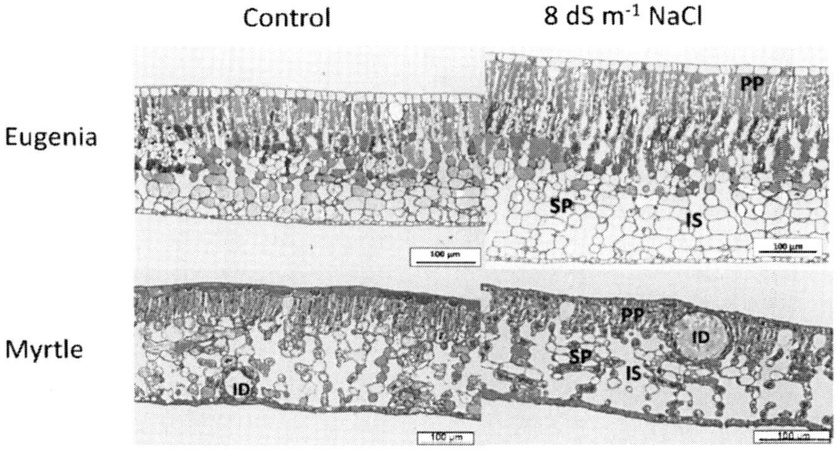

Figure 1. Light microscopy images showing the effect of NaCl on the percentage of area occupied by palisade parenchyma (PP), spongy parenchyma (SP) and intercellular spaces (IS) in leaves from *E. myrtifolia* and *M. communis* plants after 30 days of NaCl treatment. ID: idioblast. The increase in palisade parenchyma and intercellular spaces and decrease in spongy parenchyma can improve the CO_2diffusion, making it easier for CO_2 to reach the chloroplasts, which are present in a greater proportion in the palisade parenchyma. Partially reproduced from [28,29].

In general, these anatomical changes parallel to the observed decreases in P_N are not as great as the changes observed in stomatal conductance (gs). Therefore, under salt stress situations, high P_N/gs ratios (also expressed as intrinsic water use efficiency, WUEi) indicate that the leaves (more specifically, the chloroplasts), try to maintain high photosynthetic performance despite significant stomatal closure.

Damage to the chloroplast is an important aspect of the effect of salinization on leaf cells. The ultrastructure of the peroxisomes and mitochondria are also affected by salinity, albeit with less intensity. Under salinity conditions, NaCl-sensitive pea plants have been found to show leaf mitochondria with a lower number of cristae and lower electron density than the mitochondria of NaCl-tolerant pea plants [21].

The main changes that have been observed in the chloroplast of salt-irrigated plants are as follows:

(1) Changes in the number and size of chloroplasts and in the starch content.

(2) Disorganised chloroplast membranes.

(3) Changes in the number and size of plastoglobuli.

(4) Loss of the envelope and disorganisation of grana and thylakoids.

The effect of salinity on leaf ultrastructure can vary in relation to the tolerance of the plant to NaCl, as described in two rice (*Oryza sativa* L.) varieties [109]. The sensitive rice variety (Amber) showed differences in the integrity of the chloroplast compared with the tolerant variety (IR2153). Concretely, in response to high NaCl, most of the chloroplasts in the Amber variety showed signs of damage. In contrast, the chloroplasts in the tolerant variety IR2153 did not show any effect of salinity. This response correlated with a 30% decline in P_N in Amber, whereas no significant change in P_N was observed in IR2153.

The effect of NaCl on leaf chloroplast ultrastructure was studied in two different pea cultivars with different levels of sensitivity to NaCl, one salt-sensitive (cv. Challis) and the other relatively salt-tolerant (cv. Granada), up to 70 mM NaCl [22]. NaCl produced a disorganised tylakoidal structure in both pea cultivars. These authors described two different chloroplast populations that showed different densities in Percoll gradients. By electron microscopy, the authors noticed that the higher density band contained higher starch contents than the low-density band. Salinity affected the two chloroplast bands differently, and even the high-density band disappeared in the salt-treated sensitive plants [22]. In the NaCl-tolerant plants, salinity reduced the diameter of the chloroplasts from the low-density band. In both cases, an increase in the number and size of plastoglobuli was observed by the effect of NaCl, and the change was more evident in the tolerant plants [22]. Plastoglobuli are ubiquitous in chloroplasts and chromoplasts. Under different stress situations, plastoglobules increase as the thylakoids break down [110].

In the same study referred to above, Hernández et al. [22] noted that NaCl had different effects on the starch content in salt-sensitive and salt-tolerant pea plants. Whereas the percentage of starch decreased in both chloroplast populations in tolerant plants, no change occurred in sensitive plants. These data suggested that tolerant plants could use starch for different physiological processes to cope with the salt stress challenge [5].

Furthermore, as shown by Morales et al. [76], *Argyranthemum coronopifolium* is also sensitive to high salt concentrations. These authors found that salinity did not induce variations in the number of chloroplasts in these plants, although the size of the chloroplasts increased significantly in palisade and mesophyll cells due to an increase in starch content that was more evident in palisade parenchyma than in mesophyll tissue. The increased accumulation of starch in the leaf chloroplast in saline conditions could be explained by a lower accumulation of hexoses and sucrose and greater sucrose-phosphate synthase activity, leading the triose-phosphate pathway toward starch synthesis, and/or to damage of the enzymes involved in starch degradation via changes in the ionic composition in the chloroplast [111,112]. The different leaf photosynthetic rate values found by Morales et al. [76] would explain the differences in chloroplast size and starch accumulation found between palisade and mesophyll parenchyma.

Sánchez-Blanco et al. [82] studied the differences of the effects of simulated sea aerosol on the chloroplast ultrastructure of two rock-rose plants, White-leaf-rock-rose (*Cistus albidus* L.) and Montpellier rock-rose (*Cistus monspeliensis* L.). Transmission electron microscopy observation of Montpellier rock-rose control leaves showed well-organised chloroplasts with abundant starch grains and a structured cytoplasm. In plants treated only with surfactant, the ultrastructure of the organelles was not altered. When leaves were treated with sea aerosol only, the structure of some chloroplasts was altered, the starch grain decreased significantly and in some cases even disappeared, and in some of the chloroplasts, the thylakoid membranes appeared to be swollen. In leaves treated with sea aerosol plus surfactant, the chloroplasts showed the same alterations as plants treated with sea aerosol only. The substantial reduction in starch grains could be due to the fact that mesophyll cells from *C. monspeliensis* L. use soluble sugars to contribute to the osmotic adjustment induced by saline stress, which makes them more tolerant to salinity than *C. albidus* L. plants.

Navarro et al. [85] studied the effects of NaCl on the leaf ultrastructure of NaCl-sensitive *Arbutus unedo* plants. The chloroplast ultrastructure was altered and P_N was reduced by salinity. Furthermore, the number of plastoglobuli was significantly higher and the thylakoids were more dilated in the saline-treated leaves than in the control leaves.

More recently, Fernández-García et al. [94] observed changes in the ultrastructure of chloroplast from henna plants (*Lawsonia inermis* L.) irrigated with high salt levels. The control plants showed an abundant presence of starch grains and well-developed grana. Nevertheless, the chloroplast of the palisade cells had more dilated thylakoids and bigger starch grains in salt-treated plants. Moreover, the chloroplasts of the spongy mesophyll were highly altered, showing poorly developed and dilated thylakoids [94].

4. WATER RELATIONS, GAS EXCHANGE AND OSMOTIC REGULATION AND RELATED SALT TOLERANCE MECHANISMS

Plant–water relations explain the behaviour of plants in terms of how they control the hydration of their cells, which has important implications in the physiological and metabolic processes that determine the quantity and quality of plant growth [113]. Plants usually live in two different environments, the soil and the atmosphere, and these two media do not always behave in a coordinated manner. The evaporative demand of the atmosphere is practically continuous, whereas the water supply by rain or irrigation can be insufficient. Most of the water absorbed by plants from the soil is lost to the atmosphere by transpiration. In this situation, a hydraulic gradient is established in the xylem, which results in water flow from the roots to the leaves. This water movement is proportional to a driving force and in inverse proportion to resistance encountered through the soil-plant system. Transpiration creates the internal conditions under which water is absorbed by the root system and transported to the canopy [114]. As water is lost by transpiration, leaf water potential and the pressure in the xylem are reduced. Water transpiration from leaves is primarily controlled by stomata, but sustained transpiration, without sufficient replenishment, leads to progressive dehydration (loss of leaf turgidity). The degree of dehydration of plants depends on many factors: the climate and environment, the characteristics of the canopy, the growth stage of the plant, and, especially, the irrigation regime and water quality. In well-watered plants, transpiration represents the most effective means of leaf cooling [113]. Under water deficit, with the stomata closed, transpiration only occurs through the cuticle and the plant canopy. As a result, temperature tends to rise, which causes an increase in stomatal resistance to transpiration and an increase in the rate of tissue respiration. Under these conditions, photosynthetic capacity is limited and the plant uses its own photo-assimilates, resulting in decreased growth.

It is well known that the accumulation of salts in the root zone causes a decrease in osmotic potential and, consequently, a decrease in the water potential, diminishing the water available to the root zone [34,82]. Furthermore, when salts are present in the soil solution, they cause nutrient imbalances, inducing inhibition in the uptake of essential nutrients from the roots [114,115,116,117]. Other consequences of salt stress include metabolism alteration, such as hormonal changes, reduced enzyme activity and impaired photosynthesis [10,118]. In this way salinity affects plant establishment, growth and development, leading to a significant loss in productivity [45,79,89,92,119,120,121].

Nevertheless, the response to salinity varies depending on the stage of plant growth; the growing conditions, including climatic and soil factors; the agronomic and irrigation management practices; and the degree of tolerance of the species (glycophytes vs. halophytes) [42,45,77, 92,120, 122], which can even vary among the different cultivars of the same species [77,82].

Early responses to water and salt stress are very similar, because salts hinder the absorption of water through the root system due to the osmotic effect [123]. However, different plant responses have been observed in the long term [42]. Previous studies have concluded that salt-tolerant plants are not necessarily drought-tolerant plants and vice versa [121,124]. Specific mechanisms for salt avoidance either minimise the entry of salt into the plant or decrease the concentrations of salt in the cytoplasm of the cells. While halophytes have both types of mechanisms (salt exclusion and salt compartmentalisation in vacuoles), enabling them to grow for long periods of time, most glycophytes have a low ability to exclude salt and thus accumulate toxic ions in the leaves with the transpiration flow [10,15].

A plant exposed to salinity balances its water potential by losing water, which induces a decrease in osmotic potential and turgor. This creates chemical signals that can trigger subsequent adaptive responses [125,126]. During the adjustment process, an accumulation of organic and inorganic solutes that reduce cellular osmotic potential [10] and a reduction in the hydraulic conductivity of the membranes occurs, possibly by decreasing the number of water channels (aquaporins) [127]. Once turgor is recovered, growth can be restored [128].

4.1. Effects of Salinity on Leaf Water Relations and Gas Exchange

Across the soil-plant-atmosphere continuum, water flow is favoured by the gradient between soil water potential and leaf water potential. Whole plant hydraulic conductivity and individual conductance components are considered the main factors that control the movement of water through the soil-plant system and have been linked with plant transpiration and other physiological processes related with transpiration [129,130, 131,132]. Differences in root hydraulic conductance may cause differences in water transport from soil to plant, influencing the leaf water status and therefore in the growth and other physiological responses of the plant [133].

As far as plant water status is concerned, plants under saline or deficit irrigation exhibit slight dehydration, as indicated by lower water potential, due to higher difficulty in taking up water from the substrate or less available water in the substrate [89,134]. The most negative values for leaf water potential and stem water potential have been found in plants submitted to both water stress and salinity, as both passive dehydration and salt accumulation contribute to lowering the leaf water potential [135]. According to Álvarez and Sánchez-Blanco [42,44], stem water potential measured at midday is a good indicator of stress resulting from deficit irrigation or salinity. In addition, the difference between stem water potential and leaf water potential measured simultaneously in the same plant can be an indicator of the instantaneous shoot transpiration level [44].

Recently, Pérez-Pérez et al. [136] have studied the agronomic (fruit quality, yield, vegetative growth) and physiological (plant water status, gas exchange parameters, leaf mineral nutrition) responses in lemon trees using two different irrigation strategies: regulated deficit irrigation (RDI) and regulated saline irrigation (RSI). Their conclusions determine that RDI could be the best irrigation option for long water restriction periods. On the other hand, RSI would be successful for a period of only one year.

Salinity disturbs a plant's water relations due to a decreased availability of water from the soil solution as a result of lowered osmotic potential triggered by the toxic effects of the sodium and chloride ions [12]. Accordingly, an increase in resistance to water flow from soil to plant under salinity has been observed in many species [85,89]. For example, root hydraulic resistance has been found to increase in salt-stressed plants such as *Euonymus japonica*, *Phlomis purpurea*, and *Rosmarinus officinalis*, in which the lowest leaf water potential values have been recorded [45,83,89].

Salinity also causes a decrease in stomatal conductance and evaporation, and, as a result, smaller amounts of toxic ions are delivered to the root surface and into the plant. Recent research has confirmed that stomatal closure at the beginning of salinity exposure can contribute to a decline in the flow of toxic ions within the transpiration stream [137,138,139]. Decreases in stomatal conductance due to salinity have been found in several different plant species [45,90].

There is mounting evidence that stomatal regulation of vapour loss is extremely sensitive to short-term salt stress [140,141,142]. Previous comparisons of water relations in some barley cultivars with different drought resistance levels showed that lower stomatal conductance and transpiration at the beginning of salt exposure contributed to higher salt tolerance in terms of improved extension growth and less accumulation of toxic ions [139]. Other authors have reported that stomatal responses to different abiotic stress are probably regulated by root signals, as occurs in drying and re-wetting soils [143,144]. Stomatal closure in salt-stressed plants may be induced by the accumulation of abscisic acid [139,145,146]. However, Tardieu and Simonneau [147] concluded that the maintenance of water relations may be linked to an interaction between both chemical and hydraulic information.

Koyro [13] suggest that reductions in stomatal conductance represent an adaptive mechanism for coping with excessive salt, reducing the salt load of leaves and helping increase longevity by maintaining salts at subtoxic levels for longer than would occur if transpiration rates were not diminished.

Under salinity, some plants, such as *Rosmarinus officinalis*, develop tolerance and avoidance mechanisms. These mechanisms are based on stomatal closure and a reduced leaf area, in order to minimise water loss via transpiration [83]. Other resistance mechanisms include the development of osmotic adjustments, mainly involving the intake of inorganic solutes from the soil under salt stress situations to facilitate the maintenance of leaf turgor [69,123,148].

Decreases in stomatal conductance are one of the causes of photosynthesis (P_N) decline, although photoinhibition or increases in mesophyll resistance may also play an important role when stress is more severe or prolonged [149,150]. Such changes could delay plant recovery at the onset of the autumn or even cause permanent damage. Long-term reduced net CO_2 assimilation rates accompanying salinity have been attributed to stomatal closure, a decline in photosynthetic pigment content and concurrent non-stomatal factors (e.g., reduced protein concentrations) [92,151]. A decrease in P_N due to salinity stress has also been reported in many other plant species, such as *Viburnum tinus*, a salt-sensitive species [90], and *Alternathera bettzickiana*, a halophytic ornamental plant [152].

There must be a level of stress that triggers a stomatal opening control to prevent loss of water and maintain CO_2 fixation without an excessive decrease in productivity or plant growth. Thus the ratio of stomatal opening and the level of photosynthesis (photosynthesis efficiency or intrinsic water use efficiency, WUEi, [153]) are indicators of tolerance to osmotic stress. In *P. purpurea* plants submitted to a salt treatment, both P_N and gs decreased in a similar way. On the other hand, *E. myrtifolia* and *C. citrinus* plants submitted to salt stress were able to increase their WUEi throughout most of the growing season, i.e., the plants maintain higher P_N levels despite reduced stomatal opening [7,42]. This response agrees with the findings of Munns and Tester [9] and James et al. [154], who suggested that rates of P_N per unit leaf area in salt-treated plants often remain unchanged, even though gs is reduced.

When drought and heat co-occur, stomatal closure and decreased transpiration, associated with high WUEi, may lead to a dramatic increase in leaf temperature (up to 7 °C above air temperature) [155]. If this situation remains for long periods of time, leaf photodamage and/or xylem embolism may occur, leading to severe defoliation and plant death [156].

Variations in stomatal opening affect transpiration, which is a mechanism that leaves use for cooling. Accordingly, a decrease in stomatal conductance leads to an increase in leaf temperature. Under salinity conditions, as osmotic stress increases, transpiration rates and the resulting leaf latent-heat flux decrease, leading to an increased surface temperature that consequently increases the difference in temperature between salt-stressed and control plants. Accordingly, Halim et al. [157], Gerard et al. [158] and Sirault et al. [159] reported that stomatal closure in saline-stressed plants induced an increase in canopy foliage temperature. Furthermore, canopy temperature measurements in sorghum plants were found to be sensitive enough to indicate differences across the salinity treatments when soil moisture was uniform for several days following irrigation. The results of several studies have demonstrated that canopy temperature or thermal indices may be useful indicators of osmotic effects due to salt stress on myrtle, sorghum and euonymus plants [6,158,160].

4.2. Plant Osmotic Regulation

The accumulation and compartmentalisation of toxic ions and/or compatible solutes to increase osmotic tolerance are among the mechanisms involved in salt tolerance [161]. The osmotic regulation process of stomatal opening occurs via the movement of solutes, called osmolytes, which influence water movement between the cells. This water movement has several purposes linked to cell hydration inside plants, such as stomatal opening and higher water absorption via the root system. Several compounds that work in the osmotic regulation of plants are well known: carbohydrates (sucrose, sorbitol, mannitol, glycerol, arabinitol, pinitol) [162]; nitrogen compounds (proteins, betaine, glutamate, aspartate, glycine, proline, choline, putrescine, 4-gamma aminobutiric acid) [163]; and organic acids (malate and oxalate). Common responses in species exposed to saline or drought stress are an increase in osmotic adjustment, changes in the cell wall elasticity, decrease in saturation weight/dry weight ratio and increase in the percentage of water apoplastic, which minimizes the saline effects by maintaining the foliar turgidity. According to Navarro et al. [123], Zheng et al. [164] and Suárez [165] osmotic and/or elasticity adjustment result in the turgor loss point being reached at a lower leaf water potential and at a lower relative water content. The degree of osmotic adjustment differs among plant species. Furthermore, the degree of osmotic adjustment can also be affected by the speed at which the stress is imposed and the age of the plant, among other factors [13,14].

Limited osmotic adjustment has been observed in Mediterranean ornamental plants submitted to saline stress [89,123]. *C. citrinus,* for example, may behave as a typical Cl^- includer (accumulator), compartmentalising Cl^- within the leaf vacuoles, where it may be used as an osmolyte to lower the osmotic potential necessary for the maintenance of leaf turgor [13]. Such osmotic adjustment made it possible for *C. citrinus* plants to maintain turgor potential throughout saline treatment, despite only moderate osmotic adjustment (0.2 MPa). This behaviour and the osmotic adjustment values observed are within the range of those reported in other studies on Mediterranean ornamental plants submitted to saline stress [69,123,148].

Moderate salinity did not change the cell wall elasticity of the shoot tissue in *C. laevis*, however, suggesting that elastic adjustment did not play a role in the adaptation mechanism, although some changes in the cell wall composition may have occurred (such as in cell wall thickness and chemical composition, for example), which could have contributed to the decrease in growth recorded in this species in response to salt stress [166,167].

Among the organic osmolytes, proline (Pro) and glycine-betaine (GB) are the most important and efficient compatible solutes [11]. In addition to acting as an osmoprotectant, Pro has been reported to have antioxidant properties, and it can act as a molecular chaperone to protect the structure of biological macromolecules during dehydration, thus conferring plant tolerance to environmental stresses [11,168]. In plants, proline synthesis is mainly catalysed by Δ-1-pyrroline-5-carboxylate synthetase (P5CS; EC 2.7.2.11/1.2.1.41), which converts glutamate into Δ-1- pyrroline-5-carboxylate (P5C) in the cytoplasm or chloroplast. P5C reductase (P5CR) further reduces P5C to proline [169]. In addition, proline may be also synthesised in mitochondria from ornithine by ornithine δ-aminotransferase (OAT; EC 2.6.1.13), catalysing the formation of P5C, which is further reduced to proline by P5CR. Proline degradation, occurring in the mitochondria, is mediated by the action of proline dehydrogenase (ProDH; EC 1.5.99.8), which produces P5C from proline, and of P5C dehydrogenase, which converts P5C to glutamate [170].

In general, salt-tolerant species accumulate Pro after a salt-stress challenge. For example, *Eugenia myrtifolia* L. plants showed a salt-tolerant response when treated with 8 $dS{\cdot}m^{-1}$ NaCl for 30 days, as observed by the effect in plant growth, and this response correlated with a 20% increase in leaf Pro contents [30]. Furthermore, the application of Pro and/or GB to salt-stressed rice plants improved plant growth [171].

Glycine betaine is synthesised and accumulated under a variety of abiotic stresses, and it acts as an osmolyte to protect PSII during salinity [172]. It can also protect the integrity of membranes and the activity of enzymes against osmotic stress [173].

Salt stress is also associated with the toxic effect produced by the absorption of salt ions that can cause damage to the various subcellular organelles, mainly mitochondria and chloroplasts, when accumulated to high levels in the cells. These ions can also have a negative effect on the different enzymatic activities. Most unwanted ions in the plant are compartmentalised in the vacuoles, where they cease to be directly harmful, and only those that fall outside the vacuoles cause saline damage. Tolerance mechanisms in salt stress are therefore strategies aimed at the inclusion and exclusion of toxic ions.

The exclusion of salts can take place in the roots, preventing ion movement to the aerial part of the plant. This mechanism occurs in some glycophytes and in most of the halophytes (salt-tolerant). The development of saline vesicle glands in the epidermis can facilitate the excretion of salts, preventing salt accumulation in different organs such as the leaves. Another example of exclusion mechanism is the salts accumulation in the old leaves causing death and fall of these organs.

The halophyte *Plantago coronopus* L. may behave as a typical Na^+ includer that may be used as osmoticum to lower the osmotic potential necessary for the maintenance of leaf turgor [13]. This could indicate that these solutes were responsible for the osmotic adjustment described above. The cost of intracellular ion compartmentation is relatively small compared with the energy needed to synthesise organic solutes for osmotic adjustment [10]. However, the maintenance of leaf turgor is a controversial issue. Horie et al. [174] considered that salt accumulation in the leaf apoplast could be responsible for the death of leaves in plants exposed to salinity.

5. PHOTOSYNTHESIS AND CHLOROPHYLL FLUORESCENCE

Salt stress affects photosynthesis both in the short and long term. In the short term, salinity can affect photosynthesis by stomatal limitations, leading to a decrease in carbon assimilation [18]. This effect can produce rapid growth cessation, even after just a few hours of salt exposure [17].

In the long term, salt stress can also affect the photosynthetic process due to salt accumulation in young leaves [9] and decreases in chlorophyll and carotenoid concentrations even in halophyte plants [14,175,176]. The photosynthesis rate (P_N) can drop due to stomatal closure (gs), and/or other non-stomatal limitations, like the disturbance of the photosynthetic electron chain and/or the inhibition of the Calvin Cycle enzymes, such as Rubisco, phosphoenol pyruvate carboxylase (PECP), ribulose-5-phosphate kinase, glyceraldehyde-3-phosphate dehydrogenase or fructose-1,6-bisphosphatase [18,177]. A drop in gs can prevent excess water loss by transpiration, whereas proper regulation of the photosynthetic process can minimise the generation of ROS in PS II ($^1O_2$) and in the reducing side of the PSI ($O_2^{\cdot -}$ and H_2O_2) [178].

It is known that salt-tolerant species show increased or unchanged chlorophyll content under salinity conditions, whereas chlorophyll levels decrease in salt-sensitive species, suggesting that this parameter can be considered a biochemical marker of salt tolerance in plants [14,27].

In myrtle plants treated with different NaCl levels (from 4 to 8 $dS{\cdot}m^{-1}$) for 15 days, the leaf chlorophyll levels remained unchanged, whereas in eugenia plants subjected to the same treatment, the chlorophyll levels increased [28,29].

Analysing chlorophyll fluorescence is an easy and popular technique used in plant physiology studies that can provide interesting information about the state of the PSII [179]. Under saline conditions, a general decrease in photochemical quenching parameters (F_v/F_m, Y(II), qP) and in the electron transport rate (ETR) takes place, but increases in non-photochemical quenching parameters [qN, NPQ, Y(NPQ)] have also been reported [28,29,180,181,182] (Figure 2). The response to NaCl stress is correlated with decreases in PSII efficiency and increases in non-photochemical quenching parameters as a mechanism to safely dissipate excess energy. Salt stress was found to reduce ETR in a salt-sensitive rice cultivar, whereas only a slight reduction in ETR occurred in a salt-tolerant cultivar. However, the effect of NaCl stress on non-photochemical quenching parameters was somewhat different, since qN increased more substantially in the salt-tolerant cultivar than in the salt-sensitive cultivar

[180]. In salt-sensitive plants, salt stress, besides reducing photochemical-quenching parameters and ETR, also reduces non-photochemical quenching parameters [181,183].

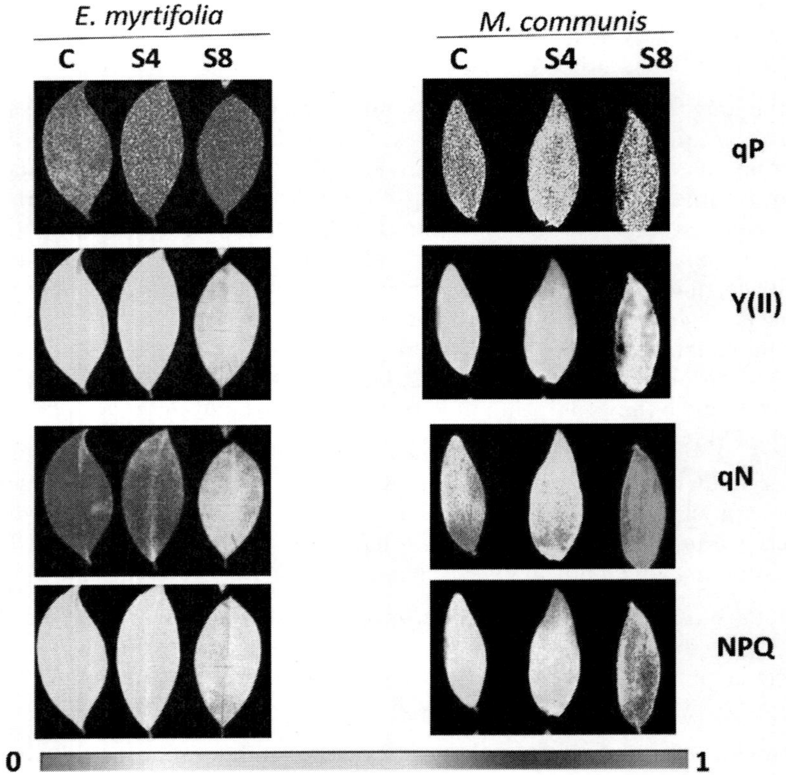

Figure 2. Chlorophyll fluorescence parameters in leaves of *E. myrtifolia* and *M. communis* plants at 30 days of NaCl stress. Images of the coefficient of photochemical quenching (qP), the effective PSII quantum yield [Y(II)], the non-photochemical quenching coefficient (qN) and non-photochemical quenching (NPQ). C, control plants; S4, 4 dS m^{-1}; S8, 8 dS m^{-1}. In the presence of salt stress, the photochemical quenching parameters increased in Eugenia plants, whereas myrtle plants showed increased non-photochemical quenching parameters, as a safe mechanism for dissipating excess light energy and minimising reactive oxygen species (ROS) generation. Partially reproduced from [28,29].

Decreases in photochemical and non-photochemical quenching parameters that occur in plants subjected to salt stress depend on the plant species and the severity of the stress. For example, in the salt-sensitive *Cucumis sativus* L. plants, whose growth was severely inhibited when subjected to 75 mM NaCl for up to 7 days [183]. After 15 days of salt treatment, eugenia plants subjected to 8 and 12 dS m^{-1} NaCl responded to the imposed stress with decreases in qP and Y(II) and a concomitant increase in the non-photochemical quenching parameters, a mechanism for safely dissipating excess light energy and minimising ROS generation [179]. At 30 days of stress, the increase in qP and Y(II) and the decrease in the non-photochemical quenching parameters observed in salt-treated plants paralleled the response observed in the gas exchange parameters, indicating an adaptive response to the imposed stress conditions (Figure 2). The recovery period was detected by plants as a new challenge, as evidenced by an alteration in the fluorescence parameters, especially in plants previously treated with 8 or 12 dS m^{-1} NaCl [29]. The observed decrease in qP as well as the drop in non-photochemical quenching parameters in this period suggested the generation of ROS in the chloroplasts as well as photooxidative damage [184], a response similar to that in NaCl-sensitive plants. In myrtle plants, no correlation between P_N and fluorescence parameters was observed at 15 days of salt stress. At 30 days of stress, however, the drop in P_N paralleled a decline in qP and Y(II) and increases in non-photochemical parameters, as a safe mechanism for dissipating excess light energy [29] (Figure 2). After the post-stress period, the partial recovery in P_N observed in myrtle plants previously treated with 4 dS m^{-1} NaCl correlated with increases in qP and Y(II), whereas in plants previously subjected to 8 dS m^{-1} NaCl, both qN and Y(NPQ) dropped [28], reflecting a decrease in the safe dissipation of excess energy that could induce photo-oxidative damage in the photosynthetic apparatus [179]. Long-term NaCl treatment (75–150 mM) also decreased Y(II) and Y(NPQ) in *Lawsonia inermis* L. plants [94].

5.1 Mechanisms for the Protection of the Photosynthetic Machinery

The photosynthetic apparatus might be damaged even under moderate light if the excess light cannot be properly dissipated, favouring the generation of ROS that can damage the photosynthetic machinery [185]. This situation may be more severe under salinity conditions. One of the mechanisms involved in the safe dissipation of excess excitation energy in PSII is the so-called xanthophyll cycle. Under stress conditions, the carotenoid violaxanthin is converted to zeaxanthin by the enzyme violaxanthin de-epoxidase [186]. This cycle has a role in the dissipation of excess excitation energy in the PSII antenna as heat by a process termed non-photochemical quenching (NPQ) [186]. In addition, zeaxanthin has been reported to play an important role as an antioxidant. Accordingly, the accumulation of zeaxanthin in transgenic tomatoes was found to alleviate salt-induced photoinhibition and photo-oxidation due to the antioxidant role of zeaxanthin in the scavenging of singlet oxygen and/or free radicals in the tilakoid membranes [186]. Furthermore, Huang et al. [187] reported the important role of ascorbate in the protection of the photosynthetic machinery. Under salt stress, the Arabidopsis *vtc-1* mutant showed a more significant decrease in P_N and in all the chlorophyll fluorescence parameters than did the wild-type plants, indicating that the decrease in P_N was due to ascorbate deficiency. Accordingly, ascorbate plays a central role in the protection of the chloroplast under salt stress conditions since it acts as a ROS scavenger and is a cofactor of violaxanthin de-epoxidase, which is involved in excess light energy dissipation in the xanthophyll cycle [185].

Photorespiration can act as a sink for excess reducing power, thereby preventing the over-reduction of the photosynthetic electron chain [188]. In addition, the photorespiratory pathway serves to convert phosphoglycolate (2PG) to 3-phosphoglycerate (3PGA), which can be metabolised to either regenerate RuBP or to make complex sugars and other carbon-based organic metabolites [189]. These effects can be even more important considering that different stress conditions can increase photorespiratory rates [190]. Drought and salinity, for example, trigger a decrease in stomatal conductance, thus decreasing the CO_2:O_2 ratio and increasing photorespiration [190]. Under salinity conditions, the photorespiratory pathway can prevent photo-oxidative damage by continuously recycling CO_2 for the chloroplast from the decarboxylation of glycine in the mitochondria. In addition, glycerate, produced inside the peroxisome, can be imported into the chloroplast and enter the Calvin cycle. These mechanisms keep the Calvin cycle working and prevent ROS generation in the electron transport chain, providing substrates for the chloroplast and decreasing the risk of photoinhibition under environmental stress conditions.

The stimulation of glycolate oxidase activity by salinity in pea leaf peroxisomes from salt-tolerant plants suggests that photorespiration can act as a molecular mechanism in NaCl-tolerance [191]. In C3 plants, both CO_2 fixation and photorespiration are the major sinks of electrons from PSII (4 electrons are required for CO_2 or O_2 reacting with Rubisco) [192].

In some salt-tolerant plants, increases in catalase activity have been recorded after an NaCl challenge, such as those described in tomato plants and myrtle, suggesting increased photorespiratory activity [25,26,28]. CAT has been found predominantly in leaf peroxisomes, where it chiefly functions to remove H_2O_2 formed in photorespiration or in the β-oxidation of fatty acids in the glyoxysomes [193]. The increase in CAT activity in salt-tolerant plants may suggest an involvement of photorespiration in the NaCl stress response. A correlation between CAT activity and photosynthesis has been described, since an increase in CAT reduces the photorespiratory loss of CO_2 [194].

Another protective mechanism for the photosynthetic machinery is the **water-water cycle** [178]. This cycle can be defined as the photoreduction of molecular oxygen to water in PSI by the electrons generated in PSII from water [178]. In this cycle, half of the electrons generated in PSII are used for the photoreduction of dioxygen to superoxide in PSI, and the other half are used for the regeneration of ascorbate (for more details of this cycle please consult the revision by Asada [178]). The primary function of the water-water cycle is the protection of the chloroplast components from the ROS. For example, different enzymes (Calvin cycle, photorespiration, chloroplast APX and SOD, etc.) can be inhibited by ROS [178]. In addition, the water-water cycle allows linear electron flow for ATP production that could be used in the photorespiratory pathway, which also dissipates excess photon energy.

6. BIOCHEMICAL ADAPTATIONS

6.1. Antioxidative Metabolism

It is well known that salinity induces oxidative stress in plants at the subcellular level [21,22,23,25,26,195]. Different works have reported that salt stress induces an accumulation of superoxide radicals (O_2) and hydrogen peroxide (H_2O_2) in different cell compartments, including chloroplasts, mitochondria and apoplastic space, which correlates with increases in some oxidative stress parameters, such as lipid peroxidation and protein oxidation [7,21,22,23,25,196,197]. In contrast, in leaf or root peroxisomes from salt-treated plants, no important changes in $O_2^{\cdot -}$ or H_2O_2 concentration have been reported [25,26,191], although lipid peroxidation levels have increased in some cases. In pea leaf peroxisomes, H_2O_2 concentrations were even statistically lower in the presence of NaCl than in control plants, in spite of the inhibition of catalase (an H_2O_2-scavenger enzyme) and an increase in the H_2O_2-producing enzyme glycollate oxidase [190].

Different papers have shown that the response of the antioxidative defences of the salt-tolerant species is somewhat different to that of salt-sensitive species. For example, some authors have described a coordinated up-regulation of the antioxidative machinery as one of the mechanisms involved in the salt tolerance response [23,25,195,196,198]. Other authors have correlated salt tolerance with higher constitutive levels of certain antioxidant enzymes [24,199,200,201]. In contrast, the antioxidant defences in salt-sensitive species show an unchanged response or even a decrease, and these species display lower constitutive antioxidant enzyme levels than salt-tolerant species [21,22,24,25,26,201] (Table 2). In different salt-tolerant species, increases in the antioxidant defences have been reported. For example, in salt-tolerant pea cultivars (cv. Granada and cv. Puget) increases have been observed in the antioxidant defences in chloroplasts and mitochondria [21,22,196]. In the wild salt-tolerant related tomato species *Lycopersicon pennellii*, salt-induced oxidative stress was alleviated by a selective up-regulation of a set of antioxidant enzymes at the subcellular level. Increases in SOD and the ASC-GSH cycle components took place in the chloroplasts, mitochondria and peroxisomes from salt-treated (100 mM NaCl) *Lycopersicon pennellii* plants, whereas in the cultivated tomato (*Solanum lycopersicom*), which is sensitive to 100 mM NaCl, a drop occurred in most of these antioxidant defences [25,26,197] (Table 3). It has been also described that NADPH regeneration systems, especially in roots, contribute to maintaining the redox state, and this seems to be essential in the biochemical mechanisms of plant defence against oxidative stress situations [202].

Furthermore, the salinity tolerance response of myrtle and eugenia plants to NaCl (44–88 mM) has been found to parallel increases in certain antioxidant defences, including MDHAR and catalase in the case of myrtle and MDHAR, GR, SOD and GSH concentrations in the case of eugenia [28,29].

In general, salt stress affects ascorbate and glutathione levels much more in sensitive plants than in tolerant plants. NaCl stress produced a 50% reduction in the total ascorbate pool in soluble fractions in an NaCl-sensitive pea genotype after 15 days of stress. This reduction was due to the loss of both the oxidized (DHA) and the reduced (ASC) forms. In an NaCl-tolerant genotype, salt stress also reduced SC and DHA levels but less so than the sensitive plants. Furthermore, no effect was observed on the ASC/DHA ratio. In contrast, less difference was observed in the glutathione concentrations. Even so, the reduced form (GSH) concentration decreased more in the NaCl-sensitive plants than in the NaCl-tolerant plants [195].

Table 2. Effect of salt stress on the activity of antioxidant enzymes in different plant models from salt-tolerant species.

Plant Specie	NaCl Treatment	Enzyme	Response	Cell Compartment	References
Pisum sativum cv. Granada	70 mM	Mn-SOD	increase	Leaf mitochondria	[21]
Pisum sativum cv. Puget [a]	90 mM	APX, MDHAR, DHAR	increase	Leaf mitochondria	[196]
Lycopersicon pennellii	100 mM	Mn-SOD, APX, MDHAR, DHAR, GPX	increase	Leaf mitochondria	[25]
		GR	decrease	Leaf mitochondria	[25]
		Mn-SOD, APX	increase	Root Mitochondria	[26]
		DHAR	decrease	Root Mitochondria	
Pisum sativum cv. Puget	90 mM	APX, DHAR, GR	increase	Chloroplasts	[196]
Pisum sativum cv. Granada	70 mM	Cu,Zn-SOD, APX	increase	Chloroplasts	[22]
Triticum durum Desf. cv. Ofanto	50 mM	stAPX, tAPX	increase	Chloroplasts	[203]
		stMDHAR, stDHAR	decrease	Chloroplasts	
Suaeda salsa	400 mM	tCu-Zn-SOD	increase	Chloroplasts	[204]
Pisum sativum cv. Puget	70–110 mM	tFe-SOD, sFe-SOD, tCuZn-SOD, sCuZn-SOD, sAPX, tAPX	increase	Chloroplasts	[205]
Pisum sativum cv. Granada	70 mM	CAT	decrease	Leaf peroxisomes	[191]
Lycopersicon pennellii	100 mM	SOD, APX, MDHAR, CAT	increase	Leaf peroxisomes	[25]
	100 mM	Mn-SOD, APX, MDHAR, CAT	increase	Root peroxisomes	[26]
Pisum sativum cv. Puget	90 mM	SOD	increase	Leaf apoplast	[23]
Triticum aestivam L. cv Gerek-79	250–500 mM	POX, SOD	increase	Leaf apoplast	[206]
Populus euphratica Oliv.	100–250 mM	SOD	Increase	Xylem sap	[207]
Brassica oleracea L.	80 mM	POX	Increase	Xylem sap	[208]
Pisum sativum cv. Granada	70 mM	APX, MDHAR, DHAR, GR	increase	Soluble fractions	[195]
Lotus japonica	150 mM	Cu,Zn-SOSD, DHAR	increase	Leaf and root crude extract	[198]
Oriza sativa cv. Pokkali	100 mM	CAT	decrease	Leaf crude extract	[181]
Oriza sativa cv. 651, cv. 632	6 and 12 dS/m	SOD, GR, APX	increase	Leaf crude extract	[180]

[a] cv Puget behaves as salt-tolerant at NaCl concentrations up to 90 mM.

Table 3. Effect of salt stress on the activity of antioxidant enzymes in different cell compartments from salt-sensitive species.

Plant Specie	NaCl Treatment	Enzyme	Response	Cell Compartment	References
Pisum sativum cv. Challis	70 mM	Mn-SOD	decrease	Leaf mitochondria	[21]
Pisum sativum cv. Puget [a]	110 mM	Mn-SOD, APX, MDHAR,	increase	Leaf mitochondria	[196]
		DHAR	decrease		
Lycopersicon esculentum L.	100 mM	Mn-SOD, MDHAR, GR	decrease	Leaf mitochondria	[25]
		Mn-SOD, MDHAR	decrease	Root Mitochondria	[26]
Pisum sativum cv. Puget	110 mM	Fe-SOD, Cu,Zn-SOD, APX, DHAR	increase	Chloroplasts	[196]
		MDHAR	decrease		
Triticum durum Desf. cv. Adamello	50 mM	tAPX	increase	Chloroplasts	[203]
		stAPX, stDHAR	decrease	Chloroplasts	
Pisum sativum cv. Puget	70–110 mM	Fe-SOD, CuZn-SOD, APX, DHAR	increase	Chloroplasts	[205]
		MDHAR	decrease		
Pisum sativum cv. Challis	70 mM	CAT	decrease	Leaf peroxisomes	[191]
Lycopersicon esculentum L.	100 mM	SOD, APX, MDHAR, DHAR	decrease	Root peroxisomes	[26]
Pisum sativum cv. Lincoln	90 mM	SOD	decrease	Leaf apoplast	[23]
Triticum aestivam L. cv Bezostaya	250–500 mM	SOD	increase	Leaf apoplast	[206]
		POX	decrease		
Populus popularis cv. 35–44	100–250 mM	SOD, APX, GR	Increase	Xylem sap	[207]
Pisum sativum cv. Challis	70 mM	CuZn-SOD I	decrease	Soluble fractions	[195]
		DHAR	increase		
Carrizo citrange	90 mM	SOD, CAT, GR, APX	increase	Leaf crude extract	[80]
Arabidopsis thaliana	100 mM	Cu,Zn-SOD, CAT	increase	Leaf crude extract	[202]
Cucumis sativus	75 mM	APX, DHAR, GR	decrease	Chloroplast	[183]
		MDHAR	increase		
Vitis vinifera	100 mM	APX	decrease	In vitro plants	[182]
		SOD, POX	increase		
Oryza sativa cv. IR-29	100 mM	APX, CAT	increase	Leaf crude extract	[181]
	6, 12 dS/m	SOD, APX, GR	decrease	Leaf crude extract	[180]

[a] cv Puget behaves as salt-sensitive at NaCl concentrations over 110 mM.

Both ASC and GSH were detected in the leaf apoplastic space of two pea cultivars with different levels of NaCl sensitivity. ASC and GSH levels did not change in either cultivar in response to NaCl, although the redox state of the ascorbate and glutathione pools decreased progressively with the severity of the stress (0 to 90 mM NaCl). Nevertheless, all these changes were, again, less pronounced in the resistant cultivar, which furthermore had higher constitutive levels of ASC and GSH in its apoplastic space than the sensitive cultivar [23]. The effect of NaCl on the ASC and GSH levels was also analysed in the leaf symplast of both pea cultivars. The data showed that ASC levels began to fall in the susceptible cultivar in the presence of 50 mM NaCl, whereas in the resistant cultivar this response occurred in the presence of higher NaCl levels (90 mM) [23]. Regarding symplastic GSH, the response was similar in both cultivars, but GSSG accumulation was only significant in the susceptible cultivar when exposed to 90 mM NaCl [23].

In grapevine plantlets, 100 mM NaCl did not affect ASC levels, but an accumulation of DHA occurred. Regarding glutathione pool, salt treatment reduced the GSH concentration but no effect was observed on GSSG levels. In both cases, a reduction in the redox state of ascorbate and glutathione occurred [182].

NaCl resulted in a decrease in the ascorbate and glutathione levels in chloroplasts from cucumber seedlings, and both forms of the two antioxidants were affected, especially the oxidized forms (DHA and GSSG). In this case, the GSH levels were much more affected than the ascorbate levels [183].

NaCl stress (100 mM) produced a decline in the ASC content in leaves of the salt-sensitive rice cultivar IR-29, but no effect was recorded in the roots of this rice cultivar or in the leaves or roots of an NaCl-tolerant rice cultivar (Pokkali) [181].

6.2. ROS and RNS Generation

Chloroplasts and mitochondria are important sources of ROS in plants subjected to salt stress. As a consequence of photosynthesis and Calvin cycle impairment, the over-reduction of the electron chain along with the lack of regeneration of the final electron acceptor in PSI ($NADP^+$) favour electron transfer from ferredoxin to oxygen to form $O_2^{\cdot -}$ (Mehler Reaction), which is disproportionated to H_2O_2 and O_2 by SOD [178]. In addition, singlet oxygen ($^1O_2$) generation is increased when the photoinhibition of PSII occurs in illuminated chloroplasts [209]. Under salt stress conditions, an increase in H_2O_2 content was reported in the chloroplasts of two pea cultivars, although this increase was much higher in the salt-sensitive cultivar than in the salt-tolerant one [22]. In another salt-tolerant pea cultivar (cv. Puget), the chloroplast H_2O_2 content increased up to 3.5-fold when plants were treated with 110 mM NaCl [196].

In plant mitochondria, two different sites of superoxide generation have been described which are located at the flavoprotein NADH dehydrogenase and at the ubiquinone-cytochrome b segments [210]. In sub-mitochondrial particles from pea plants, NaCl treatments increased NADH-dependent $O_2^{\cdot-}$ generation in an NaCl-tolerant cultivar and in an NaCl-sensitive cultivar. Nevertheless, the succinate-dependent $O_2^{\cdot-}$ generation was much higher in the NaCl-sensitive cultivar than in the resistant cultivar [21]. However, in these conditions, the Mn-SOD and Cu,Zn-SOD I isoenzymes only decreased in the mitochondria from the NaCl-sensitive cultivar, whereas a significant increase in Mn-SOD took place in the mitochondria from the tolerant cultivar. The ability of the mitochondria from the salt-sensitive cultivar to scavenge $O_2^{\cdot-}$was therefore reduced under conditions with higher $O_2^{\cdot-}$ generation. However, despite the increase in the $O_2^{\cdot-}$ rate generation in pea mitochondria under salt stress conditions, no change in the endogenous H_2O_2 concentration was detected. A similar response was reported in mitochondria from the salt-tolerant pea cultivar Puget by the effect of 70 mM NaCl. The unchanged mitochondrial H_2O_2 levels under salinity conditions could be explained by the fact that the membranes are permeable to H_2O_2, and H_2O_2 diffusion can be increased by NaCl stress [211]. However, 2- and 3-fold increases in mitochondrial H_2O_2 have been observed in the presence of 90 and 110 mM NaCl, respectively [196]. The decrease in mitochondrial antioxidant defences has been correlated with higher rates of lipid peroxidation and protein oxidation [21,196].

Peroxisomes also generate ROS as a result of their metabolism. In pea leaf peroxisomes, there are at least two sites of $O_2^{\cdot-}$generation: the reaction of the enzyme xanthine oxidase in the matrix and an NAD(P)H-dependent $O_2^{\cdot-}$ generating enzyme in the peroxisomal membrane [212]. In addition, H_2O_2 is produced at a high rate in peroxisomes by the action of different H_2O_2-producing flavin oxidases, with the photorespiratory glycolate oxydase enzyme being one of the most important H_2O_2 generators [213]. In peroxisomal membranes from pea leaves, neither the NADH-induced generation of $O_2^{\cdot-}$ nor the lipid peroxidation of membranes was found to be altered by salt stress, and results were similar in NaCl-tolerant and NaCl-sensitive pea plants [191]. In contrast, NaCl produced a decrease in the endogenous H_2O_2 concentration in leaf peroxisomes in both pea cultivars [191]. These authors suggested that H_2O_2 could react non-enzymatically with some peroxisomal metabolite whose concentration could be induced by salt stress [191]. This response has also been observed in root peroxisomes from tomato plants [26]. In this case, the authors attributed the decrease in the H_2O_2 levels in root peroxisomes, at least partially, to the fact that there were greater increases in APX and CAT activities than in SOD activities. In other words, the H_2O_2 detoxification rate is higher than the H_2O_2 production rate under salinity [26]. Alternatively, peroxisomal H_2O_2 could be potentially dangerous by leaking out, reaching the cytosol and inducing oxidative stress in this cell compartment [191]. H_2O_2 is also generated by the spontaneous and/or enzymatic dismutation of $O_2^{\cdot-}$ radicals in all cell compartments [178,214].

The effect of salt stress on ROS accumulation (mainly H_2O_2) can be also analysed by colorimetric methods in different plant tissues. By using the xylenol orange method [215] and Barba-Espín et al. [97] observed that 70 mM NaCl did not affect leaf H_2O_2 in pea leaves, but in the combined treatment using NaCl plus salicylic acid (50 or 100 μM), a 35% increase in leaf H_2O_2 was observed. Using the same methodology, García de la Garma et al. [216] analysed the effect of NaCl stress on the H_2O_2 levels in tobacco BY-2 cells at different growth phases. These authors also observed an accumulation of H_2O_2 in all the growth phases analysed. Furthermore, by using confocal laser microscopy, they found ROS accumulation in the cytoplasm and mitochondria. The effect of salinity on ROS levels ($O_2^{\cdot-}$ and H_2O_2) was also monitored in the xylem sap in two poplar cultivars with different levels of salt sensitivity [217]. After 18 days of treatment (250 mM NaCl) these authors only observed significant $O_2^{\cdot-}$ and H_2O_2 accumulation in the NaCl-sensitive plants. NaCl increased H_2O_2 in the apoplast in a concentration-dependent manner. Nevertheless, the apoplastic H_2O_2 levels were nearly 4–5-times higher in an NaCl-sensitive cultivar than in an NaCl-tolerant cultivar when plants were exposed to 70 or 90 mM NaCl [23].

Salt stress also produces increased $O_2^{\cdot-}$ generation in the chloroplast, as observed in cucumber seedling growth in the presence of 75 mM NaCl, which also correlated with higher lipid peroxidation results [183].

Salt stress was also found to increase the H_2O_2 levels in different plum lines overexpressing cytosolic Cu,Zn-SOD and/or APX. In this study, all transgenic plum lines had lower levels of H_2O_2 than non-transformed wild-type plants under salt exposure (100 mM) [207].

ROS accumulation under salt stress can be also monitored by the staining of plant tissues with diaminebenzydine (DAB, to locate H_2O_2) or with Nitroblue tetrazolium (NBT, to locate $O_2^{\cdot-}$). In pea leaves, salt stress (90 mM NaCl) was found to produce an accumulation of H_2O_2 and $O_2^{\cdot-}$ that was more evident in a salt-sensitive cultivar (cv. Lincoln) than in a relatively salt-tolerant cultivar (cv. Puget) [23]. In grapevine plantlets, 100 mM NaCl also produced ROS accumulation in leaves. In this case, a different pattern of H_2O_2 and $O_2^{\cdot-}$ accumulation was observed: the H_2O_2 staining appeared as small spots on the leaf surface, whereas the $O_2^{\cdot-}$ staining was observed in the veins [182].

Besides ROS, nitric oxide (NO) is also involved in the response of plants to salinity. Different works have reported that pre-treatment with a NO-donor (such as SNP) can improve the response of plants to NaCl-induced damage [218,219]. For example, exogenous NO application has been found to protect soybean plants and sunflower cotyledons from NaCl stress. In soybean plants, NO increased RWC, chlorophyll and ABA levels, and plants displayed better stomatal control [218]. In addition, NO can act as a regulator of glutathione metabolism, modulating the activity of GSH-dependent enzymes such as GST, GR, and GSH contents [218,219,220].

NaCl increased NO generation in tomato roots, but produced a decrease in nitrosoglutathione (GSNO), probably due to the strong drop in GSH under saline conditions [220]. GSNO is considered a NO reservoir in plants, and the balance between both molecules may act as a regulation mechanism in the response to salt stress, at least in roots [220]. It seems that under salt stress conditions, plant peroxisomes, in addition to generating ROS, are the main source of NO production [221]. These same authors also demonstrated that peroxisomes are required for NO accumulation in the cytosol, thereby participating in the production of peroxynitrite ($ONOO^-$), resulting in a nitrosative stress, moniroted by an increase in protein Tyr nitration in salt-stressed plants [221,222].

6.3. Effect of Biostimulants in the Salt-Stress Response

Different studies have demonstrated that the addition of pre-treatment with various molecules can improve or enhance the plant response to salt stress. For example, the exogenous application of H_2O_2 can induce salt tolerance in different plant species, including rice, wheat and maize [223].

Recently, Kaur and Bhatla [219] found that exogenous melatonin can improve salinity effects in sunflower cotyledons by modulating GSH metabolism or by increasing the activity of H_2O_2-scavenging enzymes (APX, CAT, POX), improving the photosynthetic rate and increasing the K^+/Na^+ ratio in *Malus hupehensis* Rehd [224].

The effects of SA treatments on the response of plants to salinity have also been studied. However, the results obtained vary depending on the plant species, growth conditions, and the SA and stress application rates. To give an example, foliar SA application significantly improved the growth of NaCl-treated tomato plants. This alleviation of NaCl toxicity by SA was related to a decrease in Na concentrations, an increase in K and Mg concentrations in the shoots and roots, and an increase in the antioxidant mechanism [225]. SA application was also found to partially protect pistachio seedlings against salt stress injuries by improving different physiological parameters such as RWC, chlorophyll content, chlorophyll fluorescence and the photosynthetic rate [226]. However, other works have found that SA treatments negatively affect the response of plants to NaCl, and that this response correlated with an imbalance in the antioxidant metabolism [97]. Borsani et al. [227] found that SA increased oxidative damage during both NaCl and osmotic stress and was critical in Arabidopsis seedling mortality under these conditions.

The protective role of polyamines (PAs) against salt stress has been also examined. For example, PA treatment improved the response of mung bean seedlings to salt stress by enhancing the levels of glutathione and ascorbate in addition to the activity of different antioxidant enzymes, including DHAR, GR, CAT and GPX [228]. The possible involvement of PAs in salt stress adaptation has also been investigated in grapevine (*Vitis vinifera* L.) plantlets. Treatment of plantlets with methylglyoxal bis(guanylhydrazone) (MGBG), a S-adenosylmethionine decarboxylase (SAMDC) inhibitor, enhanced the salt stress effects [182]. Salt and/or MGBG treatments also triggered an increase in lipid peroxidation and ROS accumulation but decreased APX activity and reduced glutathione [182]. These authors suggested that maintaining PA biosynthesis through enhanced SAMDC activity in grapevine leaf tissues under salt stress conditions could contribute to enhanced ROS scavenging activity and to protection of the photosynthetic apparatus from oxidative damage.

7. CONCLUSIONS

Salinity is one of the most significant environmental challenges limiting plant productivity. In the short term, salt stress is first perceived by the root system, inducing osmotic stress and causing reduced water availability. In the long term, salt stress induces ion toxicity due to nutrient imbalances in the cytosol.

A proper root system can ensure water and nutrient uptake in saline conditions and improve plant resistance to saline conditions. In general, salt stress produces a decrease in the aerial part of the plant, partially associated with leaf abscission mediated by ethylene production. This decrease in growth can be considered a mechanism to minimise water loss by transpiration. Moreover, an increased root to shoot ratio has been described under salinity. A bigger root proportion under salt stress can favour the retention of toxic ions, which can be an important mechanism of plant resistance/survival under saline conditions (Figure 3).

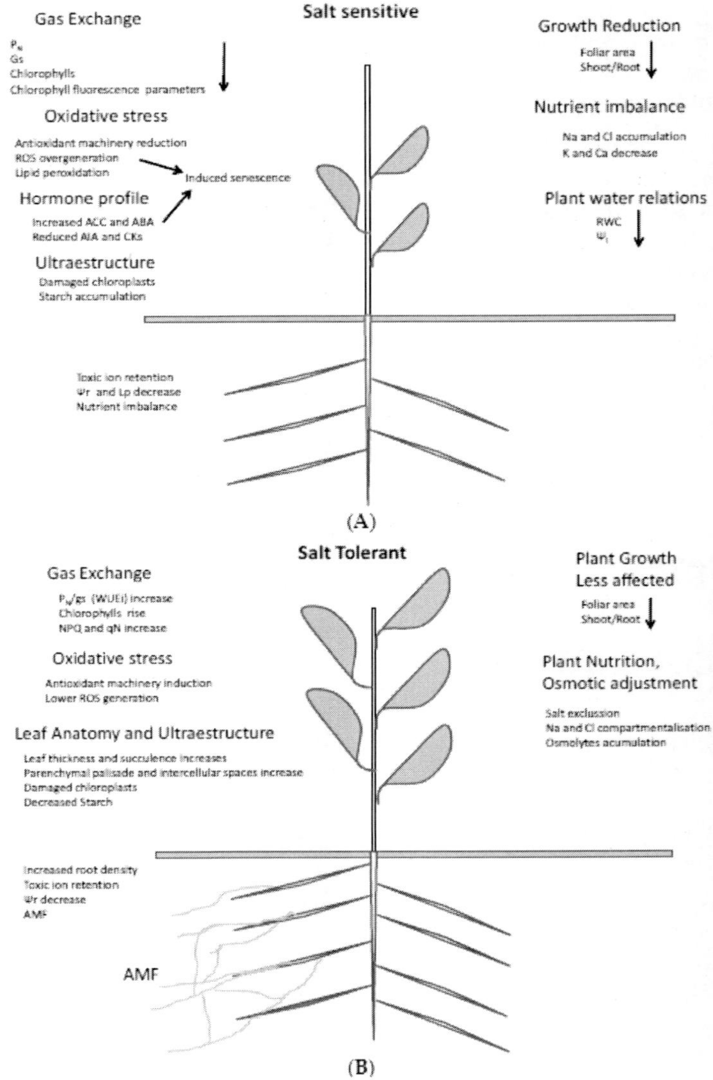

Figure 3. Schema showing the effect of salinity on salt-sensitive (**A**) and salt-tolerant (**B**) plants. Salt stress produces an important decrease in plant growth in salt-sensitive plants, especially in relation to the aerial part. The accumulation of phytotoxic ions in leaves results in a nutrient imbalance particularly reflected in decreases in K^+ and Ca^{2+}. Salinity disturbs a plant's water relations, leading to lower leaf relative water content (RWC) and leaf water potential (ψ_l). Salt stress affects the photosynthesis rate (P_N) by stomatal and non-stomatal limitations. In

addition, decreases in chlorophyll content and in chlorophyll fluorescence parameters occurs in parallel. In addition, oxidative stress at the subcellular level takes place by the effect of salt stress. In salt-stressed plants the synthesis of ethylene and abscisic acid (ABA) increases, but a decrease occurs in indol-3-acetic acid (IAA) and cytokinin (CK) that favours the progression of senescence in salinized leaves. Damage to the chloroplast is an important aspect of the effect of salinization, accompanied by starch accumulation. The accumulation of salts in the root zone causes a decrease in osmotic potential, root water potential (ψ_r) and the root hydraulic conductivity (Lp), diminishing the water available to the root zone and provoking a nutrient imbalance in roots.

The accumulation and compartmentalisation of toxic ions and/or compatible solutes is therefore one of the mechanisms involved in salt-tolerance. Toxic ion accumulation can be an energetically cheap and positive way for plants to obtain osmotic regulation, but only if they are able to compartmentalise the ions in the vacuole. Among the organic osmolytes, Pro and GB are the most important as efficient compatible solutes. In addition to acting as an osmoprotectant, Pro has been reported to have antioxidant properties, but it can also act as a molecular chaperone to protect the structure of biological macromolecules. In general, salt-tolerant species accumulate Pro after a salt-stress challenge. GB, on the other hand, can act as an osmolyte to protect PSII during salinity (Figure 3).

It is known that prolonged salt stress may cause changes in leaf anatomy. In general, tolerant species respond by increasing leaf thickness. Anatomical modifications in leaves also include an increase in palisade parenchyma and intercellular spaces and a decrease in spongy parenchyma, serving to facilitate CO_2 diffusion in a situation of reduced stomatal aperture.

Damage to the chloroplast is an important aspect of the effect of salinization on leaf cells (Figure 3). The main salt-induced changes in the chloroplast ultrastructure include changes in starch content, disarranged thylakoids and grana and increased numbers of plastoglobuli. In tolerant plants, decreases observed in starch content could suggest that the starch is used for different physiological processes to cope with the salt stress challenge.

In salt-tolerant plants, plant growth is less affected. Salt exclusion and salt compartmentalisation in vacuoles and the accumulation of osmolytes are important mechanisms for salt tolerance. PN and gs are also affected in salt-tolerant species, and an increase in the ratio P_N/gs (WUEi) indicates better photosynthetic performance. In general, salt-tolerant species display greater increases in the induction and/or constitutive levels of certain antioxidant defences than salt-sensitive plants. In general, tolerant species respond by increasing leaf thickness and succulence. Anatomical modifications in leaves also include an increase in palisade parenchyma and intercellular spaces and a decrease in spongy parenchyma, serving to facilitate CO_2 diffusion in a situation of reduced stomatal aperture. At the root zone, the accumulation of toxic ions also provokes a decrease in the root osmotic potential. In general, increases in root density takes place in tolerant species that favour the retention of toxic ions in roots, limiting the accumulation of these ions in the canopy. Finally, it is well known that both the presence of arbuscular mycorrhizal fungi (AMF) and salt-tolerant rootstocks produce a synergic effect that can alleviate the negative effects of salt stress.

Salt stress affects photosynthesis both in the short and long term (Figure 3). In the short term, salinity can affect photosynthesis by stomatal limitations (the stomatal factor), leading to a decrease in carbon assimilation. In the long term, salt stress can also affect the photosynthetic process due to salt accumulation in young leaves, leading to the loss of photosynthetic pigments and the inhibition of Calvin cycle enzymes (non-stomatal factors). As a mechanism to protect the photosynthesis process, tolerant species respond by maintaining or increasing their chlorophyll content.

In general, salt-sensitive plants show a drop in photochemical quenching parameters but an increase in non-photochemical quenching parameters. However, and depending on the plant species and the severity of the stress, a decrease in both photochemical and non-photochemical quenching parameters can also take place (Figure 3).

Salinity induces oxidative stress at the subcellular level (Figure 3), particularly in the chloroplast, which can affect the photosynthesis process. In this regards, plants have developed mechanisms for protecting their photosynthetic machinery. One of the mechanisms involved in the safe dissipation of excess excitation energy in PSII is the so-called xanthophyll cycle, whose function is to dissipate excess excitation energy in the PSII antenna as heat by a process termed non-photochemical quenching (NPQ). Photorespiration is another protective mechanism that can act as a sink for excess reducing power, thereby preventing the over-reduction of the photosynthetic electron chain. Moreover, the so-called water-water cycle is also a protective mechanism for the photosynthetic machinery. The primary function of this pathway is the protection of the chloroplast components from the ROS, but it also acts as a mechanism for dissipating excess photon energy.

As mentioned, it is well known that salinity induces oxidative stress in plants at the subcellular level, and those chloroplasts and mitochondria are important sources of ROS in plants subjected to salt stress. Some authors have described a coordinated up-regulation of the antioxidative machinery as one of the mechanisms involved in the salt tolerance response, whereas other authors have correlated salt tolerance with higher constitutive levels of certain antioxidant enzymes. In contrast, salt-sensitive species show an unchanged response or even a decrease in antioxidant defences, and they display lower constitutive antioxidant enzyme levels than salt-tolerant species (Figure 3).

ACKNOWLEDGMENTS

PDV thanks CSIC and the Spanish Ministry of Economy and Competitiveness for their 'Ramon & Cajal' research contract, co-financed by FEDER funds. We thank Spanish Ministry of Economy and Competitiveness (Project AGL2014-52563-R) and Seneca Foundation of Murcia (Project 19903/GERM/15).

REFERENCES

1. Ashraf, M.; Harris, J.C. Potential biochemical indicators of salinity tolerance in plants. *Plant Sci.* **2004**, *166*, 3–16.
2. Hussain, K.; Majeed, A.; Nawaz, K.; Khizar, H.B.; Nisar, M.F. Effect of different levels of salinity on growth and ion contents of black seeds (*Nigella sativa* L.). *Curr. Res. J. Biol. Sci.* **2009**, *1*, 135–138.
3. Colla, G.; Rouphael, Y.; Leonardi, C.; Bie, Z. Role of grafting in vegetable crops grown under saline conditions. *Sci. Hort.* **2010**, *127*, 147–155.
4. De Pascale, S.; Dalla-Costa, L.; Vallone, S.; Barbieri, G.; Maggio, A. Increasing water use effieciency in vegetable crop production: From plant to irrigation system efficiency. *HortTechnology* **2011**, *21*, 301–308.
5. Munns, R.; Gilliham, M. Salinity tolerance of crops—What is the cost? *New Phytol.* **2015**, *208*, 668–673.
6. Acosta-Motos, J.R.; Ortuño, M.F.; Álvarez, S.; López-Climent, M.F.; Gómez-Cadenas, A.; Sánchez-Blanco, M.J. Changes in growth, physiological parameters and the hormonal status of *Myrtus communis* L. plants irrigated with water with different chemical compositions. *J. Plant Physiol.* **2016**, *191*, 12–21.
7. Acosta-Motos, J.R.; Hernández, J.A.; Álvarez, S.; Barba-Espín, G.; Sánchez-Blanco, M.J. Long-term resistance mechanisms and irrigation critical threshold showed by *Eugenia myrtifolia* plants in response to saline reclaimed water and relief capacity. *Plant Physiol. Biochem.* **2017**, *111*, 244–256.
8. Cirillo, C.; Rouphael, Y.; Caputo, R.; Raimondi, G.; Sifola, M.I.; De Pascale, S. Efects of high salinity and the exogenous of an osmolyte on growth, phosynthesis and mineral composition in two ornamental shrubs. *J. Hortic. Sci. Biotechnol.* **2016**, *91*, 14–22.
9. Munns, R.; Tester, M. Mechanisms of salinity tolerance. *Ann. Rev. Plant Biol.* **2008**, *59*, 651–681.
10. Munns, R. Comparative physiology of salt and water stress. *Plant Cell Environ.* **2002**, *25*, 239–250.

11. Tang, X.; Mu, X.; Shao, H.; Wang, H.; Brestic, M. Global plant-responding mechanisms to salt stress: Physiological and molecular levels and implications in biotechnology. *Crit. Rev. Biotechnol.* **2015**, *35*, 425–437.
12. Munns, R. Genes and salt tolerance: Bringing them together. *Plant Physiol.* **2005**, *167*, 645–663.
13. Koyro, H.W. Effect of salinity of growth, photosynthesis, water relations and solute composition of the potential cash crop halophyte *Plantago coronopus* (L.). *Environ. Exp. Bot.* **2006**, *56*, 136–146.
14. Stepien, P.; Johnson, G.N. Contrasting responses of photosynthesis to salt stress in the glycophyte *Arabidopsis* and the halophyte *Thellungiella*: Role of the plastid terminal oxidase as an alternative electron sink. *Plant Physiol.* **2009**, *149*, 1154–1165.
15. Larcher, W. *Physiological Plant Ecology*, 4th ed.; Springer-Verlag: Heidelberg/Berlin, Germany; New York, NY, USA, 2003.
16. Munns, R.; Termaat, A. Whole plant response to salinity. *Aust. J. Plant Physiol.* **1986**, *13*, 143–160.
17. Hernández, J.A.; Almansa, M.S. Short-term effects of salt stress on antioxidant systems and leaf water relations of pea plants. *Physiol. Plant.* **2002**, *115*, 251–257.
18. Parida, A.K.; Das, A.B. Salt tolerance and salinity effects on plants: A review. *Ecotoxicol. Environ. Saf.* **2005**, *60*, 324–349.
19. Flowers, T.J.; Colmer, T.D. Plant salt tolerance: Adaptations in halophytes. *Ann. Bot.* **2015**, *115*, 327–331.
20. Marschner, H. *Mineral Nutrition of Higher Plants*, 2nd ed.; Academic Press: London, UK, 1995.
21. Hernández, J.A.; Corpas, F.J.; Gómez, M.; del Río, L.A.; Sevilla, F. Salt induced oxidative stress mediated by activated oxygen species in pea leaf mitochondria. *Physiol. Plant.* **1993**, *89*, 103–110.
22. Hernández, J.A.; Olmos, E.; Corpas, F.J.; Sevilla, F.; del Río, L.A. Salt-induced oxidative stress in chloroplast of pea plants. *Plant Sci.* **1995**, *105*, 151–167.
23. Hernández, J.A.; Ferrer, M.A.; Jiménez, A.; Ros-Barceló, A.; Sevilla, F. Antioxidant systems and O_2/H_2O_2 production in the apoplast of *Pisum sativum* L. leaves: Its relation with NaCl-induced necrotic lesions in minor veins. *Plant Physiol.* **2001**, *127*, 817–831
24. Hernández, J.A.; Aguilar, A.; Portillo, B.; López-Gómez, E.; Mataix-Beneyto, J.; García-Legaz, M.F. The effect of calcium on the antioxidant enzymes from salt-treated loquat and anger plants. *Funct. Plant Biol.* **2003**, *30*, 1127–1137.
25. Mittova, V.; Tal, M.; Volokita, M.; Guy, M. Up-regulation of the leaf mitochondrial and peroxisomal antioxidative systems in response to salt-induced oxidative stress in the wild salt-tolerant tomato species *Lycopersicon pennellii. Plant Cell Environ.* **2003**, *26*, 845–856.
26. Mittova, V.; Guy, M.; Tal, M.; Volokita, M. Salinity up-regulates the antioxidative system in root mitocondria and peroxisomes of the wild salt-tolerant tomato species *Lycopersicon pennellii*. *J. Exp. Bot.* **2004**, *399*, 1105–1113.
27. Ashraf, M.; Harris, P.J.C. Photosynthesis under stressful environments: An overview. *Photosynthetica* **2013**, *51*, 163–190.
28. Acosta-Motos, J.R.; Díaz-Vivancos, P.; Álvarez, S.; Fernández-García, N.; Sánchez-Blanco, M.J.; Hernández, J.A. NaCl-induced physiological and biochemical adaptative mechanism in the ornamental *Myrtus cummunis* L. plants. *J. Plant Physiol.* **2015**, *183*, 41–51.
29. Acosta-Motos, J.R.; Díaz-Vivancos, P.; Álvarez, S.; Fernández-García, N.; Sánchez-Blanco, M.J.; Hernández, J.A. Physiological and biochemical mechanisms of the ornamental *Eugenia myrtifolia* L. plants for coping with NaCl stress and recovery. *Planta* **2015**, *242*, 829–846.
30. Munns, R. A leaf elongation assay detects an unknown growth inhibitor in xylem sap from wheat and barley. *Aust. J. Plant Physiol.* **1992**, *19*, 127–135.
31. Azza Mazher, A.M.; Fatma El-Quesni, E.M.; Farahat, M.M. Responses of ornamental plants and woody trees to salinity. *World J. Agric. Sci.* **2007**, *3*, 386–395.
32. Passioura, J.B. Water Transport in and to Roots. *Ann. Rev. Plant Physiol. Plant Mol. Biol.* **1988**, *39*, 245–265.

33. Vamerali, T.; Saccomani, M.; Bona, S.; Mosca, G.; Guarise, M.; Ganis, A. A comparison of root characteristics in relation to nutrient and water stress in two maize hybrids. *Plant Soil* **2003**, *255*, 157–167.
34. Franco, J.A.; Bañón, S.; Vicente, M.J.; Miralles, J.; Martínez-Sánchez, J.J. Root development in horticultural plants grown under abiotic stress conditions—A review. *J. Hortic. Sci. Biotechnol.* **2011**, *86*, 543–556.
35. Ma, S.C.; Li, F.M.; Xu, B.C.; Huang, Z.B. Effect of lowering the root/shoot ratio by pruning roots on water use efficiency and grain yield of winter wheat. *Field Crops Res.* **2010**, *115*, 158–164.
36. Steudle, E. Water uptake by roots: Effects of water deficit. *J. Exp. Bot.* **2000**, *51*, 1531–1542.
37. Ranathunge, K.; Shao, S.; Qutob, D.; Gijzen, M.; Peterson, C.A.; Bernards, M.A. Properties of the soybean seed coat cuticle change during development. *Planta* **2010**, *231*, 1171–1188.
38. Shannon, M.C.; Grieve, C.M.; Francois, L.E. Whole-plant response to salinity. In *Plant-Environment Interactions*; Wilkinson, R.E., Ed.; Marcel Dekker, Inc.: New York, NY, USA, 1994; pp. 199–244.
39. Walker, R.R.; Sedgley, M.; Blesing, M.A.; Douglas, T.J. Anatomy, Ultrastructure and Assimilate Concentrations of Roots of Citrus Genotypes Differing in Ability for Salt Exclusion. *J. Exp. Bot.* **1984**, *35*, 1481–1494.
40. Whiley, A.W.; Schaffer, B.; Wolstenholme, B.N. *The Avocado. Botany, Production and Uses*; CAB International: Wallingford, UK, 2002.
41. Bernstein, N.; Meiri, A.; Zilberstaine, M. Root Growth of Avocado is More Sensitive to Salinity than Shoot Growth. *J. Am. Soc. Hortic. Sci.* **2004**, *129*, 188–192.
42. Álvarez, S.; Sánchez-Blanco, M.J. Long-term effect of salinity on plant quality, water relations, photosynthetic parameters and ion distribution in *Callistemon citrinus*. *Plant Biol.* **2014**, *16*, 757–764.
43. Álvarez, S.; Navarro, A.; Nicolás, E.; Sánchez-Blanco, M.J. Transpiration, photosynthetic responses, tissue water relations and dry mass partitioning in *Callistemon* plants during drought conditions. *Sci. Hort.* **2011**, *129*, 306–312.
44. Álvarez, S.; Sánchez-Blanco, M.J. Changes in growth rate, root morphology and water use efficiency of potted *Callistemon citrinus* plants in response to different levels of water deficit. *Sci. Hort.* **2013**, *156*, 54–62.
45. Gómez-Bellot, M.J.; Álvarez, S.; Castillo, M.; Bañón, S.; Ortuño, M.F.; Sánchez-Blanco, M.J. Water relations, nutrient content and developmental responses of *Euonymus* plants irrigated with water of different degrees of salinity and quality. *J. Plant Res.* **2013**, *126*, 567–576.
46. Croser, C.; Renault, S.; Franklin, J.; Zwiazek, J. The effect of salinity on the emergence and seedling growth of *Picea mariana*, *Picea glauca*, and *Pinus banksiana*. *Environ. Pollut.* **2001**, *115*, 9–16.
47. Cameron, R.W.F.; Harrison-Murray, R.S.; Atkinson, C.J.; Judd, H.L. Regulated deficit irrigation: A means to control growth in woody ornamentals. *J. Hortic. Sci. Biotech.* **2006**, *81*, 435–443.
48. Franco, J.A.; Bañón, S.; Fernández, J.A.; Leskovar, D.I. Effect of nursery regimes and establishment irrigation on root development of *Lotus creticus* seedlings following transplanting. *J. Hortic. Sci. Biotechnol.* **2001**, *76*, 174–179.
49. Franco, J.A.; Martínez-Sánchez, J.J.; Fernández, J.A.; Bañón, S. Selection and nursery production of ornamental plants for landscaping and xerogardening in semiarid environments. *J. Hortic. Sci. Biotechnol.* **2006**, *81*, 3–17.
50. Rengasamy, P.; Olsson, K.A. Irrigation and sodicity. *Aust. J. Soil Res.* **1993**, *31*, 821–837.
51. Jonathan, N.F.G.; Lehti-Shiu, M.D.; Ingram, P.A.; Deak, K.I; Biesiada, T.; Malamy, J.E. Identification of quantitative trait loci that regulate *Arabidopsis* root system size and plasticity. *Genetics* **2006**, *172*, 485–498.
52. Zobel, R.W.; Kinraide, T.B.; Baligar, V.C. Fine root diameters can change in response to changes in nutrient concentrations. *Plant Soil* **2007**, *297*, 243–254.

53. Navarro, J.M.; Gómez-Gómez, A.; Pérez-Pérez, J.G.; Botía, P. Effect of saline conditions on the maturation process of clementine Clemenules fruits on two different rootstocks. *Span. J. Agric. Res.* **2010**, *8*, 21–29.
54. Penella, C.; Nebauer, S.G.; Quiñones, A.; San Bautista, A.; López-Galarza, S.; Calatayud, A. Some rootstocks improve pepper tolerance to mild salinity through ionic regulation. *Plant Sci.* **2015**, *230*, 12–22.
55. Penella, C.; Landi, M.; Guidi, L.; Nebauer, S.G.; Pellegrini, E.; San Bautista, A.; Remorini, D.; Nali, C.; López-Galarza, S.; Calatayud, A. Salt-tolerant rootstock increases yield of pepper under salinity through maintenance of photosynthetic performance and sinks strength. *J. Plant Physiol.* **2016**, *193*, 1–11.
56. Navarro, J.M.; Pérez-Tornero, O.; Morte, A. Alleviation of salt stress in citrus seedlings inoculated with arbuscular mycorrhizal fungi depends on the rootstock salt tolerance. *J. Plant Physiol.* **2014**, *171*, 76–85.
57. Gómez-Bellot, M.J.; Ortuño, M.F.; Nortes, P.A.; Vicente-Sánchez, J.; Martín, F.F.; Bañón, S.; Sánchez-Blanco, M.J. Protective effects of *Glomus iranicum* var. *tenuihypharum* on soil and *Viburnum tinus* plants irrigated with treated wastewater under field conditions. *Mycorrhiza* **2015**, *25*, 399–409.
58. Tudela, D.; Primo-Millo, E. l-Aminocyclopropane-1-carboxylic acid transported from roots to shoots promotes leaf abscission in Cleopatra mandarin (*Citrus reshni* Hort. ex Tan.) seedlings rehydrated after water stress. *Plant Physiol.***1992**, *100*, 131–137.
59. Gómez-Cadenas, A.; Tadeo, F.R.; Talon, M.; Primo-Millo, E. Leaf abscission induced by ethylene in water stressed intact seedlings of Cleopatra mandarin requires previous abscisic acid accumulation in roots. *Plant Physiol.* **1996**, *112*, 401–408.
60. Gómez-Cadenas, A.; Tadeo, F.R.; Primo-Millo, E.; Talon, M. Involvement of abscisic acid and ethylene in the response of citrus seedlings to salt shock. *Physiol. Plant.* **1998**, *103*, 475–484.
61. Dodd, I.C. Root-to-shoot signalling: Assessing the roles of 'up' in the up and down world of long-distance signalling in planta. *Plant Soil* **2005**, *74*, 257–275.
62. Chrominski, A.; Bhat, R.B.; Weber, D.J.; Smith, B.N. Osmotic stress-dependent conversion of 1-aminocyclopropane-1-carboxylic acid (ACC) to ethylene in the halophyte, *Allenrolfea occidentalis*. *Environ. Exp. Bot.* **1988**, *28*, 171–174.
63. Albacete, A.; Ghanem, M.E.; Martínez-Andújar, C.; Acosta, M.; Sánchez-Bravo, J.; Martínez, V.; Lutts, S.; Dodd, I.C.; Pérez-Alfocea, F. Hormonal changes in relation to biomass partitioning and shoot growth impairment in salinized tomato (*Solanum lycopersicum* L.) plants. *J. Exp. Bot.* **2008**, *59*, 4119–4131.
64. Ghanem, M.E.; Albacete, A.; Martínez-Andújar, C.; Acosta, M.; Romero-Aranda, R.; Dodd, I.C.; Lutts, S.; Pérez-Alfocea, F. Hormonal changes during salinity-induced leaf senescence in tomato (*Solanum lycopersicum* L.). *J. Exp. Bot.* **2008**, *59*, 3039–3050.
65. Savé, R.; Olivella, C.; Biel, C.; Adillón, J.; Rabella, R. Seasonal patterns of water relationships, photosynthetic pigments and morphology of *Actinidia deliciosa* plants of the Haywards and Tomouri cultivars. *Agronomie* **1994**, *2*, 121–126.
66. Ruiz-Sánchez, M.C.; Domingo, R.; Torrecillas, A.; Pérez-Pastor, A. Water stress preconditioning to improve drought resistance in young apricot plants. *Plant Sci.* **2000**, *156*, 245–251.
67. Colmer, T.D.; Muñiz, R.; Flowers, T.J. Improving salt tolerance of wheat and barley: Future prospects. *Aust. J. Exp. Agric.* **2005**, *45*, 1425–1443
68. Franco, J.A.; Fernández, J.A.; Bañón, S.; González, A. Relationship between the effects of salinity on seedling leaf area and fruit yield of six muskmelons cultivars. *J. Hortic. Sci.* **1997**, *32*, 642–647.
69. Rodríguez, P.; Torrecillas, A.; Morales, M.A.; Ortuño, M.F.; Sánchez-Blanco, M.J. Effects of NaCl salinity and water stress on growth and leaf water relations of *Asteriscus maritimus* plants. *Environ. Exp. Bot.* **2005**, *53*, 113–123.
70. Hsiao, T.C.; Xu, L.K. Sensitivity of growth of roots versus leaves to water stress: Biophysical analysis and relation to water transport. *J. Exp. Bot.* **2000**, *51*, 1595–1616.

71. Cassaniti, C.; Romano, D.; Flowers, T.J. The response of ornamental plants to saline irrigation water. In *Irrigation Water Management, Pollution and Alternative Strategies*; Garcia-Garizabal, I., Ed.; InTech Europe: Rijeka, Croatia, 2012; pp. 132–158.
72. Cassaniti, C.; Leonardi, C.; Flowers, T.J. The effects of sodium chloride ornamental shrubs. *Sci. Hortic.* **2009**, *122*, 586–593.
73. Cramer, G.R. Sodium–calcium interactions under salinity stress. In *Salinity: Environment–Plants–Molecules*; Läuchli, A., Lüttge, U., Eds.; Kluwer: Dordrecht, The Netherlands, 2002; pp. 205–227.
74. Tattini, M.; Gucci, R.; Coradeschi, M.A.; Ponzio, C.; Everard, J.D. Growth, gas exchange and ion content in *Olea europaea* plants during salinity stress and subsequent relief. *Physiol. Plant.* **1995**, *95*, 203–210.
75. Zekri, M.; Parsons, L.R. Growth and root hydraulic conductivity of several citrus rootstocks under salt and polyethylene glycol stresses. *Physiol. Plant.* **1989**, *77*, 99–106.
76. Morales, M.A.; Sánchez-Blanco, M.J.; Olmos, E.; Torrecillas, A.; Alarcón, J.J. Changes in the growth, leaf water relations and cell ultraestructure in *Argyranthemum coronopifolium* plants under saline conditions. *J. Plant Physiol.* **1998**, *153*, 174–180.
77. Alarcón, J.J.; Morales, M.A.; Torrecillas, A.; Sánchez-Blanco, M.J. Growth, water relations and accumulation of organic and inorganic solutes in the halophyte *Limonium latifolium* cv. *Avignon* and its interspecific hybrid *Limonium caspia* x *Limonium latifolium* cv. *Beltlaard* during salt stress. *J. Plant Physiol.* **1999**, *154*, 795–801.
78. Hernández, J.A.; Campillo, A.; Jiménez, A.; Alarcon, J.J.; Sevilla, F. Response of antioxidant systems and leaf water relations to NaCl stress in pea plants. *New Phytol.* **1999**, *141*, 241–251.
79. Chartzoulakis, K.; Loupassaki, M.; Bertaki, M.; Androulakis, I. Effects of NaCl salinity on growth, ion content and CO_2 assimilation rate of six olive cultivars. *Sci. Hort.* **2002**, *96*, 235–247.
80. Arbona, V.; Flors, V.; García-Agustín, P.; Jacas, J.; Gómez-Cadenas, A. Enzymatic and non-enzymatic antioxidant responses of Carrizo citrange, a salt-sensitive citrus rootstock, to different levels of salinity. *Plant Cell Physiol.* **2003**, *44*, 388–394.
81. Torrecillas, A.; Rodríguez, P.; Sánchez-Blanco, M.J. Comparison of growth, leaf water relations and gas exchange of *Cistus albidus* and *C. monspliensis* plants irrigated with water of different NaCl salinity levels. *Sci. Hort.* **2003**, *97*, 353–368.
82. Sánchez-Blanco, M.J.; Rodríguez, P.; Olmos, E.; Morales, M.A.; Torrecillas, A. Differences in the Effects of Simulated Sea Aerosol on Water Relations, Salt Content, and Leaf Ultrastructure of Rock-Rose Plants. *J. Environ. Qual.* **2004**, *33*, 1369–1375.
83. Alarcón, J.J.; Morales, M.A.; Ferrández, T.; Sánchez-Blanco, M.J. Effects of water and salt stresses on growth, water relations and gas exchange in *Rosmarinus officinalis*. *J. Hortic. Sci. Biotechnol.* **2006**, *81*, 845–853.
84. Niu, G.; Rodriguez, D.S. Relative Salt Tolerance of Five Herbaceous Perennials. *HortScience* **2006**, *41*, 1493–1497.
85. Navarro, A.; Bañón, S.; Olmos, E.; Sánchez-Blanco, M.J. Effects of sodium chloride on water potential components, hydraulic conductivity, gas exchange and leaf ultrastructure of *Arbutus unedo* plants. *Plant Sci.* **2007**, *172*, 473–480.
86. López-Climent, M.F.; Arbona, V.; Pérez-Clemente, R.M.; Gómez-Cadenas, A. Relationship between salt tolerance and photosynthetic machinery performance in citrus. *Environ. Exp. Bot.* **2008**, *62*, 176–184.
87. Bañón, S.; Miralles, J.; Franco, J.A.; Ochoa, R.; Sánchez-Blanco, M.J. Effects of diluted and pure treated wastewater on the growth, physiological status and visual quality of potted lantana and polygala plants. *Sci. Hortic.* **2011**, *129*, 869–876.
88. Valdez-Aguilar, L.A.; Grieve, C.M.; Razak-Mahar, A.; McGiffen, M.; Merhaut, D.J. Growth and ion distribution is affected by irrigation with saline water in selected landscape species grown in two consecutive growing-seasons: Spring-summer and fall-winter. *HortScience* **2011**, *46*, 632–642
89. Álvarez, S.; Gómez-Bellot, M.J.; Castillo, M.; Bañón, S.; Sánchez-Blanco, M.J. Osmotic and saline effect on growth, water relations, and ion uptake and translocation in *Phlomis purpurea* plants. *Environ. Exp. Bot.* **2012**, *78*, 138–145.

90. Bañón, S.; Miralles, J.; Ochoa, J.; Sánchez-Blanco, M.J. The effect of salinity and high boron on growth, photosynthetic activity and mineral contents of two ornamental shrubs. *Hort. Sci.* **2012**, *39*, 188–194.
91. Sabra, A.; Daayf, F.; Renault, S. Differential physiological and biochemical responses of three *Echinaceae* species to salinity stress. *Sci. Hort.* **2012**, *135*, 23–31.
92. Gónmez-Bellot, M.J.; Álvarez, S.; Bañón, S.; Ortuño, M.F.; Sánchez-Blanco, M.J. Physiological mechanisms involved in the recovery of *euonymus* and *laurustinus* subjected to saline waters. *Agric. Water Manag.* **2013**, *128*, 131–139.
93. Acosta-Motos, J.R.; Álvarez, S.; Barba-Espín, G.; Hernández, J.A.; Sánchez-Blanco, M.J. Salts and nutrients present in regenerated waters induce changes in water relations, antioxidative metabolism, ion accumulation and restricted ion uptake in *Myrtus communis* L. plants. *Plant. Physiol. Biochem.* **2014**, *85*, 41–50.
94. Fernández-García, N.; Olmos, E.; Bardisi, E.; García-De la Garma, J.; López-Berenguer, C.; Rubio-Asensio, J.S. Intrinsic water use efficiency controls the adaptation to high salinity in a semi-arid adapted plant, henna (*Lawsonia inermis* L.). *J. Plant Physiol.* **2014**, *171*, 64–75.
95. Álvarez, S.; Sánchez-Blanco, M.J. Comparison of individual and combined effects of salinity and deficit irrigation on physiological, nutritional and ornamental aspects of tolerance in *Callistemon laevis* plants. *J. Plant Physiol.* **2015**, *185*, 65–74.
96. Salachna, P.; Piechocki, R. Efects of sodium chloride on growth and mineral nutrition of purpletop vervain. *J. Ecol. Eng.* **2016**, *17*, 148–152.
97. Barba-Espín, G.; Clemente-Moreno, M.J.; Álvarez, S.; García-Legaz, M.F.; Hernández, J.A.; Díaz-Vivancos, P. Salicylic acid negatively affects the response to salt stress in pea plants: Effects on PR1b and MAPK expression. *Plant Biol.***2011**, *13*, 909–917.
98. Van Genuchten, M.T. *Research Report No. 120*; U.S. Salinity Laboratory, USDA-ARS: Riverside, CA, USA, 1983; p. 50.
99. Zhen, A.; Bie, Z.; Huang, Y.; Liu, Z.; Li, Q. Effects of scion and rootstock genotypes on the anti-oxidant defense systems of grafted cucumber seedlings under NaCl stress. *Soil Sci. Plant Nutr.* **2010**, *56*, 263–271.
100. Beadle, C.L. Growth analysis. In *Photosynthesis and Production in a Changing Environment*; Hall, D.O., Scurlock, J.M.O., Bolhar-Nordentrampf, H.R., Leegod, R.C., Long, S.P., Eds.; Chapman and Hall: London, UK, 1993; pp. 36–45.
101. Curtis, P.S.; Läuchli, A. The role of leaf area development and photosynthetic capacity in determining growth of kenaf under moderate salt stress. *Aust. J. Plant Physiol.* **1986**, *18*, 553–565.
102. Cramer, G.R.; Epstein, E.; Läuchli, A. Effects of sodium, potassium and calcium on salt-stressed barley. 1. Growth analysis. *Physiol. Plant.* **1990**, *80*, 83–88.
103. Olmos, E.; Sánchez-Blanco, M.J.; Ferrández, T.; Alarcón, J.J. Subcellular effects of drought stress in *Rosmarinus officinalis*. *Plant Biol.* **2007**, *9*, 77–84.
104. Longstreth, D.J.; Nobel, P.S. Salinity effects of leaf anatomy. *Plant Physiol.* **1979**, *63*, 700–703.
105. Romero-Aranda, R.; Moya, J.L.; Tadeo, F.R.; Legaz, F.; Primo-Millo, E.; Talon, M. Physiological and anatomical disturbances induced by chloride salts in sensitive and tolerant citrus: Beneficial and detrimental effects of cations. *Plant Cell Environ.* **1998**, *21*, 1243–1253.
106. Franco-Navarro, J.D.; Brumós, J.; Rosales, M.A.; Cubero-Font, P.; Talón, M.; Colmenero-Flores, J.M. Chloride regulates leaf cell size and water relations in tobacco plants. *J. Exp. Bot.* **2016**, *67*, 873–891.
107. Gómez-Bellot, M.J.; Nortes, P.A.; Ortuño, M.F.; Romero-Trigueros, C.; Fernández-García, N.; Sánchez-Blanco, M.J. Influence of arbuscular mycorrizal fungi and treated wastewater on water relations and leaf structure alterations of *Viburnum tinus* L. plants during both saline and recovery periods. *J. Plant Physiol.* **2015**, *188*, 96–105.
108. Rouphael, Y.; De Micco, V.; Arena, C.; Raimondi, G.; Colla, G.; De Pascale, S. Effect of *Ecklonia maxima* seaweed extract on yield, mineral composition, gas exchange, and leaf anatomy of zucchini squash grown under saline conditions. *J. Appl. Phycol.* **2016**.

109. Flowers, T.J.; Duque, E.; Hajibagheri, M.A.; McGlonigle, T.P.; Yeo, A.R. The effect of salinity on leaf ultrastructure and net photosynthesis of two varieties of rice: Further evidence for a cellular component of salt-resistance. *New Phytol.* **1985**, *100*, 37–43.
110. Austin, J.R.; Frost, E.; Vidi, P.A.; Kessler, F.; Staehelin, A. Plastoglobules are lipoprotein subcompartments of the chloroplast that are permanently coupled to thylakoid membranes and contain biosynthetic enzymes. *Plant Cell* **2006**, *18*, 1693–1703.
111. Szabo-Nagi, A.; Galiba, G.; Erdei, L. Induction of soluble phosphatases under ionic and non-ionic osmotic stresses in wheat. *J. Plant Physiol.* **1992**, *140*, 629–633.
112. Balibrea, M.E.; Dell'Amico, J.; Bolarín, M.C.; Pérez-Alfocea, F. Carbon partitioning and sucrose metabolism in tomato plants growing under salinity. *Physiol. Plant.* **2000**, *110*, 503–511.
113. Passioura, J.B. Scaling up: The essence of effective agricultural research. *Funct. Plant Biol.* **2010**, *37*, 585–591.
114. Chavarria, G.; dos Santos, H.P. Plant water relations: Absorption, transport and control mechanisms. *InTechOpen***2012**, *5*, 105–132.
115. Zhu, J.K. Plant salt tolerance. *Trends Plant Sci.* **2001**, *6*, 66–71.
116. Ashraf, M. Some important physiological selection criteria for salt tolerance in plants. *Flora* **2004**, *199*, 361–376.
117. Meloni, D.A.; Gulotta, M.R.; Martínez, C.A. Salinity tolerance in *Schinopsis quebracho colorado*: Seed germination, growth, ion relations and metabolic responses. *J. Arid. Environ.* **2008**, *72*, 1785–1792.
118. Acosta-Motos, J.R.; Álvarez, S.; Hernández, J.A.; Sánchez-Blanco, M.J. Irrigation of *Myrtus communis* L. plants with reclaimed water: Morphological and physiological responses to different levels of salinity. *J. Hortic. Sci. Biotechnol.***2014**, *89*, 487–494.
119. Giri, B.; Kapoor, R.; Mukerji, K.G. Influence of arbuscular mycorrhizal fungi and salinity on growth, biomass, and mineral nutrition of *Acacia auriculiformis*. *Biol. Fert. Soils* **2003**, *38*, 170–175.
120. Katerji, N.; van Hoorn, J.W.; Mastrorilli, M.; Hamdy, A. Crop sensitivity to salinity. In *Non-Conventional Water Use: WASAMED Project*; Hamdy, A., ElGamal, F., Lamaddalena, N., Bogliotti, C., Guelloubi, R., Eds.; CIHEAM/EU DG Research: Bari, Italy, 2005; pp. 43–51.
121. Bhatnagar-Mathur, P.; Devi, M.J.; Reddy, D.S.; Lavanya, M.; Vadez, V.; Serraj, R.; Yamaguchi-Shinozaki, K.; Sharma, K.K. Stress inducible expression of At DREB1A in transgenic peanut (*Arachis hypogaea* L.) increases transpiration efficiency under water limiting conditions. *Plant Cell Rep.* **2007**, *26*, 2071–2082.
122. Alarcón, J.J.; Sánchez-Blanco, M.J.; Bolarín, M.C.; Torrecillas, A. Growth and osmotic adjustment of two tomato cultivars during and after saline stress. *Plant Soil* **1994**, *166*, 75–82.
123. Navarro, A.; Bañón, S.; Conejero, W.; Sánchez-Blanco, M.J. Ornamental characters, ion accumulation and water status in *Arbutus unedo* seedlings irrigated with saline water and subsequent relief and transplanting. *Environ. Exp. Bot.***2008**, *62*, 364–370.
124. Kefu, Z.; Hai, F.; San, Z.; Jie, S. Study on the salt and drought tolerance of *Suaeda salsa* and *Kalanchoe claigremontiana* under iso-osmotic salt and water stress. *Plant Sci.* **2003**, *165*, 837–844.
125. Hasegawa, P.M.; Bressan, R.A.; Zhu, J.K.; Bhnert, H.J. Plant cellular and molecular responses to high salinity. *Ann. Rev. Plant Physiol. Plant Mol. Biol.* **2000**, *51*, 463–499.
126. Abdul-Jaleel, C.; Manivannan, P.; Sankar, B.; Kishorekumar, A.; Gopi, R.; Somasundaram, R.; Panneerselvam, R. Water deficit stress mitigation by calcium chloride in *Catharanthus roseus*: Effects on oxidative stress, proline metabolism and indole alkaloid accumulation. *Colloids Surf. B.* **2007**, *60*, 110–116.
127. Kaldenhoff, R.; Ribas-Carbo, M.; Flexas, J.; Lovisolo, C.; Heckwolf, M.; Uehlein, N. Aquaporins and plant water balance. *Plant Cell Environ.* **2008**, *31*, 658–666.
128. Jacoby, B. Mechanisms envolved in salt tolerante by plants. In *Handbook of Plant and Crop Stress*; Pessarakli, M., Ed.; Marcel Dekker: New York, NY, USA, 1994; pp. 97–123.
129. Kriedmann, P.E.; Barrs, H.D. Citrus orchards. In *Water Deficits and Plant Growth*; Kozlowski, T.T., Ed.; VI Academic Press: New York, NY, USA, 1981; pp. 325–417.

130. Stiller, V.; Lafitte, H.R.; Sperry, J.S. Hydraulic Properties of Rice and the Response of Gas Exchange to Water Stress. *Plant Physiol.* **2003**, *132*, 1698–1706.
131. Tyree, M.T. Hydraulic limits on tree performance: Transpiration, carbon gain and growth of trees. *Trees Struct. Funct.***2003**, *17*, 95–100.
132. Brodribb, T.J. Xylem hydraulic physiology: The functional backbone of terrestrial plant productivity. *Plant Sci.* **2009**, *177*, 245–251.
133. Lo Gullo, M.A.; Trifilò, P.; Raimondo, F. Hydraulic characteristics and water relations in pigment-less mutant shoots of an orange tree. *Tree Physiol.* **2007**, *27*, 209–217.
134. Sánchez-Blanco, M.J.; Rodríguez, P.; Morales, M.A.; Ortuño, M.F.; Torrecillas, A. Comparative growth and water relations of *Cistus albidus* and *Cistus monspeliensis* plants during water deficit conditions and recovery. *Plant Sci.* **2002**, *162*, 107–113.
135. Slama, I.; Ghnaya, T.; Savouré, A.; Abdelly, C. Combined effects of long-term salinity and soil drying on growth, water relations, nutrient status and proline accumulation of *Sesuvium portulacastrum*. *C. R. Biol.* **2008**, *331*, 442–451.
136. Pérez-Pérez, J.G.; Robles, J.M.; García-Sánchez, F.; Botía, P. Comparison of deficit and saline irrigation strategies to confront water restriction in lemon trees grown in semi-arid regions. *Agric. Water Manag.* **2016**, *164*, 46–57.
137. Kerstiens, G.; Tych, W.; Robinson, M.F.; Mansfield, T.A. Sodium-related partial stomatal closure and salt tolerance of *Aster tripolium*. *New Phytol.* **2002**, *153*, 509–515.
138. Veselov, D.S.; Sharipova, G.V.; Veselov, S.U.; Kudoyarova, G.R. The Effects of NaCl Treatment on Water Relations, Growth, and ABA Content in Barley Cultivars Differing in Drought Tolerance. *J. Plant Growth Regul.* **2008**, *27*, 380.
139. Vysotskaya, L.; Hedley, P.E.; Sharipova, G.; Veselov, D.; Kudoyarova, G.; Morris, J.; Jones, H.G. Effect of salinity on water relations of wild barley plants differing in salt tolerance. *AoB Plants* **2010**.
140. Cochard, H.; Coll, L.; Roux, X.L.; Ameglio, T. Unraveling the effects of plant hydraulics on stomatal closure during water stress in walnut. *Plant Phys.* **2002**, *128*, 282–290.
141. Meinzer, F.C. Co-ordination of vapour and liquid phase water transport properties in plants. *Plant Cell Environ.* **2002**, *25*, 265–274.
142. Bunce, J.A. How do leaf hydraulics limit stomatal conductance at high water vapour pressure deficits? *Plant Cell Environ.* **2006**, *29*, 1644–1650.
143. Davies, W.J.; Kudoyarova, G.; Hartung, W. Long-distance ABA signaling and its relation to other signaling pathways in the detection of soil drying and the mediation of the plant´s response to drought. *J. Plant Growth Regul.* **2005**, *24*, 285–295.
144. Dodd, I.C.; Puértolas, J.; Huber, K.; Pérez-Pérez, J.G.; Wright, H.R.; Blackwell, M.S.A. The importance of soil drying and re-wetting in crop phytohormonal and nutritional responses to deficit irrigation. *J. Exp. Bot.* **2015**, *66*, 2239–2252.
145. Mulholland, B.J.; Taylor, I.B.; Jackson, A.C.; Thompson, A.J. Can ABA mediate responses of salinity stressed tomato? *Environ. Exp. Bot.* **2003**, *50*, 17–28.
146. Fricke, W.; Akhiyarova, G.; Veselov, D.; Kudoyarova, G. Rapid and tissue-specific changes in ABA and in growth rate in response to salinity in barley leaves. *J. Exp. Bot.* **2004**, *55*, 1115–1123.
147. Tardieu, F.; Simonneau, T. Variability among species of stomatal control under fluctuating soil water status and evaporative demand: Modeling isohydric and anisohydric behaviours. *J. Exp. Bot.* **1998**, *49*, 419–432.
148. Sánchez-Blanco, M.J.; Morales, M.A.; Torrecillas, A.; Alarcón, J.J. Diurnal and seasonal osmotic potential changes in *Lotus creticus* plants grown under saline stress. *Plant Sci.* **1998**, *136*, 1–10.
149. Flexas, J.; Bota, J.; Loreto, F.; Cornic, G.; Sharkey, T.D. Diffusive and metabolic limitations to photosynthesis under drought and salinity in C3 plants. *Plant Biol.* **2004**, *6*, 269–279.
150. Galle, A.; Florez-Sarasa, I.; Tomás, M.; Pou, A.; Medrano, H.; Ribas-Carbó, M.; Flexas, J. The role of mesophyll conductance during water stress and recovery in tobacco (*Nicotiana sylvestris*): Acclimation or limitation? *J. Exp. Bot.***2009**, *60*, 2379–2390.
151. Mugnai, S.; Ferrante, A.; Petrognani, L.; Serra, G.; Vernieri, P. Stres-Induced variation in leaf gas exchange and chlorophyll a fluorescence in *Callistemon* plants. *Res. J. Biol. Sci.* **2009**, *4*, 913–921.

152. Ali, A.; Iqbal, N.; Ali, F.; Afzal, B. *Alternanthera bettzickiana* (Regel.) G. Nicholson, a potential halophytic ornamental plant: Growth and physiological adaptations. *Flora* **2012**, *207*, 318–321.
153. Chaves, M.M.; Osorio, J.; Pereira, J.S. Water use efficiency and photosynthesis. In *Water Use Efficiency in Plant Biology*; Bacon, M., Ed.; Blackwell Publishing: Oxford, UK, 2004; pp. 42–74.
154. James, R.A.; Rivelli, A.R.; Munns, R.; von Caemmerer, S. Factors affecting CO_2 assimilation, leaf injury and growth in salt-stressed durum wheat. *Funct. Plant Biol.* **2002**, *29*, 1393–1403.
155. Blum, A. Towards a conceptual ABA ideotype in plant breeding for water limited environments. *Funct. Plant Biol.* **2015**, *42*, 502–513.
156. Chaves, M.M.; Costa, J.M.; Zarrouka, O.; Pinheiro, C.; Lopes, C.M.; Pereira, J.S. Controlling stomatal aperture in semi-arid regions—The dilemma of saving water or being cool? *Plant Sci.* **2016**, *251*, 54–64.
157. Halim, R.A.; Buxton, D.R.; Hattendorf, M.J.; Carlson, R.E. Crop water-stress index and forage quality relationships in alfalfa. *Agron. J.* **1990**, *82*, 906–909.
158. Gerard, J.; Kluitenberg, J.; Biggar, W. Canopy temperature as a measure of salinity stress on sorghum. *Irrig. Sci.* **1992**, *13*, 115–121.
159. Sirault, X.R.R.; James, R.A.; Furbank, R.T. A new screening method for osmotic component of salinity tolerance in cereals using infrared thermography. *Funct. Plant Biol.* **2009**, *36*, 970–977.
160. Gómez-Bellot, M.J.; Nortes, P.A.; Sánchez-Blanco, M.J.; Ortuño, M.F. Sensitivity of thermal imaging and infrared thermometry to detect water status changes in *Euonymus japonica* plants irrigated with saline reclaimedwater. *Biosyst. Eng.* **2015**, *133*, 21–32.
161. Türkan, I.; Demiral, T. Recent developments in understanding salinity tolerance. *Environ. Exp. Bot.* **2009**, *67*, 2–9.
162. Hare, P.D.; Cress, W.A.; van Staden, J. Proline synthesis and degradation: A model system for elucidating stress-related signal transduction. *J. Exp. Bot.* **1999**, *50*, 413–434.
163. Kinnersley, A.M.; Turano, F.J. Gamma aminobutyric acid (GABA) and plant responses to stress. *CRC Crit. Rev. Plant Sci.* **2000**, *19*, 479–509.
164. Zheng, Q.; Liu, Z.; Chen, G.; Gao, Y.; Li, Q.; Wang, J. Comparison of osmotic regulation in dehydration- and salinity-stressed sunflower seedlings. *J. Plant Nutr.* **2010**, *33*, 966–981.
165. Suárez, N. Comparative leaf anatomy and pressure-volume analysis in plants of Ipomoea pes-caprae experimenting saline and/or drought stress. *Int J. Bot.* **2011**, *7*, 53–62.
166. Mustard, J.; Renault, S. Effects of NaCl on water relations and cell wall elasticity and composition of red-osier dogwood (*Cornus stolonifera*) seedlings. *Physiol. Plant.* **2004**, *121*, 265–271.
167. Sassi, S.; Aydi, S.; Hessini, K.; Gonzalez, E.M.; Arrese-Igor, C.; Abdelly, C. Long-term manitol-induced osmotic stress leads to stomatal closure, carbohydrate accumulation and changes in leaf elasticity in *Phaseolus vulgaris* leaves. *Afr. J. Biotechnol.* **2010**, *9*, 6061–6069.
168. Ashraf, M.; Fooland, M.R. Roles of glycine betaine and proline in improving plant abiotic stress resistance. *Environ. Exp. Bot.* **2007**, *59*, 206–216.
169. Planchet, E.; Verdu, I.; Delahaie, J.; Cukier, C.; Girard, C.; Morère-Le Paven, M.C.; Limami, A.M. Abscisic acid-induced nitric oxide and proline accumulation in independent pathways under water-deficit stress during seedling establishment in *Medicago truncatula*. *J. Exp. Bot.* **2014**, *65*, 2161–2170.
170. Szabados, L.; Savouré, A. Proline: A multifunctional amino acid. *Trends Plant Sci.* **2010**, *15*, 89–97.
171. Wutipraditkul, N.; Wongwean, P.; Buaboocha, T. Alleviation of salt-induced oxidative stress in rice seedlings by proline and/or glycinebetaine. *Biol. Plant.* **2015**, *59*, 547–553.
172. Jagendorf, A.T.; Takabe, T. Inducers of glycinebetaine synthesis in barley. *Plant Physiol.* **2001**, *127*, 1827–1835.
173. Mäkelä, P.; Kärkkäinen, J.; Somersalo, S. Effect of glycinebetaine on chloroplast ultrastructure, chlorophyll and protein content, and Rubisco activities in tomato grown under drought or salinity. *Biol. Plant* **2000**, *43*, 471–475.

174. Horei, T.; Karahara, I.; Katsuhara, M. Salinity tolerance mechanisms in glycophytes and overview central focus on rice plants. *Rice* **2012**, *5*, 11–18.
175. Parida, A.; Das, A.B.; Das, P. NaCl stress causes changes in photosynthetic pigments, proteins, and other metabolic components in the leaves of a true mangrove, *Bruguiera parviflora*, in hydroponic cultures. *J. Plant Biol.* **2002**, *45*, 28–36.
176. Duarte, B.; Santos, D.; Marques, J.C.; Caçador, I. Ecophysiological adaptations of two halophytes to salt stress: Photosynthesis, PS II photochemistry and anti-oxidant feedback—Implications for resilience in climate change. *Plant Physiol. Biochem.* **2013**, *67*, 178–188.
177. Chaves, M.M.; Flexas, J.; Pinheiro, C. Photosynthesis under drought and salt stress: Regulation mechanisms from whole plant to cell. *Ann. Bot.* **2009**, *103*, 551–560.
178. Asada, K. The water–water cycle in chloroplasts: Scavenging of active oxygen and dissipation of excess photons. *Ann. Rev. Plant Physiol. Plant Mol. Biol.* **1999**, *50*, 601–639.
179. Maxwell, K.; Johnson, G.N. Chlorophyll fluorescence: A practical guide. *J. Exp. Bot.* **2000**, *51*, 659–668.
180. Moradi, F.; Ismail, A.M. Responses of photosynthesis, chlorophyll fluorescence and ROS-scavenging systems to salt stress during seedling and reproductive stages in rice. *Ann. Bot.* **2007**, *99*, 1161–1179.
181. Lee, M.H.; Cho, E.J.; Wi, S.G.; Bae, H.; Kim, J.E.; Cho, J.Y.; Lee, S.; Kim, J.H.; Chung, B.Y. Divergences in morphological changes and antioxidant responses in salt-tolerant and salt-sensitive rice seedlings after salt stress. *Plant Physiol. Biochem.* **2013**, *70*, 325–335.
182. Ikbal, F.; Hernández, J.A.; Barba-Espín, G.; Koussa, T.; Aziz, A.; Faize, M.; Diaz-Vivancos, P. Enhanced salt-induced antioxidative responses involve a contribution of polyamine biosynthesis in grapevine plants. *J. Plant Physiol.* **2014**, *171*, 779–788.
183. Shu, S.; Yuan, L.Y.; Guo, S.R.; Sun, J.; Yuan, Y.H. Effects of exogenous spermine on chlorophyll fluorescence, antioxidant system and ultrastructure of chloroplasts in *Cucumis sativus* L. under salt stress. *Plant Physiol. Biochem.***2013**, *63*, 209–216.
184. Foyer, C.H.; Harbison, J. Oxygen metabolism and the regulation of photosynthetic electron transport. In *Causes of Photooxidative Stresses and Amelioration of Defense Systems in Plants*; Foyer, C.H., Mullineaux, P., Eds.; CRC Press: Boca Raton, FL, USA, 1994; pp. 1–42.
185. Demmig-Adams, B.; Adams, W.W., III. Photoprotection and other responses of plants to high light stress. *Annu. Rev. Plant Physiol. Plant Mol. Biol.* **1992**, *43*, 599–626.
186. Zhang, Q.Y.; Wang, L.Y.; Kong, F.Y.; Deng, Y.S.; Li, B.; Meng, Q.W. Constitutive accumulation of zeaxanthin in tomato alleviates salt stress-induced photoinhibition and photooxidation. *Physiol. Plant.* **2012**, *146*, 363–373.
187. Huang, C.; He, W.; Guo, J.; Chang, X.; Su, P.; Zhang, L. Increased sensitivity to salt stress in an ascorbate-deficient Arabidopsis mutant. *J. Exp. Bot.* **2005**, *422*, 3041–3049.
188. Halliwell, B.; Gutteridge, J.M.C. *Free Radicals in Biology and Medicine*; Oxford University Press: London, UK, 2000.
189. Hodges, M.; Dellero, Y.; Keech, O.; Betti, M.; Raghavendra, A.S.; Sage, R.; Zhu, X.G.; Allen, D.K.; Weber, P.M. Perspectives for a better understanding of the metabolic integration of photorespiration within a complex plant primary metabolism network. *J. Exp. Bot.* **2016**, *67*, 3015–3026.
190. Kangasjärvi, S.; Neukermans, J.; Li, S.; Aro, E.-M.; Noctor, G. Photosynthesis, photorespiration, and light signalling in defence responses. *J. Exp. Bot.* **2012**, *63*, 1619–1636.
191. Corpas, F.J.; Gomez, M.; Hernández, J.A.; del Río, L.A. Metabolism of activated oxygen in peroxisomes from two *Pisum sativum* L. cultivars with different sensitivity to sodium chloride. *J. Plant Physiol.* **1993**, *141*, 160–165.
192. Krall, J.P.; Edwards, G.E. Relationship between photosystem II activity and CO_2 fixation in leaves. *Physiol. Plant.***1992**, *86*, 180–187.
193. Dat, J.; Vandenabeele, S.; Vranova, E.; Van Montagu, M.; Inzé, D.; Van Breusegem, F. Dual action of the active oxygen species during plant stress responses. *Cell. Mol. Life Sci.* **2000**, *57*, 779–795.
194. Brisson, L.F.; Zelitch, I.; Havir, E.A. Manipulation of catalase levels produces altered photosynthesis in transgenic tobacco plants. *Plant Physiol.* **1998**, *116*, 259–269.

195. Hernández, J.A.; Jiménez, A.; Mullineaux, P.M.; Sevilla, F. Tolerance of pea (*Pisum sativum* L.) to long-term salt stress is associated with induction of antioxidant defenses. *Plant Cell Environ.* **2000**, *23*, 853–862.
196. Gómez, J.M.; Hernández, J.A.; Jiménez, A.; del Río, L.A.; Sevilla, F. Differential response of antioxidative enzymes of chloroplasts and mitochondria to long-term NaCl stress of pea plants. *Free Rad. Res.* **1999**, *31*, S11–S18.
197. Mittova, V.; Tal, M.; Volokita, M.; Guy, M. Salt stress induces up-regulation of an efficient chloroplast antioxidant system in the salt-tolerant wild tomato species *Lycopersicon pennellii* but not in the cultivated species. *Physiol. Plant.* **2002**, *115*, 393–400.
198. Rubio, M.C.; Bustos-Sanmaded, P.; Clemente, R.M.; Becana, M. Effects of salt stress on the expression of antioxidant genes and proteins in the model legume *Lotus japonicus*. *New Phytol.* **2009**, *181*, 851–859.
199. Gueta-Dahan, Y.; Yaniv, Z.; Zilinskas, B.A.; Ben-Hayyim, G. Salt and oxidative stress: Similar and specific responses and their relation to salt tolerance in Citrus. *Planta* **1997**, *203*, 460–469.
200. Tsugane, K.; Kobayashi, K.; Niwa, Y.; Ohba, Y.; Wada, K. A recessive *Arabidopsis* mutant that grows photoautotrophically under salt stress shows enhanced active oxygen detoxification. *Plant Cell* **1999**, *11*, 1195–1206.
201. López-Gómez, E.; Sanjuán, M.A.; Diaz-Vivancos, P.; Mataix-Beneyto, J.; García-Legaz, M.F.; Hernández, J.A. Effect of salinity and rootstocks on antioxidant systems of loquat plants (*Eriobotrya japonica* Lindl.): Response to supplementary boron addition. *Environ. Exp. Bot.* **2007**, *160*, 151–158.
202. Leterrier, M.; Barroso, J.B.; Valderrama, R.; Palma, J.M.; Corpas, F.J. NADP-Dependent Isocitrate Dehydrogenase from *Arabidopsis* Roots Contributes in the Mechanism of Defence against the Nitro-Oxidative Stress Induced by Salinity. *Sci. World. J.* **2012**, *2012*, 694740.
203. Meneguzzo, S.; Sgherri, C.L.M.; Navari-Izzo, F.; Izzo, R. Stromal and thylakoid-bound ascorbate peroxidase in NaCl-treated wheat. *Physiol. Plant.* **1998**, *104*, 735–740.
204. Qiu-Fang, Z.; Yuan-Yuan, L.; Cai-Hong, P.; Cong-Ming, L.; Bao-Shan, W. NaCl enhances thylakoid-bound SOD activity in the leaves of C3 halophyte *Suaeda salsa* L. *Plant Sci.* **2005**, *168*, 423–430.
205. Gómez, J.M.; Jiménez, A.; Olmos, E.; Sevilla, F. Location and effects of long-term NaCl stress on superoxide dismutase and ascorbate peroxidase isoenzymes of pea (*Pisum sativum* cv. Puget) chloroplasts. *J. Exp. Bot.* **2004**, *55*, 119–130.
206. Mutlu, S.; Atici, Ö.; Nalbantoglu, B. Effects of salicylic acid and salinity on apoplastic antioxidant enzymes in two wheat cultivars differing in salt tolerance. *Biol. Plant.* **2009**, *53*, 334–338.
207. Diaz-Vivancos, P.; Faize, M.; Barba-Espin, G.; Faize, L.; Petri, C.; Hernández, J.A.; Burgos, L. Ectopic expression of cytosolic superoxide dismutase and ascorbate per-oxidase leads to salt stress tolerance in transgenic plums. *Plant Biotechnol.* **2013**, *11*, 976–985.
208. Fernández-García, N.; Hernández, M.; Casado-Vela, J.; Bru, R.; Elortza, F.; Hedden, P.; Olmos, E. Changes to the proteome and targeted metabolites of xylem sap in *Brassica oleracea*. *Plant Cell Environ.* **2011**, *34*, 821–836.
209. Hideg, E.; Barta, C.; Kalai, T.; Vass, M.; Hideg, K.; Asada, K. Detection of singlet oxygen and superoxide with fluorescence sensors in leaves under stress by photoinhibition or UV radiation. *Plant Cell Physiol.* **2002**, *43*, 1154–1164.
210. Boveris, A. Determination of the production of superoxide radicals and hydrogen peroxide in mitochondria. *Methods Enzymol.* **1984**, *105*, 429–435.
211. Schmitt, N.; Dizengremel, P. Effect of osmotic stress on mitochondria isolated from etiolated mung bean and sorghum seedlings. *Plant Physiol. Biochem.* **1989**, *27*, 17–26.
212. del Río, L.A.; Pastori, G.M.; Palma, J.M.; Sandalio, L.M.; Sevilla, F.; Corpas, F.J.; Jiménez, A.; López-Huertas, E.; Hernández, J.A. The activated oxygen role of peroxisomes in senescence. *Plant Physiol.* **1998**, *116*, 1195–1200.
213. Zelitch, I.; Schultes, N.P.; Peterson, R.B.; Brown, P.; Brutnell, T.P. High glycolate oxidase activity is required for survival of maize in normal air. *Plant Physiol.* **2009**, *149*, 195–204.

214. del Río, L.A.; Sandalio, L.M.; Altomare, D.A.; Zilinskas, B.A. Mitochondrial and peroxisomal manganese superoxide dismutase: Differential expression during leaf senescence. *J. Exp. Bot.* **2003**, *54*, 923–933.
215. Bellicampi, D.; Dipierro, N.; Salvi, G.; Cervone, F.; De Lorenzo, G. Extracellular H_2O_2 induced by oligogalacturonides is not involved in the inhibition of the auxin-regulated rolB gene expression in tobacco leaf explants. *Plant Physiol.***2000**, *122*, 1379–1385.
216. García de la Garma, J.; Fernández.-García, N.; Bardasi, E.; Pallol, B.; Asensio-Rubio, J.S.; Olmos, E. New insights into plant salt acclimation: The roles of vesicle trafficking and reactive oxygen species signalling in mitochondria and the endomembrane system. *New Phytol.* **2015**, *205*, 216–239.
217. Wang, R.; Chen, S.; Zhou, X.; Shen, X.; Deng, L.; Zhu, H.; Shao, J.; Shi, Y.; Dai, S.; Fritz, E.; Hüttermann, A.; Polle, A. Ionic homeostasis and reactive oxygen species control in leaves and xylem sap of two poplars subjected to NaCl stress. *Tree Physiol.* **2008**, *28*, 947–957.
218. Dinler, B.S.; Antoniou, C.; Fotopoulos, V. Interplay between GST and nitric oxide in the early response of soybean (*Glycine max* L.) plants to salinity stress. *J. Plant Physiol.* **2014**, *171*, 1740–1747.
219. Kaur, H.; Bhatla, S.C. Melatonin and nitric oxide modulate glutathione content and glutathione reductase activity in sunflower seedling cotyledons accompanying salt stress. *Nitric Oxide* **2016**, *59*, 42–53.
220. Manai, J.; Gouia, H.; Corpas, F.J. Redox and nitric oxide homeostasis are affected in tomato (*Solanum lycopersicum*) roots under salinity-induced oxidative stress. *J. Plant Physiol.* **2014**, *171*, 1028–1035.
221. Corpas, F.J.; Hayashi, M.; Mano, S.; Nishimura, M.; Barroso, J.B. Peroxisomes are required for in vivo nitric oxide accumulation in the cytosol following salinity stress of Arabidopsis plants. *Plant Physiol.* **2009**, *151*, 2083–2094.
222. Valderrama, R.; Corpas, F.J.; Carreras, A.; Fernández-Ocaña, A.; Chaki, M.; Luque, F.; Gómez-Rodríguez, M.V.; Colmenero-Varea, P.; del Río, L.A.; Barroso, J.B. Nitrosative stress in plants. *FEBS Lett.* **2007**, *581*, 453–461.
223. Hossain, M.A.; Bhattacharjee, S.; Armin, S.M.; Qian, P.; Xin, W.; Li, H.Y.; Burritt, D.J.; Fujita, M.; Tran, L.S.P. Hydrogen peroxide priming modulates abiotic oxidative stress tolerance: Insights from ROS detoxification and scavenging. *Front. Plant Sci.* **2015**, *6*, 420.
224. Li, C.; Wang, P.; Wei, Z.; Liang, D.; Liu, C.; Yin, L.; Jia, D.; Fu, M.; Ma, F. The mitigation effects of exogenous melatonin on salinity-induced stress in *Malus hupehensis*. *J. Pineal Res.* **2012**, *53*, 298–306.
225. He, Y.; Zhu, Z.J. Exogenous salicylic acid alleviates NaCl toxicity and increases antioxidative enzyme activity in *Lycopersicon esculentum*. *Biol. Plant.* **2008**, *52*, 792–795.
226. Bastam, N.; Baninasab, B.; Ghobadi, C. Improving salt tolerance by exogenous application of salicylic acid in seedlings of pistachio. *Plant Growth Regul.* **2013**, *69*, 275–284.
227. Borsani, O.; Valpuesta, V.; Botella, M.A. Evidence for a role of salicylic acid in the oxidative damage generated by NaCl and osmotic stress in Arabidopsis seedlings. *Plant Physiol.* **2001**, *126*, 1024–1030.
228. Nahar, K.; Hasanuzzaman, M.; Rahman, A.; Alam, M.M.; Mahmud, J.A.; Suzuki, T.; Fujita, M. Polyamines confer salt tolerance in mung bean (*Vigna radiata* L.) by reducing sodium uptake, improving nutrient homeostasis, antioxidant defense, and methylglyoxal detoxification systems. *Front. Plant Sci.* **2016**, *7*, 1104.

CHAPTER 5

Agro-Ecology for Potential Adaptation of Horticultural Systems to Climate Change: Agronomic and Energetic Performance Evaluation

Mariangela Diacono *, Alessandro Persiani, Angelo Fiore, Francesco Montemurro and Stefano Canali

Consiglio per la ricerca in agricoltura e l'analisi dell'economia agraria—CREA-AA, Research Centre for Agriculture and Environment, Via Celso Ulpiani 5, 70125 Bari, Italy

AUTHOR CONTRIBUTIONS

M.D., A.F., F.M. and S.C. conceived and designed the experiments; A.F. and A.P. performed the experiments; M.D. and A.P. analyzed the data; F.M. and S.C. contributed reagents/materials/analysis tools; M.D. and A.P. wrote the paper."

ABSTRACT

Adaptation can be a key factor that will shape the future severity of climate change impacts on food production. The objective of this study was to assess the suitability of an agro-ecological approach based on various techniques as potential adaptation strategy in organic horticultural systems. A long-term field experiment was set up in Southern Italy, combining: (i) appropriate soil surface shaping; (ii) cash crop rotation; (iii) agro-ecological service crops (ASC) introduction as living mulch and complementary crops; (iv) tailored organic fertilization; and (v) alternative tillage strategies. In this paper, the first two-year results on cauliflower (*Brassica oleracea* L.) and tomato (*Solanum lycopersicum* L.) crops, as well as energy consumptions through the Energy Analysis (EA) method are reported. Due to the climatic conditions that occurred, which were characterized by the absence of extreme climatic events (particularly rainfall), it was not possible to verify if the designed experimental device was able to mitigate the impact of climate change, whereas the EA indicated that total energy inputs were lower when ASC are introduced in cropping systems.

Keywords: agro-ecological service crops; crop rotations; energy analysis; energy productivity; living mulch; organic fertilization; roller crimper

1. INTRODUCTION

Agriculture is highly sensitive to climate variations, which are the dominant source of the overall interannual variability of production in many regions and a continuing source of disruption to ecosystem services [1]. In particular, according to Altieri et al. [2], predicted changes in temperature and increased frequency of extreme events (e.g., droughts and floods) will lead in some areas to reduced crop yields and yield stability, thus likely affecting food security.

Indeed, there are many potential ways to introduce management-level adaptation options for cropping systems, thus dealing with projected climate changes [3,4,5]. These adaptations include strategies such as altering varieties/species to those with more appropriate vernalization requirements and/or with resistance to heat shock and drought. As a matter of fact, plant breeding could provide later/earlier-maturing hybrids and cultivars, which would be beneficial for crop productivity, due to shorter vegetative growth period under expected climate change. Soil organic matter management and use of technologies to harvest water and conserve soil moisture are other options for adaptation, the last one being related to changes in planting dates especially for spring crops, which could be sown earlier under the climate change scenarios. This could be carried out in order to reduce the yield loss and allowing crops to develop during an earlier period of the year with ample soil water.

Moreover, according to Alexandrov et al. [4], switching from monocultures, which are more vulnerable to pest and diseases, to more diversified agricultural production systems, such as appropriate crop rotation systems, will help farmers to cope with climate variation from year to year. This type of agronomic adaptation can be considered at the farm system level, therefore, in the present study it was considered the most important option. Agricultural biodiversity and knowledge preserved by traditional farming systems as well as the innovation developed by effective research are essential to strengthen the resilience of agro-ecosystems in response to a potential change in the climate. In low input and organic farming systems, proper rotations, agro-ecological service crops (ASC) introduction as buffer zones or break crops, reduced tillage and fertilization management can provide beneficial services to the agro-ecosystems, contributing to weed, pest and diseases management, NO_3^- leaching reduction and improving soil water retention and crop tolerance to drought occurrences [6].

In addition, efficient energy use by the agriculture sector is one of the conditions for sustainable agriculture, because it allows financial savings, fossil resources preservation and air pollution decrease [7], and, for this reason, in a climate change context, energy input-output balance can be a win-win tool to study the efficiency and the environmental impacts of agricultural systems in which adaptation strategies are being used.

The research objectives were, therefore, to assess the agronomic and environmental performance as well as the suitability of a set of agro-ecological techniques (and their best synergistic combination) as potential strategies for organic horticulture adaptation to climate change. Cauliflower and tomato crops were identified as model cash crops of our experimental device, since they are usually grown in the horticultural systems of the area in accordance to an autumn-winter and spring-summer cropping cycle, respectively. Their yield and energetic performances were investigated in the innovative MITIORG long-term experimental device, to get an insight into the overall system performances.

2. RESULTS

2.1. Cash Crops Agronomic Results

Significant main effects of living mulch (LM) sowing time and fertilizer treatments on cauliflower head yield and diameter, as well as significant two-way interactions, were found(Table 1). Also, significant main effects of ASC treatment, fertilizer and ASC termination management on tomato marketable yield and fruit dry matter and significant two-way interactions were found.

Table 1. Analysis of variance for the treatments effect on yield performance in cauliflower and tomato crop.

Treatments	Cauliflower Head Yield	Cauliflower Head Diameter	Tomato Marketable Yield	Tomato Fruit Dry Matter
Living mulch (LM)	***	***	-	-
ASC species (ASC)	-	-	***	**
Fertilizer (FERT)	*	*	***	***
Termination (TERM)	-	-	***	***
LM × FERT	**	*	-	-
ASC × FERT	-	-	***	***
ASC × TERM	-	-	***	***
FERT × TERM	-	-	***	***
ASC × FERT × TERM	-	-	n.s.	n.s.

*, **, ***: Significant at the $p < 0.05$, 0.01 and 0.001, respectively. n.s. = not significant.

The highest cauliflower head yield value was obtained in NMC × AD combination, which was significantly different from all ES × fertilizer treatment interactions and the other interactions with NMC treatment (Table 2). In particular, NMC × AD value was significantly higher by 108% thanES × MSW. Among fertilizers, the best results were found for AD, which showed head yield values in combination with NMC and CS that were comparable with CS × ORG.

Table 2. Interactions between living mulch (LM) sowing times and fertilizer treatments on cauliflower head yield and head diameter.

Treatment	Cauliflower Head Yield (t ha^{-1})				Head Diameter (cm)			
	Fertilizer				Fertilizer			
Living mulch (LM)	AD [2]	MSW	ORG	NFC	AD	MSW	ORG	NFC
ES [1]	16.1 bd	11.8 d	16.1 bd	14.7 cd	16.6 bd	15.4 d	16.13 cd	15.9 cd
CS	19.4 ac	19.5 ac	21.6 ab	22.4 ab	17.7 ac	17.6 ac	18.4 ab	18.5 ab
NMC	24.7 a	17.6 bd	15.9 bd	17.3 bd	18.9 a	16.8 ad	16.0 cd	17.8 ac

[1] ES: early sowing LM, 20 days before cauliflower transplanting; CS: at cauliflower transplanting; NMC: no LM control; [2] AD: anaerobic digestate fertilizer; MSW: composted municipal solid organic wastes; ORG: commercial humified organic fertilizer; NFC: unfertilized control. Means followed by different letters within raw are significantly different ($p \leq 0.05$ probability level).

The highest head diameter value was obtained by no living mulch control (NMC) × anaerobic digestate (AD) combination (Table 2), whereas the lowest one was determined by early sowing LM (ES) × composted municipal solid organic wastes (MSW), which was not significantly different both from all the ES × fertilizer combinations and NMC × commercial humified organic fertilizer (ORG). Moreover, concurrent sowing LM (CS) significantly increased head diameter on average by 12.5% than ES treatment.

In Figure 1, effects by fertilizers (a) and termination methods (b) of ASC treatments (vetch, barley, mixture and control) on tomato marketable yield are reported.

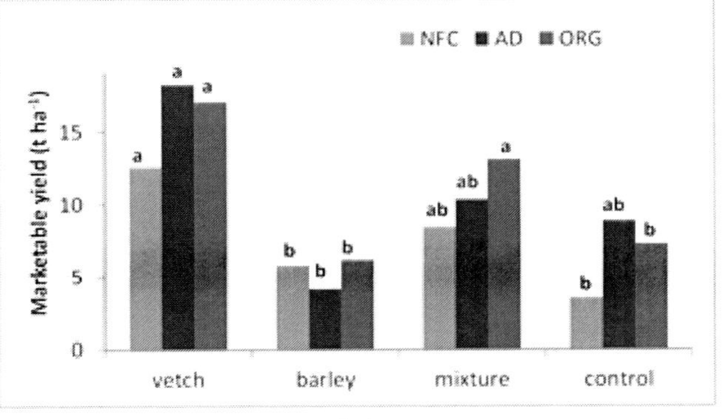

(a)

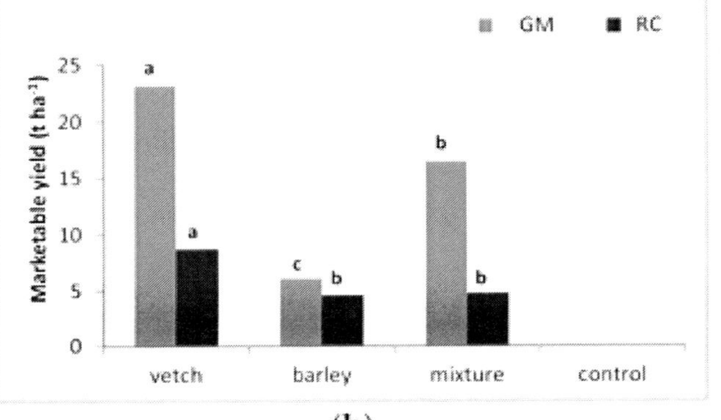

(b)

Figure 1. Effects by fertilizers (a) and termination methods (b) of Agro-ecological Service Crops treatments (vetch, barley, mixture and control) on tomato marketable yield (t ha^{-1}). AD: anaerobic digestate fertilizer; ORG: commercial humified organic fertilizer; NFC: unfertilized control; GM: green manure by incorporating ASC biomass into soil; RC: no till and ASC flattening by roller crimper.

As regards fertilizer effect, AD fertilization determined the significantly highest value in vetch plot and the lowest in barley, whereas the ASC mixture and the control had comparable and intermediate marketable yield values (Figure 1a). The ORG fertilizer determined the highest marketable yield in vetch and ASC mixture, which were higher on average by 144% and 106% than the combination with barley and control, respectively. Best results were obtained in the vetch ASC by NFC treatment, which was not significantly different from the value for mixture, whereas the lowest values were obtained by barley and the double control. On the whole, the green manure (GM) treatment determined higher marketable yield results than roller crimper (RC) treatment. The GM termination method showed the highest value in vetch, the lowest in barley and intermediate value in the mixture (Figure 1b). Conversely, the RC determined the best result with vetch, being the marketable yield higher by 89% and 81% than the combination with barley and the mixture, respectively.

In Figure 2, effects by fertilizers (a) and termination methods (b) of ASC treatments (vetch, barley, mixture and control) on tomato fruit dry matter are reported.

As regards fertilizer effect, fruit dry matter results were similar to those of marketable yield (Figure 2a), except for ORG, which determined the highest value in vetch plots, which was comparable to mixture value and higher by 185% and 71% as compared to the dry matter in barley and control plots, respectively. Dry matter results by termination methods were perfectly comparable to marketable yield ones (Figure 2b).

2.2. Cropping Systems Energetic Performance

The energy analysis showed that the LM treatments (ES and CS) consumed low energy inputs per hectare than the NMC in cauliflower production (Table 3), whereas GM termination consumed more energy inputs than RC management in tomato production.

The highest energy consuming input was fertilizer (on average, 82% and 65% for cauliflower and tomato, respectively), followed by fuels (10% and 17% for cauliflower and tomato, respectively) and water (10% for tomato).

In cauliflower, 88% on average of total energy input resulted from renewable as compared to 12% from non-renewable energy, as well as 12% derived from direct and 87% from indirect energy inputs. In tomato, 78% of total energy input resulted from renewable as compared to 22% from non-renewable energy, and 19% derived from direct and 80% from indirect energy inputs.

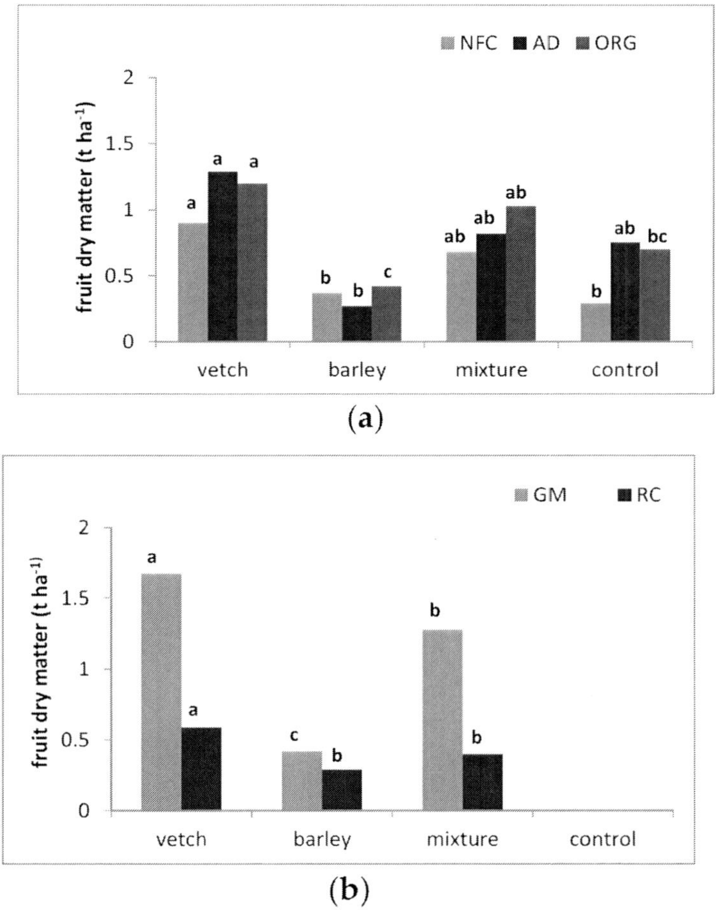

Figure 2. Effects by fertilizers (a) and termination methods (b) of Agro-ecological Service Crops treatments (vetch, barley, mixture and control) on tomato fruit dry matter (t ha^{-1}) (AD: anaerobic digestate fertilizer; ORG: commercial humified organic fertilizer; NFC: unfertilized control; GM: green manure by incorporating ASC biomass into soil; RC: no till and ASC flattening by roller crimper.

As regards the amount of energy productivity, in cauliflower production this parameter was higher for CS than both ES (43%) and NMC (86%) treatments. In tomato crop, energy productivity was higher for GM than RC (125%) and fallow-control (CT) (203%). Moreover, on the whole, the parameter was lower in tomato than in cauliflower crop.

Table 3. Amounts of each energy input and output, total inputs and production outputs, energy productivity and energy inputs divided by categories, both in cauliflower and tomato cropping cycles (MJ ha^{-1}).

	Cauliflower Crop				Tomato Crop		
	ES [1]	CS	NMC		GM [2]	RC	CT
Human labor	387	504	483	Human labor	542	299	303
Machinery	255	255	238	Machinery	504	510	238.2
Fuels	2148	1735	1736	Fuels	4077	2970	1981
Fertilizers	11,461	11,461	22,922	Fertilizers	8595	8595	17,191
Chemicals	180	180	180	Chemicals	480	480	480
Water	428	428	428	Water	1734	1734	1734
Seeds	30	30	0	Seeds	140	140	0
Total inputs	14,889	14,594	25,988	Total inputs	16,074	14,729	21,928
Outputs	24,990	35,190	33,490	Outputs	16,960	6992	7584
Energy Productivity	1.68	2.41	1.29	Energy productivity	1.06	0.47	0.35
Renewable energy	12,306	12,423	23,834	Renewable	11,012	10,769	19,229
Non-renewable energy	2583	2171	2154	Non-renewable energy	5062	3960	2699
Direct energy	2535	2239	2219	Direct energy	4620	3269	2284
Indirect energy	12,354	12,354	23,768	Indirect energy	11,454	11,460	19,644

[1] ES: early sowing LM, 20 days before cauliflower transplanting; CS: at cauliflower transplanting; NMC: no LM control; [2] GM: green manure by incorporating ASC biomass into soil; RC: no till and ASC flattening by roller crimper; CT: fallow-control.

Table 4. Amounts of energy inputs (MJ ha^{-1}) in cauliflower and tomato production divided by crop management operations.

Cauliflower				Tomato			
	Operations				Operations		
	ES [1]	CS	NMC		GM [2]	RC	CT
Soil hydraulic arrangement	503	503	503	Soil hydraulic arrangement	503	503	503
Tillage	589	589	589	Tillage	589	589	589
Seedbed preparation	574	574	574	Seedbed preparation	574	574	574
Sowing	90	90	42	Sowing	200	200	0
ASC/weeds mowing	126	126	126	ASC flattering	0	562	0
Planting	78	78	78	Strip tillage for transplanting	0	650	0
Spraying	444	235	235	ASC biomass chopped	1468	0	0
Irrigation	481	481	481	ASC plowed into the soil	846	0	0
Fertilization	11,718	11,515	22,976	Planting	97	97	97
Harvest	287	404	384	Irrigation	1786	1786	1786
				Fertilization	8853	8853	17449
				Spraying	744	744	744
				Harvest	413	170	185

[1] ES: early sowing LM, 20 days before cauliflower transplanting; CS: at cauliflower transplanting; NMC: no LM control; [2] GM: green manure by incorporating ASC biomass into soil; RC: no till and ASC flattening by roller crimper; CT: fallow-control.

As regards the energy consumption for each crop operation done (Table 4), fertilization was the most energy consuming and impacting phase in both analyzed systems. Moreover, in cauliflower soil preparation operations consumed a consistent rate of energy, as well as GM operations and irrigation in tomato crop.

3. DISCUSSION

3.1. Cauliflower and Tomato Agronomic Performance

In the MITIORG long-term experimental device, soil surface shaping was carried out to increase rooting depth layer, allowing crop survival in the event of flooding, and to make easier the lateral outflow of excess water. This technique was combined with soil organic matter management, by using organic fertilizers and amendments, that improves soil's water retention capacity and infiltration [8], and farming system diversification to provide agro-ecological services. In particular, on the top of two of the three planned soil convex strips, a leguminous ASC was intercropped as LM with the winter cash crop (cauliflower), in comparison with a no-living mulch control on the third strip. Time of sowing of the ASC, in respect to the transplanting of the cash crop, may determine the effectiveness of this technique in providing agro-ecosystem services [9]. In addition, in the concave soil strips the ASC cultivated in the winter period, as break crops between two consecutive summer cash crops, need to be terminated before cash crop transplanting to avoid competition. Therefore, according to Montemurro et al. [10], to allow the provision of agro-ecological services, ASC are incorporated into the soil by using the traditional green manure method [11], or can be flattened by the innovative roller crimper technique [12].

Although due to the absence of extreme rain events during the trial, which are not completely predictable, it was not possible to verify the effects on climate change adaptation of these combined agro-ecological practices, it was possible to test the agronomic and energy performance of the crop rotation in the innovative experimental device. Cauliflower production and quality (head yield and head diameter) were better with CS than ES (Table 2), in agreement with Kolota and Adamczewska-Sowińska's [13] results, indicating the role of LM sowing period. This result also suggests that no competition was determined by ASC in CS treatment. The comparable yield results of AD and ORG (in combination with both ES and CS), indicated that both are viable options for organic farmers, in agreement with other studies in Mediterranean area [14,15].

Similarly, as regards tomato crop, AD determined comparable results with commercial fertilizer for both marketable yield and fruit dry matter (Figure 1a and Figure 2a). However, for tomato crops, the combination of AD with barley had a marketable yield lower by 77% in comparison with vetch, maybe because of competitiveness of ASC with the vegetable cash crop. This explanation is further supported by considering that a higher absolute value in the AD × control combination was found. In the absence of fertilization, particularly in the double control (no ASC—no fertilizer) the yield performance was very low, indicating that this option of management is not sustainable. Fruit dry matter had on the whole similar results than marketable yield, suggesting poor effectiveness of the combination of fertilizer with barley ASC.

The GM results both for tomato marketable yield and fruit dry matter (Figure 1b and Figure 2b), being higher than RC, confirmed other studies [16]. This is likely because of higher availability in GM of total N by decomposition of aboveground biomass in soil, unlike RC treatment that can readily supply only a small part of the N contained in the leguminous ASC. In fact, the GM termination method showed the highest value in vetch and ASC mixture.

3.2. Energy Analysis Evaluation

Regarding the energy analysis, the most energy consuming inputs were fertilizers, fuels and water (Table 3), in agreement with others studies [17,18,19]. In addition, fertilization was the most impacting phase in both analyzed systems (Table 4). The great reduction of total energy inputs that was detected in ES and CS in comparison with NMC for cauliflower, and in GM and RC as compared to CT for tomato, would indicate that leguminous ASC can be useful not only in providing agro-ecological services [20,21], but also in maintaining cropping system sustainability. In fact, the energy productivity was higher in ASC treatments than in control ones, as well as CS seemed to be the best solution likely due to the reduction of ASC competitiveness with the cash crop. The energy productivity results also suggest that GM termination was more efficient than RC in tomato crop.

In general, under the experimental conditions of our study, the energy consumption in the innovative organic horticultural system managed with agro-ecological practices was lower than the energy used in conventional systems founded in literature [18].

The synergistic combination of hydraulic arrangement, conservation tillage and ASC introduction and management can help farmers to increase energy use efficiency by decreasing energy consumption and sustaining crop yield.

In conclusion, combined application of agro-ecological techniques is a viable strategy, sustaining yield of cash crops in rotation, and helping in adaptation of horticultural systems to the climate changes. This is due to a reduction of energy consumption (and consequently of GHG emissions) and increase of energy productivity.

However, the study period could not be sufficient to draw general conclusions, therefore, more data will be analyzed in this ongoing research to validate the reported preliminary results.

4. MATERIALS AND METHODS

4.1. Study Site and Experimental Device

The study has been conducted in the innovative MITIORG organic field experiment (*Long-term climatic change adaptation in organic farming: synergistic combination of hydraulic arrangement, crop rotations, agro-ecological service crops and agronomic techniques*) on the research farm 'Azienda Sperimentale Metaponto' of the Research Centre for Agriculture and Environment, Council for Agricultural Research and Economics (CREA-AA) (lat. 40°24' N; long. 16°48' E).

Soils are classified as Typic Epiaquerts [22]. The climate is classified as "accentuated thermomediterranean" according to the UNESCO-FAO soil classification system [23], with mean monthly temperatures of 8.8 °C in the winter and 24.4 °C in the summer. The site is generally characterized by winter temperatures which can fall below 0 °C, and summer temperatures which can rise above 40 °C.

During the cauliflower cropping cycle, over the period August-December 2014, total rainfall was lower than the 30-year long-term average (221 mm and 267 mm, respectively). A single peak was reached at the beginning of October with 63 mm of rainfall in 24 h. Mean temperature was higher (18.6 °C) than the long-term average (17.1 °C).

Conversely, during tomato cropping cycle (April-August 2015 period), both the total rainfall (116 mm) and mean temperature (21.8 °C) were very close to the 30-year long-term averages(117 mm and 20.8 °C, respectively).

The field experiment was designed combining a suite of functionally integrated techniques (conceptually identified as "layers"), namely: (i) soil surface shaping; (ii) crop rotations; (iii) ASC introduction; iv) ASC termination techniques; and (v) organic fertilization.

The base layer is the soil hydraulic arrangement by means of soil surface shaping as a kind of ridge system. Vegetable crops are cultivated both above the raised bed (*ridges* 2.5 m wide) and in the 2.5 m flat area (or *strips*) between them. Rotation (the second conceptual layer) is designed to avoid cash crops cultivation during the winter-rainy period of the year in the flat strips, which can be waterlogged in the case of heavy rain and/or temporary flooding, as frequently have occurred in the area in the last years. Moreover, in order to protect the soil from erosion and provide N to the system *via* biological fixation, a next conceptual layer based on tailored introduction of ASC is implemented. On the top of the ridges, a leguminous ASC (burr medic, *Medicago polymorpha* L.) is living mulched (LM) in the winter vegetable crop and maintained as a living ground cover throughout its cycle, controlling time by time its growth by mowing. In the flat soil strips, pure ASC or mixtures of different proportions of legume and non-legume crops, potentially resistant to temporary water excess, are cultivated in the winter-rainy period as break crops. The break ASC in the furrow strips are terminated by the no-till roller crimper technique, before the transplant of the next summer cash crops. The last layer consists of an organic fertilization strategy, which is implemented by using commercial and/or novel fertilizers and amendments, to maintain or increase long-term soil organic matter and fertility.

4.2. Experimental Setup and Treatments

The research here reported was carried out during the 2014–2015 season comparing winter cauliflower (*Brassica oleracea* L. cv. Triunphan) cultivated on the ridges with different management variant of the legume LM, and tomato (*Solanum lycopersicum* L. cv. Donald) cultivated in the furrow strips after different composition variant of the break ASC. In more detail, the cauliflower experiment was carried out according to a strip-plot design, where two factors and three replications were tested [24]. The first factor was LM management comparing three treatments: (i) early sowing LM (ES; 20 days before cauliflower transplanting); (ii) concurrent sowing LM (CS; at cauliflower transplanting) and (iii) no living mulch control (NMC). Each main plot was divided, as a randomized complete block design, into four sub-plots to test the second factor corresponding to the following organic fertilizers (F), allowed in organic farming [25]: (i) local anaerobic digestate fertilizer, based on cattle slurry (AD); (ii) composted municipal solid organic wastes (MSW; *Fertileva*-Progeva s.r.l., Laterza, TA, Italy); (iii) commercial humified organic fertilizer (ORG), based on dried cattle manure (Italpollina; CRAI s.r.l., Rivoli Veronese, VR, Italy); all compared to (iv) an unfertilized control (NFC). The organic materials were applied to soil one month before cauliflower transplanting, at 100 kg N ha^{-1} rate. The fertilization applied in ES and CS was compared to an application rate of 200 kg N ha^{-1} in the NMC plots, considering the potential contribution of LM biological N fixation in the first two treatments. Each LM sowing times × fertilizer plot (intersection plot) resulted in a 24 m^2 area. The cauliflower was manually transplanted on 2 September 2014(1 × 0.60 m; 17,000 plants ha^{-1}) and it was harvested four times at commercial maturity, from 28 November to 3 December 2014.

As regards tomato crop, the experiment was carried out according to strip-split-plot design where three factors were tested. The ASC composition was the first factor and vetch (*Vicia sativa* L.), barley (*Hordeum vulgare* L.) and a mixture of them were cultivated in the winter period, before cash crop transplanting (compared to a no-ASC control; NAC). The second factor was the soil tillage-ASC termination technique and the following variants were tested: (i) tillage/ASC green manure obtained incorporating into soil the ASC biomass by rotary hoe (to a 15–20 cm in depth) at the end of ASC flowering (GM); (ii) no till and ASC flattening by roller crimper (RC), in which the obtained thick mulch layer remained in place covering the soil surface and, after tomato harvest, it was incorporated in the soil by rotary tiller (15 cm in depth). These treatments were compared to (iii) a fallow-control, in which ASC was not sown (CT). The third factor was the organic fertilization and the following treatments were compared: (i) AD; (ii) ORG; compared to (iii) NFC. The organic materials were applied four days before ASC termination, at the rate of 75 kg N ha^{-1}. To account for the potential contribution of ASC for N, the fertilization applied in AD and ORG was compared to an application rate of 150 kg N ha^{-1} in the NAC. Each intersection plot resulted in a 30 m^2 area. Tomato crop was manually transplanted on 27 April 2015 (1 × 0.40 m; 25,000 plants ha^{-1}) and it was harvested three times at the crop commercial maturity, from 14 July to 26 August 2015.

4.3. Crop Measurements and Statistical Analysis

Cauliflower heads were collected from three randomly selected plants in each plot during the cash crop harvest and head yield (t ha^{-1}) and head diameter (cm) were determined. Similarly, tomato marketable yield (t ha^{-1}) was obtained from three randomly selected plants in each plot, by collecting marketable red fruits. Fruit dry matter (t ha^{-1}) was also determined after measuring fruit dry weight at 70 °C for 48 h.

Analysis of variance (ANOVA) was carried out, with ASC management strategies and cultivar/fertilizer as factors. To compare the differences obtained, means were further analyzed by Tukey HSD test and SNK test ($p < 0.05$) for cauliflower and tomato, respectively.

4.4. Energy Analysis

To assess the environmental impact of introduction of the above-described agro-ecological strategies, an Energy Analysis (EA) was carried out. Effects of energy input on cauliflower and tomato crops management were determined according to the Namdari et al. [26] methodology. Total energy inputs and outputs for production unit (hectare), reported in MJ, were established by multiplying each input with its own coefficient of equivalent energy as specified in Table 5. The energy equivalents per unit were found in the literature [18,27,28].

Table 5. Energy equivalents of each input and output for the agricultural production, according to [18,27,28].

Input	Unit	Energy Equivalent (MJ unit^{-1})
Human labor	h	1.96
Machinery	kg	80.0
Fuels		
diesel	kg	46.2
lubricant	kg	78.1
Fertilizers		
N	kg	66.1
P_2O_5	kg	12.4
K_2O	kg	11.1
Chemicals	kg	120
Irrigation water	m^3	1.02
Seeds	kg	1.00
Output		
Cauliflower	kg	1.70
Tomato	kg	0.80

Moreover, 'energy productivity' parameter was calculated by means of the following formula by Mohammadi et al. [29]:

$$\text{Energy ratio} = \text{Energy output (MJ ha}^{-1}\text{)/Energy input (MJ ha}^{-1}\text{)} \quad (1)$$

The energy inputs were further identified as direct and indirect intakes. They were also divided into renewable and non-renewable energy categories [30]. In particular, both human labor and fossil fuels were considered as direct energy, whereas seeds, water, chemicals, fertilizers and machinery were accounted for as indirect energy. Renewable energy consisted of human labor, seeds, water and organic fertilizers, and non-renewable energy was represented by machinery, fuels and chemicals. The EA was used for the living mulch strategy and different ASC management, whereas the fertilization strategies were not analyzed because only the balance for total N, P_2O_5 and K_2O amounts was determined.

ACKNOWLEDGMENTS

This paper is a result of the AGROCAMBIO (Sistemi e tecniche AGROnomiche di adattamento ai CAMbiamenti climatici in sistemi agricoli BIOlogici) and RETIBIO (Attività di supporto nel settore dell'agricoltura biologica per il mantenimento dei dispositivi sperimentali di lungo termine e il rafforzamento delle reti di relazioni esistenti a livello nazionale e internazionale) research projects, funded by the Organic Farming Office of the Italian Ministry of Agriculture.

REFERENCES

1. Howden, M.; Soussana, J.F.; Tubiello, F.; Chhetri, N.; Dunlop, M.; Meinke, H. Adapting agriculture to climate change. *Proc. Natl. Acad. Sic.* 2007, *104*, 19691–19696.
2. Altieri, M.A.; Nicholls, C.I.; Henao, A.; Lana, M.A. Agroecology and the design of climate change-resilient farming systems. *Agron. Sustain. Dev.* 2015, *35*, 869–890.
3. Easterling, W.; Aggarwal, P.; Batima, P.; Brander, K.; Erda, L.; Howden, M.; Kirilenko, A.; Morton, J.; Soussana, J.F.; Schmidhuber, J.; et al. *Climate Change 2007: Impacts, Adaptation and Vulnerability*; Cambridge University Press: Cambridge, UK, 2007; pp. 273–313.
4. Alexandrov, V.; Eitzinger, J.; Cajic, V.; Oberforster, M. Potential impact of climate change on selected agricultural crops in north-eastern Austria. *Glob. Chang. Biol.* 2002, *8*, 372–389.
5. Jarvis, D.I.; Khaka, E.; Pert, P.L.; Thiombiano, L.; Boelee, E. Managing agroecosystem services. In *Managing Water and Agroecosystems for Food Security—Comprehensive Assessment of Water Management in Agriculture Series 10*; Boelee, E., Ed.; CABI: Wallingford, UK, 2013; pp. 124–141.
6. Canali, S.; Diacono, M.; Campanelli, G.; Montemurro, F. Organic no-till with roller crimpers: Agro-ecosystem services and applications in organic Mediterranean vegetable productions. *Sustain. Agric. Res.* 2015, *4*, 66–75.
7. Pervanchon, F.; Bockstaller, C.; Girardin, P. Assessment of energy use in arable farming systems by means of an agro-ecological indicator: The energy indicator. *Agric. Syst.* 2001, *72*, 149–172.
8. Diacono, M.; Montemurro, F. Long-term effects of organic amendments on soil fertility. A review. *Agron. Sustain. Dev.* 2010, *30*, 401–422.
9. Canali, S.; Campanelli, G.; Ciaccia, C.; Diacono, M.; Leteo, F.; Fiore, A.; Montemurro, F. Living mulch strategy for organic cauliflower (*Brassica oleracea* L.) production in central and southern Italy. *Ital. J. Agron.* 2015, *10*, 90–96.

10. Montemurro, F.; Fiore, A.; Campanelli, G.; Tittarelli, F.; Ledda, L.; Canali, S. Organic fertilization, green manure, and vetch mulch to improve organic zucchini yield and quality. *HortSci.* 2013, *48*, 1027–1033.
11. Watson, C.A.; Atkinson, D.; Gosling, P.; Jackson, L.R.; Rayns, F.W. Managing soil fertility in organic farming systems. *Soil Use Manag.* 2002, *18*, 239–247.
12. Teasdale, J.R.; Mirsky, S.B.; Spargo, J.T.; Cavigelli, M.A.; Maul, J.E. Reduced-tillage organic corn production in a hairy vetch cover crop. *Agron. J.* 2012, *104*, 621–628.
13. Kolota, E.; Adamczewska-Sowińska, K. The effects of living mulches on yield, overwintering and biological value of leek. *Acta Hortic.* 2004, *638*, 209–214.
14. Montemurro, F.; Tittarelli, F.; Lopedota, O.; Verrastro, V.; Diacono, M. Agronomic performance of experimental fertilisers on spinach (*Spinacia oleracea* L.) in organic farming. *Nutr. Cycl. Agroecosyst.* 2015, *102*, 227–241.
15. Montemurro, F.; Ciaccia, C.; Leogrande, R.; Ceglie, F.; Diacono, M. Suitability of different organic amendments from agro-industrial wastes in organic lettuce crops. *Nutr. Cycl. Agroecosyst.* 2015, *102*, 243–252.
16. Ciaccia, C.; Montemurro, F.; Campanelli, G.; Diacono, M.; Fiore, A.; Canali, S. Legume cover crop management and organic amendments application: Effects on organic zucchini performance and weed competition. *Sci. Hortic.* 2015, *185*, 48–58.
17. Ozkan, B.; Akcaoz, H.; Karadeniz, F. Energy requirement and economic analysis of citrus production in Turkey. *Energy Convers. Manag.* 2003, *45*, 1821–1830.
18. Jadidi, M.R.; Sabuni, M.S.; Homayounifar, M.; Mohammadi, A. Assessment of energy use pattern for tomato production in Iran: A case study from the Marand region. *Res. Agric. Eng.* 2012, *58*, 50–56.
19. Rasouli, M.; Namdari, M.; Mousavi-Avval, S.H. Modeling and analysis of energy efficiency in grape production of Iran. *J. Agric. Technol.* 2014, *10*, 517–532.
20. Gilmour, J.T.; Mauromostaukos, A.; Gale, P.M.; Norman, R.J. Kinetics of cropresidue decomposition: Variability among crops and years. *Soil Sci. Soc. Am. J.* 1998, *62*, 750–755.
21. Davis, S.A. Cover-crop roller-crimper contributes to weed management in no-till soybean. *Weed Sci.* 2010, *58*, 300–309.
22. Soil Survey Staff. Soil Survey Staff. Soil taxonomy. In *Agriculture Handbook 436*; USDA-NRCS: Washington, DC, USA, 1999.
23. Unesco-FAO. Etude écologique de la zone Méditerranéenne. Carte bioclimatique de la zone Méditerranéenne. Available online: http://unesdoc.unesco.org/images/0013/001372/137255fo.pdf (accessed on 5 April 2017).
24. Gomez, K.A.; Gomez, A. *Statistical Procedures for Agricultural Research*, 2nd ed.; John Wiley and Sons: New York, NY, USA, 1984.
25. Decree, L. Riordino e revisione della disciplina in materia di fertilizzanti, a norma dell'articolo 13 della legge 7 luglio 2009, n. 88. G.U. n. 121 26/05/2010. 26 May 2010.
26. Namdari, M.; Kangarshahi, A.A.; Amiri, N.A. Input-output energy analysis of citrus production in Mazandaran province of Iran. *Afr. J. Agric. Res.* 2011, *6*, 2558–2564.
27. Page, G. An Environmentally-Based Systems Approach to Sustainability Analysis of Organic Fruit Production Systems in New Zealand. Ph.D. Thesis, Massey University, Pamerston North, New Zealand, 2009.
28. Pimentel, D.; Pimentel, M. *Food, Energy and Society*; John Wiley and Sons: New York, NY, USA, 1979.
29. Mohammadi-Barsari, A.; Firouzi, S.; Aminpanah, H. Energy-use pattern and carbon footprint of rain-fed. *Inf. Process. Agric.* 2016, *3*, 69–75.
30. Khojastehpour, M.; Nikkhah, A.; Hashemabadi, D.A. Comparative study of energy use and greenhouse gas emissions of canola production. *Int. J. Agric. Manag.* 2015, *5*, 51–58.

CHAPTER 6

Applying New Technologies to Transform Blueberry Harvesting

Fumiomi Takeda [1,*], Wei Q. Yang [2], Changying Li [3], Andris Freivalds [4], Kiseok Sung [4], Rui Xu [3], Bo Hu [5], Jeffrey Williamson [6] and Steven Sargent [6]

[1] *United States Department of Agriculture—Agricultural Research Service, Appalachian Fruit Research Station, Kearneysville, WV 25430, USA*

[2] *North Willamette Research and Extension Center, Oregon State University, Aurora, OR 97002, USA*

[3] *College of Engineering, University of Georgia, Athens, GA 30602, USA*

[4] *Harold and Inge Marcus Department of Industrial and Manufacturing Engineering, The Pennsylvania State University, State College, PA 16802, USA*

[5] *Qingdao Blueberry Engineering Center, Jiaonan City, Qingdao 266400, China*

[6] *Horticultural Sciences Department, University of Florida, Gainesville, FL 32611, USA*

ABSTRACT

The growth of the blueberry industry in the past three decades has been remarkably robust. However, a labor shortage for hand harvesting, increasingly higher labor costs, and low harvest efficiencies are becoming bottlenecks for sustainable development of the fresh market blueberry production. In this study, we evaluated semi-mechanical harvesting systems consisting of a harvest-aid platform with soft fruit catching surfaces that collected the fruit detached by portable, hand-held, pneumatic shakers. The softer fruit catching surfaces were not glued to the hard sub-surfaces of the harvest-aid platform, but suspended over them. Also, the ergonomic aspect of operating powered harvesting equipment was determined. The pneumatic shakers removed 3.5 to 15 times more fruit (g/min) than by hand. Soft fruit catching surfaces reduced impact force and bruise damage. Fruit firmness was higher in fruit harvested by hand compared to that by pneumatic shakers in some cultivars. The bruise area was less than 8% in fruit harvested by hand and with semi-mechanical harvesting system. The percentage of blue, packable fruit harvested by pneumatic shakers comprised as much as 90% of the total, but less than that of hand-harvested fruit. The ergonomic analysis by electromyography showed that muscle strain in the back, shoulders, and forearms was low in workers operating the light-weight, pneumatic shakers that were tethered to the platform with a tool balancer. The new harvesting method can reduce the labor requirement to about 100 hour/hectare/year and help to mitigate the rising labor cost and shortage of workers for harvesting fresh-market quality blueberries.

Keywords: fruit detachment; harvest efficiency; blueberry; shaking equipment; mechanical harvesting; ergonomics; instrumented sphere; acceleration; impact damage; impact recording device

1. INTRODUCTION

The highbush blueberry (Vaccinium corymbosum hybrids) industry in the United States (U.S.), South America, China, and elsewhere has experienced a phenomenal growth in the past three decades [1]. Per capita consumption of blueberries during this period doubled as consumers have become more aware of health benefits of blueberries. In 2011, blueberry production in the U.S. was 231 MT, about 57% of world production. In the U.S. about 60% of production is packed for the fresh market, of which the majority is still hand harvested [2]. Outside the U.S., much of the blueberry is hand harvested and for fresh market distribution to North American, Asian, and European markets. By 2014, blueberry production across five continents had risen to 450 MT. By 2017, the global blueberry production is projected to approach 680 MT, with about half coming from North America [3].

As the blueberry production acreage continues to increase, growers foresee a shortage of labor for hand harvesting, increasingly high labor cost, and low harvest efficiencies becoming bottlenecks for sustainable development of the fresh market blueberry production. Hand harvesting blueberries is labor-intensive and requires up to 1200 h of labor per hectare annually [4].

In contrast, commercially available over-the-row (OTR) mechanical harvesters (Figure 1) can cut harvest labor to 25 h/hectare for blueberries destined for processing and reduce the cost of harvest from $2.80/kg by hand to $0.26/kg [4,5]. Although OTR harvesters can significantly cut harvest labor, they now cost as much as $240,000 per unit making them unaffordable for small- and medium-size blueberry farms.

Figure 1. A commercial over-the-row (OTR) berry harvester manufactured by Littau Harvester (Stayton, OR, USA). OTR berry harvesters are self-propelled with the driver positioned on the top platform. The frame is sufficiently sized to straddle blueberry plants that could be >2.5-m tall and >1.5-m wide. Inside the tunnel of the harvester frame, two vertically-oriented rotary, drum shakers are mounted. Closer to the ground, there are catcher plates and conveyor belts installed to collect and transfer detached fruit to the rear and some units to the top platform. Fruit then passes through a blower to separate the trash and debris (e.g., leaves, small berries, and stems) before dropping into a lug.

Research has shown that blueberries harvested by OTR machines have significantly reduced shelf life [4,6,7,8], and as much as 25% of fruit is lost to the ground, thus reducing the packable yield [8]. Also, blueberries harvested with conventional OTR harvesters exhibit excessive bruise damage and have lower fruit firmness (g/mm) than hand-harvested blueberries, and their quality deteriorates during >3 week postharvest cold storage required for marketing and for late-season distribution [4,7,9]. This is especially the case in production areas such as the Southeast U.S. (e.g., Florida and Georgia) where hot temperatures prevail during the harvest period, and certainly mechanical harvesters cannot be used to harvest blueberries in South America for marine transport for distant markets in Asia and North America which may take as long as 4 to 5 weeks before reaching the consumers.

Much of quality loss in fruit harvested by OTR machines comes from the direct contact between the machine-detached fruit and harvester's hard surfaces such as the plastic beater rods, plastic or fiberglass catcher plates when the detached fruit drop from the bush onto the catch pans and to the conveyor belt, and when the fruit drop from the conveyor belt into the fruit lugs. Yu et al. [10,11] and Xu et al. [12] measured the mechanical impacts using a miniaturized blueberry impact recording device (BIRD). These studies showed that the impacts created by rotary harvester generated not only smaller number but also lower magnitude than the impacts created by the sway and slapper harvesters. However, in all three harvesters, the real-time impact analysis revealed that the maximum impact value was recorded when the detached fruit and sensor contacted the catch plates. Also, a close-up video recording of the harvest operation showed that blueberry fruit was more likely to collide with the catch plates than with the tunnel wall [10,11].

Over the years, mechanical harvesters have been developed in attempts to harvest blueberries for the fresh market, but with a limited success [4,9,13]. Also, there has been research to determine the effect of modifying fruit catching surfaces on blueberry quality. The team led by Peterson [13,14] designed a blueberry harvester called the V45 blueberry harvester, which had two shaker drums that were angled 45° from horizontal and a prow at the front of the harvester. With this design, the fruit-laden canes were pushed away from plant's crown and the shaking tines engaged the plants from the top. An intermediate fruit catching surface located about 30 cm above the catcher plates on each side decreased the initial falling height for detached blueberries. The padding on the intermediate fruit catching surface reduced the mechanical impact force to the fruit landing on it. Also, ground loss was reduced because the bending of canes outward positioned much of fruit away from the crown [13]. With conventional OTR machines (Figure 1), in contrast, the canes are fed between two vertically oriented drums and fruit-laden canes are held upright and whorls of tines engage canes from the side. Detached fruit falls as much as 1.5 m through the vibrating tines and lands on catcher plates located at the bottom of the harvester. Brown et al. [4] and Xu et al. [12], have shown blueberry fruit that fall more than 60 cm onto plastic catch plates sustain extensive bruise damage. However, to date, harvesters that minimize drop height have not been commercially available. Only about six V45 blueberry harvesters were ever built because it was not widely accepted by growers due to (a) limited operating speed; (b) non-durable fruit catching surface on the V45 harvester; and (c) special cane training and pruning required to minimize plant damage [9].

Attempts have been made to harvest blueberries with semi-mechanical harvesting systems [15,16,17,18]. In these studies, hand-held, vibrating shakers (e.g., Model H1, BEI Inc., South Haven, MI, USA) and Hook Shaker (Campagnola SRL, Zola Predosa-Bologna, Italy) were used to detach blueberries from the plant and two portable fruit catching frames placed around two plants were used to collect the fruit (Figure 2). The frame is manually positioned against the plant and required additional handling to pour harvested fruit into lugs. The use of a netted fabric on the portable catch frame did result in mechanically detached fruit with little or no bruise damage [16]. However, the BEI's H1 vibrator weighs 2.4 kg and has a pistol grip attached to the shaker head. Holding the shaker by one hand made it difficult for a worker to use it continuously for a long period [15]. The hook shaker used in Chile [17] engaged one cane at a time limited harvest efficiency not much greater than the level achieved by hand harvesting. In a recent nation-wide grower survey conducted as part of the U.S. Department of Agriculture, National Institute of Agriculture and Food, Specialty Crop Research Initiative Project No. 2014-07897, over 300 blueberry growers responded that they would welcome a semi-mechanical harvesting system that would improve harvest efficiency, capacity, and fruit quality, and at the same time minimize ground loss and be less expensive than currently available commercial OTR blueberry harvesters.

It is critical to optimize the operating parameters for mechanization and hand-held shakers (e.g., vibration frequency, magnitude, angle at which the shaker is engaged with the blueberry bush) for increasing harvest efficiency and capacity, and to reduce labor cost. Shaker optimization can be achieved by better understanding how the vibrational force from the shaker is transmitted to the bush and its branches and fruit as well as to the operator. Such analysis will document the potentially excessive frequencies and high forces transmitted from harvesting equipment to the operator and whether the effect of worker's limb postures could result in work-related musculoskeletal disorders (WRMSDs) [19]. In agricultural work, WRMSDs can account for as much as 28% of the ~200,000 lost-time injuries on U.S. farms, whereby hand/wrist problems were a significant proportion of these injuries [20]. In addition to improving harvest efficiency compared to hand harvesting, semi-mechanical harvesting systems should reduce worker fatigue and stress level when the weight of the shaking apparatus is counter-balanced with a tool balancer and modifications are made to the shaker handle to reduce the transmission of vibration to operator's hand.

(a)

(b)

Figure 2. (**a**) A hand-held, pneumatic shaker (Golia model) engaging a blueberry plant with mature fruit. Two portable catch frames each with netted fabric were placed tightly against the plants for collecting detached blueberries; (**b**) The shaking heads of Golia olive shaker (Campagnola, Bologna, Italy) used in the study shown with short (12.5-cm long (left)) and long (23-cm long (right)) tines.

The objectives of this study were to (i) compare the quality and quantity of blueberries harvested by hand and with a pneumatically-powered olive shaker; (ii) develop better understanding of the interaction between machine parts and the fruit; (iii) develop a modified fruit-catching surface; and (iv) perform time-motion analysis and physiological status of workers operating hand-held blueberry harvesting equipment.

2. RESULTS

2.1. Shaker Performance and Fruit Quality

The pneumatic shakers removed as much as 6 times more fruit (g/min) than by hand in 'Draper' and 'Legacy' and nearly 16 times in 'Liberty' blueberries (Table 1). All samples lost weight (1% to 5%) during 4 weeks in cold storage, but the difference between harvest parameters for pneumatic shakers was small. Brix was slightly lower (<0.5 °Brix) and TA was higher (<0.4%) in fruit harvested with pneumatic shakers than by hand or OTR harvester. The operating pressure and tine length has little effect on fruit quality parameters, but more fruit (3× to 5×) was harvested with long tines on the shaker operated at 454 KPa compared to a shaker with short tines and operated at 310 KPa. Fruit removal efficiency was quite good at all shaker settings. More than 90% of fruit of 'Draper', 'Legacy' and 'Liberty' NHB blueberry detached by the Golia shaker were sorted as blue and acceptable for packaging.

Among the NHB blueberry cultivars harvested in Oregon in 2015, fruit firmness was higher in those harvested by hand compared to that by pneumatic shakers in 'Liberty', but lower in 'Draper' and 'Legacy' blueberry fruit (Table 2). The average bruised area and percentage of fruit with >19% bruised was less in 'Draper' than in 'Legacy' and 'Liberty'. There was a slight difference in bruising due to shaking parameters, however, the percentage of sliced surface area indicating bruise damage was not affected by harvesting methods. Compared to >25% bruise in blueberries harvested by the OTR harvester as previous studies have reported [4,7], pneumatic shakers and modified catching surface limited bruise damage to less than 8% (Table 2).

In 2016, we evaluated a semi-mechanical harvesting system consisting of a tractor-pulled harvest-aid platform with fruit catching surfaces (Figure 3), coupled with portable, hand-held, pneumatic shakers to detach fruit from several SHB and NHB blueberry cultivars (Table 3). Hitching the harvesting platform to a tractor eliminated the manual push-pull operation to move the platform down the plant row. The pneumatic shakers were operated with long and short nylon tines and at 310 and 454 KPa. Detached fruit was collected on suspended net and foam pad surfaces installed on the harvest-assist platform. Previous study on mechanical harvesting of SHB blueberries [7] showed that when 'Farthing' and 'Meadowlark' SHB plants were harvested with a commercial OTR harvester (Model 8000, Oxbo International, Lynden, WA, USA), marketable yield was significantly less and the percentage of fruit rated "soft" was significantly higher compared to hand harvested fruit.

Table 1. The effect of shaker parameters on fruit detachment of 'Draper', 'Legacy', and 'Liberty' northern highbush blueberry. [x]

Shaker Setting		Cultivar					
		Draper		Legacy		Liberty	
Pressure (psi)	Tine Length (cm)	Harvest Rate (kg/min)	Blue Fruit (%)	Harvest Rate (kg/min)	Blue Fruit (%)	Harvest Rate (kg/min)	Blue Fruit (%)
45	12.5	0.80 b [y]	81 b	2.47 a	94 b	1.09 ab	96 a
45	22.5	1.50 a	78 b	2.97 a	93 b	1.93 a	91 a
65	12.5	0.97 b	83 b	0.95 b	91 b	1.47 a	97 a
65	22.5	1.54 a	82 b	2.55 a	85 c	2.03 a	91 a
Hand harvest		0.25 c	89 a	0.48 c	98 a	0.13 b	92 a

[x] Blueberry plants were either shaken with hand-held shaker for 1 min or hand harvested for 5 min to calculate harvest rate (kg/min). Harvested fruit was sorted into blue (ripe) and red/green (immature) groups to calculate the percent of blue fruit detached by different harvesting methods. [y] Means within a column followed by the same letter are not significantly different at $p = 0.05$.

Table 2. Fruit firmness and bruise damage 'Draper', 'Legacy' and 'Liberty' northern highbush blueberries harvested by hand into a pail and with a hand held semi-mechanical shaker and fruit captured on a netted portable fruit catching frame. [x]

Cultivar	Harvest Method	Fruit Firmness (g/mm)	% Bruised Area
Draper	Hand	220 b [y]	3.0 a
	Shaker	239 a	1.5 a
Legacy	Hand	166 b	8.0 a
	Shaker	178 b	10.0 a
Liberty	Hand	174 a	- [z]
	Shaker	152 b	-

[x] Fruit firmness was determined with a FirmTech II instrument. Bruise assessment was performed24 h after harvest by visual means to quantify the bruised area in fruit sliced through the equator. [y] Means within a column followed by the same letter are not significantly different at $p = 0.05$. [z] No data collected.

Figure 3. Harvesting blueberries with a semi-mechanical harvest-assist system consisting of hand-held pneumatic shakers to detach fruit and platforms to collect detached fruit. The frame is manually pulled/pushed over plants and the fruit catching platforms are pulled out while positioning the frame over the plants to be harvested and pushed towards each other to seal around the base of blueberry plants. Shown here are two workers harvesting blueberries with hand-held shakers. The yellow hoses carry pressurized air from a compressor to shakers. Yellow fruit lugs are positioned below the catching surface. The inclined catching surface diverted detached fruit into lugs.

In a preliminary study on three SHB cultivars ('Chickadee', 'Flicker' and 'Springhigh') (Table 4) bruise assessment evaluated the suitability of these cultivars for mechanical harvesting. In general, the fruit of 'Chickadee' and 'Flicker' developed less bruise damage than 'Springhigh'. The fruit of the latter cultivar is described as "soft" by the University of Florida blueberry breeding program [21] using sparkleberry as a genetic source for machine harvest traits for southern highbush blueberry. In addition, the study provided information about the effectiveness of two suspended fruit collecting surfaces in reducing mechanical impact damage. Also, it showed that fruit dropped from greater heights exhibited more impact damage.

Table 3. The main effects of operating pressures of the hand-held fruit shaker and fruit catch surface materials installed on the harvest-assist platform on impact bruise in 'Draper' blueberry harvested by semi-mechanical harvesting system to detach fruit and modified fruit collecting surfaces in 2016 in Oregon.

Treatment	Mean Bruised Area (%)	Percent of Fruit with >10% Bruised [x]	Percent of Fruit with >20% Bruised [x]
Operating pressure (psi)			
0 (hand harvest)	0.5 c [y]	0.0 c	0.0 c
50	2.0 b	6.5 b	1.0 c
60	3.4 a	13.5 a	5.2 b
Catching surface material			
Hand harvest	0.5 ns	0.0 b	0.0 ns
Foam pad	2.6	9.5 b	2.5
Fabric net	2.8	10.5 a	3.8
$p > F$-value			
Operating pressure	0.006	0.005	0.004
Catching surface material	0.619	0.406	0.359
Pressure × Material	0.660	0.278	0.359

[x] Analysis was performed using ARCSIN (square root) transformed data; un-transformed lsmeans were presented in table. [y] Means within a column followed by the same letter are not significantly different at $p = 0.05$. ns = not significantly different at $p = 0.05$.

Table 4. Determination of suitability of southern highbush blueberry cultivars ('Chickadee', 'Flicker', and 'Springhigh') for mechanical harvesting system consisting of hand-held shaker and harvest-assist platform. [x]

Treatment	Average Bruise Rating (%)
Cultivar	
Chickadee	4.3 a [y]
Flicker	4.3 a
Springhigh	6.7 b
Catching surface	
None/Hand harvest control	2.2 a
Metal sheet	9.0 c
Suspended foam pad	3.7 b
Suspended woven fabric net	4.1 c
Drop height	
0 cm	2.2 a
60 cm	4.9 b
90 cm	6.3 c
$p > F$-value	
CV	0.0682
Surface	<0.0001
CV × Surface	0.5065
Height	0.0002
Height × Surface	<0.0001
CV × Height	0.3679
CV × Surface × Height	0.0019

[x] The fruit collection surfaces were metal sheet riveted to the frame, and unsupported foam pad and woven fabric net positioned 6 cm above the metal sheet. Hand harvested fruit was dropped from heights of 60 and 90 cm. [y] Means within a column followed by the same letter are not significantly different at $p = 0.05$.

Harvesting of SHB blueberries ('Chickadee', 'Flicker', and 'Kestrel') by a semi-mechanical harvesting system with modified fruit collecting surfaces limited bruise damage to about 8.0% or less in all three cultivars (Table 5, Table 6 and Table 7). In contrast, in the fruit collected on the metal surface the bruise damage ranged from about 11 to 17% (Figure 4). Also, the percentage of fruit with several bruise damage (e.g., >10% and >20% of cut surfaces indicating damage) was quite high among the fruit collected on the metal surface in all three cultivars. Percentage of severely bruised fruit was relatively low in 'Chickadee' and 'Kestrel' blueberry when the fruit was collected on modified fruit collecting surfaces. In the case of 'Flicker' blueberry, impact damage was high even for fruit collected on softer, modified fruit collecting surface materials.

The study, however, showed that pneumatic shakers were not as efficient as hand picking in limiting the detachment of immature fruit in SHB blueberries. The percentage of green and red fruit comprised as low as 13% in 'Chickadee', but 42% in 'Kestrel' blueberry. At the time of harvest, it was estimated that about 70% of fruit on 'Chickadee' and 'Flicker' plants were ripe and ready for harvesting while in 'Kestrel' the ripe fruit comprised only 40% of fruit still on the plant.

Table 5. The influence of catching surface on bruising of 'Chickadee' blueberries harvested with harvest aide in 2016.

Catching Surface	Average Bruise Area (%)	% Berries $\geq$ 10% Bruised Area	% Berries $\geq$ 20% Bruised Area
Hand harvest	1.2 a [x]	2.7 a	1.3 a
Metal sheet	10.9 b	41.3 b	24.0 b
Foam sheet	2.9 a	8.0 a	2.7 a
Net fabric	2.8 a	6.7 a	4.0 a
$p > F$-value			
Surface	<0.0001	0.0085	0.0222

[x] Means within a column followed by the same letter are not significantly different at $p = 0.05$.

Table 6. The influence of catching surface on bruising of 'Flicker' blueberries harvested with harvest aide in 2016.

Catching Surface	Average Bruise Area (%)	% Berries $\geq$ 10% Bruised Area	% Berries $\geq$ 20% Bruised Area
Hand harvest	1.3 a [x]	2.7 a	0.0 a
Metal sheet	17.5 c	69.7 c	33.3 c
Foam sheet	5.5 b	26.7 b	9.3 b
Net fabric	8.3 b	30.7 b	16.0 b
$p > F$-value			
Surface	<0.0001	0.0004	0.0018

[x] Means within a column followed by the same letter are not significantly different at $p = 0.05$.

Table 7. The influence of catching surface on bruising of 'Kestrel' blueberries harvested with harvest aid in 2016.

Catching Surface	Average Bruise Area (%)	% Berries ≥ 10% Bruised Area	% Berries ≥ 20% Bruised Area
Hand Harvest	1.3 a [x]	2.7 a	1.3 a
Metal Sheet	10.9 c	52.0 c	25.3 c
Foam sheet	3.3 ab	12.0 b	6.7 b
Fabric net	5.4 b	17.3 b	9.3 b
$p > F$-value			
Surface	<0.0001	0.0008	0.0042

[x] Means within a column followed by the same letter are not significantly different at $p = 0.05$.

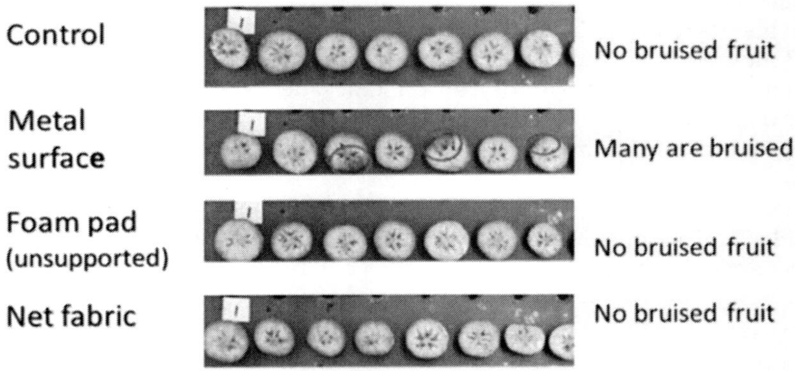

Figure 4. The effect of fruit catching surface on extent of bruise damage in the fruit of southern highbush blueberry 'Flicker'. Fruit was harvested by hand (Control) and semi-mechanical harvesting system using a harvest-assist platform. Hand harvested fruit was dropped into a 5-L plastic pail. With the semi-mechanical system, fruit was detached with a hand-held 'Golia' olive shaker and detached fruit landed on platform's metal surface or onto a pad or net fabric suspended over the metal surface. A fruit damage by impact was initially indicated by water soaked tissues and then after about 1 h the damaged area turned darker (circled in red).

2.2. Modification of Fruit Catching Surface

Modifying the fruit collection surfaces using suspended foam sheet and woven net fabric (Figure 5) significantly reduced impact (Figure 6). BIRD sensor recorded impact values >200 g when it dropped even from a height of less than 30 cm. Glued-on foam sheet over the metal surface reduced impact slightly, but as the drop height increased to nearly 45 cm, the impact value reached 200 g. A significant reduction in impact was achieved by the two modified fruit collecting surfaces used in the study. The suspended foam sheet maintained impact value to <200 even when the drop height was increased to 1.2 m. The lowest impact values were recorded on the suspended woven net fabric.

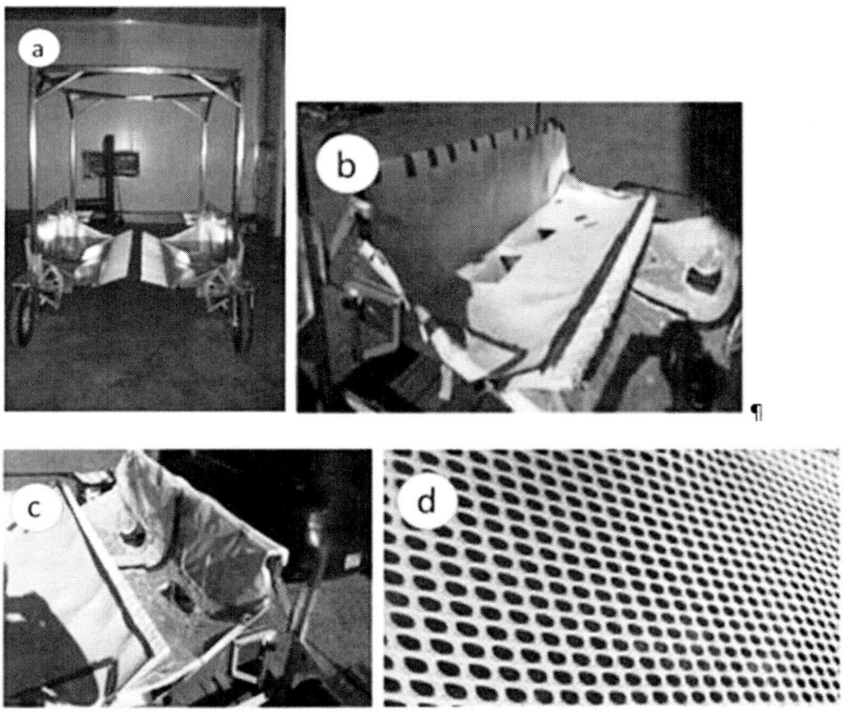

Figure 5. Harvest-assist platform for semi-mechanical harvesting of blueberries. (a) A harvest-assist platform (BBC Technologies, Fresno, CA, USA) shown with left-side and right-side stainless steel fruit catching surface and brush bristle in the middle to seal around individual blueberry plants; (b) A close-up of the modified fruit catching surface constructed with foam pad installed about 3 cm over the stainless steel surface; (c) Shown is a woven net fabric suspended over the other catching surface; (d) A close of the netted woven fabric installed on one side of the harvest-assist platform.

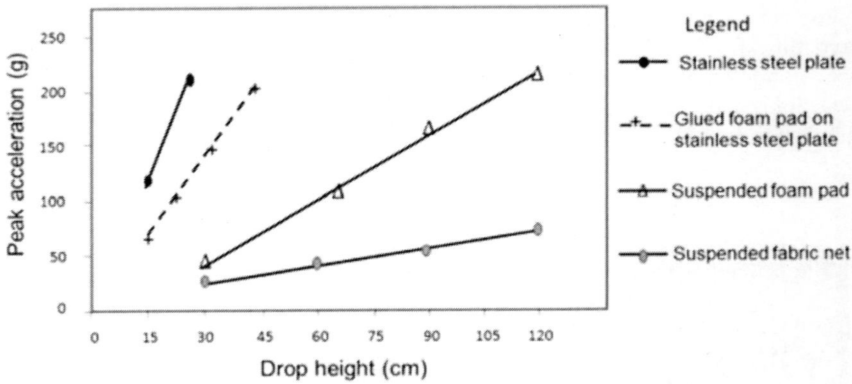

Figure 6. The relationship between impact levels and contact surface materials. The BIRD sensor was dropped on four different contact surface materials installed on the harvest-assist platform: Stainless steel plate, foam pad glued onto the stainless steel plate, unsupported foam pad installed over the stainless steel plate, and woven net fabric suspended over the stainless steel plate. Data points are the means of 10 recordings by the BIRD sensor for each drop height and only the first impact of each drop was collected for further data analysis. The BIRD sensor recorded impacts in three orthogonal directions. The largest acceleration value from each of the x, y, and z accelerometers captured during the first impact was summed to calculate the peak acceleration. Peak acceleration reflects the magnitude of the impact.

2.3. Vibration to Blueberry

Amplitude analysis of vibration transmitted from the shaking equipment (e.g., Golia olive shaker) through the limb to BIRD sensors attached at three different distances was performed (Figure 7). The findings indicated that the vibration amplitude was highest near the shake site and decreased gradually. The amplitude at the top section of the limb was only 30% of that recorded at the bottom section which was closest to the engagement point. Higher amplitude of vibration was recorded with increasing operating pressure for the shaker. It was also observed that when a large diameter limb was shaken much lower levels of vibration were recorded at all three locations along the limb.

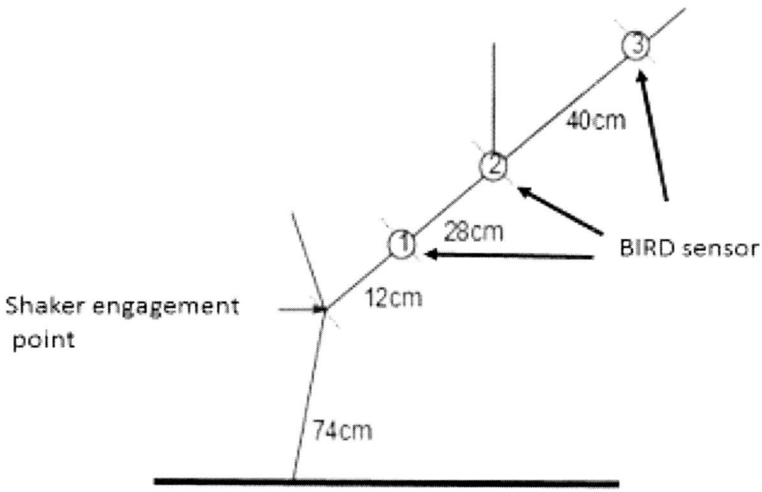

Figure 7. A drawing of the blueberry cane denoting the locations of BIRD sensor attachments and where the pneumatic shaker engaged the plant.

2.4. Ergonomic Study

Ergonomic evaluation by electromyography (EMG) was conducted on workers operating the portable, hand-held blueberry harvesting equipment (Figure 8). In this study, muscle activity (μV) was recorded with electrode sensors attached at seven upper body regions locations (left and right forearms, biceps, shoulders, and lower back (Figure 9). EMG results showed that the forearm and bicep muscles experienced the least amount of strain regardless of the operating posture (e.g., measured μV output of <60). The only exception was with the BEI H1 berry stripper held high above the shoulders. The high posture activated the forearm muscles, especially with the heavier BEI shaker. The overhead posture activated the lateral deltoid muscles as well with the heavier BEI shaker generating greater EMG than lighter shakers. The highest muscle activity was recorded in the lower back region. When the shakers were operated in a bent position, EMG registered well above 1600 μV compared to about 1000 μV operating the shaker in standing, high, and overhead operating postures.

In another study, both light-weight and heavier shaking equipment were tethered to the harvesting platform with a tool balancer which showed that the heavy equipment with a support balancer required the minimum effort and it had less workload then the untethered light-weight shaking equipment.

Figure 8. Ergonomic evaluation of portable hand-held shaker used for harvesting blueberries. Muscle activity was measured using electromyography. The electrode sensors were attached on multiple body regions: Forearm, upper arm, shoulder, and lower back. The shaker was operated with the operator in bent (a), standing (b), arched backward (c), and standing (d) postures.

EMG data were used to perform rapid upper limb assessment (RULA) for operating postures and weight of shaking equipment. RULA assists in determining the potential risk for work-related musculoskeletal disorders (WRMSDs) for operating hand-held shakers for harvesting blueberries. A RULA score of 5 to 6 indicates that design changes are required and a score from 3 to 4 warrants further investigation. A score of >6 indicates that the operating posture and equipment usage should discontinue and strongly suggests that necessary modification to the equipment and/or operating technique is a prerequisite to reducing unacceptable level of physiological and biochemical workload imposed on the operator. The results of RULA analysis showed that when light-weight shaker was operated without tethering it to the platform the RULA score for both upper limbs was 6. By tethering the light- and heavy-weight shakers to the platform with a tool balancer, the RULA score dropped to 4 or less.

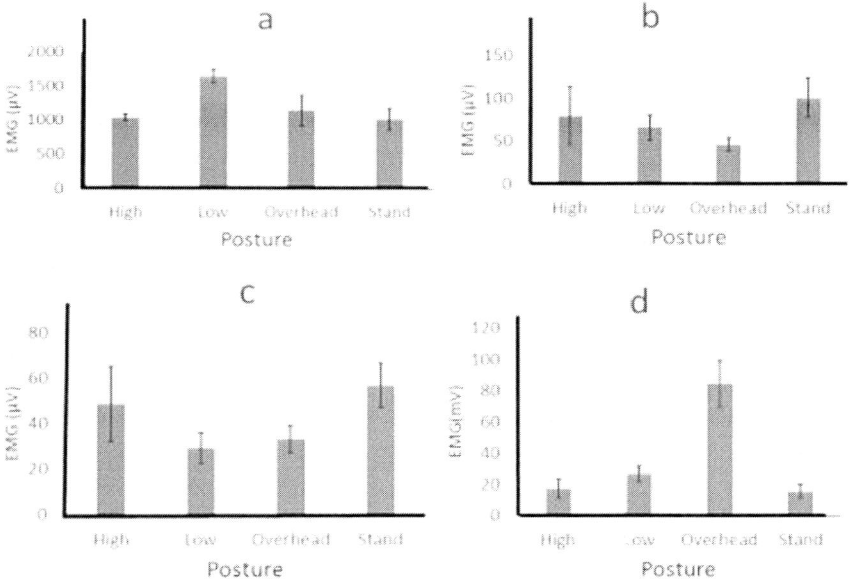

Figure 9. Muscle activity measured by electromyography as EMG (µV) was recorded by electrode sensors attached to the lower back (**a**), bicep (**b**), forearm (**c**), and shoulder (lower deltoid) (**d**) while human subjects operated a hand-held blueberry shaker in four postures: High (arched back with shaker head above shoulder height), Low (bent at waist with shaker head below waist), Overhead (straight back with both arms raised), and Standing (straight back and shaker at waist height). The vertical lines at the top of the bars represent the standard error of the mean.

Hand-held blueberry harvesting equipment is operated with a quick push-pull motion with the shaking head engaging the blueberry plant in about a 1-s pulsing action. From the RULA postural analysis, lower score was obtained when using a heavy shaker that was tethered to the platform, no difference was observed between pushing and pulling motions. Vibration data from BIRD II sensor attached to the handle showed that the heavy shaker, if tethered to the harvesting platform, transmitted less vibration to operator's hands than untethered light-weight shaker. A simple shocking absorbing pad wrapped on the handle reduced the vibration on the hand from the shaking head was transmitted through the handle to operator's left and right hands.

3. DISCUSSION

3.1. Mechanical Harvesting System

According to Brown et al. [4], 78% of the blueberry harvested by commercial mechanical harvesters had severe bruise damage compared to only 23% in hand harvested fruit. The results of this study clearly showed that harvesting blueberries with hand-held shakers and capturing detached fruit on softer surfaces was gentler than OTR harvesters and resulted in less bruise damage. Still, more technological improvements are needed to increase harvest capacity, reduce bruising, limit percent ground loss of fruit, and reduce detachment of immature (green or red) fruit. In addition to these key technical limitations, OTR mechanical harvesters (~$140,000 to $240,000) are generally not affordable to small- and medium-sized growers. In some of the major blueberry producing states such as Oregon [22], Florida [23], and North Carolina [24], the average blueberry farm is less than 20 hectares in size.

As an alternative to hand harvesting, portable hand-held shaking devices (e.g., H model, BEI Inc., South Haven, MI, USA [24] and pneumatic limb shaker (Campagnola SRL, Bologna, Italy) have been used to harvest fresh market quality fruit onto a small, manually positioned catch frame [15,16,17]. However, the hand-held shaker such as the H1 berry stripper which weighs ~2.4 kg has two major problems that limit its adoption: low harvest capacity and potential worker fatigue and injury. The H-model hand-held shaker is powered by electromechanical or pneumatic system and is quite heavy. The workers must position the catch frame under the plant, move the battery or air compressor, shake the fruit from the plant, and then handle the fruit collection box or have other workers maneuver the catch frame and collection boxes, and then move the battery or compressor. Field observations indicated that the workers are actively engaged in an assigned task of fruit harvesting less than half of the time. Thus, the pace of fruit harvesting is slow and tedious when a portable fruit catching frame is used. In addition, the workers can easily get fatigued by operating the handheld shaker because of vibration and the weight of the shaker and other laborious operations such as constantly moving the catch frame and handing empty and filled lugs. It is well documented that any repetitive job having potentially excessive frequencies, high forces and poor limb postures resulted in work-related musculoskeletal disorders (WRMSDs) [19]. In agricultural work, WRMSDs have accounted for over 60,000 lost-time injuries on U.S. farms. Hand and wrist problems were a significant proportion of these injuries [20]. Due to these limitations, the handheld shaker devices such as the BEI H-Model have not been widely adopted among blueberry growers. For instance, a recent survey of the blueberry industry in Oregon showed that only 10% of growers had used handheld shakers and many expressed that their experiences with them was not favorable. Harvest inefficiency and fatigue after prolong use of hand-held shakers contributed to their decision to discontinue the use of portable, hand-held shakers.

The new harvesting method can potentially reduce labor requirement to less than 100 hour/hectare/year and help to mitigate the rising labor cost and shortage of workers for harvesting currently faced by the blueberry industry. The semi-mechanical system studied here harvested high quality blueberries with little or no damage, and the fruit was acceptable for the fresh-fruit market even after several weeks in cold storage. The semi-mechanical harvesting platform (Figure 5) is expected to cost about $11,000 for a tractor-pulled unit (BBC Technologies, Fresno, CA, USA) and up to $60,000 for a self-propelled unit with powered fruit conveyance system (GK Machine, Donald, OR, USA). The system is scale neutral and its performance will be especially appealing to small- and medium-size growers.

3.2. Designing Mechanical Harvesters to Reduce Impact Damage

In terms of mechanical impact to blueberry fruit, research has shown that softening and bruise damage in machine-harvested fruit can be reduced if the drop height can be maintained at less than 30 cm [4]. The V45 blueberry harvester (BEI; South Haven, MI, USA) had design features to minimize drop height, and reduce ground loss [13]. However, the V45 blueberry harvester has not been widely accepted commercially and by 2005 all V45 blueberry harvesters sold about 10 years earlier were retro-fitted with two vertical rotary or slapper fruit shakers [9]. Other problems with mechanical harvesting of blueberries have stimulated innovations in cultural practices for improving machine harvest efficiency. These practices are focused primarily on minimizing ground loss and reducing the number of mature fruit missed by the machine [9,13,21]. Strik and Buller [25] in Oregon showed that supporting 'Bluecrop', an NHB blueberry, with a simple two-wire trellis system increased harvest efficiency significantly, mainly by reducing the number of mature fruit missed by the machine. The maximum proportion of fruit remaining after the last machine harvest was reduced 30.8% in non-trellised plots to 15.5% in trellised plots with canes kept more upright by trellis wires.

Blueberry growers are interested in using mechanical harvesting equipment for harvesting blueberries destined for fresh market because of financial and labor concerns [22,23,24]. Breeders are considering bush architecture, ease of detachment of mature fruit but not immature fruit, loose fruit clusters, small, dry-stem scar, firm fruit, and a concentrated ripening period to develop cultivars suitable for machine harvest [21]. In this study, the research focused on machine harvesting commercially grown SHB and NHB blueberry cultivars. Hand-held, powered olive shakers and modified collecting surface materials (Figure 3 and Figure 5) were used together with a harvest-assist platform in an attempt to increase harvest capacity, fruit removal efficiency, and maintaining fruit quality. Several blueberry cultivars were identified to be suitable for harvesting by this mechanically-assisted fruit picking system. Based on high bruise potential and low harvest efficiency, 'Springhigh' and 'Kestrel' SHB blueberries, respectively, were not suitable for machine harvest. Among the NHB blueberries reported here 'Draper' and 'Legacy' appear to be good candidates for further investigation.

The harvesting system used in this study appears to be a promising platform for further mechanization effort. An international blueberry harvesting equipment manufacturer has incorporated a number of harvesting system components developed in this study into a self-propelled ORT harvesting platform (e.g., Model 7420, Oxbo International, Lynden, WA, USA). This harvesting equipment will be used in 2017 and 2018 season in major blueberry producing states (e.g., FL, CA, OR, and WA, USA). Additional design changes in the harvesting system, such as an improved fruit decelerating, intermediate fruit catching surface, are being incorporated into the Model 7420 blueberry harvester (Figure 10). On the modified OTR harvester, the hard catch plates have been replaced with patent-pending plates with soft surfaces to reduce impact damage. Additional collection/deflection surfaces have been mounted 30 cm above the catch plates to decrease drop height and prevent detached fruit from directly falling >1 m onto catch plates and conveyance belts located below. The powered conveyance system has eliminated the need to manually handle harvested fruit, as is the case with current harvest-assist platforms.

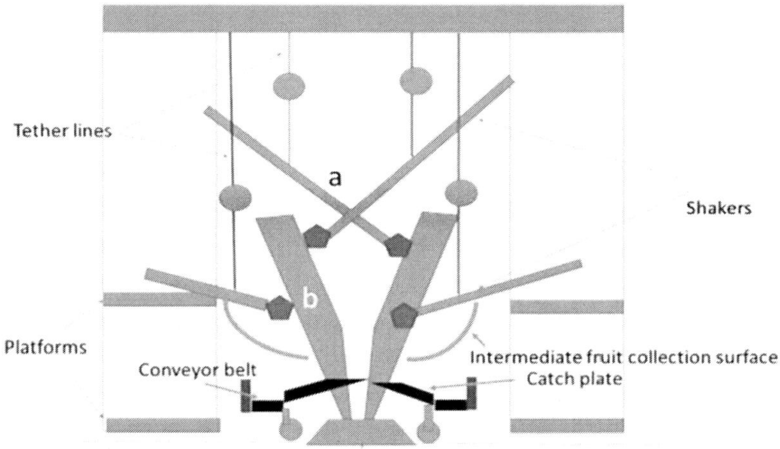

Figure 10. A drawing of the proposed mobile, semi-mechanical blueberry harvesting platform with hand-held shaker (a) tethered to the platform. The harvesting platform will have two levels on each side to position shaker operators. The lower level for the operator to engage the lower section of the blueberry plant (b) and the upper level for engaging the upper portion of the plant. Stool seats (not shown) will be installed on the platforms for workers to set and operate the shakers that are tethered to the harvesting platform with a tool balancer. The detached fruit is collected on the intermediate fruit collecting surface and soft catch plates. Decelerated fruit will roll onto the conveyor belt and transferred to fruit lugs.

3.3. Needs to Reduce Work-Related Musculoskeletal Disorders

Ergonomic analysis suggested that more modifications are required on harvesting equipment, and also how workers use hand-held equipment to reduce potential work-related musculoskeletal disorders. A self-propelled, mobile, harvest-assist platform is envisioned for future studies(Figure 10). The platform will have two levels for operators to work the shaking equipment. The worker positioned on the upper level can engage the upper half of the plant canopy while the worker at the lower level engages the lower half of plant canopy. This worker configuration would reduce the need to raise the shaker above the shoulders (Figure 8). A Stool seat can also be installed on each platform allowing workers to set and be positioned lower in relation to the plant and minimize the need to bend. Properly positioning the worker relative to the blueberry plant could contribute to relieving muscle strain prevent and potentially reduce WRMSDs in workers operating blueberry harvesting equipment. For continuous operation of harvesting equipment, it appears that having the harvesting equipment tethered to the platform with a tool balancer is an important consideration.

4. MATERIALS AND METHODS

4.1. 2015 Study Site and Harvesting System Descriptions

In 2015 summer, a preliminary blueberry harvesting study was conducted in Oregon (OR) using mature plants of northern highbush (NHB) blueberry plants ('Draper', 'Legacy', and 'Liberty'), most widely grown cultivars in the Pacific Northwest. For semi-mechanical harvesting method, a pneumatic (air-powered) Campagnola 'Golia' olive shaker (Brewt Power Systems, Merced, CA, USA) was used to detach fruit (Figure 2). Two different tines (10-cm-long and 20-cm-long) were installed into the head of a Golia shaker. The shaker was operated at 45 psi (310 KPa) or 65 psi (454 KPa). Two portable frames constructed of metal pipes and lined with netting (Haven Harvesters, South Haven, MI, USA) were placed around the base of the plants for collecting the detached fruit (Figure 2). Fruit collected on the net were manually poured into lugs. Semi-mechanical shaking time ranged from 1 to 6 min to harvest each of 4 plants for a given treatment. Hand-harvested plots were manually hand harvested for 5 min. Hand-harvested fruit were gently dropped into a 4-L plastic pail. In both hand and semi-mechanical harvesting methods, the harvested fruit from four replicate plots were weighed and sorted into blue, and red/green categories (Figure 11). The fruit in each category were weighed to calculate the percentage of green, red, and blue fruit. For comparison of fruit quality with those harvested with an OTR harvester, 'Draper' fruit harvested by a rotary harvester (Littau Harvester, Stayton, OR, USA) were included in the analysis.

Twenty-five blue-colored fruit per plot were used to determine fruit firmness using aFirmTech II (BioWorks, Inc., Wamego, KS, USA), and frozen for later analysis of total soluble solids (°Brix) and total titratable acidity (TAA) determinations. Blue fruit for cold storage were poured into clamshells and placed in a cold room maintained at 1 °C for 6 weeks. Fruit damage and quality evaluation was performed 1 day after harvest and after 6-week cold storage, plus 1 day at room temperature using the methods previously described for fruit evaluation [4,7,11,12]. Fruit were sliced through the equator with a razor blade and the bruised area was estimated from cut surface area indicated by tissue discoloration (Figure 4).

Figure 11. The fruit harvested by the experimental semi-mechanical harvesting system prior to sorting. Fruit consisted mostly of blue-colored fruit and some green-colored (shown circled in green) and red-colored (shown circled in red) fruit. The latter two categories are culled manually and by electronic sorting machines on the packing line and only the blue-colored fruit are packaged. Occasionally, machine-harvested fruit have over-ripe and smashed berries (not shown).

4.2. 2016 Study Sites and Harvesting System Descriptions

In 2016, blueberry harvesting studies were conducted in Florida (FL) and Oregon (OR). In FL, three southern highbush (SHB) blueberry cultivars (Chickadee, Flicker, and Kestrel), recently released blueberries from the University of Florida blueberry breeding program were either hand harvested or harvested with a semi-mechanical harvesting system (Figure 3 and Figure 9). In Oregon, the NHB cultivar Draper was harvested by hand or with a semi-mechanical harvesting system (Figure 9). The pneumatic olive shaker was operated at either 50 or 60 psi and the platform was set up with an unsupported foam sheet on one side and with a net fabric suspended over the platform frame on the other side (Figure 9). Harvested fruit was weighed and sub-samples were taken from each harvested plot and sorted into green-, red-, and blue-colored fruit. Fifty blue fruit from each plot were held at room temperature for 24 h and then tested for firmness using a FirmTech II instrument and sliced through the equator to assess for bruise damage (Figure 4). Bruise assessment consisted of slicing the fruit along the equator and photographing the sliced fruit. The images were used to estimate the discolored area (% of cut surface area with bruise damage) by two evaluators [4,7,11,12].

4.3. Impact Measurements, Blueberry Drop Tests and Bruise Evaluation

The mechanical impacts generated by fruit catching surfaces were measured with the BIRD (berry impact recording device) sensor [10,11,12]. The BIRD sensor is a microcontroller-based data logger and measures mechanical impacts using a tri-axial accelerometer (Figure 12). The sensor has been miniaturized to a size and weight approaching a very large blueberry fruit, and it has the sufficient sensing range and precision to measure the impacts generated on the packing lines and by harvesting equipment. It has an internal memory and power supply, enabling it to be operated independently in the field to quantify mechanical impacts. The BIRD sensor records impacts in three orthogonal directions with sensing range of ± 346 g (gravitational acceleration, g = 9.8 m/s^2) in the vector summation and maximum 2 kHz sampling frequency. One mechanical impact recorded by the BIRS sensor resembles a bell curve shape. The maximum acceleration value of the impact is considered the peak acceleration [10,11,12]. PeakG which is the largest summation acceleration in one impact curve and directly reflects the magnitude of the impact [12].

The drop test consisted of dropping the BIRD sensor 7 times onto the metal surface of the harvest-assist platform (BBC Technologies, Fresno, CA, USA) and two suspended fruit catching surfaces (Figure 8), consisting of 6.35 mm thick 'No Bruze' foam sheet (PORON Urethane Foam, Rogers Corporation, Woodstock, CT, USA) or 0.8-mm-thick woven net fabric (Haven Harvester, South Haven, MI, USA) from heights of 15, 30, 60 and 90 cm. Four replicates of 50 hand-harvested berry samples of 'Chickadee', 'Draper', 'Flicker', and Kestrel' were dropped individually onto the metal surface and suspended surfaces. Dropped fruit was placed in a clamshell and held at 21 °C. For each genotype, additional 50 fruits were transferred into a clamshell and served as control samples. After 24 h, both the dropped and control berries were sliced through the equator and photographed. Afterwards, two evaluators read the images of the sliced fruits along the equator for surface area of discolored zones and estimated the discolored area (% of cut surface area with bruise damage), following the protocol described by Brown et al. [4].

4.4. Shaker Tests

In a field test, energy transfer from the shaker to various parts of blueberry plants was determined with the BIRD sensor. One tall fruit-bearing cane was selected on 4 mature blueberry plants. BIRD sensors were attached to the cane at 3 positions (bottom, middle, and top) with electrical tape (Figure 5). The canes were shaken one at a time with a Golia olive shaker and hook shaker that engaged the cane just below the location of the bottom sensor at 50 psi (344 KPa), 80 psi (551 KPa), and 110 psi (758 KPa). After recording the acceleration the maximum acceleration of each pulse was calculated and the average and standard deviation of the maximum acceleration was tabulated.

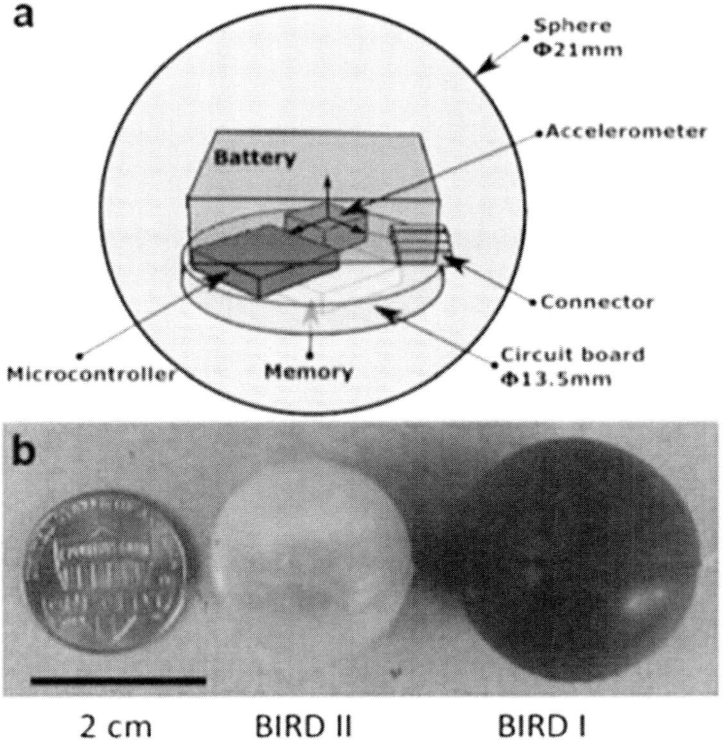

Figure 12. The blueberry impact recording device (BIRD) sensor (a). The electronic circuit board was cast in silicon rubber. The BIRD II sensor (b) used in this study is a an instrumented sphere with a diameter of 22 mm and mass of 6 grams and housing a microcontroller-based data logger which measures mechanical impacts in three orthogonal directions with a tri-axial accelerometer with sensing range of ± 346 *g* (gravitational acceleration, g = 9.8 m/s^2) in the vector summation and maximum of 2 KHz sampling frequency. The sensor was fully calibrated using a centrifuge and an impact table. The accuracy of the sensor was between 0.50% and 0.63% and the precision error was within 0.06% to 0.23%. For size comparison, the BIRD II sensor placed alongside a US penny coin (left) and the original BIRD I sensor (right).

4.5. Ergonomic Studies

Ergonomic analyses used in this study were previously utilized in analyzing appleharvesting [26,27,28]. An independent comparison of hand vs. semi-mechanical harvesting was performed utilizing a number of physiological approaches: electromyograms (EMG), rapid upper limb assessment (RULA) was conducted for estimating the postural risks of work-related upper limb injury [29]. EMG is a measure of the electrical activity of a muscle that is directly proportional to the force being produced by the muscle [30]. The EMG signal from four specific muscles was recorded with a Flex-Comp Infiniti physiological recording system (NexGen Ergonomics): trapezius and latissimus dorsi for shoulder strain and the flexor digitorum profundus and superficialis for forearm strain. Heart rate was recorded with a Polar Heart Rate Monitor and checked to see if it was within reasonable levels according to the Brouha criteria and monitor heart rate creep which indicated fatigue [31]. Again, these measurements were taken on several workers in a laboratory setting and at two field sites. Electrode sensors were attached on multiple body regions: Forearms, upper arms, shoulders, and lower back. The shaker was operated with the operator in bent with hands holding the shaker below waist, standing with shaker held at waist height, high with hands at shoulder height, and overhead postures with hands over the shoulder height and lifting the shaker (Figure 11). The shakers used in the ergonomic analysis included the Campagnola 'Golia' olive shaker with long or short tines (Figure 2), attached to a long handle that weighed 2 kg. The BEI H1 berry stripper was attached to a long handle and weighed 4 kg as shown in Figure 6. Also, a Campagnola hook shaker on a long pole [17] that weighed 1.5 kg (not shown) has been evaluated in Chile for harvesting blueberries.

4.6. Statistical Analysis

Data were analyzed by using SAS General Linear Model analysis of variance (ANOVA) (SAS Institute, Cary, NC, USA) procedure to summarize the data. When differences were detected at $p \leq 0.05$, the data were subjected to mean separation by differences of least-square means using SAS Proc Mixed (SAS v. 9.1, SAS Institute, Cary, NC, USA).

5. CONCLUSIONS

The semi-mechanical, platform-based harvest aid system performed reasonably well in terms of harvested amount, detachment selectivity, and quality of fruit after harvest and storage. In some cultivars, the fruit harvested by semi-mechanical harvesting system was as good as that harvested by hand and maintained good quality in cold storage. The semi-mechanical harvesting system offers the following benefits: (1) Improved harvest efficiency compared with hand-harvest. The workers do not need to position the catch frame under the plant, move the air compressor, shake the fruit off the plant, and then handle the fruit collection box. Because of the elimination of these steps, harvest efficiency is expected to be improved 10–20 times compared with hand-harvest; (2) Reduced fatigue because the workers do not need to support the full weight of the shaker all the time. The study showed that a number of refinements in shaker design and operating protocol are required for maintaining the well-being of workers who will operate hand-held harvesting equipment for hours, instead of minutes, on the mobile, harvesting platform.

FUTURE PERSPECTIVES

A harvesting system consisting of portable pneumatically–powered portable shakers and collecting the detached fruit on soft fruit catching surfaces and conveyance belts mounted on a mobile platform is promising.

ACKNOWLEDGMENTS

This work was financially supported by the U.S. Department of Agriculture, National Institute of Food and Agriculture, Specialty Crop Research Initiative research program (Award No.: 2014-51181-22383). The authors would like to acknowledge Scott Wolford, Renee Allen-Holland, Ann Rose, Jedediah Smith, Heather Andrews, Eric Ostmark, Eunsik Kim, and Adrian Berry for assistance with equipment fabrication, ergonomic analysis, field and laboratory work, and fruit quality assessment.

REFERENCES

1. Statistics Division FAO of the United Nations. Crops (Production). 2015. Available online: http://www.fao.org/faostat/en/#data/QC/visualize (accessed on 15 January 2017).
2. Eklund, B. 2014 Blueberry Statistics; National Agricultural Statistics Service, United States Department of Agriculture: Washington, DC, USA, 2016; p. 9. Available online: http://www.nass.usda.gov/Statistics_by_Subject/result.php?B31557DC-DDD9–3C5D-9FA2-C7F176ADE896 sector = CROPS group = FRUIT (accessed on 15 January 2017).
3. Brazelton, C. 2014 World Blueberry Statistics and Global Market Analysis; U.S. Highbush Blueberry Council: Folsom, CA, USA, 2015; p. 105.

4. Brown, G.K.; Schulte, N.L.; Timm, E.J.; Beaudry, R.M.; Peterson, D.L.; Hancock, J.F.; Takeda, F. Estimates of mechanization effects on fresh blueberry quality. Appl. Eng. Agric. 1996, 12, 21–26.
5. Dale, A.; Hanson, E.J.; Yardborough, D.E.; McNicol, R.J.; Stang, E.J.; Brennan, R.; Morris, J.R.; Hergert, G.B. Mechanical harvesting of berry crop. In Horticultural Reviews; Janick, J., Ed.; AVI Publishing: Westport, CT, USA, 1996; Volume 16, pp. 255–382.
6. Mehra, L.K.; MacLean, D.D.; Savelle, A.T.; Scherm, H. Postharvest disease development on southern highbush blueberry fruit in relation to berry flesh type and harvest method. Plant Dis. 2013, 97, 213–221.
7. Takeda, F.; Krewer, G.; Li, C.; MacLean, D.; Olmstead, J.W. Techniques for increasing machine-harvest efficiency in blueberries. HortTechnology 2013, 23, 430–436.
8. Van Dalfsen, K.B.; Gaye, M.M. Yield from hand and mechanical harvesting of highbush blueberries in British Columbia. Appl. Eng. Agric. 1999, 15, 393–398.
9. Takeda, F.; Krewer, G.; Andrews, E.L.; Mullinix, B.; Peterson, D.L. Assessment of the V45 blueberry harvester on rabbiteye and southern highbush blueberries pruned to V-shaped canopy. HortTechnology 2008, 18, 130–138.
10. Yu, P.; Li, C.; Takeda, F.; Krewer, G.; Rains, G.; Hamrita, T. Measurement of mechanical impacts created by rotary, slapper, and sway blueberry mechanical harvesters. Comput. Electron. Agric. 2014, 101, 84–92.
11. Yu, P.; Li, C.; Takeda, F.; Krewer, G. Visual bruise assessment and analysis of mechanical impact measurement in southern highbush blueberries. Appl. Eng. Agric. 2014, 30, 29–37.
12. Xu, R.; Takeda, F.; Krewer, G.; Li, C. Measure of mechanical impacts in commercial blueberry packing lines and potential damage to blueberry fruit. Postharvest Biol. Technol. 2015, 110, 103–113.
13. Peterson, D.L.; Wolford, S.D.; Timm, E.J.; Takeda, F. Fresh market quality blueberry harvester. Trans. Am. Soc. Agric. Eng. 1997, 40, 535–540.
14. Peterson, D.L.; Brown, G.K. Mechanical harvester for fresh market quality blueberries. Trans. Am. Soc. Agric. Eng. 1996, 39, 823–827.
15. Casamali, B.; Williamson, J.G.; Kovaleski, A.P.; Sargent, S.A.; Darnell, R.L. Mechanical harvesting and postharvest storage of two southern highbush blueberry cultivars grafted onto Vaccinium arboretum rootstocks. HortScience 2016, 51, 1–8.
16. Hu, B.; Yang, W.Q.; Andrews, H.; Li, C.; Takeda, F. Towards a Semi-Mechanical Harvesting Platform System for Harvesting Blueberries with Fresh-Market Quality. Acta Hortic. (in press).
17. Lobos, G.; Moggia, C.; Retamales, J.B.; Sanchez, C. Effect of mechanized (self-propelled or shaker) vs. hand harvest on fruit quality of blueberries (Vaccinium corymbosum L.) in postharvest. Acta Hortic. 2014, 1017, 141–145.
18. McKibben, H.E.; Jones, P.F. Oscillating Blueberry Stripper. U.S. Patent No. 3,023,565, 6 March 1962.
19. National Research Council and Institute of Medicine. Musculoskeletal Disorders and the Workplace: Low Back and Upper Extremities; National Academy Press: Washington, DC, USA, 2001; p. 492.
20. Gomez, M.I.; Hwang, S.; Stark, A.D.; May, J.J.; Hallman, E.M.; Pantea, C.I. An analysis of self-reported joint pain among New York farmers. J. Agric. Saf. Health 2003, 9, 143–157.
21. Olmstead, J.W.; Armenta, H.P.R.; Lyrene, P.M. Using sparkleberry as a genetic source for machine harvest traits for southern highbush blueberry. HortTechnology 2013, 23, 419–424.
22. Yang, W.Q.; Oregon State University, Aurora, OR, USA. Personal communication, 2014.
23. Williamson, J.; University of Florida, Gainesville, FL, USA. Personal communication, 2014.
24. Mainland, C.M.; North Carolina State University, Castle Hayne, NC, USA. Personal communication, 2014.
25. Strik, B.; Buller, G. Improving yield and machine harvest efficiency of 'Bluecrop' through high density planting and trellising. Acta Hortic. 2002, 574, 227–231.
26. Earle-Richardson, G.; Jenkins, P.L.; Freivalds, A.; Burdick, P.; Park, S.; Lee, C.; Mason, C.; May, J.J. Laboratory evaluation of belt usage with apple buckets. Am. J. Ind. Med. 2006, 49, 21–33.

27. Freivalds, A.; Park, S.; Lee, C.; Earle-Richardson, G.; Mason, C.; May, J.J. Effect of belt/bucket interface in apple harvesting. Int. J. Ind. Ergon. 2006, 36, 1005–1010.
28. Earle-Richardson, G.; Jenkins, P.L.; Strogatz, D.; Bell, E.M.; Freivalds, A.; Sorensen, J.A.; May, J.J. Electromyographic assessment of apple bucket intervention designed to reduce back strain. Ergonomics 2008, 51, 902–919.
29. Borg, G.A.V. Psychophysical bases of perceived exertion. Med. Sci Sports Exerc. 1982, 14, 377–381.
30. Freivalds, A. Biomechanics of the Upper Limbs, Mechanics, Modeling and Musculoskeletal Injuries, 2nd ed.; CRC Press: Boca Raton, FL, USA, 2011; p. 564.
31. Freivalds, A. Niebel's Methods, Standards and Work Design, 13th ed.; McGraw-Hill: New York, NY, USA, 2014; p. 747.

CHAPTER 7

Effects of coated and non-coated ZnO nano particles on cucumber seedlings grown in gel chamber

Sahar Moghaddasia, Amir Fotovata, Fathallah Karimzadehb, Hamid Reza Khazaeic, Reza Khorassania and Amir Lakziana

[a] Department of Soil Science, Ferdowsi University of Mashhad, Mashhad, Iran

b Department of Materials Engineering, Isfahan University of Technology, Isfahan, Iran

c Department of Crop Science, Ferdowsi University of Mashhad, Mashhad, Iran

ABSTRACT

Zinc oxide-engineered nanoparticles (ZnO ENPs) have received the most attention in recent years. This increasing interest has been directed towards studying the environmental fate and effects of ZnO ENPs on ecological terrestrial species. In this study, ZnO NPs were synthesized by atmospheric pressure solution evaporation method and were coated or uncoated with humic acid (HA). The root uptakes of uncoated and HAcoated ZnO NPs and zinc (Zn) were investigated by gel-grown cucumber. Two ZnO levels (1 and 200 μM) were applied in the form of coated (T_3) and non-coated (T_2) NPs or bulk particles (T_1). The results showed that coating NPs by HA increases zeta potential of NPs and decreases their aggregation size due to the increase in the repulsion forces among the particles. Addition of 1 mgL^{-1} ZnO into gel chamber enhanced root and shoot biomass; however, the shoot growth was higher in the presence of NPs compared to its bulk counterpart. Moreover, greater phytotoxicity of ZnO from the source of NPs than bulk particles in shoot was observed.Scanning electron microscopy results showed a clear evidence of the penetration of NPs into root cells.

1. INTRODUCTION

In recent years, nanotechnology has appeared in all fields for production of novel materials. The properties of engineered nanoparticles (ENPs) are vastly different from their corresponding bulk materials due to their extremely small size and high-specific surface area. However, the fate and the effects of ENPs on ecological terrestrial species including plants are unknown and require precise attention and study.

The unique properties of zinc oxide including its high electrical conductivity, antimicrobial activity and biocompatibility make it one of the most applicable of ENPs in many areas from optoelectronics and sensor development to sunscreen and glass products (Dhobale et al. 2008; Lupan et al. 2007; Salvati-Niasari et al. 2011). Therefore, much attention has been attracted toward its potential harm on the environment. Acute toxicity of ZnO NPs in bacteria and plant has been suggested (Brayner et al. 2006; Lin & Xing 2007). Meanwhile, Zn is one of the essential elements for plant growth and can be supplied by Zn nano fertilizers. Uptake and translocation of NPs into plant cells may increase their efficiency when used as nano fertilizers (Carpita et al. 1979; Remya et al. 2010). The translocation of ZnO NPs from roots to shoots of ryegrass has been reported by Lin and Xing (2008). Pathways trough cuticle (for sizes less than 5 nm) and stomata (for micrometer sizes and smaller) are two possible conduits for foliar uptake of NPs (Eichert et al. 2008). In the literature, both positive and negative effects of NPs on seed germination, root growth and plant yield have been suggested. For example, NPs of Al had no significant effects on the growth of California red kidney bean and ryegrass (Lin & Xing 2008) while they inhibited the root growth of corn, soybean, cabbage, cucumber, and carrot (Yang & Watts 2005). Severe detrimental effects of high concentrations of ZnO NPs on root elongation of ryegrass, radish, and rape, have been reported. However, Moghaddasi et al. (2013) showed that applying rubber ash NPs as a source of Zn can increase dry biomass of cucumber. It can be concluded that the type of NP and its size, dose, or method of application as well as the plant species are all factors that determine the effects of NPs on a plant (Lin & Xing 2007; Torney et al. 2007; Zhang et al. 2007).

Baalousha et al. (2008) showed that depending on the zero-point charge of NP and pH of the solution, NPs can be coated by humic substance sand give overall negative charge. Therefore, by increasing the electric double layer repulsive energy between NPs, the stability of NPs and thus toxicity of NPs can be increased. Accordingly, Zn could be released from ZnO NPs at high pH levels (9.0–11.0) and the presence of humic acids (HA) increase the concentration of the dissolved Zn (Kanel & Al-Abed 2011). Therefore, it seems that coating NPs by an organic matter like HA may affect the fate and impacts of ZnO NPs on plants. However, up to now, no report has been documented about the effect of surface coating on the ZnO NP behavior in gel. In most previous researches, NPs were generally applied to seeds or roots under hydroponic conditions. In this study we synthesized ZnO NPs and for the first time, the fate and effects of uncoated and HA-coated NP on growth and zinc, uptake by cucumber plants grown in gel chamber were investigated.

2. MATERIAL AND METHODS

2.1 Synthesis of coated and uncoated Zinc oxide nanoparticles

The atmospheric pressure solution evaporation method, suggested by Hosseini et al. (2011), was used for synthesis of ZnO NPs. $Zn(CH_3COO)_2$ $2H_2O$ (99.5% purity) and ethylene glycol (EG) (90% purity) was purchased from MERK company. In the atmospheric pressure solution evaporation method, 5 g of zinc precursor (zinc acetate) reached a particular reaction temperature (80 °C). After stabilization of the temperature of the precursor solution, 450 mL of stoichiometric solution of ethylene glycol was added to the suspension on a drop-by-drop basis. The stirring and heating of the reaction mixture was continued for 3–4 h. The suspension was then dried in an oven at 100°C for 12 h, and then crystallized at 600 °C for 1 h. To prepare the complexes, HA solution (500 mgL^{-1}) was obtained by dissolving solid HA (HA technical, purchased from SIGMA-ALDRICH) with 0.1 M of NaOH. The pH of the solution was adjusted to 5.0 by adding 0.1 M HCl. Nano-oxide (5 g) was then mixed with 1 L of HA solutions in a bottle and shaken for 2 days. After that, the suspension was centrifuged at 3500 × g for 30 min. The precipitated material was rinsed with deionized water, freeze dried (Model FD-4, Pishtaz engineering co., Iran), and ground gently.

The white synthesized zinc NPs and coated NPs were characterized by different methods. The crystal phase of ZnO NPs was determined by X-ray diffraction (XRD) (Model Philips Expert-MPD, Netherlands). The morphology and size of the coated and uncoated NP samples were examined by transmission electron microscopy (TEM) (Model EM10 C-80KV, Germany). The primary particle size distributions were obtained by randomly sampling 50 particles in their corresponding TEM images. The hydrodynamic (agglomerate) particle sizes of NP aqueous suspensions were measured by dynamic light scattering (DLS) method using Zetasizer Nano ZS system (Malvern Instruments Ltd.). Fourier transform infrared (FTIR) spectra were recorded with a Perkin-Elmer Spectrum One FTIR spectrometer (Nicolet iS10). Five mg of HA, nano-oxide, or HA-coated nano-oxide was mixed gently with 95 mg of KBr as a background matrix using a pestle and mortar and analyzed. FTIR spectra were recorded from 400 to 4000 cm^{-1} at 8 cm^{-1} resolution over 200 averaged scans. The FTIR spectra of nano-oxide-bound HA were obtained by subtracting that of the nano-oxides before HA coating from that of the HA-coated nano-oxides.

2.2 Growth system

Seeds of cucumbers Cucumis sativus L. cv. Dominus were placed in rows on wet blotting paper to germinate for two days. At same time the gel chambers were prepared, using methods described in Bengough et al. (2004). Chambers were constructed from two plates (one black polyvinylchloride and one transparent perspex), each size 215 × 300 × 3 mm. Strips of Perspex (3 mm thickness) were used as spacers around each plate, giving a plate separation of 6 mm, leaving three gaps, each approximately of 25 mm length, along the top surface, to allow gas exchange with the surrounding atmosphere, and unimpeded shoot extension. The agar-agar gel (Merck company) solution at 1 g100 g^{-1} water was used in the experiments sandwich approach to ensure that roots were growing in a uniform environment. Two concentrations of Zn (1 and 200 mg L^{-1}) that were respectively the optimum and toxic concentrations in our previous research in hydroponic (Moghaddasi et al. 2013), were added into the liquid gel in the form of bulk ZnO or its coated and uncoated NPs. NPs were added into the gel, followed by water bath ultrasonic treatment (25°C, 100 W, 40 kHz) for 15 min immediately before closing gel. Free-ZnO source was used as the blank treatment. Four uniformly germinated cucumber seeds were transferred to two gel chambers. The grains were spaced uniformly on the black gel plate with the seam facing the top plate – 80 mm from the top of the plate. The transparent gel plate was kept in 12 h dark/light at 18°C for 10 days in a growth cabinet with radicles pointing downward at an angle of 80° (to cause the roots to grow against the top plate). The chambers were scanned and images saved every 3 days and data were used to calculate the root length. After10 days, the plants were extracted from the plates, any gel was rinsed off, and shoots and roots were oven dried at 60°C for 48 h and weighed. Root and shoot materials were digested by 2 M HCl and Zn concentrations in the extracts were measured by atomic absorption spectroscopy.

2.3 Microscopy observations

To observe the effect of ZnO NPs on root tips of cucumber by LM (Model E100, German), when harvesting, the first 4 cm rootlets were longitudinally sliced (500 nm thick), washed with deionized water and prefixed with alcohol (70%) and glycerin. Sliced root samples were washed in sodium hypochlorite (20%) and acetic acid and embedded in carmine for 15 min. The samples were then washed with deionized water and embedded in green methyl for 30 s.

Root and stem samples were washed by running water and carefully rinsed with distilled water before SEM microscopy observations. For this reason, small cross section pieces of (prepared from 0–2 cm) roots and stems, were dehydrated using freeze dryer (Model FD-4, Pishtaz engineering co., Iran) and coated with gold for 60 s by a Sputter Coater (Model SCDOOS-Baltec, Switzerland). SEM observations were carried out using a Philips XL30 Robeahi SEM-EDS unit, with an accelerating voltage of 20 KV.

2.4 Statistical analysis

The gel chamber culture experiment was set up in a completely randomized design with three treatments and three replicates. Treatment effects were analyzed by analyzing the variance using the general linear model procedure. Means were compared using the least significant differences (LSD) at $P < 0.05$ (25). JMicroVision was used for determining the root length.

3. RESULTS AND DISCUSSION

3.1 Characterization of coated and uncoated Zinc oxide nanoparticles

XRD patterns of ZnO NPs are illustrated in Figure 1. All peaks can be well indexed to the zincite phase, which are consistent with the values in standard card (JCPDS 36–1451). No peaks from other ZnO phases or impurity can be observed. This indicates the high purity of the obtained ZnO NPs.

TEM images (Figure 2) indicated the successful production of nano-size particles of ZnO and its coating. In the TEM image, thin layers of HA shells have coated ZnO NPs. It should be noted that most of the times, instead of a single particle, agglomerated NPs are coated with HA. Therefore, the average primary size of NPs may increase to about 100 nm. In addition, when coating NPs, the thickness of the shell of HA should be added to the diagonal of ZnO. However, the hydrodynamic size of the coated NPs is a result of the initial particle size and zeta potential of ZnO and HA. DLS data (Table 1) indicate the hydrodynamic size of ZnO NPs is much larger than the observation size in TEM images. One possible explanation might be that by adding dry NPs to water, high local concentrations lead to an enhancement in collision frequencies and rapid aggregation. This is also a thermodynamic-driven process in which the surface energy of particles is reduced by forming larger aggregates (French et al. 2009). Based on the results, when coating NPs with HA, their zeta potential increases whereas their aggregation size decreases due to the increases in the repulsion forces between these particles. The FTIR spectra of HA and nano-oxide-bound HA are presented in Figure 3. FTIR spectra of HA-coated ZnO NPs showed that the HA crust is successfully formed around ZnO NPs. Compared with bulk HA, the peak intensity of bands at 3700–2500 cm^{-1}diminishes largely after being adsorbed by nano-oxides (Figure 3). This indicates the release of hydroxyl groups (Tatzber et al. 2007). The COOH peak at 1662 cm^{-1} also significantly decreased due to the strong interactions of COOH with surfaces of nano-ZnO. Yang et al. (2009) showed that HA is adsorbed by ZnO by the COOH groups whereas no FTIR spectra of nano-SiO_2- bound HA were obtained because of the insignificant adsorption. A previous report (Hajdu et al. 2009) has shown that TiO_2, Al_2O_3, and ZnO NPs were bounded to HA by electrostatic attraction and ligand exchange.

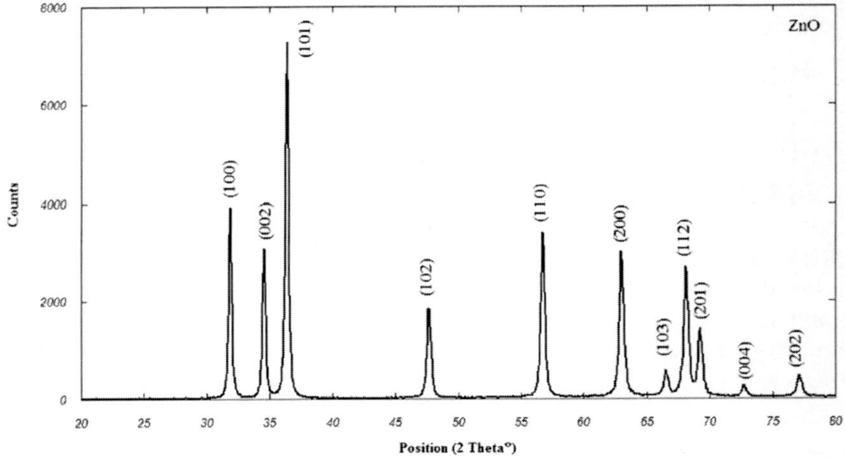

Figure 1. XRD patterns of ZnO nanoparticles calcined at 600 C for 2 h.

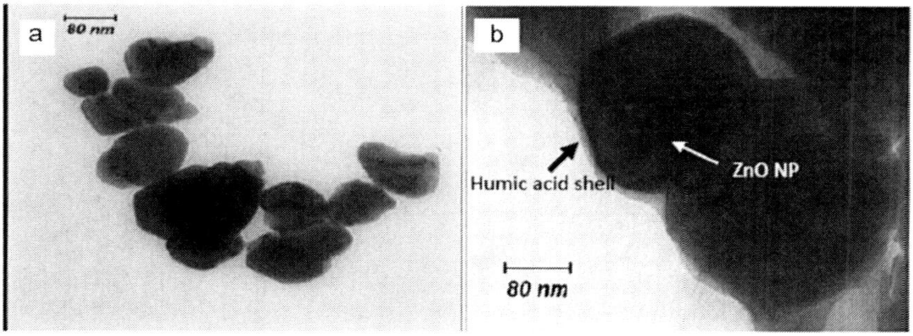

Figure 2. TEM images of non-coated (a) and coated (b) ZnO NPs.

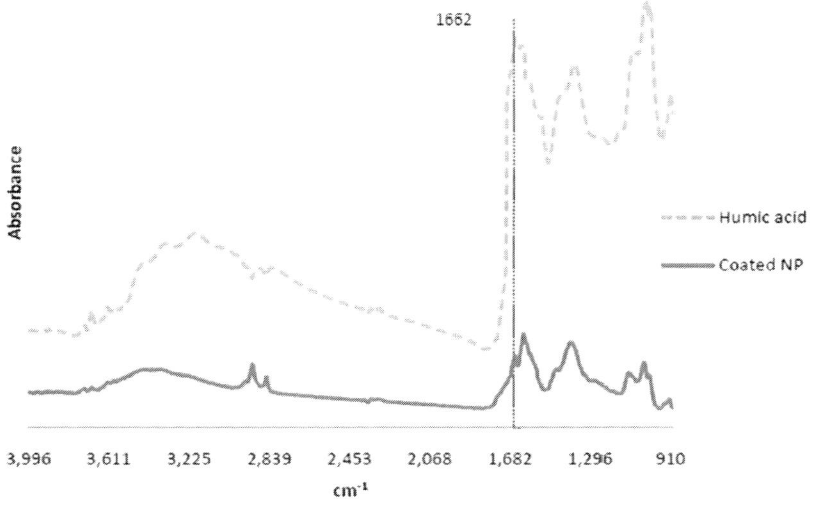

Figure 3. FTIR spectra of bulk HA and nano-zinc-bound HA was obtained by subtraction of the IR spectra of pure nano-oxide from that of the HA-coated nano-oxides after freeze drying.

Table 1. Size and zeta potential of nanoparticles (the concentration of NPs suspension for hydrodynamic size and ζ- potential was 20 mgL−1 dionize water).

Particle	ζ- potential (mV)	TEM diameter (nm)	Hydrodynamic diameter (nm)	pH
ZnO NPs	3.1	75	441	6.95
Coated ZnO NPs	−5.4	100	328	6.87

3.2 Effect of coated and uncoated nanoparticle on cucumber biomass and Zn concentration

Addition of 1 mg L^{-1} ZnO in gel chamber enhanced the root and shoot biomass; however, increment of shoot plant growth was greater due to NPs than its bulk counterpart (Table 2). This can be attributed to higher efficiency of NPs in supplying Zn to plant. Higher efficiency of ground rubber ash NPs compared to Zn-sulfate in supplying Zn to plants has earlier been reported (Moghaddasi et al. 2013). Colman et al. (2013) also observed that addition of $AgNO_3$ at a fourfold higher concentration (to act as a positive control for plants and microbes) had never greater impact than that of the AgNP treatment. Data showed greater phytotoxicity of ZnO from the source of NPs than bulk particles (Table 2) in shoot. This result indicates the possibility of penetration and transport of ZnO NPs in plants. Figure 5 showed depletion of particles in root area at high applied levels (200 mg L^{-1}) of coated and non-coated NPs. However, this depletion was not seen in bulk ZnO treatments that can show impossibility of penetration of them to root cells and it is an interesting aspect of this study. There was no significant difference between root biomass in plant treated with ZnO NPs and their bulk counterpart at concentration of 1 mg L^{-1} (Table 2). Therefore, the greater positive effect of ZnO NPs on gel-grown cucumber may be associated with uptake and transport of NPs from roots to shoots. The entrance of ZnO NPs and carbon nano tubes into the cells of living organisms has also been previously reported (Wong et al. 2005; Zhu et al. 2007; Lin & Xing 2008). In addition, carbon nanomaterials were found as black aggregates in the following chronological order: Seed > root > stem > leaf, which may show their translocation in the plant. This observation was proved by the transmutation of C70 to the next generation of the plant which was further confirmed from IR and Raman spectral data (Husen & Siddiqi 2014). It was shown that by decreasing the particle size from micro to nano, the difference in Zn concentration of the root and shoot was decreased. The difference in Zn concentration of the root and shoot for bulk ZnO treatment was 76.6% while for coated and uncoated NPs was 33.3 and 45.5%, respectively. It seems that NPs penetrate into cell roots and transport from root to shoot. This was investigated in more details trough SEM studies on part of the penetration of NPs into the plant cells.

There was no significant difference between total biomass of plants grown in gel treated with coated and uncoated NPs (Table 2). Total concentration of Zn in plants treated with coated NPs was also greater than those treated with uncoated ZnO NPs. Aggregation of NPs in nutrient solution results in changes of root uptake and root-to-shoot translocation of these particles and thereby affects their phytotoxicity and uptake efficiency (Lin & Xing 2008). Zhao et al. (2012) showed that the surface coating plays an important role in uptake of NPs by plants at 200 mg kg^{-1} and above. In fact, he reported that surface coating increases the cerium accumulation of corn root tissues. Our results demonstrated that coating NPs by HA increases the zeta potential of NPs and decreases their aggregation size due to the increase in repulsive forces between the particles and as a result higher Zn concentrations in plants treated with coated NPs are predictable. In treatments with coated NPs, compared with those with uncoated NPs, higher transportation of Zn from root to shoot might be the reason for lower difference in Zn concentrations of the root and the shoot. However, in the case of soluble NPs, it is difficult to distinguish between the ion uptake and NP uptake (Stampoulis et al. 2009).

Table 2. Root and shoot dry biomass and total Zn concentration of cucumber as affected by different sources of Zn with concentration of 1 and 200 mg kg^{-1} in the gel chamber (T_0: Control, Zn free-treatment; T_1: bulk ZnO; T_2: non coated ZnO and T_3: ZnO coated by HA).

			T_0	T_1	T_2	T_3
Zn concentration in gel chamber (mg Kg^{-1})	1	Root biomass (mg)	4.3^{d}	6.4^{ab}	7.1^{a}	6.1^{abc}
		Shoot biomass (mg)	9.1^{d}	11.4^{c}	24.2^{a}	19.8^{ab}
		Root Zn concentration (mg kg^{-1})	32^{d}	243^{c}	584^{b}	857^{a}
		Shoot Zn concentration (mg kg^{-1})	8.24^{d}	56.8^{c}	317^{b}	571^{a}
	200	Root biomass (mg)	4.3^{a}	0.04^{b}	0^{bc}	0^{bc}
		Shoot biomass (mg)	9.1^{a}	0.8^{b}	0^{bc}	0^{bc}

Means with the same letter are not significantly different at <0.05.

Root growth is known to be more sensitive than shoot growth to metal amount (Ma et al. 2010). Xiang et al. (2015) suggested that the roots are in direct contact with the NP and thus are the first tissue to be exposed to excess concentrations of nano-ZnO. Root length increasing rate of plant treated with coated ZnO NPs is much more than that treated with uncoated NPs (Figure 4). It is predicted that the plant biomass treated with coated NPs might also increase if experiment time duration continued. However, this idea needs further investigations. According to Figure 4, all of the plants grown in presence of ZnO had significantly a longer total root length than the control, which indicates the positive effect of Zn^{2+} on root elongation. Furthermore, plant treated with coated NPs (T_3) did not differ in total root length with control, but after first 6 days, the root length rate immediately increased and was higher than this delay in the increase of the root length may be because of the HA shell around ZnO which may decrease the release rate of Zn^{2+} at early stages. This indicates that Zn^{2+} release from coated NPs occurs with delay. More recently, Waalewijn-Kool et al. (2013) showed that Zn concentration in pure water increased with time for all ZnO forms but peaked at intermediate concentrations for uncoated ZnO-NP and non-nano ZnO, while for coated ZnO-NP this dominant peak was only seen after 12 months. Navarro et al. (2012) reported that Has enhance the accumulation of CdSe/ZnS NPs onto the root surface of Arabidopsis thaliana. However, the mechanism is not clear and needs to be further studied.

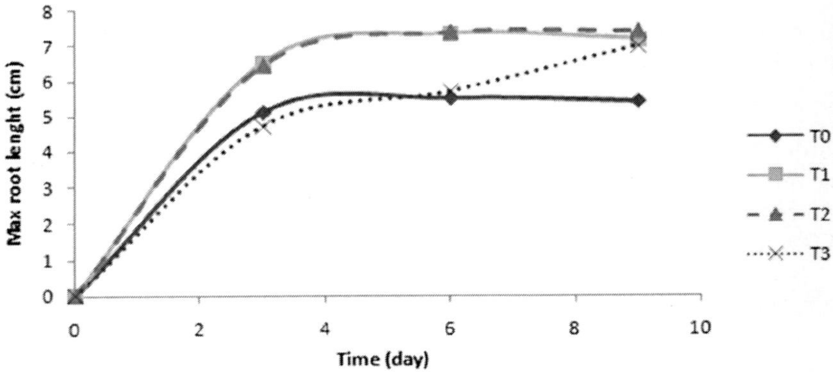

Figure 4. Total Root length of cucumber as affected by time and different sources of Zn with concentration of 1 mgL−1 in the gel chamber (T_0: Control, Zn free treatment; T_1: bulk ZnO; T_2: non-coated ZnO and T_3: ZnO coated by HA).

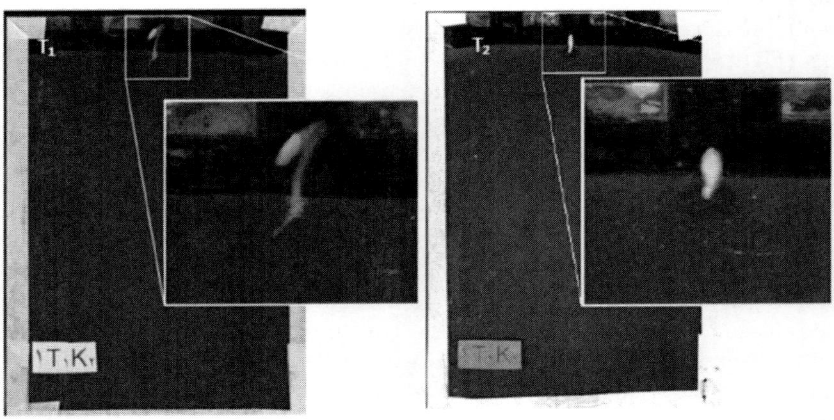

Figure 5. Toxicity to cucumber of 200 μM Zn supplied in the form of bulk ZnO (T_1) and ZnO nano particles (T_2).

3.3 Zinc oxide nanoparticles impact on root tip

The number of xylem vessels is effective to ameliorate water stress effects when plants are subjected to water stress (Zimmermann et al. 1993). Lok et al. (2013) showed that differential metaxylem counts were observed upon exposure to metal NPs or their respective ionic salts. The metaxylem counts ranged from 5 to 6 in control, but increased to 7 upon plants exposure to $ZnSO_4$ (1000 μg mL^{-1}) or $AgNO^3$ (200 μg mL^{-1}). This study results showed that there was no difference between xylem vessel numbers even when plants exposed to concentration of 200 μg mL^{-1} ZnO (data not shown). This may be due to the different concentration of Zn which was used in previous research. Light microscopy (LA) observations demonstrated that ZnO can disturb the root tip (Figure 6). It has been previously reported that in the presence of 1000 mg Zn $L^{-1}$of Zn in the form of ZnO NPs and Zn-sulfate (Zn^{2+}) in the nutrient solution, the ryegrass root tips are deformed ZnO-NP this dominant peak was only seen after 12 months. Navarro et al. (2012) reported that Has enhance the accumulation of CdSe/ZnS NPs onto the root surface of Arabidopsis thaliana. However, the mechanism is not clear and needs to be further studied. Zinc oxide nanoparticles impact on root tip The number of xylem vessels is effective to ameliorate water stress effects when plants are subjected to water stress (Zimmermann et al. 1993). Lok et al. (2013) showed that differential metaxylem counts were observed upon exposure to metal NPs or their respective ionic salts. The metaxylem counts ranged from 5 to 6 in control, but increased to 7 upon plants exposure to $ZnSO_4$ (1000 μg mL^{-1}) or $AgNO_3$ (200 μg mL^{-1}). This study results showed that there was no difference between xylem vessel numbers even when plants exposed to concentration of 200 μg mL^{-1} ZnO (data not shown). This may be due to the different concentration of Zn which was used in previous research. Light microscopy (LA) observations demonstrated that ZnO can disturb the root tip (Figure 6). It has been previously reported that in the presence of 1000 mg Zn $L^{-1}$of Zn in the form of ZnO NPs and Zn-sulfate (Zn^{2+}) in the nutrient solution, the ryegrass root tips are deformed (Lin & Xing 2008). The highest level of Zn used in present study was much lower than that used by Lin and Xing (2008) but little disturbance can be distinguished in the LA pictures. In this study also, coated ZnO NPs (Zn^{2+}) resulted in deformation of ryegrass root tips compared to control. Moreover, at high concentrations of uncoated ZnO NPs, the root tip was disturbed. Lin and Xing (2008) observed the severe impact of ZnO NPs and Zn^{2+} ions in the shrank morphology of the root tips. In the presence of 1000 mg L^{-1} of ZnO NPs or Zn^{2+}, the epidermis and the root cap were broken, the cortical cells were highly vacuolated and collapsed, and the vascular cylinder also shrank (Lin & Xing 2008).

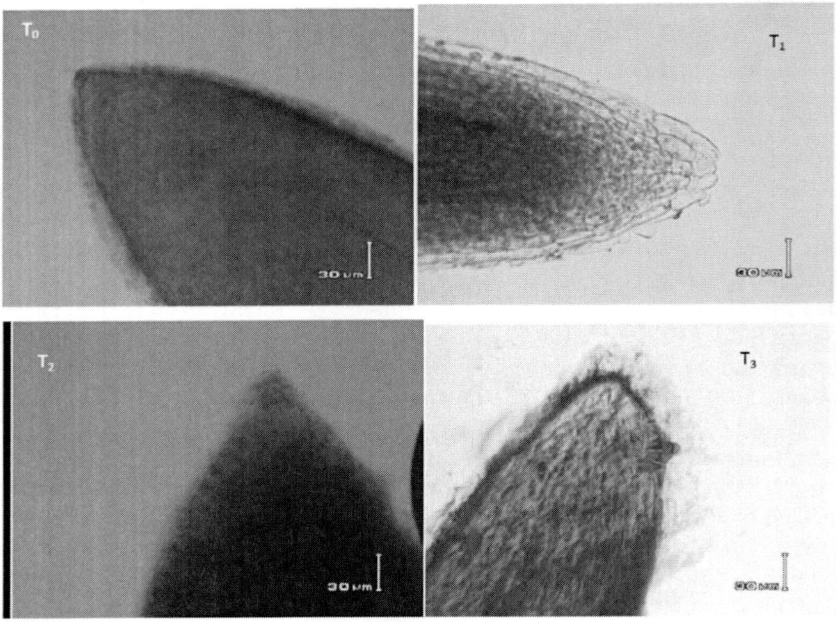

Figure 6. Light microscopy (LM) images of roots indicating destruction and deformation of root tip of cucumber plants treated with control (T_0), bulk ZnO (T_1), coated (T_3) and non-coated ZnO nano particles (T_2) and at concentration of 200 mgL^{-1} Zn^{2+}.

3.4 The penetration of coated and non-coated nanoparticles into the plant cells

The penetration of NPs into the cells of living organisms has been considered by several researchers (Lin & Xing 2008; Lin et al. 2010; Moghaddasi et al. 2015). The entrance of ZnO NPs into the cells of living organisms has also been reported (Lin & Xing 2008). They also showed the aggregation of NPs of ZnO on the root surface by using SEM images. In the present study, the SEM images also showed the aggregation of the ZnO NPs on the root surface (Figure 7). The root tip and the root hair could secrete a large amount of mucilage which might contribute to attachment of ZnO NPs on the root surface. It is assumed that ZnO NPs that covered the root surface could influence the water and nutrient uptake as well as the whole process of plant growth development. Based on the SEM images, it seems that larger NPs cannot penetrate into the root cells. ZnO NPs which entered the plant cells were different in size. It seems that aggregates (120 nm) of smaller NPs in the root cells were formed (Figure 7). ZnO NPs can pass through the cell membrane similar to the mechanisms of nutrient ions and molecules. The fate of these particles and the possible ionization or dissolution inside the plant cells is still unknown and further investigation is required. However, it seems that this entrance may affect the Zn^{2+} concentration in the root and shoot and of course in the plant's dry weight (Tables 2).

As there were no particles in control plant and the plant that treated with bulk ZnO, this hypothesis that the penetration of NPs to plant cell is the main reason of higher concentration of Zn^{2+} in nano treatments can be further supported (Figure 8). Accumulation of NPs may intensify the Zn toxicity stress to the cucumber. The presence of NPs in stem showed that NPs can transport from root to shoot. However, it seems that little ZnO NPs can transport from root to shoot (Figure 8) but Zn concentration in the shoot biomass proved that this amount is much more than the amount of Zn^{2+} provided by bulk ZnO. This can justify the higher toxicity of ZnO NPs compared to the bulk corporation. It is known that apoplastic and symplastic pathways are the main routes to transport NPs from epidermis to stele. Zhao et al. (2012) by analysis of longitudinal sections of roots showed the presence of NP aggregates mostly in cell walls of the cortex tissues and only a few of them are in the xylem. This suggest that, CeO_2 NP aggregates moved cell-to-cell cells were formed (Figure 7). ZnO NPs can pass through the cell membrane similar to the mechanisms of nutrient ions and molecules. The fate of these particles and the possible ionization or dissolution inside the plant cells is still unknown and further investigation is required. However, it seems that this entrance may affect the Zn^{2+} concentration in the root and shoot and of course in the plant's dry weight (Tables 2) from epidermis to endodermis by the apoplastic pathway. Sedimentation of NPs in this study showed that ZnO NPs can be similarly transported within the plant. Such results explain the low translocation rate from roots to shoots in cucumber plants. The presence of stained NP aggregates around vascular vessels suggests that NP aggregates have found their way to the transport system and moved through the xylem driven by transpiration. However, the mechanisms by which NPs are taken up by plants are not understood yet, but they may be present in xylem and phloem sap and be mobile within the plant (Wang et al. 2012). Zn does not exist in the SEM-EDX micrograph of control (Figure 9). In treatment plants with coated and uncoated ZnO NPs and at an intensity of 1 KeV the characteristic peak of Zn was seen inside the root cells (Figure 9) indicating that these NPs can be absorbed and transported into the plant cells. The fate of these particles and the possible ionization or dissolution inside the plant cells are still unknown and further investigation is required. Our further studies will focus on the behavior of these NPs in soil and the effect of them on plant growth. In brief, penetration of ZnO NPs into the root cells can be the reason for the difference observed in the manner of this NPs in comparison to their larger co-partners.

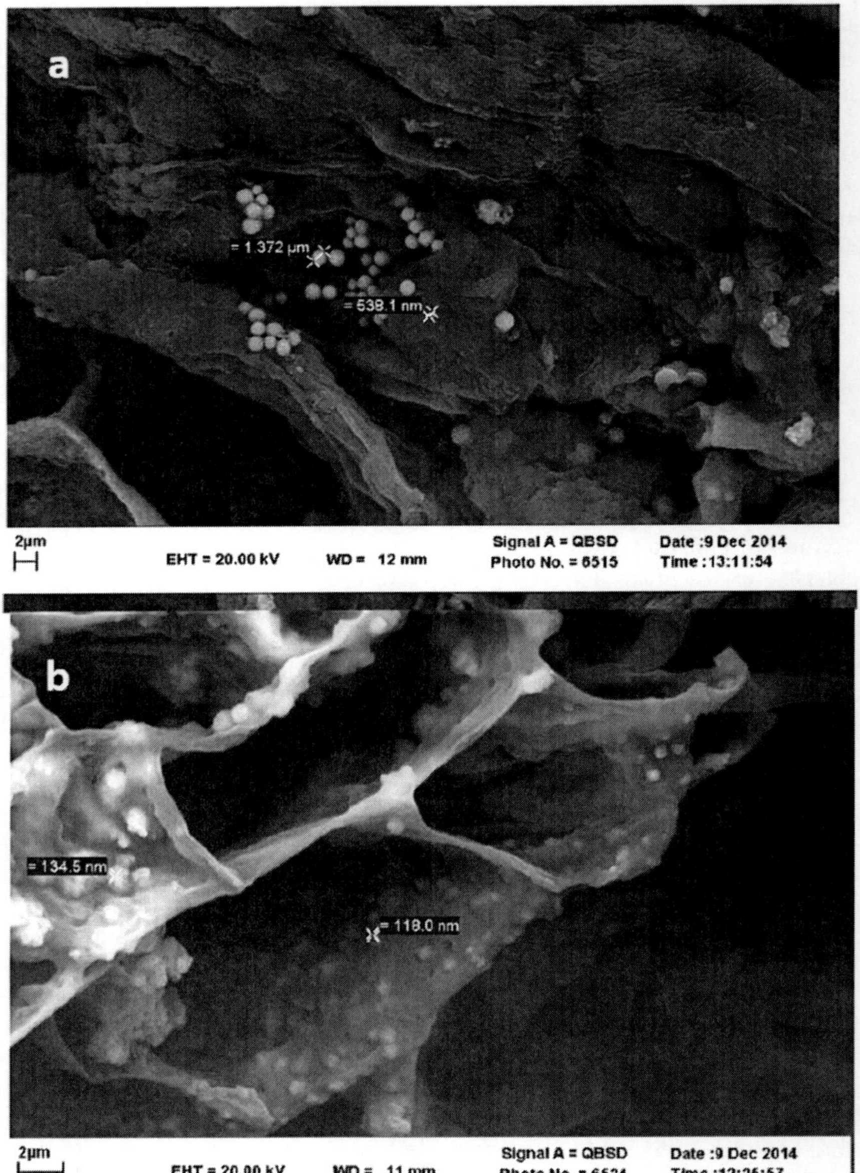

Figure 7. The SEM images of root surfaces of cucumber plants (a) and root cells (b) supplied with 200 mg L^{-1} Zn in the form of ZnO NPs.

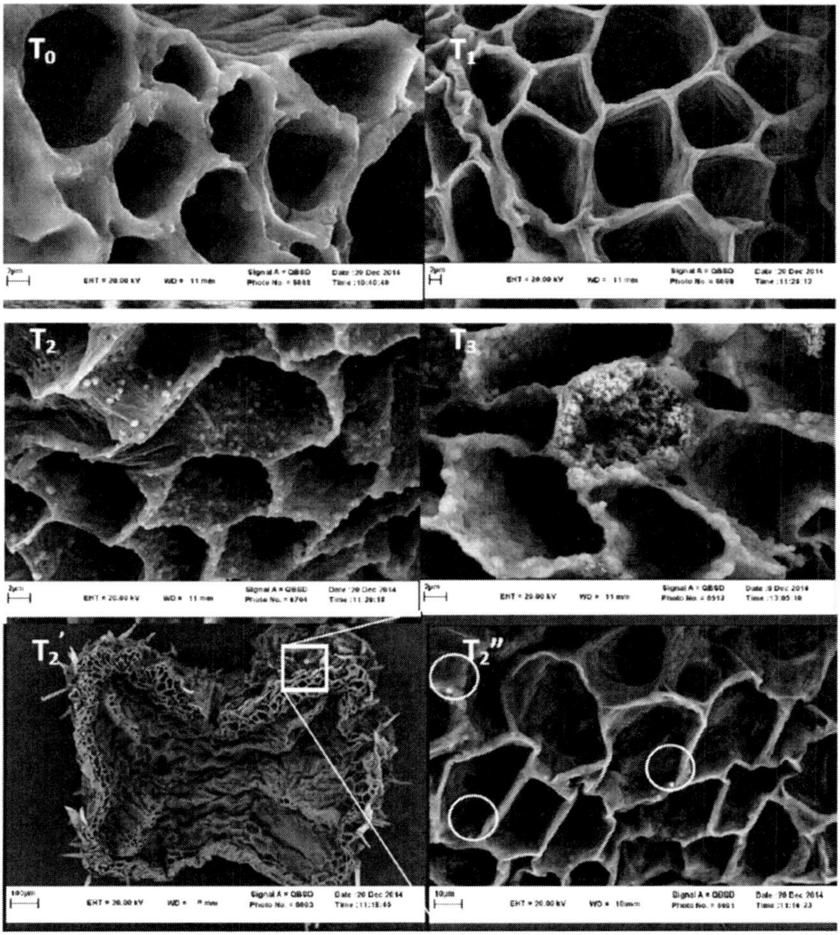

Figure 8. The SEM images of cucumber root cross sections (T_0: control, T_1: bulk ZnO treatment, T_2: ZnO treatment and T_3: coated ZnO treatment) and stem cross sections (T_2' and T_2'': ZnO treatment) at the 200 mg L^{-1} Zn treatment indicates the penetration of ZnO NPs into the root and stem cells.

4. CONCLUSION

The study results revealed that a micro to nano decrease in the particle size can increase the cucumber growth, the Zn concentration in root and shoot and the Zn transport from root to shoot. This might be because NPs can penetrate into cell roots and transport from root to shoot. In addition, this uptake into root cells makes greater phytotoxicity of ZnO from the source of NPs compared to the bulk particles at high concentrations.

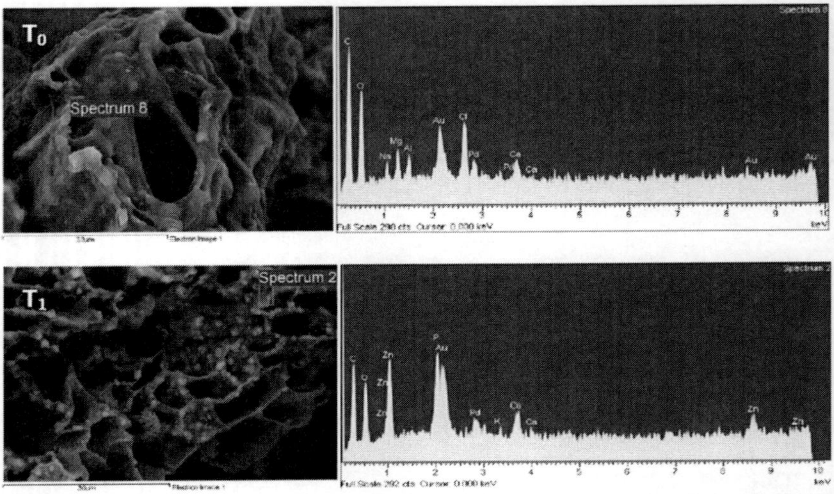

Figure 9. Scaning electron image and energy-dispray X-ray spectra of cucumber root cross sections: control treatment (T_0) and ZnO nano particle treatment (T_2).

When coating NPs with HA, the zeta potential of NPs is increased and their aggregation size is decreased due to increase in the repulsion forces among the particles. Higher Zn concentrations are therefore predictable for plants treated with coated NPs. Treatments with coated NPs are suggested for getting lower differences in Zn concentrations of the root and shoot. This is due to the higher transport of Zn from root to shoot in coated NPs. Further research is needed for investigating the root-to-shoot translocation of coated and uncoated ZnO NPs.

ACKNOWLEDGMENT

This research has been fully supported by Ferdowsi University of Mashhad.

REFERENCES

1. Baalousha M, Manciulea A, Cumberland S, Kendall K, Lead JR. 2008. Aggregation and surface properties or iron oxide nanoparticles: influence of pH and natural organic matter. Environ Toxicol Chem. 27:1875–1882.
2. Bengough AG, Gordon DC, Al-Menaie H, Ellis RP, Allan D, Keith R. 2004. Gel observation chamber for rapid screening of root traits in cereal seedlings. Plant Soil. 262:63–70.
3. Brayner R, Ferrari-Iliou R, Brivois N, Djediat S, Benedetti MF, Fiévet F. 2006. Toxicological impact studies based on Escherichia coli bacteria in Ultrafine ZnO nanoparticles colloidal medium. Nano Lett. 6:866–870.
4. Carpita N, Sabularse D, Montezinos D, Delmer DP. 1979. Determination of the pore size of cell walls of living plant cells. Science. 205:1144–1147.
5. Colman BP, Arnaout CL, Anciaux S, Gunsch CK, Hochella JR, Kim B, Lowry GV, McGill BM, Reinsch BC, Richardson CJ, et al. 2013. Low concentrations of silver nanoparticles in biosolids cause adverse ecosystem responses under realistic field scenario. Plos One. 8:57189.
6. Dhobale S, Thite T, Laware SL, Rode CV, Koppikar SJ, Ghanekar RK. 2008. Zinc oxide nanoparticles as novel alphaamylase inhibitors. J Appl Phys. 104:904–907.
7. Eichert T, Kurtz A, Steiner U, Goldbach HE. 2008. Size exclusion limits and lateral heterogeneity of the stomatal foliar uptake pathway for aqueous solutes and water-suspended nanoparticles. Physiol Plantarum. 134:151–160.
8. French R, Jacobson A, Kin B, Isley S, Penn R, Baveye P. 2009. Influence of ionic strength, pH, and cation valence on aggregation kinetics of titanium dioxide nanoparticles. Environ Sci Technol. 43:1354–1359.
9. Hajdu A, Elles E, Tombacz E, Borbath I. 2009. Surface charging, polyanionic coating and colloid stability of magnetite nanoparticles. Colloids Surf Physicochem Eng Aspects. 347:104–108.
10. Hosseini SS, Roudbar Mohammadi SH, Joshaghani HR, Eskandari M. 2011. Antifungal effect of sodium dodecil sulphate and nano particle ZnO on growth inhibition of standard strain of Candida albicans. J Gorgan Uni Med Sci. 12:64–69.
11. Husen A, Siddiqi KS. 2014. Carbon and fullerene nanomaterials in plant system. J Nanobiotech. 12:16.
12. Kanel SR, Al-Abed SR. 2011. Influence of pH on the transport of nanoscale zinc oxide in saturated porous media. J Nanopart Res. 13:4035–4047.
13. Lin DH, Xing BS. 2007. Phytotoxicity of nanoparticles: inhibition of seed germination and root elongation. Environ Pollut. 150:243–250.
14. Lin DH, Xing BS. 2008. Root uptake and phytotoxicity of ZnO nanoparticles. Environ Sci Technol. 42:5580–5585.
15. Lin J, Zhang H, Chen Z, Zheng Y. 2010. Penetration of lipid membranes by gold nanoparticles: insights into cellular uptake, cytotoxicity, and their relationship. ACS Nano. 4:5421–5429.
16. Lok R, Pokhre LA, Brajesh Dubey B. 2013. Evaluation of developmental responses of two crop plants exposed to silver and zinc oxide nanoparticles. Sci Total Environ. 452–453:321–332.
17. Lupan O, Chow L, Chai G, Roldan B, Naitabdi A, Schulte A. 2007. Nano fabrication and characterization of ZnO nanorod arrays and branched microrods by aqueous solution route and rapid thermal processing. Mater Sci Eng. 145:57–66.
18. Ma YH, Kuang LL, He X, Ding YY, Zhang ZY, Chai ZF. 2010. Effects of rare earth oxide nanoparticles on root elongation of plants. Chemosphere. 78:273–279.
19. Moghaddasi S, Khoshgoftarmanesh AH, Karimzadeh F, Chaney RL. 2013. Preparation of nano-particles from waste tire rubber and evaluation of their effectiveness as zinc source for cucumber in nutrient solution culture. Sci Hort. 160:398–403.
20. Moghaddasi S, Khoshgoftarmanesh AH, Karimzadeh F, Chaney RL. 2015. Fate and effect of tire rubber ash nanoparticles (RANPs) in cucumber. Ecotox Environ Safe. 115:137–143.

21. Navarro DA, Bisson MA, Aga DS. 2012. Investigating uptake of water-dispersible CDSe/ZnS quantum dot nanoparticles by Arabidopsis thaliana. J Hazard Mater. 211–212:427–435.
22. Remya N, Saino HV, Baiju G, Maekawa T, Yoshida Y, Sakthi Kumar D. 2010. Nanoparticulate material delivery to plant. Plant Sci. 179:154–163.
23. Salvati-Niasari M, Davar F, Khansari A. 2011. Nanosphericals and nanobundles of ZnO: synthesis and characterization. J Alloys Compd. 509:61–65.
24. Stampoulis D, Sinha SK, White JC. 2009. Assay-dependent phytotoxicity of nanoparticles to plants. Environ Sci Technol. 43:9473–9479.
25. Tatzber M, Stemmer M, Spiegel H, Katzlberger C, Haberhauer G, Mentler A, Gerzabek MH. 2007. FTIR-spectroscopic characterization of humic acids and humin fractions obtained by advanced NaOH, Na4P2O7, and Na2CO3 extraction procedures. J Plant Nutr Soil Sci. 170:522–529.
26. Torney F, Trewyn GB, Lin VSY, Wang K. 2007. Mesoporous silica nanoparticles deliver DNA and chemicals into plants. Nanotechnol. 2:295–300.
27. Waalewijn-Kool PL, Ortiz MD, Lofts S, Van Gestel CAM. 2013. The effect of pH on the toxicity of zinc oxide nanoparticles to Folsomia candida in amended field soil. Environ Toxicol Chem. 32:2349–2355.
28. Wang ZY, Xie XY, Zhao J, Liu XY, Feng WQ, White JC, Xing BS. 2012. Xylem- and phloem-based transport of CuO nanoparticles in maize (Zea mays L.). Environ Sci Technol. 46:4434–4441.
29. Wong N, Kam S, Dai HJ. 2005. Carbon nanotubes as intracellular protein transporters: generality and biological functionality. J Am Chem Soc. 127:6021–6026.
30. Xiang L, Zhao HM, Li YW, Huang XP, Wu XL, Zhai T, Yuan Y, Cai QY, Mo CH. 2015. Effects of the size and morphology of zinc oxide nanoparticles on the germination of Chinese cabbage seeds. Environ Sci Pollut Res. 22:10452–10462.
31. Yang K, Lin D, Xing B. 2009. Interactions of humic acid with nanosized inorganic oxides. Langmuir. 25:3571–3576.
32. Yang L, Watts DJ. 2005. Particle surface characteristics may play an important role in phytotoxicity of alumina nanoparticles. Toxicol Lett. 158:122–132.
33. Zhang L, Su M, Liu C, Chen L, Huang H, Wu X, Liu X, Yang F, Gao F, Hong F. 2007. Antioxidant stress is promoted by nano-anatase in spinach chloroplasts under UV-B radiation. Biol Trace Elem Res. 109:68.
34. Zhao L, Peralta-Videa JR, Varela-Ramirez A, Castillo-Michel H, Li C, Zhang J, Aguilera RJ, Keller AA, Gardea-Torresdey JL. 2012. Effect of surface coating and organic matter on the uptake of CeO2 NPs by corn plants grown in soil: Insight into the uptake mechanism. J Hazard Mater. 225:131–138.
35. Zhu L, Chang DW, Dai LM, Hong YL. 2007. DNA damage induced by multi walled carbon nanotubes in mouse embryonic stem cells. Nano Lett. 7:3592–3597.
36. Zimmermann U, Benkert R, Schneider H, Rygol J, Zhu JJ, Zimmermann G. 1993. Xylem pressure and transport in higher plants and tall trees. In: Smith, JAC, Griffiths, H, editors. Water deficits – Plant responses from cell to community. Oxford (UK): Bios Scientific Publishers; p. 87–108.

CHAPTER 8

Cropping Systems and Agronomic Management Practices in Smallholder Farms in South Africa: Constraints, Challenges and

Goitsemodimo Moswetsi, Morris Fanadzo and Bongani Ncube

Department of Agriculture, Faculty of Applied Sciences, Cape Peninsula University of Technology, Private Bag X8, Wellington 7654, South Africa

ABSTRACT

Many studies have been conducted to assess challenges faced by smallholder crop producers in South Africa but few have focused on agronomic constraints. This study is a review of agronomic constraints faced by smallholder farmers under irrigation and dryland farming in South Africa. Constraints include choice of cultivars, planting dates and densities, tillage operations, water and fertilizer management and cropping patterns. Water availability is a major constraint under dryland farming and studies showed that the impacts can be very severe. Interaction of a variety of constraints under irrigation resulted in below average yields. Strategies for increasing productivity in dryland agriculture include capturing more water and allowing it to infiltrate to the root zone and the use of available water more efficiently. Yields under irrigation can improve when micronutrients are blended with macronutrients in relatively affordable blends. Use of green manure as an alternative fertilizer can also improve yields. Soil and water management technologies that improve soil fertility and productivity were as important as those that prevent soil erosion and water loss. It was recommended that practices such as supplementary irrigation and rainwater harvesting technologies take priority in efforts to address dryland water problems. As for farmers with access to irrigation water, practices that deal with improvement in planting dates and populations and water and fertilizer management can have positive impacts on crop yields. Research focused on smallholder agriculture should also start focusing more on water productivity and improving agronomic practices.

1. INTRODUCTION

The total land area of South Africa is 127 million ha of which 82% (100 million ha) is agricultural land. Only 14% (14 million ha) of the latter receives sufficient rainfall for arable crop production and periodic droughts affect the dryland areas[1]. South Africa is currently facing one of the worst droughts in history. Less than 10% (1.35 million ha) of the arable area is irrigated but the irrigated area accounts for a substantial proportion of the value of the country's total agricultural output[1]. The commercial farming sector produces more than 95% of the total marketed agricultural output, while the smallholder sector produces only 5%.

South African agriculture is dualistic in nature, consisting of the less developed smallholder sector and a well-developed commercial sector. The number of commercial farmers is estimated at between 50,000 and 60,000. The large-scale commercial sector, consisting of mostly white farmers, provide food for the whole nation while the smallholder farmers are characterized by the overwhelming large number of the rural population who play an insignificant role in national food production. Oettle *et al.*[2] reported that 60,000 white commercial farmers occupy 102 million ha while 1.2 million black farmers share about 17 million ha of agricultural land. Recent studies such as Cousins[3] showed however increased figures of people involved in smallholder agriculture to be 4 million and Hart[4] reported a decreased number of commercial farmers by at least 23%. Armour[5] reported that there are 3.3 million smallholder farmers and 30,000 commercial farmers in South Africa. A report produced by World Wide Fund-SA[6] pointed out that a decline in commercial farming is attributed to decreased productivity and drought. Smallholder agriculture is therefore beginning to play a significant role in the country's agriculture economy.

Smallholder agriculture plays an important role in terms of food security, employment creation and income generation in many countries. In Sub-

Saharan Africa, over 60% of the rural population depends on dryland agriculture[7] and in South Africa; approximately 1.3 million rural households are involved in dryland agriculture[8]. The majority of smallholder farmers in South Africa are located in the Eastern Cape and Limpopo provinces where 52 and 60% of the households are poor, respectively[9]. Rural livelihoods in most parts of South Africa depend on agriculture and over 30% rural households in Eastern Cape and Limpopo provinces are involved in agriculture[10].

South Africa has incorporated smallholder agriculture into various support and development programs of both the public and private sectors to ensure viability and sustainability in farming practices. The aims of the smallholder agricultural development programs in South Africa include increasing crop productivity, enhancing sustainable resource use and facilitating economic growth in rural communities[11]. The government has created several support programs to improve agricultural productivity of smallholders in rural areas. Programs like the Land Reform for Agricultural Development (LRAD) are specifically meant to empower smallholder farmers. However, as Hart[4] pointed out, some of these programs have tended to make matters worse than improve them.

Despite government efforts, poverty is still pervasive in rural areas, particularly in the former homelands where smallholder farmers are the majority. Problems of unemployment, low income levels and food insecurity continue to be a challenge for the government. Alleviating food insecurity therefore requires more effort to be directed towards developing smallholder agriculture. In contrast to the international scene where irrigated agriculture is recommended as an appropriate way of addressing rural poverty and unemployment in areas where sustained dryland production of crops is limited by **WATER DEFICIT**s[12], smallholder irrigation schemes in South Africa have not performed optimally to achieve the desired goal of increased crop productivity.

Currently there are 302 smallholder irrigation schemes with 206 still operational and the remainder non-operational[13]. The majority of the schemes are in Limpopo, KwaZulu-Natal and the Eastern Cape provinces. Researchers in smallholder irrigation schemes have found that much of the schemes are operating below their potential[13-16]. An analysis of yields of main crops grown under irrigation in the Eastern Cape indicated that large gaps existed between yields achieved by farmers and those achieved with good management[15]. Yields of maize (*Zea mays*) and butternut (*Cucurbita moschata*) were 24 and 22% of maximum economic yields obtained in on-farm experiments conducted in the scheme during the same period. The fact that the yields from on-farm experiments were comparable to commercial yields suggests that other factors were responsible for the low crop productivity experienced by farmers.

Constraints and challenges faced by smallholder dryland farmers include those linked with tillage operations, cropping systems and patterns, cultivar choices, planting populations and times, plant stands, soil nutrients/fertilizer and **WATER MANAGEMENT**. However, before suggesting solutions to the problems there is need to understand what constitutes smallholder farming and an understanding of the constraints and challenges faced by the farming systems. The main objectives of this study were to (1) Review the constraints and challenges faced by smallholder farmers in South Africa, (2) Assess the impact of the agronomic constraints on smallholder cropping systems, (3) Explore the relationship between agronomic constraints and crop yields and (4) Suggest some of the interventions, options and opportunities that South Africa can utilise to improve productivity in the smallholder farming sector.

There is no standard definition of a smallholder farmer but the term is generally used in the South African context for producers who are black and otherwise distinct from the dominant (and white dominated) large-scale commercial sector[17]. No clear distinctions can be drawn between categories such as smallholder, small-scale, subsistence, communal or emergent or peasant. In the South African context and in this study, smallholder farmers are defined as black farmers most of whom reside in the former homelands. Smallholder farmers are usually affected by prices, subsidies and markets but the input and output markets, which are not fully formed, remain localised to some extent[18].

2. MAJOR AGRONOMIC CONSTRAINTS TO IMPROVED CROP PRODUCTIVITY ON SMALLHOLDER FARMS

Smallholder farmers in South Africa are generally not farming at an optimum level. This is because the constraints and challenges facing smallholder farmers have not been addressed well, hence the persistence of low productivity levels. Challenges encountered by smallholders have been linked to historical, socio-economic, financial, natural and agronomic factors[19,20,15,21]. While challenges such as socio-economic and financial factors have support programs and institutions to assist the farmers, in the case of agronomic constraints there is very little progress on efforts to help farmers. Agronomic constraints that farmers are battling with include tillage practices, nutrient (fertilizer) management, water (irrigation and rainfall) management, cultivar choice, planting dates and planting densities and resultant plant stands[15, 16].

In South Africa, the smallholder crop producer challenges and constraints have been identified and studied but to date not many studies have found real solutions to the constraints. Generally, smallholder farmers are not efficiently producing or at least sustaining the production. Fanadzo *et al.*[15,22] pointed out that limited research has been done to evaluate the relationship between agronomic constraints or farmer management practices and the resulting crop productivity.

Tillage: There is clear evidence on how conventional tillage causes and contributes to soil degradation, erosion and desiccation. Frequent soil disturbance with implements such as ploughs and hoes, which is common in smallholder production in the Southern African region, does more harm than good[7]. One obvious negative effect of tilling the soil is the creation of the restrictive soil layers commonly known as hard pans. The restrictive hard pans created by frequent tillage impede soil infiltration and root penetration and causes accelerated oxidation of **ORGANIC MATTER**[7]. Conventional tillage is labor demanding in smallholder production and it is associated with poor timing of farming operations[23]. Resource poor farmers who depend on ox-traction, which is not uncommon in many smallholder productions for ploughing operations, are affected.

Kosgei *et al.*[24] reported that dryland smallholders maximised tillage because of perceptions that it resulted in increased yields with results from Potshini Catchment supporting their statement. However, that is untrue according to Rockstrom and Falkenmark[25]. Perennial cultivation of cereals under conventional tillage practices has been linked to yields below 1 t ha^{-1} where otherwise more than 3 t ha^{-1} can be achieved. This is partly because the restrictive layers delay seed emergence and compromise plant growth[25]. Johansen *et al.*[26] noted a reduction in soil particle infiltration and the resulting slow water infiltration as a result of continuous conventional tillage. Aeration is also reduced with increased compaction and the plant's roots are not distributed effectively under such conditions.

Wani *et al.*[27] reported reduced porosity and increased surface runoff with conventional tillage. Rockstrom[23] reported a total runoff of 10-25% in similar conditions. Runoff increases in soils where tillage has excessively destroyed the **ORGANIC MATTER** status of the soil leading to reduced **WATER HOLDING CAPACITY**[26]. In Thukela River Basin, a maize yield increase of more than 5% was achieved when the surface runoff was reduced by more than 80%[28]. In a study of tillage impacts on maize yields, Kosgei *et al.*[24] reported an increase in maize yield of 168, 133 and 120% in no-till plots compared to conventional tillage. Burning or removal of crop residues after tillage leaves the top soil exposed to adverse climatic conditions[23] such as runoff and wind, hence top soil loss which further leads to reduced soil fertility for crop production.

Water management: South Africa has an average annual rainfall of less than 500 mm. According to the United Nations Council on Combating Desertification (UNCCD) system for defining dry lands, more than 80% of South Africa's land surface is classified as **SEMI-ARID** to arid. Only 18% of is classified as dry sub-humid or subhumid[1]. Rainfed crop production is concentrated in the sub-humid zones as well as in the central and eastern reaches of the semiarid zone, where favourable soil characteristics occur[1].

Unlike farmers who practice irrigation, dryland farmers have to battle drought, which is a huge constraint to resource poor farmers without irrigation. Knowledge of drought incidence is crucial under dryland smallholder agriculture[29]. Such knowledge enables smallholder farmers to be prepared and to commence the necessary measures early to prevent crop failures. Mupangwa *et al.*[30] noted that not only drought and rainfall variability affect plant growth but difficulty in planning the planting dates, crop types and varieties increases under these conditions. Water productivity in smallholder dryland farming is a limiting factor to optimum production in many areas. In addition to drought, the cropping timeliness to coincide with rain, hydrological properties of soils and timely weed control[24]also pose major constraints. The low mean annual rainfall of 450-550 mm and high annual evaporation of 2000-2500 mm in South Africa results in severe crop water stress during most seasons[31]. These conditions were prevalent in **SEMI-ARID** areas where smallholder farmers are located with little capacity to establish conventional irrigation infrastructure to handle recurring droughts and periods of dry spells[24].

Water availability in dryland farming can be cited as the single most important challenge to crop growth[23,32-34]. Most of these studies concluded that low productivity in dryland agriculture mainly relates to management practices. According to Rockstrom *et al.*[32], physical potential was the least cause of poor productivity. Improved management practices can increase yields obtained in dryland agriculture. The challenge is to maximize water infiltration, mitigate dry spells and improve primarily soil fertility in order to increase productivity[7]. De Winnaar *et al.*[33] studied the Thukela River Basin where farmers normally grow maize and soya beans (*Glycine max*) to identify potential run-off harvesting in order to capture and efficiently use water to maintain agricultural production in an economic, environmental and sustainable manner. The authors concluded that providing information of runoff that is spatially relevant is a vital step for locating runoff-generating areas and determining areas within a catchment where surface water is generated, which is an important step in promoting runoff-harvesting technologies. Irrigation reduces or removes **WATER DEFICIT** as a limiting factor in plant growth and makes it possible to grow crops where the climate is too dry for this purpose and to increase crop yields[13]. However, research also showed that with adequate water but improper management practices yields were likely to remain poor in irrigated agriculture.

Generally, infield **WATER MANAGEMENT** in smallholder irrigation schemes in South Africa is weak. In the Eastern Cape, Fanadzo *et al.*[15]partly attributed this to the infield irrigation equipment that was in a state of dilapidation as many of the sprinkler systems used were very old and were not maintained well. Different standpipe lengths, sprinklers and nozzles were found in single laterals while many connections to the laterals often leaked due to worn-out threads (Table 1).

In KwaZulu-Natal, farmers reported problems with regards to irrigation methods; for example, furrow irrigation was reported as causing erosion and wasting water as it is difficult to measure the exact amount of water to be provided to the crops[16]. In the Eastern Cape, Fanadzo *et al.*[15] reported that farmers did not exercise irrigation scheduling but used a combination of plant observation and the feel method. Sprinkler irrigation schedules of two to four hour stand times every two to three days were common. In extreme cases, farmers irrigated overnight, resulting in over irrigation and wastage of water. Irrigation scheduling was generally constant regardless of crop type and growth stage, usually resulting in over irrigation during early crop growth stages and under-irrigation during the advanced growth stages.

Similar results were obtained in the Limpopo province where farmers applied large amounts of water at a time instead of applying reduced amounts of water at intervals that are more frequent. Over irrigation was one of the reasons for reduction in yields in Limpopo[35].

Soil fertility management: Soil fertility is defined as the quality of a soil that enables it to provide nutrients in adequate quantities and in proper balance for the growth of specified plants or crops[36]. A decline in soil fertility implies a decline in the quality of the soil. Soil fertility decline is defined as the decline in chemical soil fertility or a decrease in the levels of soil **ORGANIC MATTER**, pH, cation exchange capacity and plant nutrients[37]. Soil fertility decline thus includes nutrient depletion or nutrient decline (larger removal than addition of nutrients), nutrient mining (large removal of nutrients and no inputs), acidification (decline in pH and/or an increase in exchangeable Al), the loss of **ORGANIC MATTER** and an increase in toxic elements[37].

Table 1: Sprinkler system characteristics, uniformity and efficiency parameters at two smallholder farms in the Eastern Cape, South Africa[15]

Parameters	Nofemele farm	Kalawe farm	Norm
Sprinkler type	Rain Bird 30BH	Rain Bird 30BH	-
Nozzle sizes (mm)	3.6 and 4.0	3.6 and 4.0	-
Sprinkler spacing (m)	12	12	-
Lateral spacing (m)	12	18	-
Number of sprinklers per lateral	7	4-8	-
Average application (mm h^{-1})	6.05	2.87	-
Gross application (mm h^{-1})	10.14	5.81	≥5
Christiansen uniformity coefficient (%)	74.3	93.0	≥85
Distribution uniformity (%)	54.9	90.0	>75
Application efficiency (%)	59.6	49.3	>65
System efficiency (%)	32.7	44.4	>48
Pressure variation (%)	20.4	30.3	≤20

Nutrient removal may result in a decline of the soil fertility if replenishment with inorganic or organic nutrient inputs is inadequate. In many parts where smallholders are operating, declining soil fertility is a major production constraint[23,38]. A lack of fertilizers and manure has not only been identified as a factor contributing to low soil fertility but farmers also explained how soil erosion and poor **SOIL TYPE**s are also aggravating the situation. Even in areas of adequate rainfall distribution, crop yields remain largely low. Crop management practices and soil fertility were cited by Minde *et al.*[38] as possible contributing factors to low crop productivity by smallholders. In the Olifants River Basin of South Africa, Magombeyi and Taigbenu[39] reported that even with effective management of pests and adequate moisture, yields are expected to decrease by 50% if there is a lack of macronutrients (NPK).

Knowledge of land conditions and soil health constraints is necessary to plan management options and to apply necessary soil fertility enhancements[40]. Soil health is a huge research topic under smallholder irrigation[35, 41, 42]. Research also indicated that dryland smallholder farmers tend not to apply fertilizer, fearing risks of crop failures caused by dry spells and drought[43] and at times, the low availability and the high cost of fertilizer is an impediment[44]. Soil fertility is reported to be very low in much of the Sub-Saharan Africa region and Rockstrom[23] added that farmers in Africa generally apply 11 kg of fertilizer per every harvestable hectare, while in developed countries the average fertilizer applied for every hectare is 62 kg. Soil fertility constraints often constitute the primary limiting factor to crop growth in both irrigated and dryland agriculture. Andersson *et al.*[45] identified nitrogen as the main constraint to crop growth in the Thukela river basin.

There is a considerable relationship between water usage and fertilizer applications in crop production. A combination of supplementary irrigation and fertilizer applications leaded to improvements in yields in drylands[7,43]. However, fertilizer application alone resulted in higher yields than supplementary irrigation alone. It is not uncommon in South Africa where smallholder farmers generally use low inputs to use kraal manure as a source of fertilizer. Smallholder farmers resort to manure and other non-synthetic fertilizers due to lack of capacity to buy commercial fertilizers.

Research indicated that soils under smallholder irrigation lacks the essential nutrients and farmers tend to ignore the detrimental effects of soils cultivated without replenished fertilizers. For instance, farmers in the Eastern Cape applied fertilizers at random and the applications were only made once in two to three years[16]. Mandiringana *et al.*[46] reported that soils under irrigation schemes in the Eastern Cape are already nutrient depleted. In Limpopo province, Machethe *et al.*[35] found that the soils at some irrigation schemes had accumulated salts to toxic levels due to poor **WATER MANAGEMENT**.

Fanadzo *et al.*[47] identified poor timing and low amounts of fertilizer application as the major causes of low productivity in maize and butternut production under irrigation in the Eastern Cape (Table 2). In the same province, fertilizer application rates were usually not based on soil fertility analysis and recommendations[15]. In Limpopo province, Machethe *et al.*[35] reported that farmers applied blanket amounts of inorganic fertilizers and the rates were mainly marginal, especially for the field crops. Farmers cited a lack of information on fertilizer recommendations and funds as the main reasons for resorting to low blanket applications. A study conducted in the Limpopo province showed that 5% of the farmers lacked knowledge of fertilizer use while 48% had insufficient financial power to purchase the fertilizers[41]. Farmers are not only unable to obtain appropriate fertilizers but they are also unaware of the correct fertilizers for usage in already depleted soils[38,48]. Farmers need to know the recommended fertilizer application rates and Odhiambo and Magandini[41] stressed that while that may seem obvious, inappropriate applications of fertilizer or soil amendments can waste resources, cause water pollution and damage soils.

Farmers do not have knowledge of nutrient requirements of the crops that they plant, which can increasingly overuse the nutrient reserves. Irrigation imposes a great demand for nutrients and most crops grown under irrigation are high value vegetable crops that take up large quantities of nutrients from soils. Provision should therefore, be made to replace nutrients removed by these heavy feeders for farmers to continue realizing profitable yields. The interaction of moisture supply and nutrient supply is reciprocal such that if the farmer cannot irrigate, it is a waste to fertilize and if a farmer cannot fertilize, it is a waste to irrigate[49]. Therefore, with inadequate fertilizer management under the smallholder sector, farmers are not getting the potential yields.

Cropping patterns: Cropping patterns (yearly sequence and spatial arrangement of crops or of crops and fallow on a given area) present an advantage and opportunity to farmers, allowing them to sow crop after a crop and increase cropping intensity[26]. Research indicated that most smallholder farmers in South Africa practice monoculture of maize and this is practiced by 88% of farmers in Limpopo dryland production[38]. However, Van Duivenbooden *et al.*[34] reported intercropping with leguminous crops at low densities. Snapp *et al.*[50] explained the importance of legumes for both food and feed nutrition and their contribution to subsequent cereal productivity through biologically fixed nitrogen[44], pest build-ups/cycles breakdowns, conservation and sustainable agriculture promotion. Furthermore, Ogindo and Walker[31] found out a low surface evaporation rate in a maize-bean intercrop in the Free State province. However, the transpiration rate also amounted to 5-6% higher than a sole maize and bean. A high leaf area of the intercrop can suppress weeds and provide soil moisture retention.

Table 2: Fertility management and yields of grain maize, green maize and butternut at Zanyokwe irrigation scheme, Eastern Cape, South Africa[15]

Variable (kg ha^{-1})	Seasons			
	2005/06	2006/07	2007/08	Mean
Grain maize				
Basal N	13.2	13.2	13.2	13.2
Top-dress N	28.0	37.8	46.1	37.3
Total N	41.2	51.0	50.5	47.6
Grain yield	2266	1417	3489	2391
Green maize				
Basal N	-	9.7	11.9	10.8
Top-dress N	-	52.5	45.7	49.1
Total N	-	60.6	57.6	59.1
Percent cob sales	-	42.6	49.1	45.9
Butternut				
Basal N	13.2	10.9	11.9	12.0
Top-dress N	48.8	58.2	52.4	53.1
Total N	62.0	68.7	50.6	60.4
Total yield	6800	8100	3200	6000

Planting densities of the intercropping system depends on both rainfall and **SOIL TYPE**. A common rotational practice involves maize with either a legume crop or vegetable. In the case of vegetables, farmers are mainly using them for local market income. Magombeyi and Taigbenu[39] in a study of the Olifants River basin found the same results where farmers generally planted maize in the area for approximately 121 days and after harvest, farmers prepare their fields for winter vegetables. In addition, in many parts of Sub-Saharan Africa, the inclusion of cereal-legume such as millet (*Pennisetum glaucum*), sorghum (***SORGHUM BICOLOR***) and wheat (***TRITICUM AESTIVUM***) is common[34].

Cropping patterns generally used in many smallholder-irrigated areas involve alternating summer and winter crops, field crops and vegetables[20,35]. Depending on province, the type of crops grown under irrigation differs. Of all the crops, maize is the most common and important summer crop grown in terms of the areas devoted to crop type and number of growers[20,41,38]. In Limpopo, Machethe *et al.*[35]reported maize or cotton followed by winter wheat as the dominant rotation system, while in the Eastern Cape, Fanadzo *et al.*[15]reported dominant cropping patterns of maize monoculture, butternut-cabbage and maize-cabbage rotations. Under irrigation in KwaZulu-Natal, it is possible to cultivate three crops a year on a single plot but many farmers leave at least some plots fallow in the winter months[51].

The use of arable land is not optimum in smallholder cropping, both under irrigation and under dryland. Land-use, as expressed in terms of the number of crops that are cultivated on a particular surface area per year or season is termed cropping intensity. Whilst cropping intensities of 200% are possible under irrigation in the Eastern Cape[14], a study by Fanadzo *et al.*[15] indicated low cropping intensities averaging 48% from 2006 to 2008 (Table 3).

Similar low cropping intensity levels have been reported in smallholder irrigation schemes in other provinces[14,52,53]. Increased crop productivity with higher cropping intensities is well documented[54-56].

Lack of motivation and resources were identified as the two main factors responsible for the underutilization of land in the Eastern Cape[15]. The low cropping intensities were attributed to minimal winter cropping because of limited markets for most winter vegetables such as cabbage (*Brassica oleracea* var. *capitata*), spinach (*Spinacea oleracea*), beetroot (*Beta vulgaris*) and carrots (*Daucus carota*). Farmers relied on customers who came to buy the vegetables from the field. Thus, even at the low winter cropping intensities, some of the cabbage was observed to rot in the field after farmers failed to secure customers on time.

Table 3: Cropping patterns at Zanyokwe irrigation scheme, Eastern Cape, South Africa (Modified from Fanadzo *et al.*[15])

Parameters	Seasons		
	2005/06	2006/07	2007/08
Number of crops			
Summer	1.70	2.70	2.13
Winter	0.73	0.70	0.40
Total	2.43	3.40	2.53
Cropped area			
Summer	41.57	43.20	39.33
Winter	7.97	6.53	5.33
Cropping intensity	49.54	49.73	44.66

Cultivar choice: Literature available on crop cultivars does not give specific details of the varieties of crops used by smallholder farmers in South Africa based on their farming systems. However, Fanadzo *et al.*[15] reported that irrigation farmers in the Eastern Cape used a combination of hybrid seeds and Open Pollinated Varieties (OPVs). Van Averbeke *et al.*[13] reported that hybrid maize yielded 50% more than the OPVs commonly grown by farmers in the same irrigation scheme. With timely planting and optimum fertilizer application, long season maize cultivars were favoured over short season cultivars, while the latter was a better option with delayed planting as long as the cultivars were grown at higher densities and well fertilized under irrigation in the Eastern Cape[22].

At Tugela Ferry irrigation scheme in KwaZulu-Natal, Cousins[51] reported that almost all farmers planted early green maize to benefit from the crop's comparative market advantage. Gouse[57] surveyed the Hlabisa district of KwaZulu-Natal and found out that at the end of the 2009/2010 season none of the farmers planted *Bacillus thuringiensis* (Bt) maize and few still planted conventional maize and the rest planted Herbicide Tolerant (HT) or BR (a stacked traits of Bt and HT combination) maize[57]. Farmers seemed to be willing to pay for weed-control and labour-saving benefits of HT maize than the borer-control insurance of the Bt maize.

Although literature lacks details on cultivars grown by smallholder farmers, especially under dryland, specifics are available for the type of cultivars that farmers can utilize. According to Van Duivenbooden *et al.*[34], the short-period growth varieties that can withstand the effects of drought periods were recommended for dryland drought prone environments. Ogindo and Walker[31] gave an indication of cultivars which form base for the recommended varieties in which a short season PAN 6804 [PANNAR (Pty) Ltd., South Africa] maize cultivar maturing in 121 days was used in relation to soil **MOISTURE CONTENT**. When cultivar PAN 148 was grown together with dry bean in intercropping system, the recorded advantages were water conservation and high biomass creation due to increased transpiration. The fertilizer application rate, planting density per hectare and cultivar choice forms a link when deciding on appropriate cultivars for dryland production[31].

Planting time and densities: Research showed significant relationships between moisture, fertilizer applications, planting populations and planting dates. In a study by Fanadzo *et al.*[22], planting time and plant population interaction showed that increasing plant population from 40,000 maize plants ha^{-1} resulted in higher yields when planting was done early but significantly lower yields when planted late under irrigation. Late planting was identified by Van Averbeke *et al.*[13] as one of the major causes for low productivity in many smallholder farms. In the Eastern Cape, maize **GRAIN**

YIELD decreased by 38% when planting was delayed[15]. Results from the dry parts of South Africa showed that optimum population of maize for best yields decreased from 60,000 plants ha^{-1} with 650 mm water supply to 10,000 plants ha^{-1} with 240 mm water availability[58].

Economic crop yields arise from plant densities that minimize inter and intra-row competition, which widely depends on environmental conditions, while cereal **GRAIN YIELD** is the product of heads per unit area, kernels per head and kernel weight[34]. Factors that influence these components are seeding density, plant distribution and genotype for a given area. For grain legumes, which form part of most rotational and intercropping in many smallholders[48], an optimum seeding rate depends on the delicate balance between seedling death by fungal pathogens, extent of infestation by foliar diseases and insect pests, soil moisture status throughout the crop cycle and production of sufficient vegetative growth for supporting yield formation[26]. The recommended planting density of maize and *Sorghum* grains under dryland in southern Africa is 10,000-32,000 and 50,000-80,000 plants ha^{-1}, respectively[34]. Plant population is influenced by other prevailing conditions in achieving an optimum population. Soil fertility and moisture are known to be playing a fundamental role. In conditions where soil fertility and moisture are low, planting density is likely to be affected and dryland farmers are already facing that problem.

Planting population management is generally poor in many smallholder areas under irrigation with most studies showing results of under population. Higher yielding cultivars at optimum planting densities are important if higher yields are to be achieved. However, plant populations interact with other factors such as fertilizers particularly nitrogen, cultivar selection and planting time. With nutrients and season length non-limiting, the higher plant populations will yield more for short season cultivars and long season cultivars would yield more at low planting populations in maize[22].

Machethe *et al.*[35] reported maize planting densities of 55,000-80,000 plants ha^{-1} under irrigation in Limpopo province. However, densities were relatively high taking into consideration the fact that fertilizer especially nitrogen was marginal[35]. Fanadzo *et al.*[59]showed that the standard planting population of 40,000 plants ha^{-1} which is a norm amongst smallholder irrigation farmers in the Eastern Cape is not enough to optimize on both green and grain maize production under irrigation. As supported by various studies[60-65], for example, knowledge of optimum planting populations and determinants of the choice of planting density is crucial for achieving high yields. Manipulation of planting densities and row spaces play a vital role in grain production. Maize is the agronomic species that is most sensitive to changes in planting density, such that for each production system, there is a population that maximizes the utilization of available resources allowing for the expression of maximum attainable yield in the environment[63].

OPPORTUNITIES FOR IMPROVED PRODUCTIVITY

Research showed that water is the main constraining factor to productivity under dryland agriculture and other factors such as nutrient and pest management techniques play a lesser role[66]. Rockstrom[23]identified two broad strategies for increasing yields in dryland farming when water availability in the root zone constrains crop growth: (1) Capturing more water and allowing it to infiltrate into the root zone and (2) Using the available water more efficiently (increasing water productivity) by increasing the plant water uptake capacity and/or reducing non-productive soil evaporation. Results by Rockstrom *et al.*[67] correlated with what they had reported earlier[32] on how to mitigate periods of dry-spells in **SEMI-ARID** rainfed environment:

- Maximise plant water availability (maximise infiltration of rainfall, minimise unproductive water losses, increase soil **water holding capacity** and maximise root depth)
- Maximise plant water uptake capacity (timelines of operations, crop management and soil fertility management)
- Bridge crop **water deficit**s during dry-spells through supplementary irrigation

The challenge in dryland farming is to maximise infiltration, mitigate dry-spells and to improve primarily soil fertility management in order to increase water productivity. Approaches such as rainwater harvesting techniques are argued as means to manage drought and dry spells in dryland agriculture[33,45]. Biazin *et al.*[43] stated that the techniques for enhancing infiltration, reducing runoff and evaporation or improving soil moisture storage in the crop rooting zone are known as *in situ* rainwater harvesting. These techniques are aimed at enhancing rainfall infiltration ad reducing soil evaporation. The most commonly applied *in situ* rainwater harvesting and management practices include ridging, mulching, various types of furrowing and hoeing and conservation tillage[43].

In the Free State province, intercropping was identified as one option to maximise water usage by reducing the rate of soil water evaporation on smallholder farms[31]. The maize-bean intercrop had the lowest soil surface evaporation as a percentage of precipitation and evapotranspiration. This confirmed what has been already discussed in many studies that intercrop systems, due to an early and higher leaf area index, are able to reduce the amount of energy available for soil surface evaporation. In addition to reduction in evaporation, Nyakudya and Stroosnijder[68] reported that the use of mulch protects the soil from raindrop impact and thus reduces soil surface crusting and encourages infiltration and moderates the **HIGH TEMPERATURE**s in the upper root zone. In a study by Mutiro *et al.*[66], it was reported that conservation agriculture improved root penetration and infiltration. The yields of plots that received supplementary irrigation coupled with ripping were higher than supplementary irrigation with no ripping. The study further concluded that ripping increases water availability in the root zone. In addition to intercropping for coping in drought periods, farmers in the Olifants River basin used technology driven efforts such as a new short-duration and drought tolerant crop varieties[39]. Integrated Soil Fertility Management (ISFM) is another concept developed since the 1990s, which focuses on application of soil fertility management practices and the knowledge to adapt these to local conditions, thereby maximising fertilizer and organic resource use efficiency and crop productivity[40]. These practices include appropriate fertilizer and organic input management in combination with the utilization of improved germplasm[69]. Techniques like seed priming (soaking seeds in water before sowing, for 24 h in case of maize) which enhances crop water use efficiency by hastening germination and emergence should be promoted[70].

There is considerable scope for improving water productivity of crop, livestock and fisheries from field through to basin scale. Practices used to achieve this include water harvesting, supplemental irrigation, deficit irrigation, precision irrigation techniques and soil-water conservation practices. Practices not directly related to **WATER MANAGEMENT** impact water productivity because of interactive effects such as those derived from improvements in soil fertility, pest control, crop selection or access to better markets. However, there are several reasons to be cautious about the scope and ease of achieving water productivity gains. Crop water productivity is already quite high in highly productive regions and gains in yield (per unit of land area) do not necessarily translate into gains in water productivity. Reuse of water that takes place within an irrigated area or a basin can compensate for the perceived losses at the field-scale in terms of water quantity, though the water quality is likely to be affected. While crop breeding has played an important role in increasing water productivity in the past, especially by improving the harvest index, such large gains are not easily foreseen in the future. More importantly, enabling conditions for farmers and water managers are not in place to enhance water productivity. Improving water productivity will therefore require an understanding of the biophysical as well as the socioeconomic environments crossing scales between field, farm and basin.

Supplementary irrigation: Supplementary irrigation can mitigate the dry spells in dryland agriculture and this can be achieved through water harvesting systems that collect local surface runoff in small storage structures[7]. It is a key strategy, still underused for achieving rainfed yield potential and water productivity. Supplementary irrigation systems can also be used in small vegetable gardens during dry seasons to produce fully irrigated cash crops[67]. According to Rockstrom[23], water-harvesting systems with storage for supplementary irrigation were not very common in Eastern and Southern Africa. In the Olifants River basin, rainwater harvesting tanks for supplementary irrigation were used to irrigate maize crop in bridging the intra-seasonal dry-spells and in irrigation of winter vegetables to improve family food security[39].

Rainwater harvesting: Faures and Santini[71] defined water harvesting as a process of collecting and concentrating rainfall as runoff from a catchment area to be used in a smaller area, either for agriculture or other purposes. The purpose of rainwater harvesting was to make more efficient use of rainwater that would otherwise be lost as runoff or through evaporation[72]. Footpaths, dirt roads and compounds consist of compacted soil, often with heavy erosion crusts that produce high volumes of runoff. Road runoff can vary from a simple diversion structure directing surface water into crop fields, to deep trenches with check-dams in order to enable both flood and subsurface irrigation[73]. The available means of water harvesting include micro-catchment systems, basins, pitting systems and water storage systems. The latter can also serve domestic purposes, watering livestock and supplementary irrigation[73].

As part of **WATER MANAGEMENT** practices under dryland crop production, the Agriculture Research Council's Institute for Soil, Climate and Water (ISCW) developed what is called the in-field rainwater harvesting (IRWH) technology in South Africa with the objective of harnessing rainwater for crop production[74]. This technique (Fig. 1) has resulted in a significant increase in crop yield compared to conventional practices[74]. According to Sullivan *et al.*[75], with this technology, rainfall runned off compacted strips or bunds and was collected in rectangular basins running along crop rows and infiltrating deep into the soil beneath the surface evaporation zone. The basin areas are covered with locally available mulch to further retain soil moisture. The technique is specifically suited to smallholder farmers who farm on soils with high clay content.

Conservation agriculture: Tillage operations and the management of crop residues are important in water conservation, particularly in dry areas[34]. In **SEMI-ARID** regions, conventional tillage has mainly four purposes: To prepare the seedbed, promote infiltration, water conservation within the soil profile and to prevent wind and water erosion. According to Rockstrom *et al.*[67], a change from conventional to conservation agriculture will increase crop productivity by 20-120% and water productivity by 10-40%. However, potential disadvantages include higher costs of pest control, the cost of acquiring new management skills and investments in new ploughing equipment. Conservation agriculture covers a spectrum of non-soil-inversion practices, from zero tillage to reduced tillage, aiming at maximizing infiltration and soil productivity and minimizing water losses while simultaneously conserving energy and labour[43].

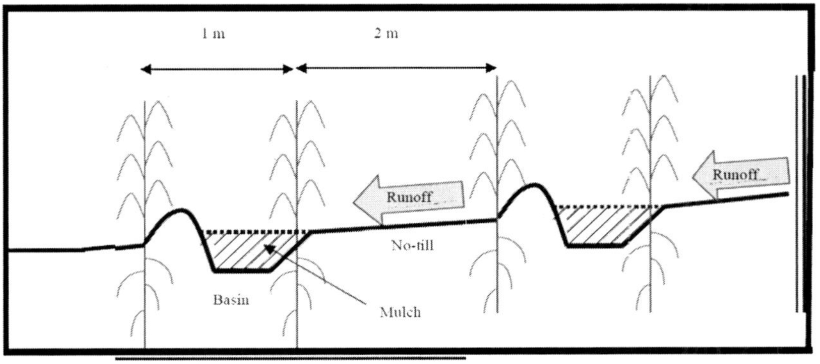

Figure. 1: Infield Rain Water Harvesting (IRWH) technique[74]

Like integrated soil fertility management, a combination of conservation agriculture with other associated technologies like the use of manure, green manure cover crops, the use of drought-tolerant and low nitrogen requiring cultivars and agro-forestry are common successful attempts to adapt the system to the local farming systems[40]. No-till is a powerful point of entry to solve the problems of soil erosion, soil fertility and soils with low **WATER HOLDING CAPACITY**[34]. Crop yields from no-tillage agriculture are usually as high as or higher than yields produced by conventional tillage. Rockstrom and Barron[76] supported these findings in that tillage influences both soil conditions at the surface, i.e. infiltration capacity and the structure in the top soil, which will affect the plant water availability and crop productivity[34].

Land and soil fertility management: Use of crop residues by farmers as fodder and none or shorter fallow periods due to a shrinking land resource base, should be balanced by the addition of chemical fertilizers and organic manure, which most smallholder farmers cannot afford. Therefore there is need to develop appropriate soil nutrient and cropping systems that minimize the need for chemical fertilizers and also find ways to integrate livestock into the farming systems[73]. In the deficient soils of smallholders in Zimbabwe, Kumwenda *et al.*[48] reported that additions of micronutrients improved the yield response to nitrogen and phosphorus. Nutrients such as zinc, boron, sulphur and magnesium can often be included relatively cheap in existing fertilizer blends and when targeted to deficient soils, these nutrients can dramatically improve fertilizer use efficiency and crop profitability.

Alternative sources of nutrients are needed in situations where soil fertility needs to be rebuilt and high costs and quantities limit inorganic fertilizer applications. Use of green manure cover crops has been suggested as one viable option because of their ability to regulate soil surface temperatures, improve soil **ORGANIC MATTER**, conserve soil moisture and suppression of weeds. High-quality organic manures provide readily available N, energy (carbon) and nutrients to the soil ecosystem and they build soil fertility and structure over the long term[48]. Research by Ncube *et al.*[77] showed evidence that there is potential to improve livelihoods of smallholder farmers through the use of small rates of manure and N under **SEMI-ARID** conditions in Zimbabwe. However, while some research studies recommend that organic fertilizers are possible and convenient way resource-poor farmers can improve their soil fertility, Minde *et al.*[38]suggested that the use of chemical fertilizers is the surest way of restoring soil fertility in the Limpopo province.

To prevent erosion, farmers in KwaZulu-Natal dug furrows by hand or opened with ploughs and then shaped by hand and placed upslope the field to re-direct water from the fields during heavy rains[78]. Physical structures (for example contour ditches, mulching and windbreaks) reduce soil erosion, especially on sloping land[77]. Technologies that improve soil fertility and productivity are as important as those that reduce soil erosion and water loss[73]. These include practices such as residue mulching, contour tillage and tied ridging, minimum tillage, sub-soiling, crop rotation, cover cropping, rotational grazing, contour ripping and direct applications of **ORGANIC MATTER**, farmyard manure and inorganic fertilizers.

Other possible opportunities: Assisting smallholders to pursue sustainable intensification of production through adoption of better seeds, technologies and other improved inputs and assisting them to use them more effectively is expected to result in improved crop productivity[79]. Transplanting is one strategy that is commonly used to establish crops when conditions are less favourable for direct seeding, such as when birds pose a threat to emerging seedlings[15]. In the Eastern Cape, Fanadzo *et al.*[59] reported that farmers would obtain higher yields and profits by increasing plant populations to 60,000 plants per hectare without necessarily having to change their inter-row spacing, although narrow rows would result in slightly higher yields and would help in weed suppression. Ensuring that farmers select the appropriate combination of crops could contribute to increased production and productivity[35], for example, incorporation of leguminous crops into the rotation would raise the nitrogen level of in the soil.

3. CONCLUSION AND RECOMMENDATIONS

Current agronomic practices make the problems already faced by the farmers even worse. Because the gap between farmer yields and the biophysical potential is very large, greater and easier gains are possible through better agronomic management practices. Programs that target emerging smallholder black farmers who in most cases have been found to possess limited farming skills will enable them to improve food security. Practices like rainwater harvesting are relatively new and farmers who are unfamiliar with the practices will need training and mentorship. All relevant stakeholders, including policy makers, private corporations and extension should assist farmers to alleviate crop production problems.

Smallholder farmers should not only be given priority in land and water provision programs but they should also be actively involved in the development of the programs that are meant for them. Training farmers and extension staff in irrigation water and crop management practices and ensuring a reliable supply of irrigation water should receive urgent attention. Sustainable technologies of improving soil fertility such as inclusion of nitrogen fixing species should be encouraged since inorganic fertilizers are expensive and organic sources often inadequate and of poor quality. Enabling policies and strategies need to be formulated to support smallholder farmers to increase productivity by providing the necessary infrastructure and support services, while at the same time fostering self-sustenance and sustainability.

REFERENCE

1. ARC., 2014. Annual report for 2013/2014. Agricultural Research Council, Pretoria, South Africa. http://www.arc.agric.za/Documents/Annual%20Reports/ ARC%20Annual%20 Report%202013%202014.pdf.
2. Andersson, J.C.M., A.J.B. Zehnder, G.P.W. Jewitt and H. Yang, 2009. Water availability, demand and reliability of *in situ* water harvesting in smallholder rain-fed agriculture in the Thukela River Basin, South Africa. Hydrol. Earth Syst. Sci., 13: 2329-2347.
3. Andersson, J.C.M., A.J.B. Zehnder, J. Rockstrom and H. Yang, 2011. Potential impacts of water harvesting and ecological sanitation on crop yield, evaporation and river flow regimes in the Thukela River basin, South Africa. Agric. Water Manage., 98: 1113-1124.
4. Armour, J., 2014. Dualism in SA agriculture. Proceedings of the FERTASA 54th Annual Congress, June 10, 2014, Johannesburg, South Africa –
5. Backeberg, G.R. and A.J. Sanewe, 2010. Towards productive water use and household food security in South Africa. Proceedings of the 6th Asian Regional Conference of the

International Commission on Irrigation and Drainage (ICID), October 10-16, 2010, Yogyakarta -.
6. Bembridge, T.J., 2000. Guidelines for rehabilitation of small-scale farmer irrigation schemes in South Africa. WRC report 891/1/00. Water Research Commission, Pretoria, South Africa.
7. Biamah, E.K., 2005. Coping with drought: Options for soil and water management in semi-arid Kenya. Ph.D. Thesis, Wageningen University, Netherlands.
8. Biazin, B., G. Sterk, M. Temesgen, A. Abdulkedir and L. Stroosnijder, 2012. Rainwater harvesting and management in rainfed agricultural systems in sub-Saharan Africa: A review. Phys. Chem. Earth Parts A/B/C, 47: 139-151.
9. Cousins, B., 2011. What is a Smallholder? Class-Analytic Perspectives on Small-Scale Farming and Agrarian Reform in South Africa. In: Reforming Land and Resource Use in South Africa: Impact on Livelihoods, Hebinck, P. and C. Shackleton (Eds.). Routledge, Abingdon, UK., ISBN: 9781136886072, pp: 86-111.
10. Cousins, B., 2013. Smallholder irrigation schemes, agrarian reform and accumulation from above and from below in South Africa. J. Agrarian Change, 13: 116-139.
11. Crosby, C.T., M. de Lange, C.M. Stimie and I. van der Stoep, 2000. A review of planning and design procedures applicable to small-scale farmer irrigation projects. WRC Report 578/2/00, Water Research Commission, Pretoria, South Africa.
12. De Winnaar, G., G.P.W. Jewitt and M. Horan, 2007. A GIS-based approach for identifying potential runoff harvesting sites in the Thukela River basin, South Africa. Phys. Chem. Earth, Parts A/B/C, 32: 1058-1067.
13. Dube, E., C. Chiduza, P. Muchaonyerwa, M. Fanadzo and T.S. Mthoko, 2012. Winter cover crops and fertiliser effects on the weed seed bank in a low-input maize-based conservation agriculture system. S. Afr. J. Plant Soil, 29: 195-197.
14. Ellis, F., 1998. Household strategies and rural livelihood diversification. J. Dev. Stud., 35: 1-38.
15. FAO., 2000. Affordable irrigation technologies for smallholders: Opportunities for technology adaptation and capacity building. Programme Formulation Report No. 6, December 2000, IPTRID Secretariat, Rome, Italy.
16. FAO., 2005. Fertilizer use by crop in South Africa. Land and Plant Nutrition Management Service, Land and Water Development Division, Food and Agriculture Organization of the United Nations, Rome. http://www.fao.org/docrep/008/y5998e/y5998e00.HTM.
17. Fanadzo, M., A.B. Mashingaidze and C. Nyakanda, 2007. Narrow rows and high maize densities decrease maize grain yield but suppress weeds under dryland conditions in Zimbabwe. J. Agron., 6: 566-570.
18. Fanadzo, M., C. Chiduza and P.N.S. Mnkeni, 2009. Investigation of agronomic factors constraining productivity of grain maize (*Zea mays* L.) at Zanyokwe irrigation scheme, Eastern Cape, South Africa. J. Applied Biosci., 17: 948-958.
19. Fanadzo, M., C. Chiduza and P.N.S. Mnkeni, 2010. Pre-plant weed control, optimum N rate and plant densities increase butternut (*Cucurbita moschata*) yield under smallholder irrigated conditions in the Eastern Cape Province of South Africa. Afr. J. Agric. Res., 5: 2192-2199.
20. Fanadzo, M., C. Chiduza and P.N.S. Mnkeni, 2010. Effect of inter-row spacing and plant population on weed dynamics and maize (*Zea mays* L.) yield at Zanyokwe irrigation scheme, Eastern Cape, South Africa. Afr. J. Agric. Res., 5: 518-523.
21. Fanadzo, M., C. Chiduza and P.N.S. Mnkeni, 2010. Overview of smallholder irrigation schemes in South Africa: Relationship between farmer crop management practices and performance. Afr. J. Agric. Res., 5: 3514-3523.
22. Fanadzo, M., C. Chiduza, P.N.S. Mnkeni, I. van der Stoep and J. Stevens, 2010. Crop production management practices as a cause for low water productivity at Zanyokwe Irrigation Scheme. Water SA, 36: 27-36.
23. Faurus, J.M. and G. Santini, 2008. Water and the rural poor: Interventions for improving livelihoods in sub-Saharan Africa. FAO Land and Water Division, Food and Agriculture Organization of the United Nations and IFAD, Rome.

24. Gouse, M., 2012. GM maize as subsistence crop: The South African smallholder experience. AgBioForum, 15: 163-174.
25. Hart, T., 2008. Reviewing 15 years of resource-poor small-scale agriculture in South Africa: Is there any way forward? HSRC: CPEG TIPS Forum, October 2008, Cape Town, South Africa, pp: 1-20.
26. Hartemink, A.E., 2006. Soil fertility decline: Definitions and assessment. Encyclopedia Soil Sci., 2: 1618-1621.
27. Hasnip, N., S. Mandal, J. Morrison, P. Pradhan and L. Smith, 2001. Contribution of irrigation to sustaining rural livelihoods. KAR Project R7879, Report OD/TN109. http://books.hrwallingford.co.uk/acatalog/od_downloads/odtn109.pdf.
28. Hensley, M., J.J. Botha, J.J. Anderson, P.P. van Staden and A. du Toit, 2000. Optimizing rainfall use efficiency for developing farmers with limited access to irrigation water. Water Research Commission Report No. 878/1/00, Water Research Commission, Pretoria, South Africa.
29. Johansen, C., M.E. Haque, R.W. Bell, C. Thierfelder and R.J. Esdaile, 2012. Conservation agriculture for small holder rainfed farming: Opportunities and constraints of new mechanized seeding systems. Field Crops Res., 132: 18-32.
30. Kosgei, J.R., G.P.W. Jewitt, V.M. Kongo and S.A. Lorentz, 2007. The influence of tillage on field scale water fluxes and maize yields in semi-arid environments: A case study of Potshini catchment, South Africa. Phys. Chem. Earth, Parts A/B/C, 32: 1117-1126.
31. Kubbinga, B., 2012. Road runoff harvesting in the drylands of sub-Saharan Africa: Its potential for assisting smallholder farmers in coping with water scarcity and climate change, based on case studies in Eastern province, Kenya. M.Sc. Thesis, Free University, Amsterdam, Netherlands.
32. Kumwenda, J.D.T., S.R. Waddington, S.S. Snapp, R.B. Jones and M.J. Blackie, 1996. Soil fertility management research for the maize cropping systems of smallholders in Southern Africa: A review. NRG Paper No. 96-02, CIMMYT, Mexico DF., USA., pp: 1-35.
33. Lahiff, E. and B. Cousins, 2005. Smallholder agriculture and land reform in South Africa. IDS Bull., 36: 127-131.
34. Lipton, M., 1996. Rural Reforms and Rural Livelihoods: The Context of International Experience. In: Land, Labour and Livelihoods in Rural South Africa, Lipton, M., M. de Klerk and M. Lipton (Eds.)., Vol. 1, Indicator Press, Durban, South Africa, ISBN: 9781868402342.
35. Machethe, C.L., N.M. Mollel, K. Ayisi, M.R. Mashatola, F.D.K. Anim and F. Vanasche, 2004. Smallholder irrigation and agricultural development in the Olifants River Basin of Limpopo province: Management transfer, productivity, profitability and food security issues. Report No. 1050/1/04, Water Research Commission (WRC), Pretoria, South Africa.
36. Magombeyi, M.S. and A.E. Taigbenu, 2008. Crop yield risk analysis and mitigation of smallholder farmers at quaternary catchment level: Case study of B72A in Olifants river basin, South Africa. Phys. Chem. Earth, Parts, A/B/C, 33: 744-756.
37. Mandiringana, O.T., P.N.S. Mnkeni, Z. Mkile, W. van Averbeke, E. van Ranst and H. Verplancke, 2005. Mineralogy and fertility status of selected soils of the Eastern Cape Province, South Africa. Commun. Soil Sci. Plant Anal., 36: 2431-2446.
38. Mashingaidze, A.B., 2004. Improving weed management and crop productivity in maize systems in Zimbabwe. Ph.D Thesis. Wageningen University, Wageningen, The Netherlands.
39. Mati, B.M., 2005. Overview of water and soil nutrient management under smallholder rainfed agriculture in East Africa. Working Paper 105, International Water Management Institute (IWMI), Colombo, Sri Lanka.
40. Minde, I., T. Pedzisa and J. Dimes, 2008. Improving access and utilization of fertilizers by smallholder farmers in the Limpopo province of South Africa. Report No. 3, International Crops Research Institute for the Semi-Arid Tropics, Bulawayo.
41. Mnkeni, P.N.S., C. Chiduza, A.T. Modi, J.B. Stevens, N. Monde, I. van der Stoep and R. Dladla, 2010. Best management practices for smallholder farming on two irrigation schemes in

the Eastern Cape and Kwazulu-Natal through participatory adaptive research. WRC Report No. TT 478/10, Water Research Commission, Pretoria, South Africa.

42. Mpandeli, S. and P. Maponya, 2014. Constraints and challenges facing the small scale farmers in Limpopo province, South Africa. J. Agric. Sci., 6: 135-143.
43. Mupangwa, W., S. Walker and S. Twomlow, 2011. Start, end and dry spells of the growing season in semi-arid southern Zimbabwe. J. Arid Environ., 75: 1097-1104.
44. Mutiro, J., H. Makurira, A. Senzanje and M.L. Mul, 2006. Water productivity analysis for smallholder rainfed systems: A case study of Makanya catchment, Tanzania. Phys. Chem. Earth Parts A/B/C, 31: 901-909.
45. Ncube, B., J.P. Dimes, M.T. van Wijk, S.J. Twomlow and K.E. Giller, 2009. Productivity and residual benefits of grain legumes to sorghum under semi-arid conditions in south-western Zimbabwe: Unravelling the effects of water and nitrogen using a simulation model. Field Crops Res., 110: 173-184.
46. Ncube, B., J.P. Dimes, S.J. Twomlow, W. Mupangwa and K.E. Giller, 2007. Raising the productivity of smallholder farms under semi-arid conditions by use of small doses of manure and nitrogen: A case of participatory research. Nutr. Cycling Agroecosyst., 77: 53-67.
47. Nyakudya, I.W. and L. Stroosnijder, 2011. Water management options based on rainfall analysis for rainfed maize (*Zea mays* L.) production in Rushinga district, Zimbabwe. Agric. Water Manage., 98: 1649-1659.
48. Odhiambo, J.J.O. and V.N. Magandini, 2008. An assessment of the use of mineral and organic fertilizers by smallholder farmers in Vhembe district, Limpopo province, South Africa. Afr. J. Agric. Res., 3: 357-362.
49. Oettle, N., S. Fakir, W. Wentzel, S. Giddings and M. Whiteside, 1998. Encouraging sustainable smallholder agriculture in South Africa. Agricultural Services Reform in Southern Africa, June 1998, pp: 80.
50. Ogindo, H.O. and S. Walker, 2005. Comparison of measured changes in seasonal soil water content by rainfed maize-bean intercrop and component cropping systems in a semi-arid region of Southern Africa. Phys. Chem. Earth, Part A/B/C, 30: 799-808.
51. Perret, S., M. Lavigne, N. Stirer, S. Yokwe and S. Dikgale, 2003. The Thabina irrigation scheme in a context of rehabilitation and management transfer: Prospective analysis and local empowerment. Final Report, Project Number: 2003-068, DWAF, CIRAD-IWMI-UP, Pretoria, South Africa. http://ageconsearch.umn.edu/bitstream/18071/1/ wp 0400 05.pdf.
52. Rockefeller Foundation, 2006. Africa's turn: A new green revolution for the 21st century. Rockefeller Foundation, New York.
53. Rockstrom, J. and M. Falkenmark, 2000. Semiarid crop production from a hydrological perspective: Gap between potential and actual yields. Crit. Rev. Plant Sci., 19: 319-346.
54. Rockstrom, J., 2000. Water resources management in smallholder farms in Eastern and Southern Africa: An overview. Phys. Chem. Earth Part B: Hydrol. Oceans Atmosphere, 25: 275-283.
55. Rockstrom, J., J. Barron and P. Fox, 2002. Rainwater management for increased productivity among small-holder farmers in drought prone environments. Phys. Chem. Earth, Parts A/B/C, 27: 949-959.
56. Rockstrom, J., J. Barron and P. Fox, 2003. Water Productivity in Rain-Fed Agriculture: Challenges and Opportunities for Smallholder Farmers in Drought-Prone Tropical Agro-Ecosystems. In: Water Productivity in Agriculture: Limits and Opportunities for Improvement, Kijne, J.W., R. Barker and D.J. Molden (Eds.)., CABI., Sri Lanka, ISBN: 9781845933395.
57. Rockstrom, J., L. Karlberg, S.P. Wani, J. Barron and N. Hatibu *et al.*, 2010. Managing water in rainfed agriculture-the need for a paradigm shift. Agric. Water Manage., 97: 543-550.
58. Rockstrom, J., and J. Barron, 2007. Water productivity in rainfed systems: overview of challenges and analysis of opportunities in water scarcity prone savannahs. Irrig. Sci., 25: 299-311.
59. SSSA., 1997. Glossary of Soil Science Terms. Soil Science Society of America, Madison, WI.

60. Sanginga, N. and P.L Woomer, 2009. Integrated Soil Fertility Management in Africa: Principles, Practices and Development Process. CIAT, Nairobi, ISBN: 978929059261, Pages: 263.
61. Sangoi, L., 2000. Understanding plant density effects on maize growth and development: An important issue to maximize grain yield. Ciencia Rural, 31: 159-168.
62. Shaxson, F.T. and R.G. Barber, 2003. Optimizing Soil Moisture for Plant Production: The Significance of Soil Porosity, Volume 1. Food and Agriculture Organization of The United Nations, Rome, ISBN: 9789251049440, Pages: 107.
63. Shumba, E.M., 1984. Yields of maize in the semi-arid regions of Zimbabwe. Zimbabwe Agric. J., 81: 91-94.
64. Snapp, S.S., P.L. Mafongoya and S. Waddington, 1998. Organic matter technologies for integrated nutrient management in smallholder cropping systems of Southern Africa. Agric. Ecosyst. Environ., 71: 185-200.
65. Sommer, R., D. Bossio, L. Desta, J. Dimes and J. Kihara *et al*., 2013. Profitable and sustainable nutrient management systems for east and southern African smallholder farming systems-challenges and opportunities. A Synthesis of the Eastern and Southern Africa Situation in Terms of Past Experiences, Present and Future Opportunities in Promoting Nutrients use in Africa, May 2013.
66. Statistics South Africa, 2011. Census 2011 agricultural households. Report No. 03-11-01 (2011). http://www.statssa.gov.za/census/census_2011/census_products/ Agricultural_ Households.pdf.
67. Sterve, H., 2010. Factors restricting adoption of sustainable agricultural practices in a smallholder agro-ecosystem: A case study of Potshini community, upper Thukela region, South Africa. M.Sc. Thesis, Stockholm University, Sweden.
68. Sullivan, A., A. Mumba, S. Hachigonta, M. Connolly and L.M. Sibanda, 2013. Appropriate climate smart technologies for smallholder farmers in Sub-Saharan Africa. Fanrpan Policy Brief 2, XIII. Pretoria, South Africa.
69. Tafesse, M., 2003. Small-scale irrigation for food security in sub-Saharan Africa. CTA Working Document No. 8031, CTA, Wageningen, The Netherlands
70. Thamaga-Chitja, J.M. and P. Morojele, 2014. The context of smallholder farming in South Africa: Towards a livelihood asset building framework. J. Hum. Ecol., 45: 147-155.
71. Tollenaar, M. and J. Wu, 1999. Yield improvement in temperate maize is attributable to greater stress tolerance. Crop Sci., 39: 1597-1604.
72. Van Averbeke, W. and J.N. Marais, 1992. Maize response to plant population and soil water supply: I. Yield of grain and total above-ground biomass. S. Afr. J. Plant Soil, 9: 186-192.
73. Van Averbeke, W., 2012. Performance of Smallholder Irrigation Schemes in the Vhembe District of South Africa. In: Problems, Perspectives and Challenges of Agricultural Water Management, Kumar, M. (Ed.). Tech Press, Rijeka.
74. Van Averbeke, W., C.K. M'Marete, C.O. Igodan and A. Belete, 1998. An investigation into food plot production at irrigation schemes in central Eastern Cape. WRC Report No. 719/1/98, Water Research Commission, Pretoria, South Africa.
75. Van Averbeke, W., J. Denison and P.N.S. Mnkeni, 2011. Smallholder irrigation schemes in South Africa: A review of knowledge generated by the Water Research Commission. Water SA, 37: 797-808.
76. Van Duivenbooden, N., M. Pala, C. Studer, C.L. Bielders and D.J. Beukes, 2000. Cropping systems and crop complementarity in dryland agriculture to increase soil water use efficiency: A review. NJAS-Wageningen J. Life Sci., 48: 213-236.
77. WWF-SA., 2012. Agriculture: Facts and trends. WWF-SA., South Africa. http:// awsassets.wwf.org.za/downloads/facts_brochure_mockup_04_b.pdf.
78. Wani, S.P., T.K. Sreedevi, J. Rockstrom and Y.S. Ramakrishna, 2009. Rainfed Agriculture-Past Trends and Future Prospects. In: Rainfed Agriculture: Unlocking the Potential, Wani, S.P., J. Rockstrom and T.Y. Oweis (Eds.). CAB International, London.
79. Whingwiri, E.E., K. Mashingaidze and M. Rukuni, 1992. Small-Scale Agriculture in Zimbabwe. Book 2. Field Crop Production. Rockwood Publisher, Harare, Zimbabwe.

CHAPTER 9

The Current Status, Problems, and Prospects of Alfalfa (*Medicago sativa* L.) Breeding in China

Shangli Shi [1,*], Lili Nan [1] and Kevin F. Smith [2]

[1] *Ministry of Education Key Laboratory of Ecosystems, College of Prataculturai Science, Gansu Agricultural University, Lanzhou 730070, Gansu, China*

[2] *Faculty of Veterinary and Agricultural Sciences, The University of Melbourne, Private Bag 105 Hamilton, Victoria 3300, Australia*

ABSTRACT

This paper reviews the current status, methodology, achievements, and prospects of alfalfa (*Medicago sativa* L.) breeding in China. There are 77 cultivars that have been registered in the country, these include 36 cultivars bred through breeding programs, 17 introduced from overseas, 5 domesticated from wild ecotypes, and 19 through regional collection/breeding programs. Cultivars have been selected for cold resistance, disease resistance, salt tolerance, grazing tolerance, high yield, and early maturity. Most of these cultivars have been developed through conventional breeding techniques, such as selective and cross breeding, and some are now being evaluated that have been developed through the application of transgenic technology. The major problems for alfalfa breeding in China include low success rate, slow progress to breed resistant varieties, lack of breeding materials and their systematic collection, storage and evaluation, lack of advanced breeding techniques, and low adoption rate of new cultivars. There are gaps in alfalfa breeding between China and the developed world. Improvement of alfalfa breeding techniques, development of cultivars with adaptations to different regions within China, and the protection and utilization of alfalfa germplasm resources have been identified as major strategies to improve the efficiency of alfalfa breeding in China.

Keywords: China; alfalfa; varieties; breeding; selection; *Medicago sativa* L.

1. INTRODUCTION

China has abundant genetic resources of *Medicago spp*. with 12 species *M. sativa* subsp. sativa L., *M. ruthenica* L. Ledeb., *M. pubescens* Sirj., *M. archiducis-nicolai Sirjaev*., *M. platycarpos* L. Trautv., *M. sativa* subsp. *falcata, M. sativa* subsp. *x varia*, *M. polymorpha* L., *M. minima* L. Grnfb, *M. lupulina* L., and *M. orbicularis* L. Bart., which are mainly distributed in northern and southwest China [1].

M. sativa L. is a perennial species of the genus, originating in Asia Minor, Transcaucasia, Iran and Turkmenistan highlands. The common name "alfalfa" refers to the main species *M. sativa* subsp. sativa but may also describe *M. sativa* subsp. falcata and *M. sativa* subsp. *x varia* (a hybrid between subsp. sativa and subsp. *falcata*) which is closely related to *M. sativa* in morphology. Among the Medicago species, *M. sativa* has the largest cultivation area in the world at present because of its numerous superior traits such as cold resistance, salt tolerance, wide adaptability, high yield, good herbage quality, resistance to frequent cutting, good persistence, soil amelioration, and economic benefit. It is therefore regarded as "the queen of forages". As the concept of "Pasture-based livestock industries" gained prominence in China in recent years, and with the application of new policies such as subsidy for grassland ecological protection, revitalization of alfalfa for dairying, and conversion of crops to forest and grassland, *M. sativa* has become the most widely used species in integrated farming systems, grazing, and ecological conservation in China. It is therefore imperative to develop new alfalfa varieties to meet production targets under local conditions, which presents both opportunities and challenges for alfalfa breeding. Although there have been previous reviews of alfalfa breeding in China [2,3,4] these have not been widely available to an international audience, this review seeks to provide an updated review of the status of alfalfa breeding in China and a comment on the future needs for research and development.

2. ALFALFA BREEDING IN CHINA

Forage breeding in China commenced about half a century later than the developed world. Until 1949, only a few scientists collected and evaluated wild forage germplasm for agronomic adaptation. Modern herbage breeding programs commenced in the 1950s and have been developed extensively since the 1980s. The National Committee for the Examination and Approval of Pasture Variety (NCEAPV) was established officially in 1986, and has promoted the selection and development of new pasture varieties, the collection and domestication of local ecotypes, and the introduction of foreign forage germplasm.

As of 2015, a total of 77 Alfalfa varieties were registered by NCEAPV, including 36 domestic varieties, 17 introduced varieties, 5 domesticated varieties, and 19 local varieties [5,6]. These do not include varieties registered by regional authorities (provinces and autonomous regions).

Varieties developed in China can be divided into eight categories based on their main characteristics (Figure 1), which have been described briefly below:

1) High yield: such as cv. Gannong 3 and 4;

2) *Disease resistance:* such as cv. Zhonglan 1, which has high resistance to downy mildew (*Peronospera trifoliorum*), and moderate resistance to brown leaf spot and rust disease, with resulting increases of herbage production of ca. 30% compared with cv. Longdong in the presence of these diseases. The resistance to downy mildew and brown leaf spot of cv. Xinmu 4 is better than cv. Xinjiang Daye;

3) *Cold resistance*: such as cvv. Longmu 801, Longmu 803, Longmu 806, Longmu 808, Caoyuan 1, Caoyuan 2, Caoyuan 3, Tumu 1, Xinmu 1, Xinmu 3, and Chicao 1. Most of these have been developed for planting in high latitude and high altitude areas in North China;

4) *Salt tolerant*: such as cvv. Zhongmu 1, Zhongmu 3, and Zhongmu 5; these have more than 10% higher yields compared with control cultivars when planted in saline-alkali soil;

5) *Grazing tolerance*: creeping-rooted types alfalfa such as cvv. Gannong 2, Zhongmu 2, and Gongnong 3. This suitability to grazing is conferred through strong root systems and greater ground cover. These cultivars are also highly suitable for conservation of water and soil, wind-breaks and sand/dune stability, soil reinforcement, and slope protection which represent other important areas of use for alfalfa in the pastoral regions of China;

6) *Early maturity*: such as cv. Xinmu 2, which matures earlier (about three to five days) than cv. Xinjiang Daye, and has high rates of regrowth following grazing;
7) *Lower fiber concentrations*: such as cv. Gannong 7 which has Neutral Detergent Fiber(NDF) and Acid Detergent Fiber (ADF) concentrations 2 percentage points lower than that in other alfalfa, and crude protein 1 percentage point higher than other cultivars, and hence improved palatability;
8) *Insect resistance*: such as cv. Gannong 5, with high levels of resistance to aphids with yields increased 14.22% [7] compared with the control cv. Golden Empress, cv. Caoyuan 4 is a variety developed for regions with serious thrip infestations.

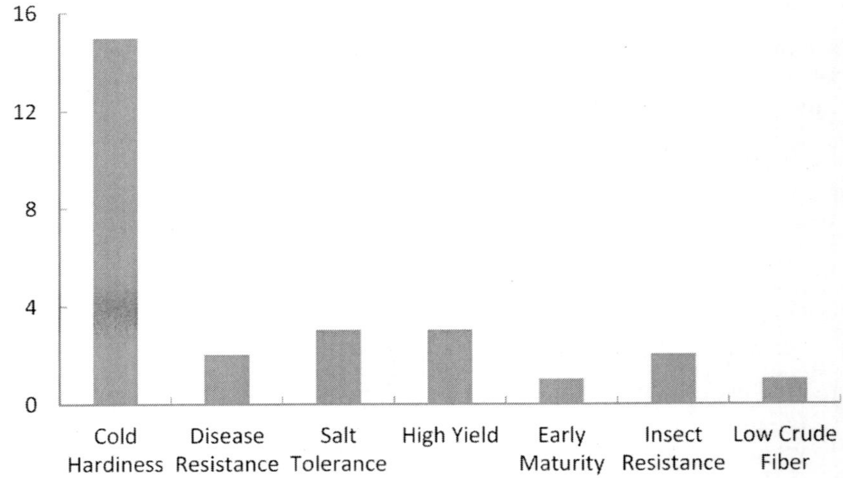

Figure 1. The number and reported general attributes of alfalfa varieties bred in China.

The introduction of new alfalfa cultivars from overseas has been increasing with the rapid development of the alfalfa industry in China. The number of alfalfa varieties imported to China has increased dramatically to a current level of approximately 400. There are presently 17 varieties registered by the NCEAPV, including cvv. AmeriGraze 401+Z, Derby, Sanditi, and WL525HQ. (Table 1). This has been attributed to the improved production and forage quality that can be directly utilized for animal production. These importations also contribute to the pool of the alfalfa germplasm resources for alfalfa breeding in China.

Table 1. Introduced Medicago sativa varieties registered by the National Committee for the Examination and Approval of Pasture Varieties.

Num.	Cultivar	Country	Main Characteristics
1	AmeriGraze 401+Z	USA	Cold tolerance, regeneration
2	Derby	France	Lodging resistance, cold tolerance, high yield
3	Eureka	Australia	Heat tolerance
4	Golden Empress	USA	Cold tolerance, drought tolerance, regeneration
5	Crown	USA	Cold tolerance, drought tolerance, regeneration
6	Sanditi	France	Lodging resistance, cold resistance, yield
7	Sitel	France	Lodging resistance, yield
8	Victor	Canada	Drought tolerance, insect resistance, heat tolerance
9	Victoria	Canada	Moderate cold tolerance, heat tolerance
10	WL232HQ	USA	Cold tolerance, regeneration
11	WL323ML	USA	Cold tolerance, yield
12	Rambler	Canada	Cold tolerance, drought tolerance, grazing tolerance
13	Caribou	Canada	Cold tolerance
14	WL525HQ	USA	Heat tolerance, yield
15	Wisdom	USA	Heat tolerance
16	WL343HQ	USA	Resistance to diseases and insect pests, cold tolerance, regeneration, frequent cutting tolerance
17	Qiuliu	Russia	Cold tolerance, drought tolerance, saline-alkali soil tolerance

Table 2. Local alfalfa varieties registered by the National Committee for the Examination and Approval of Herbage Varieties.

Num.	Cultivar	Region	Main Characteristics
1	Beijiang	Xinjiang	Drought tolerance, cold tolerance, poor regeneration
2	Xinjiang Daye	Xinjiang	Large leaves, good regeneration
3	Hexi	Gansu	Late maturing, poor regeneration
4	Longdong	Gansu	Drought tolerance, poor regeneration
5	Longzhong		Drought tolerance
6	Tianshui	Gansu	
7	Guanzhong	Shanxi	Early regrowth, early maturing
8	Shanbei	Shanxi	Late maturing, drought tolerance
9	Jinnan	Shanxi	Good regeneration, early maturing
10	Pianguan	Shanxi	Late maturing
11	Aohan	Inner Mongolia	Drought tolerance, cold tolerance, suitable for dryland cultivation
12	Junggar	Inner Mongolia	Drought tolerance, suitable for dryland cultivation
13	Cangzhou	Hebei	Comparative tolerance to saline-alkali soils
14	Weixi	Hebei	Drought tolerance
15	Baoding	Hebei	Good regeneration
16	Zhaodong	Heilongjiang	Cold tolerance
17	Huiyin	Jiangsu	Heat tolerance, early maturing
18	Chuxiong	Yunnan	Annual or biennial, high nitrogen-fixing ability, cold tolerance, drought tolerance
19	Wudi	Shandong	Highly salt tolerant

Local varieties are mainly distributed in Xinjiang, Inner Mongolia, Gansu, Shaanxi, Shanxi, Hebei, Shandong, Heilongjiang, Jiangsu, and Yunnan provinces. They have resistance to environmental stresses, for example, cv. Zhaodong and cv. Aohan are resistant to cold.

Alfalfa has been cultivated for more than 2000 years in China; a nation with vastly diverse climate, soil, and other ecological conditions across different regions. Under the long-term influence of natural conditions and farming conditions in different parts of the country, the formation of the local varieties has primarily been achieved by natural selection (Local varieties, Table 2). These local varieties play a significant role in production, are extremely valuable germplasm resources, and have strong resistance, wide adaptability, and yield stability.

The five wild varieties registered for cultivation are cvv. Aletai, Longdong, Hulunbeier, Deqin, and Qingshui. Among them, cv. Qingshui is the first rhizomatous alfalfa registered in China, and plays an important role in production for some regions. The width of the crown is 25–30 cm with well-developed underground horizontal roots, which enables the variety to produce abundant shoots (25–46 shoots/plant) and to have a strong regenerative ability following grazing.

3. BREEDING METHODS

Alfalfa is a perennial predominantly cross-pollinated plant; self-fertility is affected by many factors. Morphological features and physiological characteristics prevent alfalfa from self-pollination which results in low selfing rates. Hence, alfalfa is classified as a cross-pollinated species, and its natural cross-pollination rate is between 25% and 75%. As an autotetraploid, it has its specific requirements in breeding methods and selection of offspring that are different to diploid plant species; these factors have also contributed to the delay in utilizing some modern breeding technologies in alfalfa compared to major food crops. Currently selective breeding, cross breeding, male sterile line breeding, space breeding, and biotechnology-assisted breeding have been adopted for alfalfa breeding in China, among which selective breeding and cross breeding are the most widely applied.

3.1. Selective Breeding

Selective breeding is the selection of improved varieties and genotypes with advantageous characteristics to develop new varieties with specific breeding objectives. It is an important method to improve both existing cultivars and breed new varieties from germplasm collections or other sources. The methods of selective breeding include individual selection, mass selection, improved mass selection, group selection, and recurrent selection etc. [8]. There are a large number of new varieties developed and registered in China using this method, including cvv. Zhongmu 1, Zhongmu 2, Zhongmu 3, Gongnong 1, Gongnong 2, Gongnong 3, Xinmu 1, Xinmu 2, Xinmu 3, and Gannong 2. Selective breeding is practical, effective, manageable, and is one of the most commonly used methods in alfalfa breeding in China.

3.2. Cross Breeding

Cross breeding uses cross-pollination between different species or sub-species to breed varieties by selection from the hybrid progeny. It can recombine the parents' genes to form a variety of different types and provide abundant material for selection. Genetic recombination allows the accumulation of desirable alleles from parents that differ for major traits. Choosing the correct parents and making suitable combinations are critical for cross breeding. Among alfalfa varieties that have been bred using cross breeding and registered in China are cv. Gannong 3 and Tumu 2. *M. sativa* subsp *x. varia* cvv. Caoyuan 1, Caoyuan 2, Tumu 1, and Gannong 1 were developed through crossing subsp. sativa and subsp. *falcata*. Intergeneric reciprocal hybridization combined with radiation breeding techniques were used to develop *Melilotoides ruthenicus* Sojak hybrids cvv. Longmu 801 and Longmu 803.

3.3. Male Sterile Line Breeding

Male sterile line breeding is where the male organs in flowers do not function to produce pollen whereas the female organs function normally. No seeds of male sterile lines are produced by self-pollination but seeds can be produced when pollinated with other non-male sterile lines.

The first alfalfa male sterile line was found in Canada in 1958. Since then researchers in the USA, Russia, Hungary, Bulgaria, France, and Japan used male sterile lines in their breeding programs. Petkov discovered that the maintainer line (line to help maintain the male sterile lines) for sterile line breeding in 1979. Other researchers conducted hybridization experiments with the sterile line and obtained excellent hybrids, which showed obvious heterosis. Herbage yield increased by 30% compared with control species, with improved resistance to stress. From the 1980s, researchers have studied a range of aspects related to the basic biology and agronomic suitability of male sterility in alfalfa breeding including pollen sterility and fertility, regarding the characterization of male sterile lines, and investigated the sterility mechanism of alfalfa pollen. More recently, Hybrid force-620 that was bred with male sterile line technique by Dairyland company in the US has been widely sown by dairy farmers due to its high yield and improved quality [9].

In China, Yongfu Wu [10] from Inner Mongolia University selected and bred six male sterile lines from *M. varia* Martin. cv. Caoyuan 1; scientists from the Institute of Animal Science, Chinese Academy of Agricultural Sciences developed three male sterile lines from *M. sativa* cv. Atlantic Ocean; and scientists from the Institute of Grassland Science, Jilin Academy of Agricultural Sciences pioneered the male sterile variety test in alfalfa and obtained F1 seeds under open pollination [2]. However, there has been no practical use of male sterile line technology in the development of new cultivars of *M. sativa* in China.

3.4. Biotechnology Assisted Breeding

Biotechnology assisted breeding is the use of genomic techniques such as molecular markers and transgenesis to incorporate new traits or increase the rate of genetic gain [11,12,13,14]. Research on biotechnology assisted breeding for alfalfa in China began in late 1970s. We used tissue culture and somatic cell hybridization initially, and then moved to the development and application of transgenesis and molecular markers. At present, studies are focused on the transformation and expression of genes related to stress resistance and tolerance. Although there are few varieties developed using this technology to date, the methods adopted in China are briefly summarized below.

3.5. Transgenic Technology

Different genetic transformation methods significantly affect the efficiency of exogenous genes transfer into *M. sativa*, which is the basis for the acquisition and development of transgenic plants. There are three main methods in genetic transformation of *M. sativa*, germplasm line transformation, direct transformation, and indirect transformation. Germplasm line transformation method uses a plant's own pollen, ovary, and other germ cells to introduce exogenous genes, such as via the ovary injection and pollen tube method. Direct transformation methods use chemical or physical methods to introduce exogenous DNA into plant cells and obtain transgenic plants. Biolistic, ultrasonic, polyethelene glycol (PEG)-mediated, and microinjection are all examples of direct transformation methods. Indirect transformation methods use carriers to introduce DNA into the plant cell, and Agrobacterium-mediated transformation is the major example of this method.

In genetically modified breeding for alfalfa salt tolerance [15], Li et al. overexpressed the Alfinl transcription factor in alfalfa via Agrobacterium-mediated transformation, which enhanced the expression of endogenous gene MsPRP2 and improved the salt tolerance of transgenic alfalfa plants. Liang et al. [14] studied several factors that influence the efficiency of Agrobacterium-mediated genetic transformation in alfalfa and obtained transgenic plants expressing the betaine aldehyde dehydrogenase (BADH) gene. Yan et al. [16] showed that the BADH gene can enhance the salt tolerance of alfalfa. Wang et al. [17] used the vacuum infiltration-aided Agrobacterium-mediated method to transfer salt tolerance gene *AtNHXl* into Alfalfa and obtained salt tolerant somatic embryos. Pan et al. [18] used Agrobacterium EHA105-mediated cotyledon dissemination method to transfer *Cg-NHXl* gene into cv. Xinjiang Daye. Zhao et al. [19], transferred the *Atriplex dimorphostegia NHX* salt tolerance gene into aseptic seedling leaves and cotyledons of cv. Xinmu No.1 and cv. Xinjiang Daye, and obtained salt tolerant plants. Li et al. [20] studied genetic transformation of M. sativa and transferred *DsNRT2* (a Nitrate Transporter) that was cloned from *Dunaliella salina* into cv. Zhongmu No.1 and 7 with positive seedlings obtained. Liu et al. [21] transferred the HAL1 gene that was cloned from yeast into the embryogenic callus of cv. Longmu 803 with the Agrobacterium-mediated method and obtained 11 transgenic plants. Expression of HAL1 gene improved the salt tolerance characteristics of alfalfa. Sui [22] studied a vacuolar *Na^+/H^+antiporter* gene (*ScNHXI*) from *Suaeda corniculata*, which was introduced into *M. sativa* and improved salt tolerance of transgenic alfalfa. Sheng [23] successfully introduced the *DREB2A* gene into a range of alfalfa genotypes cvv. Zhaodong, Aoha, Gongnong 1, and Gongnong 2, primarily from Heilongjiang province with the aim of improving drought tolerance characteristics. Wang et al. [24] transferred the LEA3 gene from barley into cv. Zhongmu 1 through biolistic transformation and obtained an increased survival rate of transgenic plants under high salt stress. The salt tolerance of the transgenic plants was greater than that of the control plants, which shows that LEA3 gene has potential for application in developing drought resistance and salt tolerance for alfalfa breeding. The improvement of salt tolerance was also the aim of Bao et al. [25] where the transgenic co-expression of *tonoplast* Cation/H^+ antiporter and *H^+-pyrophosphatase* genes from the xerophytic plant *Zygophyllum xanthoxylum* increased the growth and salt tolerance of genetically modified alfalfa plants under field conditions. The use of transgenesis to introduce genes into alfalfa to confer stress tolerance outside of the range that is presented within sexually compatible species offers the potential to greatly increase the adaptation of cultivated alfalfa in stressed and degraded environments in China.

The recent advent of transcriptomics and next generation sequencing technologies offers the potential to identify genes involved in stress tolerance on a broader scale than previous technologies. Deep sequencing of the transcriptome of alfalfa was used recently to identify 5605 differentially expressed genes in the crown buds of cv. Zhaodong when grown under naturally occurring freezing stress in northern China [26]. These studies may identify further candidates for deployment using transgenic or genome editing technologies.

Due to various constraints, the further development of transgenic alfalfa in China has not been widely carried out. Although progress and achievements have been made, most are still at the laboratory experiment phase. Currently there are no transgenic alfalfa cultivars developed and adopted in practice.

3.6. Molecular Marker Technology

In stress resistance breeding, Yang et al. [27] used *M. sativa* L. cv. Zhongmu No. 1 and salt sensitive alfalfa as materials to screen from and build up the basis for research on salt tolerance genes for molecular marker development in alfalfa, using the Random Amplified Polymorphism (RAPD) technique. In disease resistance breeding, Gui et al. [28] used RAPD technique and Bulked Segregation Analysis to study molecular markers linked to resistance genes against brown spot disease in five *Medicago* species, and selected eight random primers which are able to indicate polymorphism between resistant and susceptible materials in more than three species simultaneously. Wei [29] used RAPD, Simple Sequence Repeat (SSR), and Inter Simple Sequence Repeat (ISSR) molecular marker and field experiments to study DNA fingerprints and genetic diversity of 84 alfalfa lines. Xu and Jia [30] regenerated somatic hybrid plants between alfalfa and sainfoin by protoplast fusion and culture, and extracted DNA from leaf material from regenerants of hybrid tissue; they also analyzed the recombination of the genetic materials by RAPD and Southern hybridization techniques. Yu et al. [31] combined SSR and Expressed Sequence Tag derived SSR (EST-SSR) molecular marker techniques to construct molecular marker profiles of alfalfa germplasm. Su et al. [32] established a tetraploid F2 mapping family in a hybrid between subsp. *sativa* and susbsp. *falacata*, and constructed a genetic map of this population. Fu and Pauls [33] used 20 imported alfalfa accessions and demonstrated that RAPD markers could determine the relationship between alfalfa varieties. SSR marker technologies have been used to characterize and contrast the genetic diversity within and between cultivars and landraces of Chinese origin with those from other geographic regions, thereby providing insights into the nature of population structure and variation that could be used in a genomics assisted breeding program [34] Again, there have been no bred varieties of alfalfa registered in China using these technologies to date.

3.7. Space Breeding

Space breeding, also called space technology breeding or spaceflight breeding, takes advantage of space environment which has been known to induce physiological changes in plants and heritable mutations, although the mechanisms through which these mutations occur are not fully understood [35,36]. Space breeding has been performed in China since 1987 with recoverable spacecraft allowing the recovery of plant material. Space breeding has resulted in new cultivars of rice, wheat, and soybean, and although the efforts have focused on major food crops approximately 1% of the seed accessions in China's space breeding program have been pastures [35]. Ren [37] studied alfalfa seed carried by recoverable satellite using Fourier Transform (FT)-Raman spectroscopy, which showed that the content of DNA and Ca^{2+} increased and that of sugar and lipid decreased compared with the control seed on the ground. Zhang [38] analyzed leaf microstructure of four alfalfa varieties grown from seeds carried by Shenzhou-3 spacecraft in comparison with control plants, and found that leaf thickness and palisade tissue thickness of the varieties are much greater than that of the control plants, with their leaf protuberance degree being much less than the control plants. In addition, the cell structure tightness of the four alfalfa varieties was different from the control plants. These mutations may have influenced its resistance performance, and could be used for further resistance breeding. Zhang [39] studied the space mutation of cv. Longmu 803 and cv. Zhaodong with their seeds carried by recoverable satellite. The results indicated that the types of alfalfa chromosome aberration produced by space mutation were mainly the formation of micronuclei. Visible mutations occurred in mitotic cell chromosomes due to space mutation effects.

Space breeding technology has been widely used in the breeding of other crops; however, there is no alfalfa variety registered on a national scale. Nevertheless, there are several varieties registered in provinces or autonomous regions of China.

4. PROBLEMS AND CHALLENGES

Although China has made considerable achievements in alfalfa breeding, there is room for improvement compared with other countries, and the overall level of alfalfa breeding is still low and cannot meet the requirements of development in grassland agriculture, grassland animal husbandry, and ecological reconstruction. The main problems and challenges are discussed below.

4.1. Slow Breeding Cycle

The number of varieties bred in China is relatively low, and the progress of resistance breeding is slow. Up to date, only 36 bred Alfalfa varieties have been registered by the National Committee for the Examination and Approval of Herbage Variety in China, compared with 192 alfalfa varieties registered in the USA in 2015. The breeding target is more focused on yield in China, and overwintering rates has also become one of the priority breeding targets due to northern China being one of the major regions of *M. sativa* plantation. Hence, the target characteristics of bred varieties include high yield and high overwintering rates, and to a lesser degree, cold tolerance and fall dormancy. In disease and insect resistance, the main focus is on pathogen identification and field investigation of disease conditions [40,41], disease resistance identification methods, and the influence of environment on alfalfa resistance performance. There are few varieties that have been specifically bred for disease resistance, insect resistance, and drought tolerance, and almost no varieties for herbicide resistance.

4.2. Breeding Methods

At present, alfalfa varieties registered in China are mainly bred with conventional breeding methods. These methods are simple and widely adopted, and take advantage of excellent germplasm resources in different ecological regions to develop new varieties suitable for regional conditions. However, these methods normally take a long time to develop cultivars that are used on farms. The application of genetic engineering and other modern biotechnology in alfalfa breeding has been put in place in China for only 20 years, and research is currently focused on target gene cloning, construction of expression vectors, and related laboratory experiments with the commencement of some field trials.

4.3. Low Adoption Rate of Bred Varieties

The supply chain for seed production, distribution, and marketing of seed is less well developed in China than some other countries where specialized breeding companies are often involved and responsible for this process. In China, however, alfalfa breeding resources are mainly centered in universities and research institutes. As a result, the breeding targets are not well placed to meet the market demands, and breeding research is disconnected with business development and without effective involvement from industries, thereby leading to poor adoption of new varieties [42]. Largely due to the slow process of seed production of new varieties, most enterprises choose to purchase imported varieties, further reducing the adoption rates of newly bred varieties in China.

5. OPPORTUNITIES AND FURTHER RESEARCH

In recent years, the collection of germplasm resources, innovation of breeding theory and technology, and molecular biology assisted breeding have played an important role in accelerating alfalfa breeding in China. These experiences confirmed that we could learn from the experience of developed countries and take integrated approaches in enhancing alfalfa breeding in China.

5.1. Strengthening the Research and Utilization of Alfalfa Germplasm Resources

Many developed countries give great emphasis on the collection, preservation, research, and utilization of germplasm resources. China covers a large territory with very diverse ecological environments due to the influence of latitude, elevation, and topography. This makes the country rich and unique in alfalfa germplasm resources, including conventional local and wild types, improved varieties, and new breeding materials such as alfalfa mutants, all of which form the basis of alfalfa breeding. In addition to the development of breeding methods and the adoption of new technology, a critical step for the success of modern breeding programs is to widely collect and make use of the right germplasm resources. Due to poor awareness in germplasm protection, overgrazing, and excessive exploitation, some excellent forage germplasm resources have been lost. Lack of desirable germplasm resources for screening and breeding has seriously affected the process and results for developing new varieties. Therefore, it is imperative to learn from this experience, conduct further research and identification of existing germplasm in a systematic way, screen materials with desirable characteristics, and conduct research in an innovative manner [43].

5.2. Improving Alfalfa Breeding Theory

In the developed world, great efforts have been made to develop the basic theory of breeding. Forage breeding in China is not only behind developed countries, but also behind crop breeding domestically. This is largely due to lack of advanced forage breeding theory and technology. In order to make a breakthrough in alfalfa breeding, we need to strengthen research on forage breeding theory and techniques; otherwise, slow breeding cycle/process will be unavoidable.

5.3. Enhancing the Exploration of New Techniques in Alfalfa Breeding

Conventional breeding methods are basic methods to breed new variety all over the world and will continue to play an important role in alfalfa breeding in China in the near future. With the rapid development of science and technology, molecular techniques and transgenic technology have been widely used in alfalfa breeding worldwide. Transgenic technology has been used to improve the resistance of alfalfa to cold, drought, salinity, alkalinity, disease, herbicides, and insect pests. Molecular technology has been widely used in assisted alfalfa breeding and germplasm introgression research, genetic linkage map, germplasm identification, and genetic diversity studies. The combination of conventional breeding techniques with modern breeding techniques can help to break through some of the constraints and speed up the breeding cycle, which is critical for improving alfalfa breeding technology in China.

5.4. Conservation of Distinctive and Special-Purpose Alfalfa Materials

Alfalfa breeding in China has the following characteristics: broad cultivation area, complex ecological conditions, and a wide range of uses in production. According to local conditions and uses, specific alfalfa varieties can be selected and cultivated. For example, in germplasm collection and application in western China, research should focus on germplasms that are resistant/tolerant to drought, cold, and saline soils of sandy texture. In these regions, new varieties that are able to prevent wind erosion and fix sandy soils, and to conserve water and soil resources could be the major breeding targets for alfalfa. On the other hand, improved animal production through high forage production and better forage quality could also be a high priority target. Therefore, selection and breeding of special-purpose varieties that fit both the environment/soil constraints and production requirements for various feeding systems, such as hay and silage, and different growing seasons are necessary to address regional production and sustainability issues [2]. Strategies must be worked out on a regional basis to achieve specific needs.

ACKNOWLEDGMENTS

The authors would like to thank Zhongnan Nie for his assistance during the development of this manuscript.

REFERENCES

1. Geng, H.Z. Alfalfa of China; China Agriculture Press: Beijing, China, 1995. (In Chinese)
2. Shi, S.L.; Nan, L.L.; Guo, Q.E. Achievements and prospect of alfalfa breeding in China. J. Plant Genet. Res. 2010, 11, 46–51. (In Chinese)
3. Yang, Q.C.; Sun, Y. The History, Current situation and development of alfalfa breeding in China. Chin. J. Grassl. 2011, 33, 95–101. (In Chinese)
4. Yang, Q.C.; Kang, J.M.; Zhang, T.J.; Liu, F.Q.; Long, R.C.; Sun, Y. Distribution, breeding and utilization of alfalfa germplasm resources. Chin. Sci. Bull. 2016, 61, 261–270.
5. Chinese Forage Cultivar Registration Board (CHCRB). Licensed Cultivars of Herbage Crops in China; China Agricultural University Press: Beijing, China, 1999. (In Chinese)
6. Chinese Herbage Cultivar Registration Board (CHCRB). Licensed Cultivars of Herbage Crops in China (1999–2006); China Agriculture Press: Beijing, China, 2008. (In Chinese)
7. Wu, D.G. Resistance of Alfalfa (Medicago sativa L.) to Aphid and Regional Adaptability Experiment of New Alfalfa Line HA-3. Master's Thesis, Gansu Agricultural University, Lanzhou, China, 2007.
8. Gui, Z.; Gao, J.M. Advances in research of alfalfa breeding in China. J. Tianjin Agric. Coll. 2003, 10, 37–41. (In Chinese)
9. Cai, L.Y.; Shi, F.L.; Zhang, F.S.; Gao, C.P. Research progresses in alfalfa heterosis. Chin. J. Grassl. 2010, 32, 92–97. (In Chinese)
10. Wu, Y.F. Researching on breeding Alfalfa Male Sterile Line. Chin. J. Grassl. 1980, 37–39. (In Chinese)
11. Smith, K.F.; Forster, J.W.; Spangenberg, G.C. Converting genomic discoveries into genetic solutions for diary pastures. Anim. Prod. Sci. 2007, 47, 1032–1038.
12. Kolliker, R.; Rosellini, D.; Wang, Z.-Y. Development and application of technological and molecular genetic tools. In Fodder Crops and Amenity Grasses. Vol. 5 in the Series: Handbook of Plant Breeding; Boller, B., Posselt, U., Veronesi, F., Eds.; Springer: New York, NY, USA; pp. 89–113.
13. Li, X.; Brummer, E.C. Applied genetics and genomics in alfalfa breeding. Agronomy 2012, 2, 1–44.
14. Li, H.W.; Ma, D.M.; Xu, X. Advance and prospects on salt-tolerant transgenic alfalfa. North. Hortic. 2012, 19, 184–188. (In Chinese)
15. Liang, H.-M.; Xia, Y.; Sun, Z.-X.; Wang, T.-M.; Liu, D.-X.; Wang, G.-L.; Huang, J.; Chen, S.-Y. Establishment of genetic transformation system of Medicago sativa mediated by Agrobacterium tumefaciens. J. Agric. Biotechnol. 2015, 13, 152–156. (In Chinese)
16. Yan, L.P.; Xia, Y.; Mao, X.H.; Liu, C.; Liang, H. Breeding and Salt Resistance Evaluation of BADH Transgenic Alfalfa Cultivar Shanmu 2. Chin. Bull. Bot. 2011, 46, 293–301. (In Chinese)
17. Wang, Q.L.; Wang, S.M.; Zhang, J.L.; Bao, A.K.; Chen, T.X.; Lou, J.Q.; Lu, N. Transformation Studies of Medicago sativa Mediated by Agrobacterium tumefaciens with AtNHX1 Gene. Pratacult. Sci. 2005, 23, 55–59. (In Chinese)
18. Pan, J.L. Plant Regeneration and Molecule Detection for the Transgenic Xinjiang Big Leaf Alfalfa of Cg-NHX1 Gene from Halophyte Chenopodium glaucum. Ph.D. Thesis, Xinjiang Agricultural University, Urumchi, China, 2007.

19. Zhao, H.J.; Zhang, B.; Li, P.Y.; Chen, A.P. Influencing Factors on Alfalfa Transformation with Atriplex dimorphostegia NHX Gene Mediated by Agrobacterium tumefaciens. Acta Agrestia Sin. 2007, 15, 418–422. (In Chinese)
20. Li, S.L.; Yang, G.; Yi, X.L.; Ma, G.; Ma, X.Z.; Li, Y.; Cao, Y. Transformation of Alfalfa (Medicago sativa L.) by Agrobacterium-mediated Process with DsNRT2 Gene from Dunaliella salina. J. Sichuan Univ. 2008, 45, 409–412. (In Chinese)
21. Liu, Y.Z.; Wei, Z.Y.; Xing, S.C.; Tan, H.; Gao, M.; Dong, Y.S. Transgenic Alfalfa with HAL1 Gene and Its Salt Tolerance. J. Jilin Agric. Sci. 2008, 33, 21–24. (In Chinese)
22. Sui, X. Cloning of Vacuolar Na+/H+ Antiporter Gene from Suaeda corniculata (ScNHX1) and Transformation Studies of ScNHX1 in Medicago sativa. Master's Thesis, Jilin Agricultural University, Changchun, China, 2007.
23. Sheng, H. Construction of Plant Expressive Vectors of DREB2A, SCMRP Gene and Genetic Transformation into Alfalfa. Master's Thesis, Northeast Agricultural University, Harbin, China, 2006.
24. Wang, Y.; Zhu, B.C.; Sun, Y. Transformation of Barley LEA3 Gene into Alfalfa (Medicago sativa L.). J. Nucl. Agric. Sci. 2007, 21, 249–252. (In Chinese)
25. Bao, A.K.; Du, B.Q.; Touil, L.; Kang, P.; Wang, Q.L.; Wang, S.W. Co-expression of tonoplast Cation/H+ antiporter and H+-pyrophosphatase from xerophyte Zygophyllum xanthoxylum improves alfalfa plant growth under salinity, drought and field conditions. Plant Biotech. J. 2016, 14, 964–975.
26. Song, L.; Jiang, L.; Chen, Y.; Shu, Y.; Bai, Y.; Guo, C. Deep-sequencing transcriptome analysis of field-grown Medicago sativa L. crown buds acclimated to freezing stress. Funct. Integr. Genom. 2016, 5, 495–511.
27. Yang, Q.C.; Han, J.G. A Study on Using RAPD in Alfalfa Salt Tolerant Breeding. Acta Agrestia Sin. 2003, 11, 27–32. (In Chinese)
28. Gui, Z.; Yuan, Q.H. Studies on RAPD Polymorphic in Alfalfa Resistant to Brown Blot. Acta Agrestia Sin. 2002, 10, 274–278. (In Chinese)
29. Wei, Z.W. DNA Fingerprint of Medicago sativa Variety Genomes Using SSR, ISSR and RAPD. Acta Pratacult. Sin. 2004, 13, 62–67. (In Chinese)
30. Xu, Z.Q.; Jia, J.F. Molecular Identification of Intergeneric Somatic Hybrid Plants between Alfalfa and Sainfoin. Chin. J. Biotechnol. 2000, 16, 173–178. (In Chinese)
31. Yu, L.Q.; Liu, R.X.; Su, D.; Yun, J.F. Genetic Diversities of Fall Dormancy Alfalfa Based on SSR and EST-SSR. Acta Agrestia Sin. 2009, 31, 52–58. (In Chinese)
32. Su, D.; Yu, L.Q.; Zhou, Y.L.; Sun, J.J.; Wang, F.G.; Wu, R.N. Establishment of a Tetraploid Alfalfa Family and Analysis of Its Genetic Variability. Acta Agrestia Sin. 2011, 19, 657–662. (In Chinese)
33. Fu, J.H.; Pauls, K.P. Identification of Alfalfa Populations by Using RAPD Molecular Marker Method. Crops 1995, 6, 28. (In Chinese)
34. Qiang, H.; Chen, Z.; Zhang, Z.; Wang, X.; Gao, H.; Wang, Z. Molecular diversity and population structure of a worldwide collection of cultivated tetraploid alfalfa (Medicago sativa subsp. sativa L.) germplasm as revealed by microsatellite markers. PLoS ONE 2015, 10, e0124592.
35. Chengzhi, L. Agronomy in space—China's crop breeding program. Space Policy 2011, 27, 157–164.
36. Li, Y.; Liu, M.; Cheng, Z.; Sun, Y. Space environment induced mutations prefer to occur at polymorphic sites of rice genomes. Adv. Space Res. 2007, 40, 523–527.
37. Ren, W.B.; Han, J.G.; Zhang, Y.W. A Study of the effect of space mutagenesis of grass seeds. Pratacult. Sci. 2006, 23, 72–76. (In Chinese)
38. Zhang, W.J. Effects of Four Alfalfa Cultivars on Mutagenic of Spaceflight. Master's Thesis, Gansu Agricultural University, Lanzhou, China, 2010.
39. Zhang, Y.X.; Liu, J.L.; Han, W.B.; Tang, F.; Hao, R.; Shang, C.; Du, Y.; Li, J.; Wang, C. Biological effect of aerospace environment of alfalfa. J. Nucl. Agric. Sci. 2009, 23, 266–269. (In Chinese)
40. Chen, Y.J.; Liu, X.M.; Cui, G.W.; Xi, Q.X. Research progress in root rot of alfalfa. GrassL China 2000, 22, 51–56. (In Chinese)

41. Chen, S.K.; Yao, G.J. Resistance to common leafspot in different alfalfa varieties. Pratacult. Sci. 1994, 11, 61–62. (In Chinese)
42. Yang, Q.C.; Sun, J.; Han, G.D. Studies on RAPD Polymorphic in salt-tolerant alfalfa and salt-sensitive alfalfa. Acta Agrestia Sin. 2001, 9, 83–86. (In Chinese)
43. Wang, Y.H.; Gao, H.W.; Wang, Z.; Li, Y.; Liu, G.B. Study and evaluation of drought resistance capacity of Medicago glutinosa germplasm resources at seedling stage. J. Plant Genet. Resour. 2009, 10, 443–447. (In Chinese)

CHAPTER 10

Estimation of phenotypic variability in symbiotic nitrogen fixation ability of common bean under drought stress using ^{15}N natural abundance in grain

Jose Polania[a, b,*], Charlotte Poschenrieder[b], Idupulapati Rao[a], Stephen Beebe[a]

[a]*Centro Internacional de Agricultura Tropical (CIAT), A. A. 6713 Cali, Colombiab*

[b]*Universidad Autónoma de Barcelona, Lab Fisiología Vegetal, Facultad de Biociencias, Bellaterra, Spain*

ABSTRACT

Common bean (*Phaseolus vulgaris* L.) is the most important food legume, cultivated by small farmers and is usually exposed to unfavorable conditions with minimum use of inputs. Drought and low soil fertility, especially phosphorus and nitrogen (N) deficiencies, are major limitations to bean yield in smallholder systems. Beans can derive part of their required N from the atmosphere through symbiotic nitrogen fixation (SNF). Drought stress severely limits SNF ability of plants. The main objectives of this study were to: (i) test and validate the use of ^{15}N natural abundance in grain to quantify phenotypic differences in SNF ability for its implementation in breeding programs of common bean with bush growth habit aiming to improve SNF, and (ii) quantify phenotypic differences in SNF under drought to identify superior genotypes that could serve as parents. Field studies were conducted at CIAT-Palmira, Colombia using a set of 36 bean genotypes belonging to the Middle American gene pool for evaluation in two seasons with two levels of water supply (irrigated and drought stress). We used ^{15}N natural abundance method to compare SNF ability estimated from shoot tissue sampled at mid-pod filling growth stage vs. grain tissue sampled at harvest. Our results showed positive and significant correlation between nitrogen derived from the atmosphere (%Ndfa) estimated using shoot tissue at mid-pod filling and %Ndfa estimated using grain tissue at harvest. Both methods showed phenotypic variability in SNF ability under both drought and irrigated conditions and a significant reduction in SNF ability was observed under drought stress. We suggest that the method of estimating Ndfa using grain tissue (Ndfa-G) could be applied in bean breeding programs to improve SNF ability. Using this method of Ndfa-G, we identified four bean lines (RCB 593, SEA 15, NCB 226 and BFS 29) that combine greater SNF ability with greater grain yield under drought stress and these could serve as potential parents to further improve SNF ability of common bean.

Keywords: Shoot biomass; Nitrogen derived from the atmosphere; Nitrogen derived from the soil; Nitrogen use efficiency; Terminal drought stress

HIGHLIGHTS

- New method was used to estimate symbiotic nitrogen fixation ability in bean.
- Phenotypic variation in nitrogen fixation was observed under drought stress.
- Four lines were superior in their ability to fix nitrogen under drought stress.
- Several lines were found to combine drought resistance with ability to fix nitrogen.

1. INTRODUCTION

Common bean (*Phaseolus vulgaris* L.) is the most important food legume cultivated in the tropics of Latin America and eastern and southern Africa. It is nutritionally rich in iron and protein, and is a source of fiber and carbohydrates that are essential in the nutrition of the population, especially in developing countries. The bean crop is cultivated by small farmers and it is often exposed to unfavorable conditions with minimum use of inputs (Beebe *et al.*, 2013 ; Beebe *et al.*, 2008). Bean yields are affected by various biotic and abiotic stress factors. Drought affects about 60% of the bean producing regions causing yield losses between 10 and 100%. Drought is the second most important factor in yield reduction after disease (Rao, 2014 ; Thung and Rao, 1999). In addition to drought, smallholder bean production is often affected by low soil fertility in marginal areas (Douxchamps et al., 2010) where the majority of grain legumes are cultivated (Beebe *et al.*, 2014 ; Sinclair and Vadez, 2012). About 50% of the bean production areas in Eastern and Central Africa and 60% in Southern Africa are subjected to N deficiency due to both depletion of N in the soil and application of limited N fertilizer (Beebe et al., 2014). Thus the ability of the crop to acquire N from the soil is limited. Common bean can supply at least part of its N requirement through symbiotic nitrogen fixation (SNF). However compared to other legumes, beans have lower SNF capacity (Hardarson, 2004; Hardarson *et al.*, 1993 ; Peoples *et al.*, 2009). The estimated mean value of nitrogen derived from the atmosphere (Ndfa) for common bean across different geographical regions of the world is 39% (Peoples et al., 2009). This value is in contrast with the Ndfa values of 54%–65% observed for other widely-grown legume crops and to the values recorded for soybean and faba bean which were 68% and 75%, respectively (Peoples et al., 2009).

Abiotic and biotic stress factors such as P deficiency, drought, and pest and diseases affect SNF capacity (Ramaekers et al., 2013). Among these limitations, SNF is highly sensitive to drought (Beebe *et al.*, 2014 ; Devi *et al.*, 2013), with possible interactions among stresses. Bean genotypes with resistance to drought are affected by N deficiency, and SNF capacity is affected by drought stress (Devi et al., 2013). Moreover, drought has a negative influence on both the rhizobia and the nodulation process of legumes (Sinclair and Vadez, 2012). Drought even can cause the loss of SNF activity in common bean, and other legume species that generally have low rates of N fixation even under well-watered conditions (Devi et al., 2013). Identification of parental genotypes to use in breeding that combine superior SNF ability under drought stress with other desirable traits could be a useful strategy to confront the new challenges of climate variability and to ensure food security in marginal areas.

Different techniques have been used to estimate SNF, including N_2 balance, N_2 difference, ^{15}N natural abundance, ^{15}N isotope dilution, ureide analysis, acetylene reduction assay and hydrogen evolution (Unkovich et al., 2008). The ^{15}N natural abundance method presents some advantages over other methodologies. It can be applied under both greenhouse and field conditions, allows estimation of N fixation in almost any situation where both N-fixing and non N-fixing plants are present at the same location, and it can be applied to farmers' fields, or to any field experiments where legumes and non N-fixing plants coexist (Douxchamps *et al.*, 2010 ; Unkovich *et al.*, 2008). Also, the ^{15}N natural abundance method allows to separate N derived from the atmosphere with the N derived from soil (Boddey et al., 2000). To calculate the total contribution of SNF in kg ha^{-1}, the estimation of shoot biomass in kg ha^{-1} is required (Boddey *et al.*, 2000 ; Unkovich *et al.*, 2008). This methodology is usually applied to shoot tissue of the plant harvested at different growth stages such as flowering or pod filling (Boddey *et al.*, 2000 ; Unkovich *et al.*, 2008). Generally shoot tissue is used because to use the whole plant including roots is complex. However, the fact of taking shoot biomass sample, from a breeding perspective, is equally complex because of the large sample size for destructive sampling of the plot. This could mean significant labor costs for plant breeding programs dealing with large numbers of breeding lines. For these reasons most bean breeding programs do not routinely select for better SNF ability. Developing methodologies that can estimate SNF ability using grain tissue could be easier to integrate into most breeding programs since grain is routinely harvested to estimate yield and/or nutritional quality. Therefore based on the principle that common beans mobilize much of their N from vegetative structures to the grain (Lynch and White, 1992 ; Ramaekers *et al.*, 2013), we propose that it would be much easier to apply the methodology of ^{15}N natural abundance using the grain tissue at the time of harvest.

The main objectives of this study were to: (i) test and validate the use of ^{15}N natural abundance in grain to quantify phenotypic differences in symbiotic nitrogen fixation (SNF) ability for its implementation in breeding programs aiming to improve SNF in common bean, and (ii) quantify phenotypic differences in SNF under drought stress to identify superior genotypes that could serve as parents.

2. MATERIALS AND METHODS

2.1. Experimental site and meteorological conditions

Two field trials were conducted during the dry season from June to September in two years (2012 and 2013), at the main experimental station of the International Center for Tropical Agriculture (CIAT) in Palmira, Colombia, located at 3° 29" N latitude, 76° 21" W longitude and an altitude of 965 m. Basic characteristics of this field site were described previously (Beebe et al., 2008). The soil is a Mollisol (Aquic Hapludoll) with 23.3 g kg^{-1} of organic matter and no limitations of availability of iron (Fe) and molybdenum (Mo) for the process of SNF. Common bean has been grown on this field for many years and there is adequate native *Rhizobium* in soil with 5.2×10^6 colony forming units of rhizobia per gram of soil. During the crop-growing season, maximum and minimum air temperatures in 2012 were 31.0 °C and 19.0 °C, and in 2013 were 30.2 °C and 19.2 °C, respectively. Total rainfall during the active crop growth was 85.8 mm in 2012 and 87.7 mm in 2013. The potential pan evaporation was of 385.2 mm in 2012 and 351.0 mm in 2013. Two levels of water supply (irrigated and drought) were applied through furrow irrigation (approximately 35 mm of water per irrigation). The drought stress treatment in 2012 received 3 irrigations (at 3 days before sowing and at 5 and 23 days after sowing). In 2013, irrigation was provided at 3 days before sowing and at 4 and 15 days after sowing. In both years, irrigation was suspended after the application of the third irrigation to induce terminal drought stress conditions (less water availability from flowering to physiological maturity). The irrigated control treatment received 5 irrigations in 2012 and 6 irrigations in 2013 to ensure adequate soil moisture for crop growth and development.

2.2. Plant material and experimental design

For this study 36 bush bean genotypes belonging to the Middle American gene pool were selected: twenty two elite lines of common bean (BFS 10, BFS 29, BFS 32, BFS 67, MIB 778, NCB 226, NCB 280, RCB 273, RCB 593, SCR 16, SCR 2, SCR 9, SEN 56, SER 118, SER 119, SER 125, SER 16, SER 48, SER 78, SMC 141, SMC 43 and SXB 412); five interspecific lines from the cross between elite line SER 16 and *Phaseolus coccineus* (ALB 6, ALB 60, ALB 74, ALB 88 and ALB 213); one landrace of tepary bean (*Phaseolus acutifolius*) G 40001 from Veracruz-Mexico, and two interspecific lines between tepary bean and common bean (INB 841 and INB 827 developed from five cycles of congruity backcrossing of tepary with ICA Pijao). BFS (small red) lines were developed to improve adaptation to low soil fertility and drought. SER and SCR (small red), SEN (small black) and NCB (small black) lines were developed for improved adaptation to drought. ALB (small red) lines were developed for improved adaptation to drought and aluminum toxicity. RCB (small red) lines were developed for improved yield potential, disease resistance and commercial grain. SEA 15 and BAT 477 were included as drought resistant checks, and three commercial cultivars of common bean (DOR 390, Pérola and Tio Canela) as drought sensitive materials. BAT 477 NN was included as a non-fixing bean genotype which was used as reference plant to estimate nitrogen derived from the atmosphere (Ndfa). In the two years, a 6 × 6 partially balanced lattice design with 3 replications was used. Experimental units consisted of 4 rows with 3.72 m row length with a row-to-row distance of 0.6 m and plant-to-plant spacing of 7 cm (equivalent to 24 plants m^{-2}). Trials were weeded and sprayed with insecticides and fungicides as needed.

2.3. Determination of symbiotic nitrogen fixation ability using shoot and grain

To compare and validate the method of ^{15}N natural abundance, we sampled shoot tissue at mid-pod filling and grain tissue at harvest time. We sampled a representative plant within a row of 50 cm long at mid-pod filling and also at harvest time for each genotype and from each plot of both irrigated and drought treatments for oven drying and grinding and for ^{15}N estimation. The plant was cut at the soil surface, washed with deionized water and dried in the oven at 60 °C for two days. The dried sample was finely ground using a ball-mill and was weighed using a microbalance to pack 2.5 mg of each sample in a tin capsule. These samples in tin capsules were sent to UC Davis Stable Isotope Facility in USA for ^{15}N isotope analyses. The percentage of N derived from the atmosphere (%Ndfa) was determined for both shoot and grain samples using the ^{15}N natural abundance method (Shearer and Kohl, 1986). BAT 477 NN was used as a non-fixing reference plant.

$$\%\text{Ndfa} = \frac{\delta 15\text{N non fixing reference plant} - \delta 15\text{N of N2 fixing legume}}{\delta 15\text{N non fixing reference plant} - \beta} \times 100$$

Where β is the $\delta^{15}N$ value from the nitrogen fixing bean plant grown in N free medium. The isotope discrimination occurs internally within the plant so that the different plant parts differ in $\delta^{15}N$ (Unkovich et al., 1994). Consequently, different β values were used to estimate %Ndfa for the shoot at mid-pod filling and grain at harvest. The β values used were −3.09‰ for shoot at mid-pod filling and −2.44‰ for grain at harvest for genotypes with growth habit II and −3.62‰ for shoot at mid-pod filling and −2.88‰ for grain for genotypes with growth habit III. The β values were generated from conducting a pot experiment in the greenhouse at CIAT, following the procedure of Unkovich et al. (1994). We used SMC 140 and GGR 18 as representative genotypes of growth habit II and III, respectively. Total shoot and seed N content per unit area (kg ha^{-1}) were estimated using the values of N concentration in shoot biomass and grain and dry weights of shoot biomass and grain. Total N derived from atmosphere in kg ha^{-1}(TNdfa) and total N derived from soil in kg a^{-1} (TNdfs) were estimated (Unkovich et al., 1994). Nitrogen use efficiency (NUE) was estimated as kg of grain produced per kg of shoot N uptake at mid-pod filling growth stage.

2.4. Shoot biomass and grain yield measurements

At mid-pod filling, a 50 cm segment of the row from each plot with about 7 plants was used for destructive sampling to measure shoot biomass (SB). Also at mid-pod filling, the roots of three plants per plot (selected randomly) of the non-fixing bean genotype (BAT 477 NN) were pulled from soil to check for the absence of nodules. At the time of harvest, plants in 50 cm of a row from each plot were cut and dry weights of stem, pod, seed, and pod wall were recorded. Grain was harvested from two central rows after discarding end plants in both the irrigated and drought plots. In order to compare shoot dry biomass with grain dry weight, mean values of grain yield per hectare were corrected for 0% moisture in grain.

2.5. Statistical analysis

All data were analyzed using the SAS (v 9.0) PROC MIXED and PROC CORR (SAS Institute Inc., 2008). The adjusted means for each genotype and the environment (irrigated and drought) were obtained using the mixed models theory together with the MIXED procedure considering the effects of the replications and blocks within replications as random and genotypes as fixed. Correlation coefficients were calculated by the PROC CORR. In the following text, values marked with *, ** or *** are statistically significant at probability levels of 5%, 1% and 0.1%, respectively.

3. RESULTS

3.1. Estimation of Ndfa and differences in ^{15}natural abundance in shoot and grain

Analysis of %N derived from the atmosphere (%Ndfa) estimates the proportional dependence of the biomass N on N_2 fixation. The %Ndfa in grain was compared with %Ndfa in shoot biomass to determine their relationship. If the %Ndfa in grain is closely related with %Ndfa in shoot, legume breeders would be able to select for SNF based on the grain values without the need to harvest, dry and grind large volumes of shoot biomass. A significant and positive correlation values of r = 0.81*** in 2012 and r = 0.66*** in 2013 (r = 0.83*** for combined data for two seasons, Table 1) were observed between the %Ndfa values estimated with ^{15}N natural abundance of shoot biomass at mid-pod filling growth stage and %Ndfa values estimated with ^{15}N natural abundance in the grain under irrigated conditions; the correlation values were also significant and positive under drought conditions r = 0.67*** in 2012 and r = 0.74*** in 2013 (r = 0.71*** for combined data for two seasons, Table 1).

Nodule formation was observed in both irrigated and drought treatments in both years and with all the genotypes evaluated, except for BAT477 NN which was used as a non-nodulating reference plant for estimating SNF ability. Significant differences were observed in both shoot and grain $\delta^{15}N$ between the non-fixing bean genotype and the other lines tested under both irrigated and drought conditions (Table 2). The values of $\delta^{15}N$ for shoot for the non-fixing bean genotype (BAT 477_NN) in 2012 were 5.6 and 8.7 under irrigated and drought conditions, respectively (Table 2); and in 2013 the values were 8.3 and 9.5 under irrigated and drought conditions, respectively (Table 2). Under irrigated conditions, the $\delta^{15}N$ for shoot of 35 genotypes (Excluding BAT 477_NN) ranged from −0.2 to 1.7 in 2012 and from 2.1 to 5.6 in 2013 (Table 2). Under drought conditions the $\delta^{15}N$ for shoot of 35 genotypes ranged from 3.6 to 7.5 in 2012 and from 4.3 to 8.5 in 2013 (Table 2). The values of $\delta^{15}N$ for grain for BAT 477_NN in 2012 were 5.8 and 8.6 under irrigated and drought conditions, respectively (Table 2); and in 2013 the values were 6.8 and 8.3 under irrigated and drought conditions, respectively (Table 2). Under irrigated conditions, the $\delta^{15}N$ for grain of 35 genotypes ranged from 0.0 to 1.4 in 2012 and from 1.7 to 4.8 in 2013 (Table 2). Under drought conditions, the $\delta^{15}N$ for grain of 35 genotypes ranged from 4.5 to 8.0 in 2012 and from 4.4 to 7.0 in 2013 (Table 2).

3.2. Differences in SNF ability and genotypic response to drought

An average reduction of 70% in 2012 and 38% in 2013 in SNF ability was observed under drought stress in bush Middle American genotypes using the grain method (Table 2). The lines RCB 593, BFS 32, SER 125, SMC 141 and BFS 29 maintained a relatively higher level of SNF ability under drought stress in both years (Table 2). A weak correlation was observed between %Ndfa estimates using grain samples and grain yield under irrigated conditions (Table 1). No correlation was observed between %Ndfa estimates and grain yield under drought conditions (Table 1). However, the lines RCB 593, SEA 15, NCB 226 and BFS 29 were superior in combining high values of grain yield with greater values of %Ndfa under drought stress. The line NCB 226 was superior in %Ndfa ability under both irrigated and drought conditions (Table 2). The accumulation of N (kg ha^{-1}) in grain was reduced by 55% due to drought stress in Middle American bush beans, being more sensitive the accumulation of total N in grain derived from the atmosphere (TNdfa-G) than N in grain derived from soil (TNdfs-G), with reduction of 78% and 43%, respectively (Fig. 1 ; Fig. 2). Under irrigated conditions the lines BFS 29, SCR 16, BFS 32, NCB 280 and SEN 56 presented higher total N content in grain from both TNdfs and TNdfa values compared with the other lines tested (Fig. 1 ; Fig. 2). *Phaseolus acutifolius* (G 40001) was outstanding in its ability for TNdfs, and showed a drastic decrease in TNdfa under drought (Fig. 1 ; Fig. 2). The lines SEA 15, RCB 593 and BFS 10 maintained higher total N content in grain for both TNdfs and TNdfa values, compared with the other lines tested under drought stress (Fig. 1 ; Fig. 2). The three commercial varieties (DOR 390, Tio Canela 75 and Perola) presented lower N content in grain for both TNdfs and TNdfa values under both irrigated and drought conditions (Fig. 1 ; Fig. 2). Similar tendency was observed in the relationship between shoot biomass and total Ndfa estimated using shoot tissue (TNdfa-SH) (Fig. 3). Genotypes that stood out for a higher TNdfa-G and grain yield under both irrigated and drought conditions (Fig. 1 ; Fig. 2) also exhibited higher TNdfa-SH and shoot biomass at mid-pod filling under both irrigated and drought conditions (Fig. 3). Several inbred lines were superior in their shoot biomass, grain yield and TNdfa-SH and TNdfa-G than the three commercial varieties (DOR 390, Tio Canela and Perola) under both irrigated and drought conditions (Fig. 1, Fig. 2 ; Fig. 3).

Several genotypes including BFS 29, SEN 56, NCB 226, NCB 280, SCR 16, ALB 60, RCB 593 and SER 48 presented higher total N content in grain under irrigated conditions, and also maintained N levels well under drought conditions. The lines MIB 778, Pérola, SMC 43, ALB 88, Tio Canela and the non-nodulant BAT477NN showed low grain N content under both irrigated and drought conditions. The lines SEA 15, NCB 280, BFS 10, SEN 56, BFS 29, NCB 226 and SER 16 not only showed high values in grain for TNdfs but also were outstanding in their TNdfa value using grain that resulted in greater values of grain yield under drought stress, while RCB 593 combined higher values of TNdfa and an intermediate level of Ndfs with greater grain yield under drought conditions (Fig. 1 and Fig. 2).

Table 1. Correlation coefficients (r) between% nitrogen derived from the atmosphere estimated using shoot tissue (%Ndfa-SH), % nitrogen derived from the atmosphere estimated using grain tissue (%Ndfa-G), total nitrogen derived from the atmosphere in kg ha^{-1} using grain tissue (TNdfa-G), total nitrogen derived from the soil in kg ha^{-1} using grain tissue (TNdfs-G), nitrogen use efficiency in kg of grain produced kg^{-1} of N uptake in the shoot (NUE), shoot biomass in kg ha^{-1} (SB) and grain yield in kg ha^{-1} (GY) of 36 bean genotypes of grown under irrigated and drought conditions in a Mollisol at CIAT-Palmira, Colombia. Values reported are from analysis of data collected from two seasons of evaluation (2013 and 2014).

Irrigated							
Trait	%Ndfa-SH	%Ndfa-G	TNdfa-G	TNdfs-G	NUE	SB	GY
%Ndfa-SH	1						
%Ndfa-G	0.83***	1					
TNdfa-G	0.65***	0.82***	1				
TNdfs-G	−0.53***	−0.69***	−0.29***	1			
NUE	0.06	0.07	0.32***	0.15*	1		
SB	0.07	0.05	0.20**	0.32***	−0.45***	1	
GY	0.16*	0.16*	0.61***	0.50***	0.51***	0.39***	1
Drought							
Trait	%Ndfa-SH	%Ndfa-G	TNdfa-G	TNdfs-G	NUE	SB	GY
%Ndfa-SH	1						
%Ndfa-G	**0.71*****	1					
TNdfa-G	0.56***	0.83***	1				
TNdfs-G	−0.22**	−0.37***	0.09	1			
NUE	0.20**	0.19**	0.45***	0.48***	1		
SB	−0.13	−0.16*	0.17*	0.57***	−0.12	1	
GY	0.05	0.05	0.51***	0.86***	0.61***	0.59***	1

*, **, *** Significant at the 0.05, 0.01 and 0.001 probability levels, respectively.

Table 2. Phenotypic differences in% nitrogen derived from the atmosphere estimated using shoot tissue (%Ndfa-SH), % nitrogen derived from the atmosphere estimated using grain tissue (%Ndfa-G), shoot ^{15}N natural abundance and grain ^{15}N natural abundance of 36 genotypes of common bean grown under irrigated and drought conditions in 2012 and 2013 at Palmira, Colombia.

Genotype	%Ndfa-Shoot				%Ndfa-Grain				^{15}N natural abundance in shoot (‰)				^{15}N natural abundance in grain (‰)			
	Irrigated		Drought		Irrigated		Drought		Irrigated		Drought		Irrigated		Drought	
	2012	2013	2012	2013	2012	2013	2012	2013	2012	2013	2012	2013	2012	2013	2012	2013
ALB 6	57	50	22	25	53	44	17	29	0.6	2.5	5.9	6.3	1.4	2.5	6.3	5.1
ALB 60	56	39	27	26	58	38	27	20	0.7	3.8	5.4	6.1	0.9	3.0	5.3	6.2
ALB 74	55	36	28	16	60	42	18	18	0.8	4.1	5.3	7.3	0.7	2.7	6.2	6.3
ALB 88	62	31	9	23	66	34	7	18	0.2	4.7	7.5	6.5	0.3	3.5	7.4	6.3
ALB 213	49	47	18	29	62	30	12	23	1.3	2.8	6.4	5.7	0.6	3.8	6.9	5.8
BAT 477	58	49	17	32	60	44	11	30	0.2	2.5	6.4	5.1	0.5	2.4	6.9	4.8
BAT 477_NN	0	0	0	0	0	0	0	0	5.6	8.3	8.7	9.5	5.8	6.8	8.6	8.3
BFS 10	58	34	18	28	65	40	18	22	0.5	4.4	6.4	5.8	0.4	2.9	6.1	5.8
BFS 29	59	42	24	25	70	40	25	29	0.4	3.5	5.8	6.3	0.0	2.9	5.5	5.2
BFS 32	61	45	26	23	58	39	34	25	0.2	3.1	5.5	6.5	1.0	2.9	4.5	5.6
BFS 67	61	41	29	26	62	41	12	27	0.2	3.6	5.2	6.2	0.7	2.8	6.9	5.3
DOR 390	58	38	25	23	58	42	20	19	0.6	3.9	5.6	6.5	0.9	2.7	6.0	6.2
G 40001	52	28	11	7	54	18	1	12	1.0	5.1	7.3	8.5	1.3	4.8	8.0	7.0
INB 827	60	36	25	30	68	39	18	27	0.3	4.1	5.6	5.6	0.2	3.0	6.1	5.3
INB 841	55	37	19	25	61	24	16	27	0.8	4.1	6.3	6.2	0.7	4.3	6.4	5.3
MIB 778	56	50	15	14	62	45	4	16	0.7	2.6	6.9	7.6	0.6	2.4	7.7	6.5
NCB 226	62	52	15	37	69	44	14	32	0.2	2.4	6.8	4.7	0.1	2.6	6.6	4.8
NCB 280	45	24	16	26	60	30	17	26	1.7	5.5	6.7	6.1	0.7	3.8	6.4	5.4
Pérola	57	45	21	38	59	45	23	24	0.3	2.9	6.1	4.3	0.6	2.3	5.6	5.6
RCB 273	55	48	13	25	62	36	17	23	0.8	2.9	7.0	6.3	0.6	3.3	6.3	5.7
RCB 593	45	38	35	33	55	37	31	36	1.6	3.8	4.5	5.2	1.2	3.1	4.9	4.4
SCR 2	56	23	13	34	58	22	9	25	0.6	5.6	6.9	5.1	1.0	4.5	7.1	5.6
SCR 9	63	32	23	21	65	39	19	26	0.1	4.6	5.8	6.7	0.3	3.0	6.2	5.5
SCR 16	58	35	20	25	59	37	26	27	0.5	4.3	6.2	6.2	0.9	3.2	5.4	5.4
SEA 15	55	38	34	35	60	35	19	22	0.5	3.7	4.4	4.8	0.5	3.2	6.0	5.9
SEN 56	63	41	17	28	67	30	21	20	0.1	3.6	6.6	5.8	0.3	3.8	5.9	6.1
SER 16	58	39	19	24	59	33	13	21	0.5	3.8	6.3	6.4	0.9	3.5	6.7	6.0
SER 48	66	42	27	27	61	44	23	32	−0.2	3.5	5.4	6.0	0.7	2.6	5.6	4.9
SER 78	59	31	33	25	63	31	17	23	0.4	4.6	4.7	6.2	0.6	3.6	6.3	5.8
SER 118	59	37	20	18	65	41	26	13	0.4	4.0	8.2	7.1	0.4	2.8	5.4	6.9
SER 119	66	52	26	30	66	45	16	22	−0.1	2.3	5.6	5.6	0.3	2.5	6.4	6.0
SER 125	62	40	20	30	64	39	27	28	0.1	3.7	6.2	5.6	0.5	3.0	5.2	5.3
SMC 43	56	55	21	25	64	47	13	23	0.7	2.1	6.1	6.2	0.5	2.2	6.7	5.7
SMC 141	57	51	31	20	60	53	22	28	0.6	2.4	5.0	6.8	0.8	1.7	5.7	5.3
SXB 412	47	41	17	26	51	47	18	24	1.2	3.4	6.5	5.9	1.3	2.1	6.1	5.5
Tio Canela 75	53	37	42	23	57	45	34	17	0.9	4.0	3.6	6.4	1.1	2.5	4.5	6.4
Mean	**56**	**39**	**22**	**25**	**59**	**37**	**18**	**23**	**0.7**	**3.8**	**6.0**	**6.2**	**0.8**	**3.1**	**6.2**	**5.8**
Sig. diff.	*	*	*	*	*	*	*	*	*	*	*	*	*	*	*	*

*Significant difference at 0.05 level as estimated from the MIXED procedure.

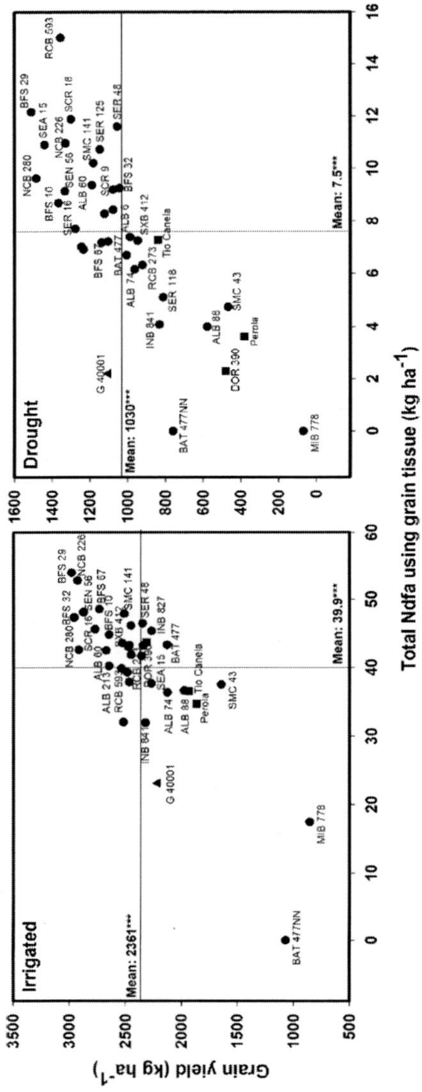

Figure. 1. Identification of genotypes that combine greater total nitrogen derived from the atmosphere in kg ha^{-1}estimated using grain tissue (TNdfa-G) with superior grain yield under irrigated and drought conditions when grown in a Mollisol at CIAT-Palmira, Colombia. Higher TNdfa-G genotypes with greater grain yield were identified in the upper, right hand quadrant. Genotypes identified with symbols of (■) are commercial varieties and with a symbol of (▲) is *P. acutifolius.*

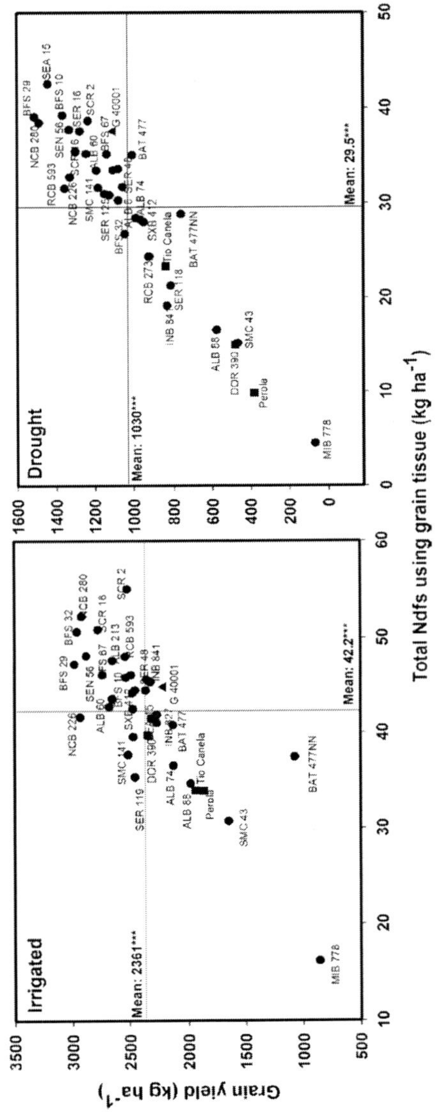

Figure. 2. Identification of genotypes that combine greater total nitrogen derived from the soil in kg ha^{-1} estimated using grain tissue (TNdfs-G) with superior grain yield under irrigated and drought conditions when grown in a Mollisol at CIAT-Palmira, Colombia. Higher TNdfs-G genotypes with greater grain yield were identified in the upper, right hand quadrant. Genotypes identified with symbols of (■) are commercial varieties and with a symbol of (▲) is *P. acutifolius.*

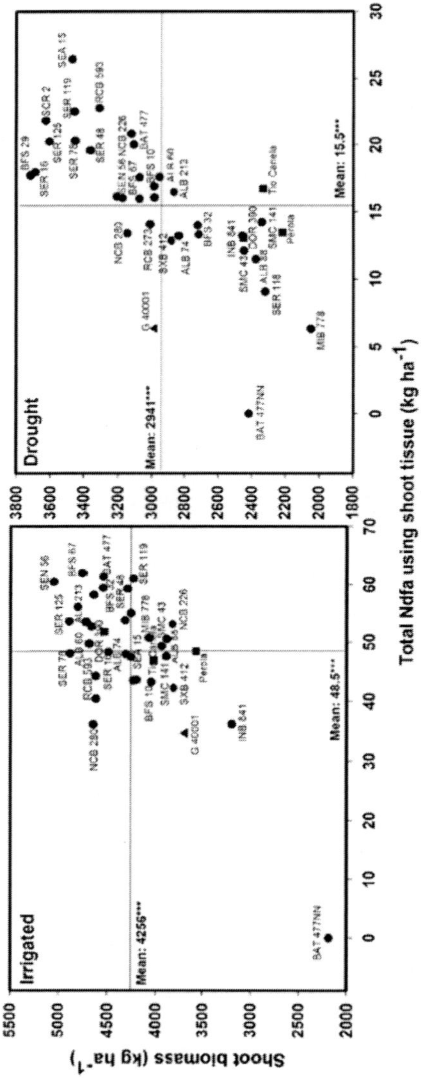

Figure. 3. Identification of genotypes that combine greater total nitrogen derived from the atmosphere in kg ha^{-1}estimated using shoot tissue (TNdfa-SH) with superior grain yield under irrigated and drought conditions when grown in a Mollisol at CIAT-Palmira, Colombia. Higher TNdfa-SH genotypes with greater grain yield were identified in the upper, right hand quadrant. Genotypes identified with symbols of (■) are commercial varieties and with a symbol of (▲) is *P. acutifolius.*

A positive and highly significant correlation was observed between grain yield and nitrogen use efficiency (NUE) under both irrigated and drought conditions (Table 1). Genotypes that combined better grain yield with higher NUE under drought conditions were RCB 593, SMC 141, BFS 32, BFS 29, SEA 15, SEN 56, NCB 280 and NCB 226. A low but significant correlation coefficient (r = 0.19**) was observed between %Ndfa and NUE under drought conditions. The lines RCB 593, SMC 141, BFS 32, BFS 29, SEA 15, SEN 56, NCB 280 and NCB 226 combined higher values of NUE and %Ndfa under drought stress (Fig. 4). The commercial cultivar Pérola showed a moderate value of %Ndfa but it was poor in its NUE and was low yielding under drought (Fig. 4).

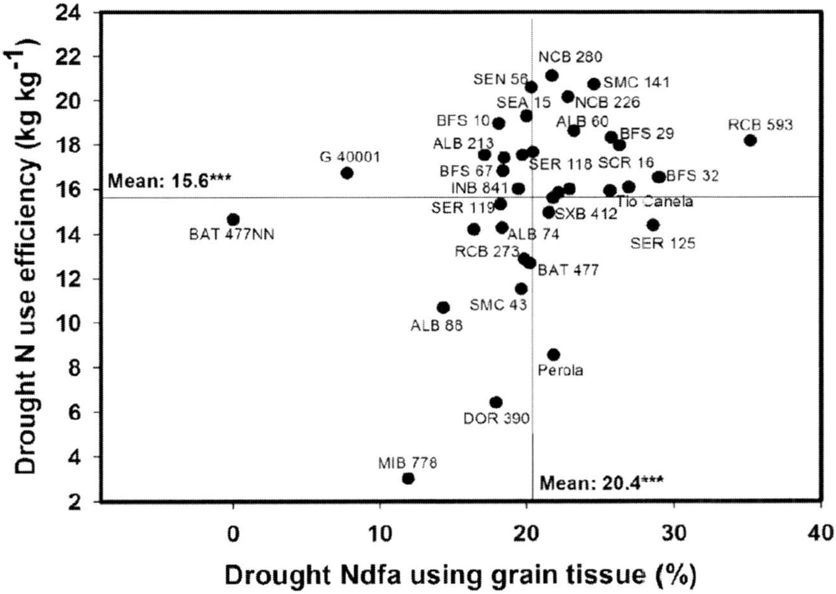

Figure. 4. Identification of genotypes that combine greater values of %nitrogen derived from the atmosphere using grain tissue (%Ndfa-G) with higher values of nitrogen use efficiency (NUE) in terms of kg of grain produced kg^{-1} of shoot N uptake under drought conditions when grown in a Mollisol at CIAT-Palmira, Colombia. Higher%Ndfa-G genotypes with greater values of NUE were identified in the upper, right hand quadrant.

4. DISCUSSION

This study allowed to compare the estimation of %Ndfa using shoot tissue (%Ndfa-SH) vs. grain tissue (%Ndfa-G), to quantify phenotypic differences in common bean for SNF ability under irrigated and drought stress conditions, and to test whether %Ndfa-G could be a useful trait in breeding programs. Results from two seasons with %Ndfa values estimated under field conditions, comparing the conventional methodology based on shoot tissue at the growth stage of mid-pod filling (%Ndfa-SH), and the method proposed here using grain tissue at the time of harvest (%Ndfa-G), showed that the latter methodology is feasible. The high correlation between both methods validates this statement. We suggest that grain samples collected at harvest can be used to quantify phenotypic differences in SNF ability in common bean using the methodology of natural abundance of ^{15}N. The proposed %Ndfa-G method, from a breeding perspective, is much easier than the conventional methodology involving shoot tissue (%Ndfa-SH) which requires destructive sampling, drying of fresh tissue, sampling of plant parts for grinding, all with more labor and therefore is less cost effective. With the proposed %Ndfa-G methodology, the breeder can take a sample of harvested grain, dry and grind it and analyze for the isotope ratios of $\delta^{13}C$ and $\delta^{15}N$, simultaneously selecting for water use efficiency (Araus *et al.*, 2002 ; Easlon *et al.*, 2014) and SNF ability (based on higher values of %Ndfa). In a recent genomic study addressing SNF ability in common bean, Kamfwa et al. (2015) using both shoot and grain samples suggested selection for high SNF ability based on grain tissue could be easier integrated into most breeding programs. Similar results for %Ndfa using the ^{15}N natural abundance technique in grains and the whole shoots have also been reported for other legume species (Bergersen et al., 1985). Discordant results reported by others can occur when grain filling during the late stage of development is highly dependent on the contribution of N fixation rather than on the extent of remobilization of N from vegetative structures (Bergersen et al., 1992).

Previous research showed that common beans are able to translocate about 80–93% of its total N to the grain (Ramaekers et al., 2013). Also it has been shown that common bean accumulates preferentially the fixed N into the grain (Dubois and Burris, 1986; Westermann *et al.*, 1985 ; Wolyn *et al.*, 1991). At such high rates of N translocation in common beans, trends in %Ndfa would be similar between shoot tissue and grain tissue. Several of the lines evaluated in this study, have high mobilization of N from shoot to grain under both irrigated and drought conditions, as can be evidenced from our results, where the genotypes with greater N accumulation in shoot at mid-pod filling also showed greater values of N accumulation in grain at harvest. The contrary was true with the genotypes that accumulated less N in shoot and grain.

This study also permitted evaluating the SNF ability in a set of elite common bean breeding lines that were recently developed for improving resistance to drought. Furthermore, we tested lines that were derived from crosses among bean races (Beebe et al., 2008), as well as interspecific crosses with introgression from *P. coccineus* (Butare et al., 2012) or *P. acutifolius* (Beebe, 2012). Several lines developed over different breeding cycles to improve drought resistance not only performed better under water shortage, but also had higher ability to fix N under these conditions. Under drought stress,both grain yield and total Ndfa in these lines doubled the values that were observed for three leading commercial cultivars grown in Latin America: DOR 390, Perola, and Tío Canela. Under unfavorable conditions, such as drought, a decrease in the effectiveness of SNF process is expected (Devi et al., 2013). The symbiosis is based on the carbon supply from the plant to the Rhizobium which provides fixed N to the plant. But under drought stress, the reduced net photosynthesis decreases the supply of photosynthates to the nodules resulting in lower values of %Ndfa (Gonzalez *et al.*, 1995 ; Sassi *et al.*, 2008). Our results confirm previous reports that SNF in common bean is especially sensitive to drought stress, as drought reduced Ndfa by 57%. SNF is a physiological process especially sensitive to soil drying (Devi et al., 2013), and the effect of drought stress on SNF varies according to the stages of development when the stress occurs. Water stress during early vegetative growth had no significant effect on SNF, but during flowering or grain filling it significantly reduced fixed N (Chalk et al., 2010). In the two growing seasons of this study the drought stress was imposed starting at the critical preflowering stage, resulting in high inhibition of SNF ability.

The superior performance of a few drought-adapted lines that combine greater grain yield with higher values of %Ndfa in grain might be due to greater carbon transport to both grain and the nodule. It is also noteworthy that the drought resistant bean lines with higher SNF ability had greater values of shoot biomass indicating the importance of plant vigor for supporting nodule development, as well as contributing to remobilization of both C and N to developing grains. Vigorous plants permit higher levels of N accumulation of both fixed N and non-fixed N, while maintaining or improving remobilization of this N to grain, contributing to increased grain yield in common bean (Wolyn et al., 1991). Higher shoot biomass and N accumulation before pod set could provide an advantage for drought adapted lines (Vadez et al., 2014). The lower sensitivity of SNF ability in drought resistant lines can be explained by the expeditious removal of N products from nodules and sequestering them in the shoot of the plant to avoid N-feedback limitation on SNF (Beebe *et al.*, 2014 ; Sinclair and Vadez, 2012).

Tepary bean (*P. acutifolius*) has been reported to present multiple traits for drought resistance including early maturity, greater photosynthate remobilization capacity, deep rooting, small leaves, and stomatal control for improved water use efficiency (Beebe *et al.*, 2014; Mohamed *et al.*, 2005 ; Rao *et al.*, 2013). But in spite of excellent drought resistance, our results indicate that *P. acutifolius* drastically reduces its SNF under drought, suggesting an internal control of SNF, possibly through decreased carbon supply to the nodules, while maintaining uptake of mineral N from soil through its fine root system. In early maturing genotypes, competition for photosynthates exists between nodules and developing pods/grain (Piha and Munns, 1987 ; Vadez *et al.*, 2014). *Phaseolus acutifolius* is early maturing, and may preferentially channel the photosynthates into pods and grains rather than destine them to the development and maintenance of the nodules. This in part could be an evolutionary response to the hot and dry environments from where tepary bean originated.

Common beans are poor N fixers compared to other grain legumes (Hardarson, 2004; Hardarson *et al.*, 1993 ; Peoples *et al.*, 2009). We conducted our evaluation in a Mollisol with adequate soil organic matter content that can limit the expression of SNF activity. Nonetheless, in the irrigated treatment, %Ndfa-G presented a mean value of 48% for two seasons or a mean value of 40 kg ha^{-1} of N fixed. Several genotypes presented %Ndfa-G values superior to 50%, representing more than 50 kg ha^{-1} of N fixed under irrigated conditions. The efficiency of major food crops in the recovery of applied N is often not more than 30% (Subbarao et al., 2013); so 50 kg N fixed through SNF would be comparable to about 150 kg N applied as chemical fertilizer. Line RCB 593 suffered less inhibition of SNF ability under drought stress when compared to irrigated conditions; moreover, and it presented the highest SNF ability under drought stress. This line could be a potential parent to improve SNF capacity under drought stress in bush beans. Other lines that were resistant to drought and showed relatively higher SNF ability included BFS 29, BFS 32, SER 48, SEA 15 and NCB 226. The drought- adapted line NCB 226 has previously been found to be superior in its SNF ability under drought stress using the methodology of acetylene reduction activity (ARA) determined with a flow-through system (Devi et al., 2013). Our results also indicated that the drought resistant check BAT 477 showed moderately high %Ndfa under drought stress. This line has previously been identified with good SNF ability both under optimal (Kipe-Nolt and Giller, 1993) and drought stress conditions (Castellanos et al., 1996). As expected, most drought sensitive lines were poor in their SNF ability, for example, DOR 390, a commercial variety used widely in at least three countries in Latin America.

This study also found that all the best N fixers under drought were drought resistant lines, confirming the tendency that was observed in previous studies with common bean (Beebe *et al.*, 2014 ; Devi *et al.*, 2013). A positive and significant correlation was observed between grain yield and TNdfa (kg ha^{-1}) in grain under both irrigated and drought conditions, indicating that the genotypes with more N accumulation from fixation presented higher grain yield under both irrigated and drought conditions. The lines RCB 593, SEA 15, BFS 29, SCR 16, NCB 280 and NCB 226 accumulated more N from symbiotic fixation, and used it for greater grain production both under drought and irrigated conditions. Several lines that performed well under drought stress, such as SMC 141, RCB 593, BFS 32, SEN 56 and NCB 226, combined greater SNF ability with more efficient use of N to produce grain. Despite the fact that SNF ability can be drastically affected by drought stress, the more drought resistant Middle American lines overcome this limitation by using the acquired N more efficiently through greater remobilization of both C and N to grain. A higher use efficiency of the acquired N can be very relevant for crop yield in environments dominated by strong droughts where the viability of *Rhizobium* and in consequence SNF is severely inhibited.

5. CONCLUSIONS

Correlations between %Ndfa using shoot tissue (%Ndfa-SH) and %Ndfa using grain tissue (%Ndfa-G) indicated that the values of %Ndfa-G can be used to quantify phenotypic differences in symbiotic nitrogen fixation (SNF) of common bean under either irrigated or drought stress conditions. Estimates of %Ndfa-G are easier to implement in a breeding program due both to less labor costs and the feasibility to determine this parameter at harvest time. Using %Ndfa-G values, we observed significant phenotypic differences in SNF ability in common bean under drought stress. We identified four bean lines RCB 593, SEA 15, NCB 226 and BFS 29 that were not only drought resistant but also were superior in their SNF ability and these lines could serve as parents in breeding programs. Our results also indicate that the drought response of the SNF ability of tepary bean (*P. acutifolius*) and common bean are different, possibly due to differences in internal regulation mechanisms of SNF.

ACKNOWLEDGEMENTS

The authors acknowledge the support of the Bill and Melinda Gates Foundation (BMGF), United States Agency for International Development (USAID) and the CGIAR research program on grain legumes for financial support of research on improving genetic adaptation of common beans to drought. Special thanks to Dr. Mark B. Peoples for his comments and suggestions to improve the quality of this manuscript. We also thank Miguel Grajales, Cesar Cajiao, Norma Barbosa, Mariela Rivera and bean breeding and physiology teams at CIAT, Colombia for their contribution.

REFERENCES

1. Araus, J.L., Slafer, G.A., Reynolds, M.P., Royo, C., 2002. Plant breeding and drought inC3 cereals: what should we breed for? Ann. Bot. 89, 925–940, http://dx.doi.org/10.1093/aob/mcf049.
2. Beebe, S., Rao, I.M., Cajiao, C., Grajales, M., 2008. Selection for drought resistance incommon bean also improves yield in phosphorus limited and favorableenvironments. Crop Sci. 48, 582, http://dx.doi.org/10.2135/cropsci2007.07.0404.
3. Beebe, S., Rao, I.M., Blair, M.W., Acosta-Gallegos, J.A., 2013. Phenotyping commonbeans for adaptation to drought. Front. Physiol. 4 (march), 1–20, http://dx.doi.org/10.3389/fphys.2013.00035.
4. Beebe, S., Rao, I.M., Devi, M., Polania, J., 2014. Common beans, biodiversity, andmultiple stress: challenges of drought resistance in tropical soils. Crop PastureSci. 65, 667–675, http://dx.doi.org/10.1071/CP13303.
5. Beebe, S., 2012. Common bean breeding in the tropics. Plant Breed. Rev. 36,357–426, http://dx.doi.org/10.1002/9781118358566 (36).
6. Bergersen, F.J., Turner, G.J., Gault, R.R., Chase, D.L., Brockwell, J., 1985. The naturalabundance of 15 N in an irrigated soybean crop and its use for the calculationof nitrogen fixation. Crop Pasture Sci. 36, 411–423.
7. Bergersen, F.J., Turner, G.L., Gault, R.R., Peoples, M.B., Morthorpe, L.J., Brockwell, J.,1992. Contributions of nitrogen in soybean crop residues to subsequent cropsand to soil. Crop Pasture Sci. 43, 155–169.
8. Boddey, R.M., Peoples, M.B., Palmer, B., Dart, P.J., 2000. Use of the 15 N naturalabundance technique to quantify biological nitrogen fixation by woodyperennials. Nutr. Cycl. Agroecosyst. 57, 235–270, http://dx.doi.org/10.1023/A:1009890514844.
9. Butare, L., Rao, I.M., Lepoivre, P., Cajiao, C., Polania, J., Cuasquer, J., Beebe, S., 2012. Phenotypic evaluation of interspecific recombinant inbred lines (RILs) ofPhaseolus species for aluminium resistance and shoot and root growthresponse to aluminium-toxic acid soil. Euphytica 186, 715–730.
10. Castellanos, J.Z., Pe˜na-Cabriales, J.J., Acosta-Gallegos, J.A., 1996. N-15-determineddinitrogen fixation capacity of common bean (Phaseolus vulgaris) cultivarsunder water stress. J. Agric. Sci. 126, 327–333.
11. Chalk, P.M., Alves, B.J.R., Boddey, R.M., Urquiaga, S., 2010. Integrated effects ofabiotic stresses on inoculant performance, legume growth and symbioticdependence estimated by 15 N dilution. Plant Soil 328, 1–16, http://dx.doi.org/10.1007/s11104-009-0187-7.
12. Devi, M., Sinclair, T.R., Beebe, S., Rao, I.M., 2013. Comparison of common bean(Phaseolus vulgaris L.) genotypes for nitrogen fixation tolerance to soil drying.Plant Soil 364, 29–37, http://dx.doi.org/10.1007/s11104-012-1330-4.
13. Douxchamps, S., Humbert, F.L., van der Hoek, R., Mena, M., Bernasconi, S.M.,Schmidt, A., Rao, I.M., Frossard, E., Oberson, A., 2010. Nitrogen balances infarmers fields under alternative uses of a cover crop legume: a case study fromNicaragua. Nutr. Cycl. Agroecosyst. 88, 447–462, http://dx.doi.org/10.1007/s10705-010-9368-2.
14. Dubois, J., Burris, R.H., 1986. Comparative study of N uptake and distribution inthree lines of common bean (Phaseolus vulgaris L.) at early pod filling stage.Plant Soil 93, 79–86, http://dx.doi.org/10.1007/BF02377147.
15. Easlon, H.M., Nemali, K.S., Richards, J.H., Hanson, D.T., Juenger, T.E., McKay, J.K.,2014. The physiological basis for genetic variation in water use efficiency andcarbon isotope

composition in Arabidopsis thaliana. Photosynth. Res. 119,119–129, http://dx.doi.org/10.1007/s11120-013-9891-5.

16. Gonzalez, E.M., Gordon, A.J., James, C.L., Arrese-Igor, C., 1995. The role of sucrosesynthase in the response of soybean nodules to drought. J. Exp. Bot. 46,1515–1523.
17. Hardarson, G., Bliss, F.A., Cigales-Rivero, M.R., Henson, R.A., Kipe-Nolt, J.A., Longeri,L., Manrique, A., Pe˜na-Cabriales, J.J., Pereira, P.A., Sanabria, C.A., Tsai, S.M., 1993.Genotypic variation in biological nitrogen fixation by common bean. Plant Soil152, 59–70, http://dx.doi.org/10.1007/BF00016333.
18. Hardarson, G., 2004. Enhancement of symbiotic nitrogen fixation in grain legumes:selected results from the FAO/IAEA program. In: Serraj, R. (Ed.), SymbioticNitrogen Fixation: Prospects for Enhanced Application in Tropical Agriculture.Oxford & IBH Publishing, New Delhi, pp. 163–171.
19. Kamfwa, K., Cichy, K.A., Kelly, J.D., 2015. Genome-wide association analysis ofsymbiotic nitrogen fixation in common bean. Theor. Appl. Genet. 128,1999–2017, http://dx.doi.org/10.1007/s00122-015-2562-5.
20. Kipe-Nolt, J.A., Giller, K.E., 1993. A field evaluation using the 15 N isotope dilutionmethod of lines of Phaseolus vulgaris L. bred for increased nitrogen fixation.Plant Soil 152, 107–114, http://dx.doi.org/10.1007/BF00016339.
21. Lynch, J., White, J.W., 1992. Shoot nitrogen dynamics in tropical common bean.Crop Sci. 32, 392–397.
22. Mohamed, F., Mohamed, M., Schmitz-Eiberger, N., Keutgen Noga, G., 2005.Comparative drought postponing and tolerance potentials of two tepary beanlines in relation to seed yield. Afr. Crop Sci. J. 13, 49–60.
23. Peoples, M.B., Brockwell, J., Herridge, D.F., Rochester, I.J., Alves, B.J.R., Urquiaga, S.,Boddey, R.M., Dakora, F.D., Bhattarai, S., Maskey, S.L., Sampet, C., Rerkasem, B.,Khan, D.F., Hauggaard-Nielsen, H., Jensen, E.S., 2009. The contributions ofnitrogen-fixing crop legumes to the productivity of agricultural systems.Symbiosis 48, 1–17, http://dx.doi.org/10.1007/BF03179980.
24. Piha, M.I., Munns, D.N., 1987. Nitrogen fixation potential of beans (Phaseolusvulgaris L.) compared with other grain legumes under controlled conditions.Plant Soil 98, 169–182.
25. Ramaekers, L., Galeano, C.H., Garzón, N., Vanderleyden, J., Blair, M.W., 2013.Identifying quantitative trait loci for symbiotic nitrogen fixation capacity andrelated traits in common bean. Mol. Breed. 31, 163–180, http://dx.doi.org/10.1007/s11032-012-9780-1.
26. Rao, I.M., Beebe, S., Polania, J., Ricaurte, J., Cajiao, C., Garcia, R., Rivera, M., 2013. Cantepary bean be a model for improvement of drought resistance in commonbean? Afr. Crop Sci. J. 21, 265–281.
27. Rao, I.M., 2014. Advances in improving adaptation of common bean and Brachiariaforage grasses to abiotic stresses in the tropics. In: Pessarakli, M. (Ed.),Handbook of Plant and Crop Physiology. CRC Press, Taylor and Francis Group,Boca Raton FL, pp. 847–889.
28. SAS Institute Inc., 2008. SAS/STAT®9.2.
29. Sassi, S., Gonzalez, E.M., Aydi, S., Arrese-Igor, C., Abdelly, C., 2008. Tolerance ofcommon bean to long-term osmotic stress is related to nodule carbon flux andantioxidant defenses: evidence from two cultivars with contrasting tolerance.Plant Soil 312, 39–48, http://dx.doi.org/10.1007/s11104-008-9613-5.
30. Shearer, G., Kohl, D.H., 1986. N2-fixationixation in field settings: estimations basedon natural 15 N abundance. Aust. J. Plant Physiol. 13, 699–756, http://dx.doi.org/10.1071/PP9860699c.
31. Sinclair, T.R., Vadez, V., 2012. The future of grain legumes in cropping systems.Crop Pasture Sci. 63, 501–512, http://dx.doi.org/10.1071/CP12128.
32. Subbarao, G.V., Sahrawat, K.L., Nakahara, K., Rao, I.M., Ishitani, M., Hash, C.T., Kishii,M., Bonnett, D.G., Berry, W.L., Lata, J.C., 2013. A paradigm shift towardslow-nitrifying production systems: the role of biological nitrification inhibition(BNI). Ann. Bot., http://dx.doi.org/10.1093/aob/mcs230.
33. Thung, M., Rao, I.M., 1999. Integrated management of abiotic stresses. In: Singh,S.P. (Ed.), Common Bean Improvement in the Twenty-First Century. Springer,Netherlands, Kimberly USA, pp. 331–370.

34. Unkovich, M.J., Pate, J.S., Sanford, P., Armstrong, E.L., 1994. Potential precision ofthe delta-N-15 natural-abundance method in-field estimates ofnitrogen-fixation by crop and pasture legumes in south-west Australia. Aust. J.Agric. Res 45, 119–132.
35. Unkovich, M.J., Herridge, D.F., Peoples, M., 2008. Measuring plant-associatednitrogen fixation in agricultural systems. Aust. Cent. Int. Agric. Res. 258(citeulike-article-id:9530718).
36. Vadez, V., Kholova, J., Medina, S., Kakkera, A., Anderberg, H., 2014. Transpirationefficiency: new insights into an old story. J. Exp. Bot. 65, 6141–6153, http://dx.doi.org/10.1093/jxb/eru040.
37. Westermann, D.T., Porter, L.K., O'Deen, W.A., 1985. Nitrogen partitioning andmobilization patterns in bean plants. Crop Sci. 25, 225–229.
38. Wolyn, D.J., Clair, D.A.S., DuBois, J., Rosas, J.C., Burris, R.H., Bliss, F.A., 1991.Distribution of nitrogen in common bean (Phaseolus vulgaris L.) genotypesselected for differences in nitrogen fixation ability. Plant Soil 138, 303–311.

CHAPTER 11

Factors Affecting the Presence and the Diversity of Bryophytes in the Petrifying Sources Habitat (7220) in Wallonia and the Brussels-Capital Region, Belgium

J.-M. Couvreur,[1] G. San Martin,[2] and A. Sotiaux[3]

[1]*Département de l'Etude du Milieu Naturel et Agricole, Service Public de Wallonie, Gembloux, Belgium*

[2]*Centre Wallon de Recherches Agronomiques, Gembloux, Belgium*

[3]*National Botanic Garden, Meise, Belgium*

ABSTRACT

Bryological composition, water chemistry, and environmental factors were characterized on 67 Belgian travertines. We explore the relationship between these environmental factors and the community composition, species richness, or presence of individual species using Redundancy Analysis with Hellinger's transformation (tb-RDA) or Generalized Linear Models (GLMs). The best variables explaining the community composition are slope, NO_3, NH_4, and PO_4. The species richness is negatively related to canopy cover and PO_4. Palustriella commutata tends to be more frequent when the slope is steeper and to a lesser degree when the canopy cover is lower. Eucladium verticillatum tends to be slightly more frequent when canopy cover and NH_4 concentrations are lower. Cratoneuron filicinum is more frequent at higher Mg concentrations and Pellia endiviifolia is more frequent at lower PO_4 concentrations and higher NO_3 concentrations. Brachythecium rivulare showed wide ecological amplitude and almost none of the tested environmental factors seem to be related to its presence. The study identifies eutrophication as the main factor responsible for habitat deterioration. Practical indications on the best ways to maintain or to enhance the quality of these petrifying sources are given.

1. INTRODUCTION

"Travertine" or "tufa" deposits include a wide variety of calcareous substrates that are characterized not only by their mineral composition but also by their morphology and processes by which they form. There are many interpretations and definitions of the terms tufa and travertine (Symoens et al. [1], Couderc [2], Pentecost [3], De Zuttere [4], Viles and Goudie [5], Ford and Pedley [6], Janssen and Swennen [7], Merz-Preiß and Riding [8], Boch et al. [9], Franco et al. [10], and Brusa and Cerabolini [11]). The term "travertine" is often used to describe calcareous substrates formed during warm periods or in warm environment including thermal springs and that contain no or very few organic material other than bacteria [6, 7]. In contrast, the term "tufa" is generally used to describe substrates formed in colder environments and containing organic material [6]. It is possible for both warm and cold travertine formation to occur simultaneously, as evidenced by several Belgian locations including the Hoyoux river near Huy and on the "Ry de Matignolles" stream near Treignes [10].

Although the European definition of the habitat [12] uses the term "tufa" we will use the term "travertine" in accordance with Pentecost [13] as the term "tufa" relates more to a soft and poorly consolidated variety of travertine.

In the framework of the mapping of Natura 2000 sites in Wallonia, south Belgium, we had the opportunity to undertake research on the ecology of the priority habitat "Petrifying springs with tufa formation (Cratoneurion) 7220." The code 7220 is specific to this priority habitat in the Interpretation Manual of European Union Habitats [12]. This priority habitat generally consists of small point or linear formations dominated by bryophytes (Cratoneurion commutati Koch, 1928, communities). The manual [12] provides a list of characteristic species for each habitat type; however, these lists have to be adapted at the national or regional scale. In Wallonia, two bryophyte species are considered strictly linked to the 7220 habitat,

Palustriella commutata (Hedw.) Ochyra and Eucladium verticillatum (With.) Bruch & Schimp. Palustriella commutata (Hedw.) Ochyra has been split into two species by some authors [14]; however, in this study we consider both as Palustriella commutata.

The following are considered to be good companion species and are also included in the Walloon definition when assessing the habitat conservation status [15]: Aneura pinguis (L.) Dumort., Conocephalum conicum (L.) Dumort., Jungermannia atrovirens Dumort., Leiocolea badensis (Gottsche.) Jörg., Pellia endiviifolia (Dicks.) Dumort., Preissia quadrata (Scop.) Nees, Brachythecium rivulare Schimp., Bryum pseudotriquetrum (Hedw.) P. Gaertn. et al., Ctenidium molluscum (Hedw.) Mitt., Cratoneuron filicinum (Hedw.) Spruce, Dichodontium pellucidum (Hedw.) Schimp., Didymodon tophaceus (Brid.) Lisa, Fissidens adianthoides Hedw., Fissidens crassipes Wilson ex Bruch & Schimp., Gymnostomum calcareum Nees & Hornsch., Hymenostylium recurvirostrum (Hedw.) Dixon, Philonotis calcarea (Bruch & Schimp.) Schimp., Plagiomnium ellipticum (Brid.) T. J. Kop., Plagiomnium rostratum (Schrad.) T. J. Kop., Plagiomnium undulatum (Hedw.) T. J. Kop., and Seligeria trifaria (Brid.) Lindb. Cratoneuron filicinum is considered a good characteristic species in other regions and countries, for example, Flanders and the Brussels-Capital Region [16], Great Britain [17], and the Netherlands [18]. However, it was not included in the small group of characteristic species as it is far more widespread in Wallonia than the two other species [19], occurring in a variety of different habitats, including man-made calcareous forest tracks. Nonetheless Cratoneuron filicinum is considered to be a good companion species for the 7220 habitat in Wallonia [15].

In Wallonia [15], the 7220 habitat is mainly associated with calcareous springs and small streams (width < 5 m) where incrustation processes occur. An atypical form of the habitat is represented in Wallonia by the "travertins" formed in the Hoyoux river consisting of calcareous deposits across the river (barrages) or its tributaries whose width can reach 10 to 20 m. The 7220*habitat in Wallonia [15] is characterized by Ca-rich waters (110–120 $mg{\cdot}L^{-1}$) and high pH values (7.5–8.5) but depending on the subregion the substrate can be mostly sandy with calcareous incrustations (sandy-loam region) or more compact calcareous rocks (Condroz, Famenne-Calestienne, Lorraine) where they are locally called "crons" or "crânières" [1, 20–24].

From a phytosociological point of view, most of the springs and small streams of Wallonia and Brussels-Capital Region can be attributed to the Montio-fontanae-Cardaminetea-amarae Braun-Blanquet et Tüxen, 1943, Class [16, 25, 26]. Following Bardat and Hauguel [26] and depending on the local situations the bryophytes communities can then be linked to the Caricion remotae Kästner, 1941, Alliance (intraforest communities of oligomesotrophic soils), to the Pellion endiviifoliae Bardat, 1998, prov.nov. Alliance (neutro-alkaline small streams), or to the Riccardio-Eucladion verticillati Bardat, 1998, prov.nov. Alliance (Ca-rich and thermophilous soils). Following Zechmeister and Mucina [16, 25], Belgian communities can be linked to the Cratoneurion commutati Koch, 1928, Alliance (sunny springs with highly oxygen-saturated waters), to the Adiantion Br.-Bl. ex Horvatic, 1934, em. Hoc. Loco Alliance (waters with high Ca contents and high summer temperatures), or to the Caricion remotae Kästner, 1941, Alliance (intraforest communities of oligotrophic waters).

The bryological aspects of travertines and their phytosociological composition are well documented in Belgium [1, 4, 16, 21, 22, 27, 28], the Netherlands [18, 27], Italy [29], France [2, 26, 30], and in a wider European context [25]. However, relatively few studies have been undertaken on datasets that consider the driving environmental variables that could influence the distribution of particular species and the species richness of these habitats [11, 31–35].

The aims of this research were (1) to learn more about the variables, both physical and chemical, that can affect bryophyte diversity in the surveyed sites; (2) to evidence the most significant variables that can affect the presence or absence of the more characteristic species; (3) to give practical indications on the best ways to maintain or to enhance the quality of this habitat both in Wallonia and the Brussels-Capital Region.

2. MATERIALS AND METHODS

2.1. Data Collection

Most of the surveyed sites lie in Wallonia, south Belgium (Figure 1), where 62 sites were identified, representing the widest variety of conditions within the 7220 habitat. Wallonia is not a homogeneous region as it consists of 5 subregions characterized by geomorphological specificities (from north to south: sandy Loess Plateau, psammitic Condroz, calcareous Famenne-Calestienne, acidic siliceous Ardennes, and calcareous Lorraine). Petrifying sources are mainly small habitats; however, they are relatively widespread in all these subregions except the Ardennes where it is virtually absent due to the acidic siliceous substrate and in the sandy Loess Plateau where there were also fewer sites. To increase the coverage, we added 5 sites within the Loess region by including some of the Brussels-Capital Region localities.

On each Walloon site a comprehensive sampling regime (random selection of all known locations) was undertaken during autumn, 2012; the bryophytes were identified in the laboratory (A. Sotiaux). Sampling of the 5 locations in the Brussels-Capital Region was undertaken during autumn, 2014, with bryophytes identified in the laboratory (J.-M. Couvreur). We collected bryophytes that are directly associated with areas of travertine and to avoid "noise" in the final dataset we also dropped before analysis species that were not directly related to the 7220 habitats. Moreover, as it is sometimes difficult to assess the rate of abundance or cover by these bryophytes especially in this kind of rather heterogeneous habitat, we decided to only mention the presence/absence of these species.

Distance from the spring was not measured but the physical and physicochemical variables, including slope, orientation, canopy cover, and water chemistry, were measured at the location of each bryophyte sample. The physical variables were measured first, including slope, orientation, and canopy cover. Slope was estimated visually and classified into three groups: low (0.5–7.5%), moderate (7.5–20%), and steep (>20%). The orientation was measured using a compass (the four cardinal points plus the four intermediate ones). The percentage of canopy cover (%) was measured visually. Water chemistry, the second group of variables, was measured at each site on the same day that the bryophytes were collected. The water sample was obtained by collecting one litre of water flushing through the site in a glass bottle. The glass bottles were kept cool by storing them in cool boxes with freeze packs, until they were transferred to a laboratory fridge. The pH was measured in the lab using a WTW197i multimeter; other physicochemical variables were measured using a DR3900 Hach Lange photometer: conductivity (μS/cm), nitrates (N-NO^3 mg·L^{-1}), ammonium (N-NH_4 mg·L^{-1}), soluble orthophosphates (P-PO_4 mg·L^{-1}), nitrites (N-NO_2 mg·L^{-1}), water hardness (dH degrees), calcium (Ca mg·L^{-1}), and magnesium (Mg mg·L^{-1}). Information including locality, date, location (and Lambert 1972 coordinates), and biogeographical region (Atlantic or Continental mentioned in the variable "Zone") was also recorded at each site.

2.2. Analysis

All of the analyses were performed with R 3.2.1 [36]. We removed one dataset from the analysis (Leffe6) for which NO_2 and NO_3 values were completely unrealistic, even in highly eutrophicated waters. Four missing values in three different explanatory variables (two for PO_4, one for Ca, and one for Mg) were modelled because the AIC approaches we intended to use are incompatible with missing values. We built three Gaussian linear models, one for each variable with missing values, with the other environmental variables as predictors; however, the species data were not used. The missing data were replaced by the predicted value from these models.

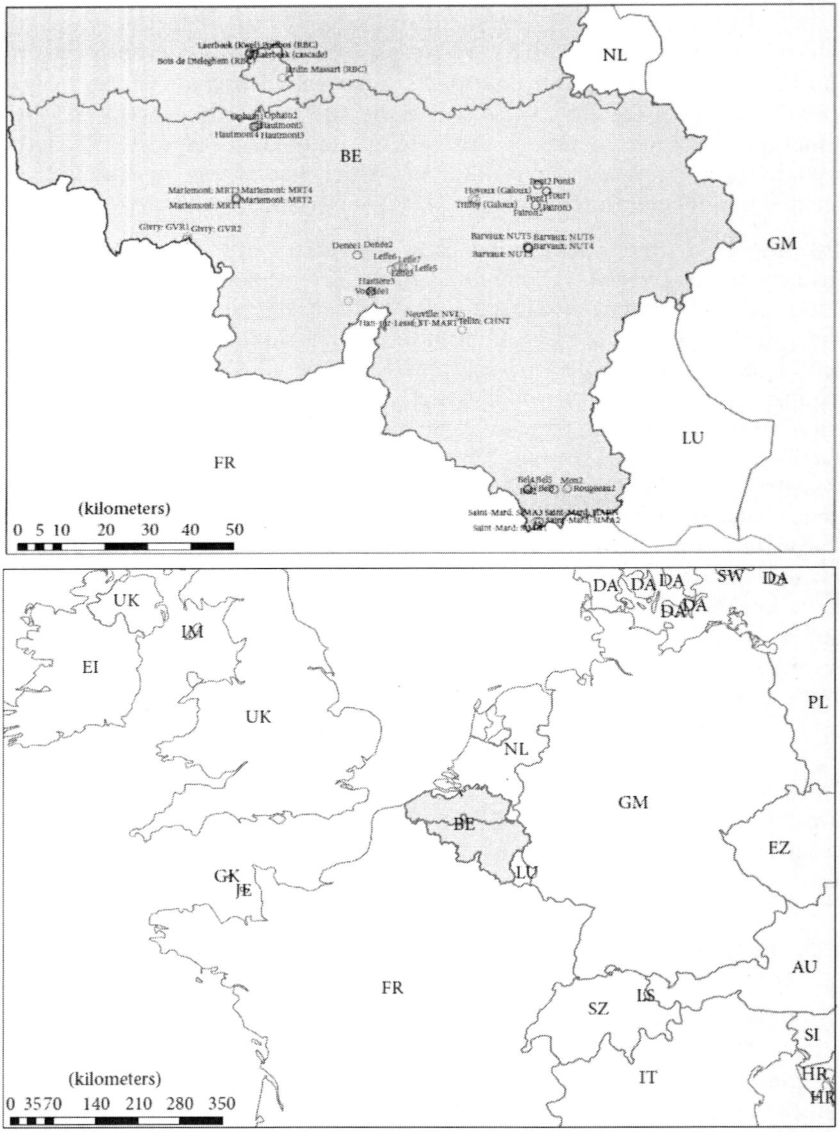

Figure 1: Location of most of the 67 surveyed petrifying sources (some locations are close to each other and cannot be represented here). Belgium (BE) is displayed in grey and the lines inside delimit the 3 administrative regions (Flemish at the north, Brussels-Capital Region just above the Flemish-Walloon border, and Walloon beneath).

The relationship between bryophyte community composition (presence/absence matrix) and the environmental factors was studied with a Hellinger transformation based Redundancy Analysis (tb-RDA [37]) performed with vegan 2.3-1 package [38]. We checked that the environmental variables were able to explain at least a part of the variance of the species matrix with a global permutation test (ANOVA function in vegan, 999 permutations). We then applied a simple forward selection procedure based on sequential permutation tests (ordistep function from vegan) to identify the environmental variables that are most related to species composition.

The relationship between species richness (total number of species) and the environmental predictors was analysed with a Poisson Generalized Linear Model (GLM). Binomial GLMs were used to characterize the presence/absence of a few individual species relative to the environmental predictors. These binomial GLMs were built only for species that were present on 20 of the 67 sites (Palustriella commutata, Eucladium verticillatum, Cratoneuron filicinum, Pellia endiviifolia, and Brachythecium rivulare).

For all GLMs, model conditions (linearity, distribution, and outliers) were checked with residuals plots and overdispersion was checked for the Poisson GLM. Multicollinearity between the explanatory variables was checked graphically and by computing Variance Inflation Factors (VIFs). If necessary, some of the explanatory variables were dropped in order to keep the VIFs < 5. We centred the pH explanatory variable on its mean value to avoid predictions at the intercept for pH = 0, that is, completely outside the range of observed values. The absence of spatial correlation of the models residuals was checked with splines correlograms.

In order to determine the most important explanatory variables for these GLMs, we applied an AICc based model selection procedure as described by Burnham and Anderson [39] and shortly summarized hereafter. GLMs with all possible combinations of explanatory variables were computed. Their AICc and AICc model weights were calculated. The AICc model weight quantify the model selection uncertainty and can be interpreted as the probability for a given model to be selected as the best model (the one with lowest AICc) in a set of models if we could resample the data. Then for each explanatory variable we computed a variable weight as the sum of the AICc weights of the models in which this explanatory variable is present. This variable weight is a measure of the relative importance of the explanatory variables. We interpreted only the explanatory variables with an AICc weight > 0.6. The shrinkage model averaged coefficients were then computed along with their unconditional standard errors. These model averaged coefficients are the mean of the coefficient of all models weighted by the model quality (AICc model weights).

For the biological interpretation of the GLMs outputs we plotted graphs, based on model averaged coefficients, for the most important explanatory variables in the range of their observed values. The explanatory variables that are not drawn on the graphs are fixed to their mean value.

3. RESULTS

3.1. Dataset

The results of the field campaign are presented in Table 1(a) (physicochemical variables) and Table 1(b) (species) as two matrices with the 67 sites as lines and the species or environmental factors as columns. The "Leffe6" site is mentioned; however, it was removed due to "out of range" values of NO_2 and NO_3. The 4 missing values were replaced by their predicted ones in the final matrix by using an explanatory GLM model based on the other variables. These values are, respectively, 0.055 and 0.058 $mg \cdot L^{-1}$ PO_4 for sites MRT1 and MRT2, respectively, 51.06 $mg \cdot L^{-1}$ Ca for site GVR2 and 25.52 $mg \cdot L^{-1}$ Mg for site MRT4.

Three pairs of explanatory variables were highly correlated: conductivity-Ca (0.8), conductivity-water hardness (0.77), and water hardness-Ca (0.85). We decided to drop the variables conductivity and water hardness which were highly correlated to the Ca concentration and to keep the last one for the next step analysis.

3.2. Key Variables of Species and Sites Assemblages

The tb-RDA analysis (Table 2) using the list of previously retained quantitative and qualitative variables (pH, NO_3, NH_4, PO_4, NO_2, Ca, Mg, slope converted into numerical values from 1 to 3, canopy cover and orientation). The first unconstrained axis (PCA1) explains a rather high amount of variation, almost comparable to the first constrained axis, RDA1. This suggests that there is another environmental variable, not measured during this study, which could influence the species composition of the communities (see Section 4). Nevertheless the variables explain 36.5% of the variation and a Monte Carlo permutation test executed on this RDA delivers a highly significant value for the model ($P < 0.001$) that confirms the global model is relevant.

Inside the constrained variance the first axis explained 44.9% (= eigenvalue RDA1 = 0.4282/0.9524) of this constrained variance and the second axis 18.1% (= eigenvalue RDA2 = 0.1721/0.9524). The use of the same dataset without the two outliers yields approximately the same results (38.3% of constrained variation, 42.7% for the first axis, and 15.4% for the second axis). As the parameter orientation is not retained by the forward selection model and because it produces a poorly readable figure, we also tested the same model without the orientation variable (Figure 2). The results are the following: 28.3% of constrained variation, 55.9% for the first axis, and 17.7% for the second axis.

The RDA analysis is potted both with and without the orientation variable (Figure 2). The scaling 2 option [37] allows the following comments and shows that slope, NO_3, NH_4, PO_4, and canopy cover play an important role in the dispersion of the sites along the first axis.

The two variables NH_4 and PO_4 are very closely correlated, suggesting that they can act in a similar way on the species assemblages and between factors themselves; however, NO_3 is negatively correlated with both NH_4 and PO_4. The two characteristic species Eucladium verticillatum and Palustriella commutata are close together suggesting they appear both on sites with higher slope and lower cover of canopy and lower NH_4 and PO_4 concentrations. Pellia endiviifolia shows a positive relationship with NO_3 whereas Brachythecium rivulare and Oxyrrhynchium hians (Hedw.) Loeske are weakly correlated with PO_4, NH_4, and high pH values. Bryum pseudotriquetrum is negatively correlated with canopy cover. Most other species are clustered together away from these extremes. They show mostly shorter projections, indicating that they are either present in most of the sites or related to intermediate ecological conditions; however, they could also be related to some variables on third or higher axis which is not displayed.

The result of the forward model selection of the RDA retained a model with (sorted by decreasing importance) slope, Mg, Ca, and canopy cover (spe ~ slope + Mg + Ca + cover). The same analysis without the outliers ("Haumont5" and "Givry2") generates a slightly different model with slope, canopy cover, Mg, and Ca as the most significant factors (spe ~ slope + cover + Mg + Ca).

Table 1: (**a**) Dataset of 67 surveyed locations with their measured physicochemical variables. Conductivity is in μS; hardness is in dH degrees; slope and canopy cover are in percentages; NO_3, NH_4, PO_4, NO_2, Ca, and Mg are in mgL^{-1}. (**b**) Dataset of 67 surveyed locations with the presence/absence data of the bryophytes species (characteristic and companion species). Species are: Aneura pinguis, Brachythecium rivulare, Bryum pseudotriquetrum, Conocephalum conicum, Cratoneuron filicinum, Ctenidium molluscum, Dichodontium pellucidum, Didymodon tophaceus, Eucladium verticillatum, Fissidens adianthoides, Fissidens crassipes, Fissidens taxifolius, Leiocolea badensis, Gymnostomum calcareum,

Palustriella commutata, Pellia endiviifolia, Philonotis calcarea, Plagiomnium ellipticum, Plagiomnium rostratum, Plagiomnium undulatum, Platyhypnidium riparioides, Preissia quadrata, and Oxyrrhynchium hians.

(a)

Site Nb	Zone	Location	pH	Conductivity	NO_3	NH_4	PO_4	NO_2	Hardness	Ca	Mg	Slope	Orientation	Canopy cover
1	atl	Hautmont1	7.61	778	7.97	0	0.053	0.011	25.4	157	14.8	Moderate	NW	80
2	atl	Hautmont2	7.88	749	0.704	0.028	0.089	0.011	23.2	145	12.7	Low	NE	90
3	atl	Hautmont3	8.04	731	7.46	0.033	0.165	0.045	23.6	138	18.5	Low	NE	80
4	atl	Hautmont4	8.08	665	2.5	0.053	0.078	0.028	21.1	131	11.7	Low	NW	70
5	atl	Hautmont5	7.87	530	2.31	0.194	0.461	0.264	16.1	92.6	13.3	Low	N	60
6	atl	Ophain1	7.07	756	7.55	0.015	0.04	0.011	25.1	157	13.3	Low	N	90
7	atl	Ophain2	7.92	790	5.52	0.026	0.042	0.013	21	132	10.7	Low	W	90
8	cont	Hastière1	7.52	799	9	0.021	0	0.007	26.7	183	4.35	Steep	NW	0
9	cont	Hastière2	8.25	594	8.86	0.016	0.055	0.007	19.7	140	0	Moderate	NW	0
10	cont	Hastière3	8.16	505	12.3	0.016	0	0.006	15.4	93.5	9.72	Steep	W	50
11	cont	Hastière4	8.21	555	10.5	0.017	0.014	0.009	17.5	98.4	15.9	Moderate	SW	100
13	cont	Leffe1	8.27	592	4.34	0.026	0.076	4.35	19	114	13	Low	W	50
14	cont	Leffe3	8.13	595	0.074	0.022	0.0262	0.076	18.7	109	14.8	Low	W	100
15	cont	Leffe5	7.51	636	0.03	0.073	0.161	0.022	21.6	124	18	Low	E	0
16	cont	Leffe6	8.18	560	19	0.018	0.005	19.1	17.5	93.3	19.3	Steep	S	50
17	cont	Leffe7	8.22	599	4.95	0.032	0.057	5	18.6	113	11.5	Low	W	0
18	cont	Vodelée1	7.13	720	6.19	0.017	0.064	6.22	24.4	148	15.8	Steep	NW	30
19	cont	Denée1	8.2	634	4.83	0.02	0.06	4.86	20.1	117	16.2	Moderate	SE	100
20	cont	Denée2	8.07	700	5.16	0.027	0.115	5.18	26.1	151	21.5	Moderate	S	100
21	atl	Mariemont: MRT1	7.76	907	4.98	0.03	NA	0.013	30.6	188	18.2	Low	S	100
22	atl	Mariemont: MRT2	8.12	879	2.16	0.035	NA	0.015	27.8	173	15.1	Moderate	SE	100
23	atl	Mariemont: MRT3	8.19	672	1.7	0.021	0.092	0.014	22.2	143	9.36	Low	S	100
24	atl	Mariemont: MRT4	8.09	659	0.779	0.033	0.029	0.02	19.1	136	NA	Low	SE	100
25	atl	Givry: GVR1	8.01	658	6.69	0.042	0.042	0.015	20.9	119	18.2	Low	W	100
26	atl	Givry: GVR2	8.06	618	2.84	0.446	0.061	0.009	4.09	NA	17.7	Low	W	100
26	atl	Givry: GVR2	8.06	618	2.84	0.446	0.061	0.009	4.09	NA	17.7	Low	W	100
27	cont	Han-sur-Lesse: ST-MART	7.23	673	2.15	0.019	0.118	0.009	23.9	134	22.6	Low	SW	100
28	cont	Tellin: CHNT	7.78	773	7.82	0.021	0.014	0.009	26.7	150	24.3	Moderate	S	100
29	cont	Neuville: NVL	7.22	687	5.94	0.019	0.009	0.014	25.1	150	17.8	Steep	E	100
30	cont	Fairon1	8.08	466	6.84	0.016	0	0.011	16.4	83.4	20.3	Steep	SE	100
31	cont	Fairon2	7.2	497	7.43	0.004	0.003	0.012	15.6	87.7	14.2	Steep	SE	100
32	cont	Fairon3	6.22	463	6.27	0.012	0	0.012	15.3	84.4	15.2	Steep	SE	100
33	cont	Pont1	8.09	733	9.33	0.01	0.002	0.006	25.1	136	26.1	Steep	NE	50
34	cont	Pont2	8.32	578	11.1	0.013	0	0.014	18.1	86.6	25.7	Steep	NE	50
35	cont	Pont3	8	630	12.1	0.007	0.004	0	20.4	89.4	34	Steep	SE	50
36	cont	Tour1	7.42	728	4.57	0.009	0	0.013	24.3	123	30.3	Low	NE	100
37	cont	Oneux1	8.1	562	5.11	0.065	0.083	0.069	16.9	87.3	22.5	Moderate	NE	10
38	cont	Oneux2	7.42	607	5	0.007	0.011	0.009	19.5	95.9	26.1	Moderate	SE	80
39	cont	Oneux3	7.57	684	5.96	0.009	0	0	22.7	108	32.6	Steep	S	100

40	cont	Barvaux: NUT1	7.88	670	0.831	0.018	0.015	0.017	22.9	131	19.3	Low	NW	80
41	cont	Barvaux: NUT2	7.63	585	0.67	0.004	0.007	0.018	19.2	109	17.3	Moderate	NW	80
42	cont	Barvaux: NUT3	7.66	675	0.385	0.018	0.01	0.017	22.2	122	22.1	Low	NW	100
43	cont	Barvaux: NUT4	7.91	621	1.33	0.013	0.005	0.015	20.4	121	14.5	Moderate	NW	50
44	cont	Barvaux: NUT5	8.02	582	0.529	0.019	0.037	0.016	19.1	108	17	Low	W	20
45	cont	Barvaux: NUT6	8.03	653	0.5	0.025	0.021	0.02	7.55	27.8	15.8	Low	W	100
46	cont	Montauban1	7.78	416	1.1	0	0.004	0.013	15.6	93.7	10.8	Steep	SW	0
47	cont	Mon2	7.81	418	1.26	0	0.003	0.012	14.6	89.6	8.77	Steep	S	0
48	cont	Rougeeau1	7.93	437	0.936	0	0.003	0.013	15.3	99	6.25	Steep	SE	0
49	cont	Rougeeau2	7.93	426	0.908	0	0.009	0.012	15.9	96	10.7	Steep	SE	0
50	cont	Huombois1	7.95	450	4.15	0	0.003	0.012	15.2	91.1	10.4	Steep	W	10
51	cont	Bel1	7.82	418	0.364	0.012	0	0.01	14.6	88.1	9.71	Steep	SE	0
52	cont	Bel2	7.99	499	3.9	0.015	0.005	0.011	17	106	9.36	Steep	SE	0
53	cont	Bel3	7.41	593	5.26	0.015	0.003	0.012	21.6	136	10.7	Steep	S	0
54	cont	Bel4	7.54	452	2.63	0.009	0.004	0.01	16.2	101	8.57	Steep	SE	80
55	cont	Bel5	7.73	460	2.91	0.009	0.015	0.013	16.6	103	9.43	Steep	N	0
56	cont	Saint-Mard: StMA1	8.11	391	1.15	0.026	0.059	0.021	13.1	83.8	5.75	Low	W	50
57	cont	Saint-Mard: StMA2	8.06	365	0.986	0.012	0.027	0.017	12.3	76.5	6.68	Low	NW	90
58	cont	Saint-Mard: StMA3	7.86	397	1.06	0.014	0.019	0.016	13.6	83.6	7.96	Moderate	NW	100
59	cont	Saint-Mard: PtCAM1	7.23	503	0.242	0.003	0.132	0.005	17.7	112	8.54	Low	NW	70
60	cont	Saint-Mard: PtCAM2	7.83	494	0.307	0.031	0.072	0.018	18.3	107	14.3	Low	SW	80
61	cont	Saint-Mard: HARN	8.09	377	0.224	0.006	0.027	0.019	13.9	81.9	10.5	Steep	N	80
62	atl	Poelbos (RBC)	8.2	908	0.236	0.031	0.042	0.013	30.1	163	31.5	Moderate	S	80
63	atl	Laerbeek (cascade)	7.94	892	0.488	0.034	0.021	0.004	28.7	158	28.4	Low	S	80
64	atl	Laerbeek (Kwel)	7.89	1085	0.598	0.047	0.022	0.011	32.3	172	65.5	Low	S	70
65	atl	Bois de Dieleghem (RBC)	8.07	1007	1.36	0.041	0.043	0.019	46.1	251	47.1	Low	S	80
66	atl	Jardin Massart (RBC)	7.8	1058	1.69	0.046	0.053	0.005	30.9	170	30.6	Moderate	NW	70
67	cont	Hoyoux (Galoux)	8.26	557	5.083	0.02	0.031	0.007	32.6	97.4	20.1	Low	NW	60
68	cont	Triffoy (Galoux)	7.83	608	5.144	0.026	0.016	0.009	34.7	96.3	25.9	Moderate	NE	60

(b)

Site Nb	Zone	Location	Aneu_pin	Brach_riv	Bryu_pse	Cono_con	Crat_fil	Cten_mol	Dich_pel	Did_toph	Eucl_vert	Fiss_adi	Fiss_cra	Fiss_tax	Leio_bad	Gymn_cal	Pal_com	Pell_end	phi_calc	Pla_ell	Plag_ros	Plag_und	Plat_rip	Pre_qua	Oxy_hians
1	atl	Hautmont1	0	1	0	0	1	0	0	0	0	0	0	1	0	0	1	1	0	0	0	0	0	0	0
2	atl	Hautmont2	0	0	0	0	0	0	0	0	0	0	0	0	0	0	0	0	0	0	0	0	0	0	0
3	atl	Hautmont3	0	0	0	0	0	0	0	0	0	0	0	0	0	0	0	0	0	0	0	0	0	0	0
4	atl	Hautmont4	0	0	0	0	0	0	0	0	0	0	0	0	0	0	0	0	0	0	0	0	0	0	0
5	atl	Hautmont5	0	0	0	0	0	0	0	0	0	0	0	0	0	0	0	0	0	0	0	0	0	0	0
6	atl	Ophain1	0	1	0	0	1	0	0	0	0	0	0	0	0	0	0	1	0	0	0	0	0	0	0
7	atl	Ophain2	0	1	0	0	1	0	0	0	0	0	0	0	0	0	0	1	0	0	0	0	0	0	0
8	cont	Hastière1	1	1	1	0	1	1	0	0	1	1	0	0	0	1	1	1	1	1	0	1	0	1	0
9	cont	Hastière2	0	0	0	0	0	0	0	0	0	0	0	0	0	0	0	0	0	0	0	0	0	0	0
10	cont	Hastière3	0	1	0	0	0	1	0	0	0	1	0	0	0	0	1	1	0	0	0	1	0	0	0
11	cont	Hastière4	0	0	0	0	1	0	0	0	0	0	0	0	0	0	0	1	0	0	0	0	1	0	0
13	cont	Leffe1	0	1	0	0	1	0	0	0	1	0	1	0	0	0	0	1	0	0	0	1	0	0	0
14	cont	Leffe3	0	0	0	0	0	0	0	0	0	0	0	0	0	0	0	0	0	0	0	0	0	0	0
15	cont	Leffe5	0	0	0	0	0	0	0	0	0	0	0	0	0	0	0	0	0	0	0	0	0	0	0
16	cont	Leffe6	1	0	1	1	1	1	0	0	1	0	1	1	0	0	0	0	0	0	0	0	1	0	1
17	cont	Leffe7	0	0	0	0	0	0	0	0	0	0	0	0	0	0	0	0	0	0	0	0	0	0	0
18	cont	Vodelée1	0	0	1	0	1	0	0	0	1	0	0	0	0	0	1	1	0	0	0	0	0	0	0
19	cont	Denée1	0	1	0	0	1	0	0	0	0	0	0	0	0	0	0	1	0	0	0	0	0	0	0
20	cont	Denée2	0	0	0	0	0	0	0	0	0	0	0	0	0	0	0	0	0	0	0	0	0	0	0
21	atl	Mariemont: MRT1	0	1	0	0	0	0	0	0	0	0	0	0	0	0	0	0	0	0	0	0	0	0	1
22	atl	Mariemont: MRT2	0	1	0	0	1	0	0	0	0	0	0	0	0	0	0	0	0	0	0	0	0	0	0
23	atl	Mariemont: MRT3	0	0	0	0	0	0	0	0	0	0	0	0	0	0	0	0	0	0	0	0	0	0	0
24	atl	Mariemont: MRT4	0	1	0	0	0	0	0	0	0	0	0	1	0	0	0	1	0	0	0	0	0	0	0
25	atl	Givry: GVR1	0	0	0	0	1	0	0	0	0	0	0	0	0	0	0	1	0	0	0	0	0	0	0
26	atl	Givry: GVR2	0	0	0	0	0	0	0	0	0	0	0	0	0	0	0	0	0	0	0	0	0	0	0
27	cont	Han-sur-Lesse: ST-MART	0	1	0	0	1	0	0	0	0	0	0	0	0	0	0	0	0	0	1	0	0	0	0
28	cont	Tellin: CHNT	0	0	0	0	0	0	0	0	0	1	0	0	0	0	0	0	0	0	0	0	0	0	0
29	cont	Neuville: NVL	0	0	0	0	0	1	0	0	0	0	0	0	0	0	1	0	0	0	0	1	0	0	0
30	cont	Fairon1	0	0	0	0	1	0	0	0	1	0	0	1	0	0	1	1	0	0	0	0	1	0	0
31	cont	Fairon2	0	0	0	0	0	0	0	0	0	0	0	0	0	0	0	0	0	0	0	0	0	0	0
32	cont	Fairon3	0	0	0	0	1	0	0	0	1	0	0	1	0	0	1	1	0	0	0	0	0	0	0
33	cont	Pont1	0	0	0	1	1	1	0	1	1	0	0	0	1	1	1	1	0	0	0	0	0	0	0
34	cont	Pont2	0	1	0	0	1	0	0	0	0	0	0	0	0	0	1	1	0	0	0	0	0	0	1
35	cont	Pont3	0	0	0	0	1	0	0	1	1	0	0	0	0	0	1	1	0	0	0	0	0	0	0

36	cont	Tour1	0	1	0	1	1	0	0	0	0	0	0	0	0	0	0	1	0	0	1	1	0	0	1
37	cont	Oneux1	0	0	0	0	1	0	1	0	0	0	0	0	0	0	1	1	0	0	0	1	1	0	0
38	cont	Oneux2	0	0	0	1	1	0	0	0	0	0	0	0	0	0	1	1	0	0	1	0	1	0	0
39	cont	Oneux3	0	0	0	0	1	0	0	0	0	0	0	0	0	0	0	1	0	0	0	0	1	0	0
40	cont	Barvaux: NUT1	0	0	0	0	0	0	0	0	0	0	0	0	0	0	0	0	0	0	0	0	0	0	0
41	cont	Barvaux: NUT2	0	1	0	0	1	0	0	0	1	0	0	0	0	0	0	1	0	0	0	1	0	0	0
42	cont	Barvaux: NUT3	1	1	0	0	1	0	0	0	0	1	0	0	0	0	0	0	0	0	1	0	0	0	0
43	cont	Barvaux: NUT4	0	1	0	0	1	0	0	0	0	0	0	0	0	0	0	1	0	0	0	1	0	0	1
44	cont	Barvaux: NUT5	0	0	0	0	1	0	0	0	1	0	0	0	0	0	0	0	0	0	0	0	0	0	1
45	cont	Barvaux: NUT6	0	0	0	0	1	0	0	0	0	0	0	1	0	0	0	1	0	0	0	0	0	0	0
46	cont	Montauban1	1	0	1	1	1	1	1	0	1	1	0	1	0	0	1	1	0	0	1	0	0	0	0
47	cont	Mon2	0	0	0	0	1	0	0	0	1	0	0	1	0	0	1	1	0	0	0	0	0	0	0
48	cont	Rougeeau1	0	0	1	0	0	1	0	0	1	1	0	0	0	0	1	1	0	0	0	0	0	0	0
49	cont	Rougeeau2	0	0	1	0	0	0	0	0	1	0	0	0	0	0	1	0	0	0	0	0	0	0	0
50	cont	Huombois1	0	0	0	0	0	0	0	0	1	0	0	0	0	0	1	0	0	0	0	0	0	0	0
51	cont	Bel1	1	0	1	1	1	1	0	0	1	1	0	0	1	0	1	1	0	0	0	0	0	0	0
52	cont	Bel2	0	1	0	0	1	0	0	0	0	0	0	0	0	0	0	1	0	0	0	1	0	0	0
53	cont	Bel3	1	0	1	0	0	0	0	0	1	1	0	0	0	0	1	1	0	0	0	0	0	0	0
54	cont	Bel4	0	0	1	0	0	0	0	0	1	0	0	0	1	0	1	1	0	0	0	0	0	1	0
55	cont	Bel5	0	0	1	0	0	0	0	0	1	1	0	0	0	0	1	1	0	0	0	0	0	1	0
56	cont	Saint-Mard: StMA1	0	1	0	0	1	0	0	0	0	0	0	0	0	0	1	0	0	0	0	0	0	0	0
57	cont	Saint-Mard: StMA2	0	1	0	0	0	0	0	0	0	0	0	1	0	0	0	0	0	0	1	0	0	0	1
58	cont	Saint-Mard: StMA3	0	0	0	0	0	0	0	0	0	0	0	0	0	0	0	0	0	0	0	1	0	0	0
59	cont	Saint-Mard: PtCAM1	0	0	0	1	1	0	0	0	0	0	0	1	0	0	1	0	0	0	1	0	0	0	1
60	cont	Saint-Mard: PtCAM2	0	0	0	0	0	0	0	0	0	0	0	1	0	0	0	0	0	0	0	1	0	0	0
61	cont	Saint-Mard: HARN	0	0	0	0	1	0	0	0	0	0	0	0	0	0	0	1	0	0	0	0	0	0	0
62	atl	Poelbos (RBC)	0	1	0	0	1	0	0	0	0	0	0	1	0	0	0	0	0	0	0	0	0	0	1
63	atl	Laerbeek (cascade)	0	0	0	0	1	0	0	0	0	0	0	0	0	0	0	0	0	0	0	0	0	0	1
64	atl	Laerbeek (Kwel)	0	1	0	1	1	0	0	0	0	0	0	1	0	0	0	0	0	0	0	0	0	0	0
65	atl	Bois de Dieleghem (RBC)	0	1	0	0	1	0	0	0	0	0	0	1	0	0	0	0	0	0	0	0	0	0	1
66	atl	Jardin Massart (RBC)	0	1	0	0	1	0	0	0	0	0	0	1	0	0	0	1	0	0	0	0	0	0	1
67	cont	Hoyoux (Galoux)	0	1	0	1	1	0	0	0	0	0	1	0	0	0	0	1	0	0	1	1	1	0	0
68	cont	Triffoy (Galoux)	0	0	0	0	1	0	0	0	0	0	1	0	0	0	0	1	0	0	1	0	1	0	0

Table 2: Results of the tb-RDA analysis.

Dataset	Percentage of variation explained by the variables (constrained variation)	Percentage explained by first axis inside constrained variation	Percentage explained by second axis inside constrained variation
With outliers and orientation variable	36.5	44.9	18.1
Without outliers and with orientation variable	38.3	42.8	15.4
With outliers and without orientation variable	28.3	55.9	17.7

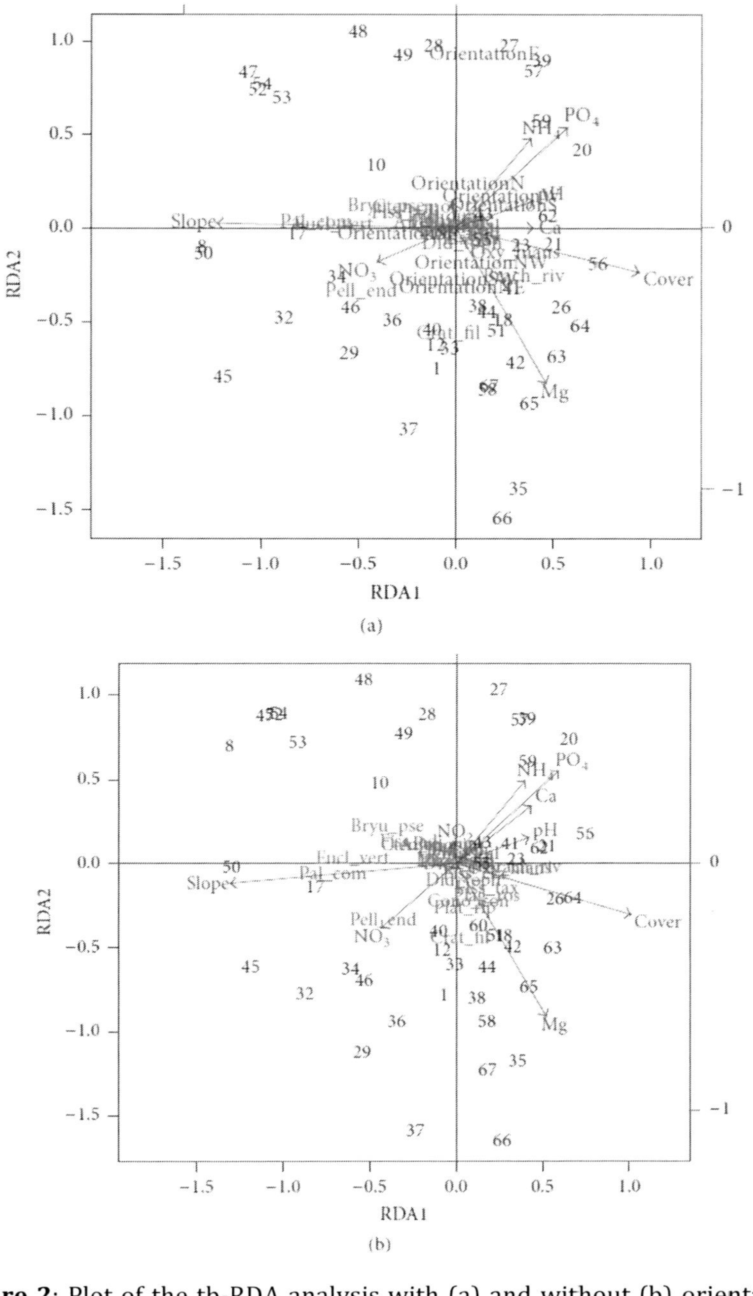

Figure 2: Plot of the tb-RDA analysis with (a) and without (b) orientation variable.

3.3. Species Richness

The results show (Table 3) that only two explanatory variables (canopy cover and PO_4) are strongly supported by the data (variable weight = 0.992 and 0.960, resp.) while two other ones are moderately supported by the data (NH_4 and Mg) with a weight of, respectively, 0.709 and 0.650. All but the Mg variables are negatively correlated with species richness.

To check the potential influence of outliers we performed the same analysis after excluding Haumont5 and Givry2 sites which have very high PO_4 and NH_4 values. The model selection results are comparable to the first analysis including outliers; AIC retains only canopy cover and PO_4 as the main explanatory variables (Table 3) (columns on the right for each parameter).

Figures 3 and 4 show the results of two predicted scenarios based on the complete model (2 outliers included) with the mean values of other variables. Figure 3 represents the relation between the PO_4 concentration and the species richness and Figure 4 represents the relation between the canopy cover and the species richness. It can be seen that for the canopy cover the expected number of species rises by 2 or 3 species (going approximately from 2 to 4 or even 5) when the cover falls from 100 to 0%. In the case of the PO_4 concentration the expected number of species falls to zero when the PO_4 concentration rises from 0 to 0.5 $mg \cdot L^{-1}$ or more.

3.4. Presence/Absence of Characteristic and Companion Species

For each species we began by excluding the variables that were highly correlated using the VIF (Variance Inflation Factor) values. The only variable to exclude from the five species was orientation and thus the binomial GLM analysis included the following explanatory variables: zone, NO_2, Mg, pH_mean, Ca, NH_4, NO_3, canopy cover, slope, and PO_4.

3.4.1. Palustriella commutata

The most important variables delivered by the AIC procedure are "steep slope" and "moderate slope" and to a lesser extend "canopy cover" (Table 4). The frequency of Palustriella commutata tends to increase when the slope is steeper and decreases when canopy cover is higher (Figure 5).

Table 3: Results of the AIC stepwise model selection for species richness showing the best significant variables selected based on their weight using the complete dataset (columns on the left) or the data without 2 possible outliers (columns on the right).

Variable	Weight of the variable		Averaged coefficient		Unconditional standard error	
	Full model	Model without possible outliers	Full model	Model without possible outliers	Full model	Model without possible outliers
Cover	0.992	0.992	−0.007	−0.006	0.002	0.002
PO_4	0.960	0.983	−7.510	−8.361	2.572	2.479
NH_4	0.709	0.300	−5.615	−1.689	4.140	2.414
Mg	0.650	0.548	0.009	0.007	0.006	0.005
pH_mean	0.587	0.616	−0.186	−0.200	0.128	0.131
Zonecont	0.309	0.300	0.065	0.058	0.086	0.080
Ca	0.300	0.269	0.001	0.000	0.001	0.001
NO_3	0.274	0.276	0.004	0.004	0.006	0.006
NO_2	0.230	0.229	−0.002	−0.001	0.014	0.013
Slope moderate	0.087	0.090	0.007	0.008	0.018	0.019
Slope steep	0.087	0.09	0.005	0.008	0.021	0.023

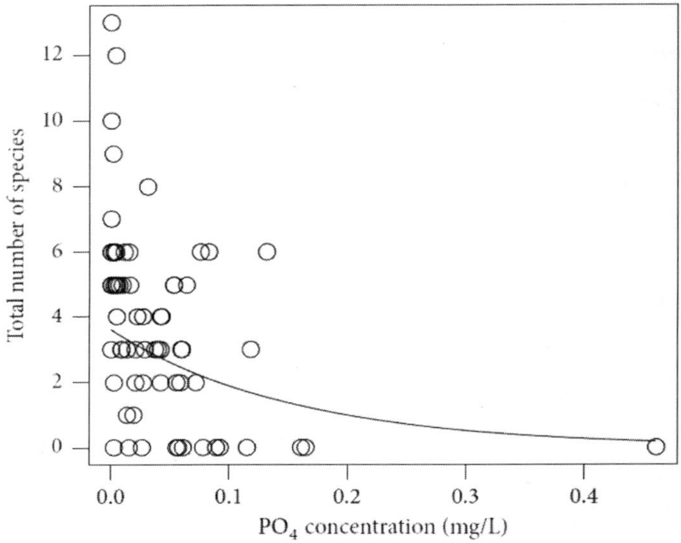

Figure 3: Prediction of species richness in relation to PO4 concentration based on the coefficients of the best models selected by AIC stepwise selection.

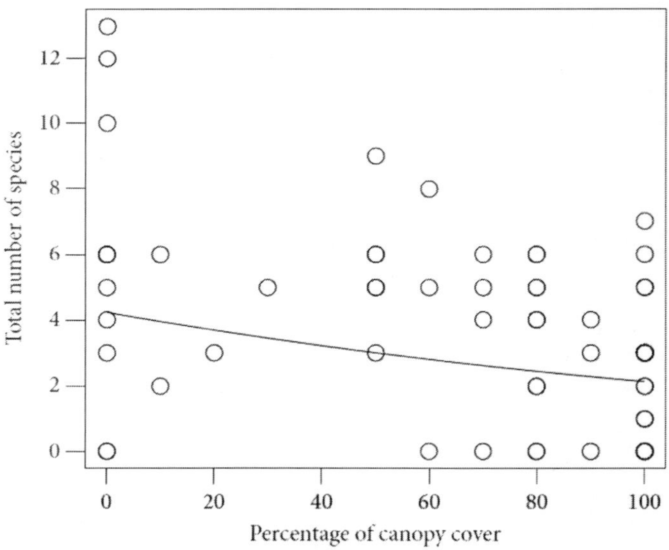

Figure 4: Prediction of species richness in relation to canopy cover based on the coefficients of the best models selected by AIC stepwise selection.

Table 4: Results of the AIC stepwise model selection for probability of presence of Palustriella commutata showing the best significant variables selected based on their weight using the complete dataset.

Variable	Weight of the variable	Averaged coefficient	Unconditional standard error
Moderate slope	0.902	1.116	1.002
Steep slope	0.902	3.152	1.010
Cover	0.797	−0.022	0.011
pH_mean	0.592	−1.293	0.947
NH_4	0.387	−13.210	14.475
NO_3	0.379	0.063	0.066
Ca	0.336	−0.006	0.007
PO_4	0.283	1.874	3.610
NO_2	0.254	−0.048	0.091
Mg	0.248	−0.004	0.014
Zonecont	0.243	0.0097	0.357

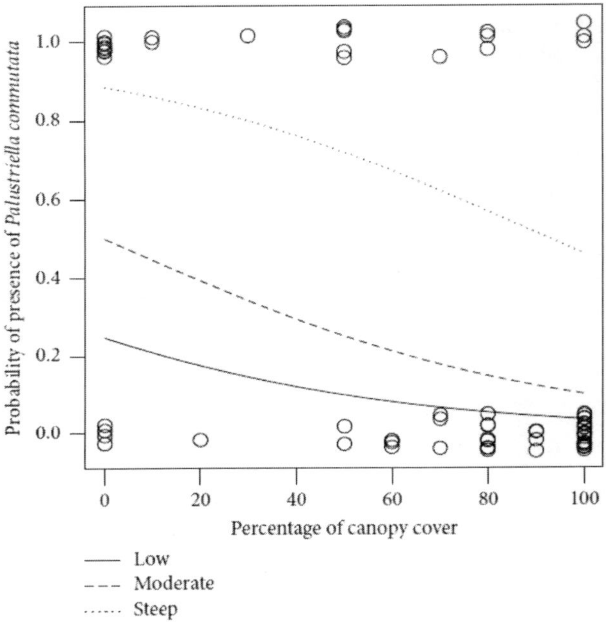

Figure 5: Prediction of presence of Palustriella commutata in relation to canopy cover based on the coefficients of the best models selected by AIC stepwise selection for different slopes: low slope (solid line), moderate slope (dashed line), and steep slope (dotted line).

3.4.2. Eucladium verticillatum

The most important variables delivered by the AIC procedure are canopy cover and NH_4, both are negatively correlated with the probability of the presence of the species (Table 5). The frequency of Eucladium verticillatum tends to decrease when the canopy cover increases and for higher concentrations of NH_4, however the size of the effect is relatively limited as shown on Figure 6.

3.4.3. Cratoneuron filicinum

The only important variable delivered by the AIC procedure is the Mg concentration (positive relationship) and to a lesser degree NH_4 (negative relationship) (Table 6, Figure 7). The results remain similar after excluding the "Laarbeek (Kwel)" site whose outlying value (65.5 $mg \cdot L^{-1}$) could have unduly influenced the correlation.

Table 5: Results of the AIC stepwise model selection for probability of presence of Eucladium verticillatum showing the best significant variables selected based on their weight using the complete dataset.

Variable	Weight of the variable	Averaged coefficient	Unconditional standard error
Canopy cover	0.931	−0.031	0.013
NH_4	0.714	−80.324	52.086
PO_4	0.523	−17.784	15.728
NO_2	0.516	0.311	0.259
pH_mean	0.453	−0.684	0.625
Slope moderate	0.445	−0.256	0.675
Slope steep	0.445	0.862	0.772
Zonecont	0.383	6.353	878.235
NO_3	0.277	−0.023	0.041
Mg	0.250	0.003	0.015
Ca	0.241	0.000	0.005

Table 6: Results of the AIC stepwise model selection for probability of presence of Cratoneuron filicinum showing the best significant variables selected based on their weight using the complete dataset.

Variable	Weight of the variable	Averaged coefficient	Unconditional standard error
Mg	0.986	0.127	0.049
NH_4	0.734	−21.408	17.105
PO_4	0.339	−2.269	3.217
Ca	0.324	−0.003	0.004
Cover	0.322	−0.002	0.003
pH_mean	0.261	−0.110	0.227
Zonecont	0.257	−0.019	0.231
NO_2	0.239	0.012	0.049
NO_3	0.238	0.005	0.022
Slope moderate	0.135	0.09	0.133
Slope steep	0.135	−0.018	0.121

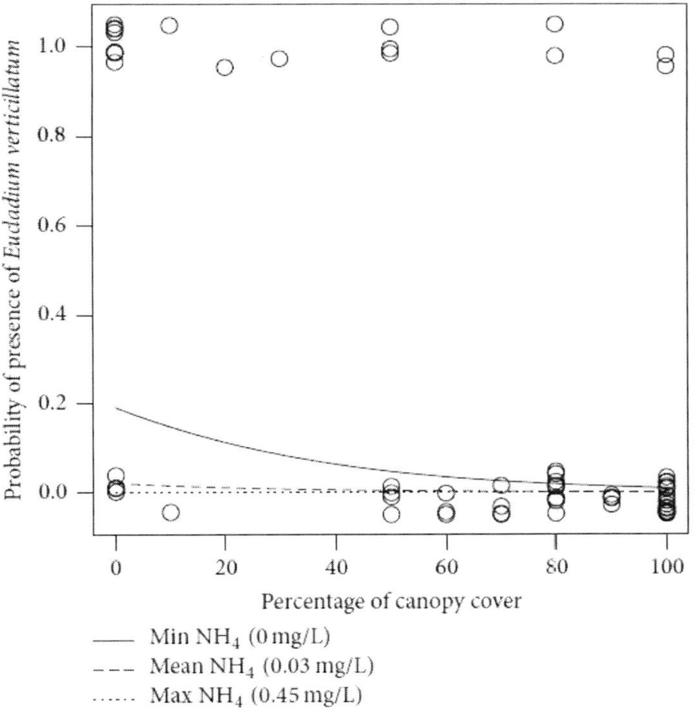

Figure 6: Prediction of presence of Eucladium verticillatum in relation to canopy cover based on the coefficients of the best models selected by AIC stepwise selection for different levels of NH4: maximum concentration in the dataset (dotted line), mean concentration in the dataset (dashed line), and minimum concentration in the dataset (solid line).

3.4.4. Pellia endiviifolia

The most important variables delivered by the AIC procedure are PO_4 (negative relationship) and NO_3 (positive relationship) (Table 7; Figure 8).

3.4.5. Brachythecium rivulare

The only important variable delivered by the AIC procedure is the Ca concentration (positive relationship) (Table 8). However Figure 9 shows that this correlation is weak and is based almost entirely on the outlying value (251 mg·L−1) of the "Bois de Dieleghem" site. It would thus be sensible to not pay too much attention to this correlation.

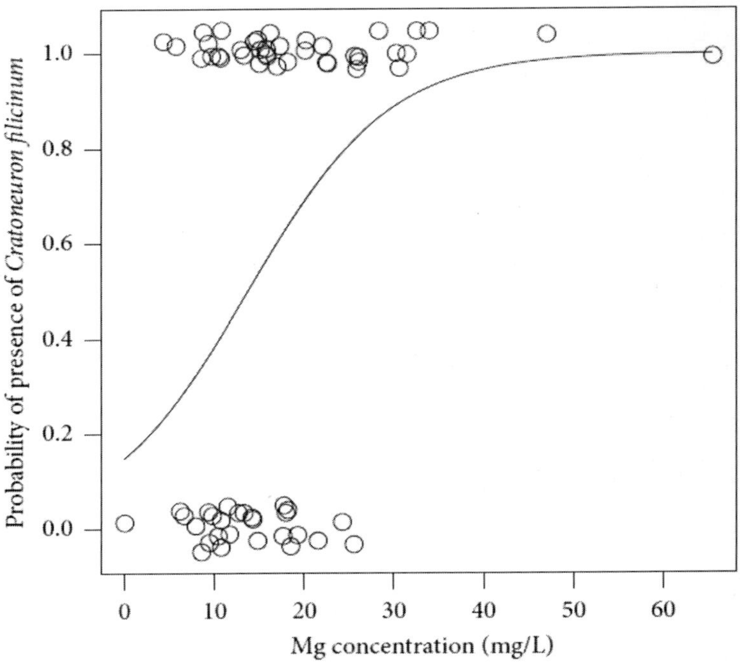

Figure 7: Prediction of presence of Cratoneuron filicinum in relation to Mg concentration based on the coefficients of the best models selected by AIC stepwise selection.

4. DISCUSSION

This study aimed to identify the driving variables that explain the species richness and the distribution of the bryophyte assemblages within the 7220 habitat. The environmental driving factors discussed within this study apply only to the 7220 habitat of travertine forming springs and watercourses, and it would not be advisable to apply these findings outside of this habitat such as (less calcareous) types of water courses in Wallonia and Brussels-Capital Region. The springs and watercourses in this study are characterized by medium to high levels of Ca (60–300 $mg \cdot L^{-1}$) [40] and high pH values (7.5–8.5). The extreme values in our samples range from 76.5 to 251 $mg \cdot L^{-1}$ for Ca (the 27.8 value seeming doubtful) and pH from 6.2 to 8.2. In comparison Brusa and Cerabolini [11] recorded Ca values between 33.2 and 60.2 $mg \cdot L^{-1}$ and pH values between 7.5 and 8.4.

Table 7: Results of the AIC stepwise model selection for probability of presence of Pellia endiviifolia showing the best significant variables selected based on their weight using the complete dataset.

Variable	Weight of the variable	Averaged coefficient	Unconditional standard error
PO_4	0.922	−28.250	13.310
NO_3	0.885	0.251	0.113
Ca	0.650	−0.015	0.010
NO_2	0.448	0.168	0.157
NH_4	0.422	−4.857	5.898
Zonecont	0.419	−0.660	0.677
Cover	0.339	−0.003	0.004
Mg	0.257	0.004	0.010
pH_mean	0.240	−0.095	0.273
Slope moderate	0.233	0.252	0.269
Slope steep	0.233	0.296	0.319

Table 8: Results of the AIC stepwise model selection for probability of presence of Brachythecium rivulare showing the best significant variables selected based on their weight.

Variable	Weight of the variable	Averaged coefficient	Unconditional standard error
Ca	0.894	0.023	0.010
PO_4	0.579	−9.225	7.229
Slope moderate	0.468	−0.193	0.369
Slope steep	0.468	−0.897	0.650
pH_mean	0.354	0.354	0.414
NH_4	0.351	−7.123	10.000
Cover	0.292	0.002	0.003
Zonecont	0.285	−0.149	0.293
Mg	0.262	0.004	0.010
NO_3	0.251	0.009	0.025
NO_2	0.240	0.007	0.054

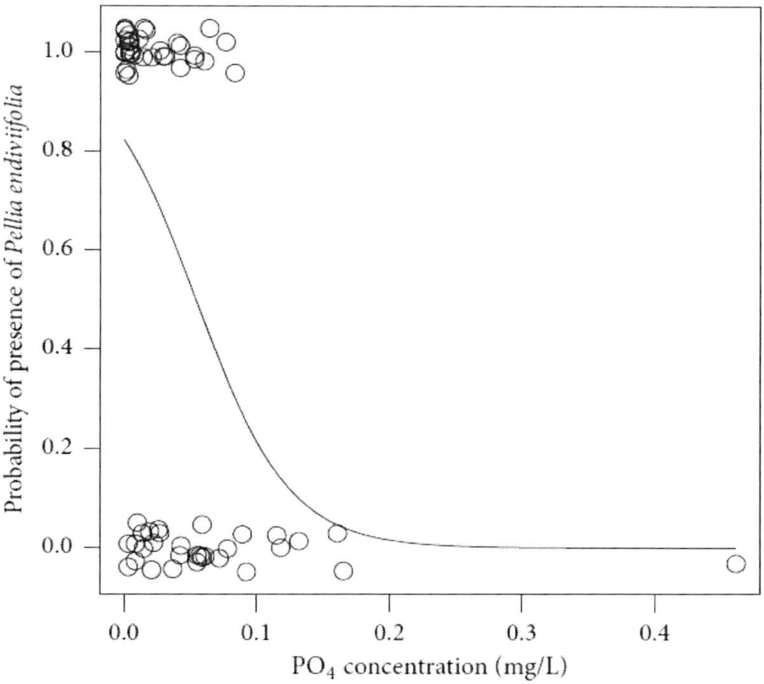

Figure 8: Prediction of presence of Pellia endiviifolia in relation to PO4 concentration based on the coefficients of the best models selected by AIC stepwise selection with the full dataset.

4.1. Global Analysis

The results of the tb-RDA showed that only 36.5% of the total variance in the dataset can be explained by the environmental variables, considered within this study. This suggests that other variables not measured here could explain better the observed species assemblages. Following other authors [41, 42], we suggest that factors such as the substrate itself (sand, loam, and rock), the micro topography [11, 16, 18, 33, 43], the water turbulence [2, 8, 44], the distance from the spring [8, 13], the presence and characteristics of travertine-productive microorganisms like cyanobacteria [8, 16, 45], and the local relative humidity could play an important role in the species assemblages even if they would probably be difficult to quantify.

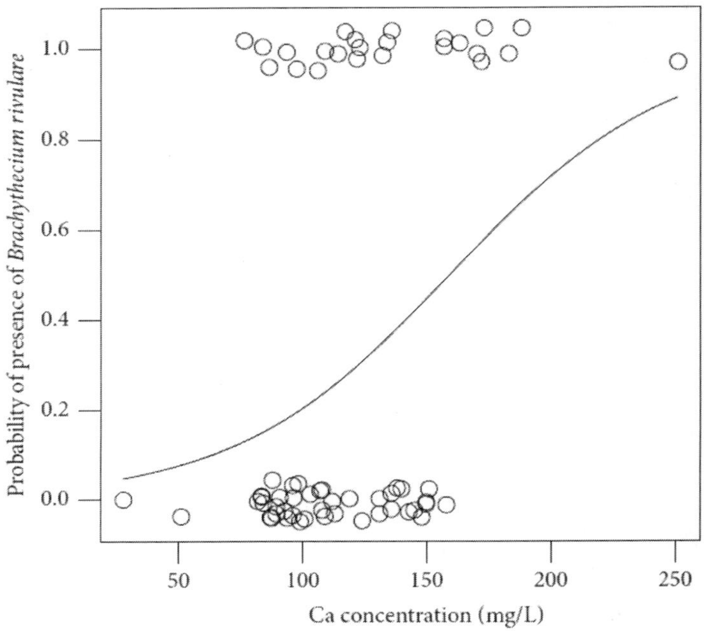

Figure 9: Prediction of presence of Brachythecium rivulare in relation to Ca concentration based on the coefficients of the best models selected by AIC stepwise selection.

The results show that the most important variables explaining the rest of the variance are the slope, NO_3, NH_4, and PO_4 concentrations (Figure 2). The slope has been identified by many authors as a key factor for the presence of the travertine communities [18] mainly because a steep slope generates more turbulence that enhances CO_2 degassing and thus calcite precipitation that accounts for more than 80% of the precipitation in travertine [13, 45]. The negative relationship between the PO_4 concentration in the spring water and the presence of mosses specific to petrifying sources should be further explored but could possibly be explained by the fact that PO_4 has an inhibitive role in the formation of calcite [3, 46, 47]. Pentecost [13] also highlights the negative role of phosphate pollution on the deposits formation and on the travertine communities. These results are partly confirmed by the forward selection that has similarly retained slope and canopy cover but also Mg and Ca (as well as NH_4 when the 2 possible outliers are dropped). That means that the relief, the canopy cover, and the degree of eutrophication play an important role among the variables. It is not surprising that NH_4 and PO_4 variables act in the same way (Figure 2). More curiously though is the opposite relation between these two variables and the NO_3 concentration though a similar observation has been made by Denys and Oosterlynck [48] in Flanders. The second axis showed that the canopy cover also plays an important role in the dispersion of the species assemblages. This result is expected since we know that the exposure and the degree of relative humidity play a key role in the ecology of most of bryophytes species [49].

The analysis also indicates that some species are more specifically linked to some variables considered in this study. Eucladium verticillatum and Palustriella commutata appear mostly in the same kind of locations characterized by a moderate or steep slope and with no or very low levels of eutrophication. This result is similar to analysis in Italy, using the same multivariate analysis [11]. Canopy cover is also negatively correlated with these sites indicating that these species occur more frequently at enlightened sites. Pellia endiviifolia shows a tendency to grow in NO_3-rich sites, which is confirmed by other observation that the species can grow in oligotrophic and eutrophic waters [40]. Finally Brachythecium rivulare and Oxyrrhynchium hians are more likely to be found in more eutrophicated sites (determined by PO_4 and NH_4 values).

4.2. Species Richness

The Poisson GLM analysis and the AIC have retained two major significant variables concerning the species richness, that is, canopy cover and PO_4 concentration. However other variables that have not been included in this analysis, such as soil type and water turbulence, should be considered in future studies. For example, Pentecost [13] has stated that two other variables, site size and heterogeneity, are also important factors affecting species richness. It can be seen in Figure 4 that all other registered variables being equal there could be a gain of only 2 or 3 species when clearing completely a theoretical site. This gain does not predict if the gained species would be characteristic species or other species.

The negative relation with PO_4, and with NH_4 when keeping the outliers in the dataset, indicates that the more oligotrophic the conditions the more rich the community. Other studies indicate that this kind of relationship is not always attributed to eutrophication, and human impacts are sometimes positively correlated to species richness [48]. These studies may address a wider range of conditions than in the present study and suggest that this relationship is scale-dependent. The majority of the richest sites were characterized by PO_4 concentration $< 0.1\ mg{\cdot}L^{-1}$ which is very close to the $0.05\ mg{\cdot}L^{-1}$ threshold determined by Galoux et al. [41] for the reference conditions of Walloon watercourses. Brusa and Cerabolini [11] in Italy using multivariate analysis found that the 3 major factors explaining their species assemblages were the permanency/nonpermanency of the water flow, the slope, and the canopy cover. In this study the slope is not retained in the species richness analysis; however, it is an important variable for Palustriella commutata.

4.3. Characteristic and Companion Species

4.3.1. Palustriella commutata

Our analysis shows that slope and to a lesser extend canopy cover are important variables explaining the presence of Palustriella commutata in the surveyed sites. The absence of canopy combined with a steep slope increases the probability of presence of Palustriella commutata (Figure 5). The slope has been detected as the only significant variable by Brusa and Cerabolini [11]. The fact that slope seems to play an important role in the probability of presence of the species seems contradictory with the well-known occurrence of the species in calcareous meadows in Belgium [16, 28]. It is possible that the properties of the water including highly mineralized or calcareous waters and permanently wet surfaces could be even more important than slope [11, 19, 28, 50]. It has been mentioned by many authors [13, 43, 51, 52] that there are two subspecies or species [14] Palustriella commutata and P. falcata whose ecological preferences are slightly different. According to these authors P. commutata occurs relatively more often on boggy sites and around small springs and seepages with a higher pH and higher Ca concentrations, and it tolerates higher rates of nitrates and grows almost exclusively on limestone. P. falcata seems more restricted to open habitat of turbulent water at higher altitude and in less bases-rich waters. Concerning the canopy cover other authors [1, 13, 18, 50] mention the fact that this factor must play a role even if the species tolerates much shade. The species seems also to tolerate a certain amount of eutrophication [18, 34, 50] even if it grows mostly in oligotrophic circumstances.

4.3.2. Eucladium verticillatum

Although canopy cover and NH*4* are both negatively linked with the probability of presence of the species these variables seem not to play a major role. It is probable that other variables could also influence species richness including less permanently wet substrate by capillarity or because of being splashed. Indeed we found the species abundant around a spring on a steep sandy substrate in the Rouge-Cloître in the Brussels-Capital Region that is not splashed at all but is permanently wet due to capillarity and high relative humidity. The relative weak negative relationship with canopy cover is mentioned by some authors [2, 24, 50] that found the species in much shaded habitats.

4.3.3. Cratoneuron filicinum

This species is not recognized in Wallonia as a characteristic species of the 7220 habitat as it is widely distributed [19], occupying calcareous meadows, calcareous springs, and wet rocks, as well as forest tracks where dolomite gravels have been added to harden the soil. In the Walloon watercourses Cratoneuron filicinum has been identified as one of the few indicative species of more or less calcareous streams [41]. In our sites the species has a relatively wide ecological amplitude as regards canopy cover and light to moderate eutrophication. The tolerance to a certain level of eutrophication is also mentioned by Bailly et al. [50] and by Sossey-Alaoui and Rosillon [40]. Mg proves to be a key-element in our sites and this observation should be related to the fact that the species is especially common on dolomite substrates, such as forests tracks that are rich in Mg, as well as on marl substrates in Lorraine [19].

4.3.4. Pellia endiviifolia

This is a widespread species in Wallonia and Brussels [19, 53] commonly associated with neutral to basic waters. It is also known to tolerate shade as well as moderately eutrophic waters [50]. It is therefore not surprising that none of these variables show clear relationships with the probability of presence in our sites. The apparent contradictory relationships with PO_4 and NO_3 are hard to clarify. Sossey-Alaoui and Rosillon [40] indicate that the species can be an indicator species both in oligotrophic streams and in nitrates-rich streams, providing that the waters are mineral-rich (high conductivity/high Ca concentration).

4.3.5. Brachythecium rivulare

Together with Pellia endiviifolia this species is relatively widespread in Wallonia and Brussels-Capital Region [19, 53]. It occurs along water courses and in marshes, though it is less linked to calcareous waters. In Wallonia it has been demonstrated that it has very wide ecological amplitude that is common in every natural region independent of the substrate [41]. It has also wide ecological amplitude occurring in a range of trophic levels and tolerating shade but preferring open sites [50]. This high ecological amplitude mirrors in the quasi-absence of significant variables explaining the presence of the species in our dataset.

4.4. Management Measures

One of the main objectives of this study was to define management objectives that could be implemented to maintain or improve the conservation status of the habitat 7220. As only relatively small proportion of the observed variability can be explained by the surveyed variables; the inclusion of further variables is recommended for future studies. Additionally among the surveyed variables some cannot be managed at all like the slope and Mg concentrations. Nevertheless, our results provide information on key variables that should be considered when managing these sites. The management measures may be different depending on their intention and may vary if the aim is to enhance biodiversity or favouring particular species. Generally speaking, we have seen that one of the most prominent results is that the species assemblage characteristic of the 7220 habitat (Palustriella commutata and Eucladium verticillatum) favours very low PO_4 concentration ($PO_4 < 0.2$ mg·L−1) and low canopy cover. This is also the case but to a lesser degree with NH_4. Graham and Farr [34] have stated that these two species are also mostly present when PO_4 concentration is less than 0.05 mg·L^{-1}. Brachythecium rivulare, Oxyrrhynchium hians, and even Pellia endiviifolia can occur in eutrophic conditions; however, in these conditions there is a lower probability of presence of the two characteristic species.

The more open the site the more diverse the moss assemblages will be even if the expected number of additional species is only 2 or 3 (Figure 4); however, the favoured species would not automatically be characteristic species of the 7220 habitat. If canopy cover plays a role in favouring species richness, it would be advisable not to cut down all existing trees and bushes but to undertake thinning to avoid possible radical negative changes in relative humidity and temperature on the bryophyte communities.

We recommend that regular monitoring of the water chemistry is the best way to detect abnormal changes especially eutrophication (PO_4, NH_4, NO_3). Monitoring regimes are already in place for most of the Walloon and Brussels sites but should be extended to all 7220 known sites.

4.5. Suggestions for Future Studies

To try to intercept other significant factors in the distribution and the presence/absence of the most characteristic species of the 7220 habitat, further studies should seek to take into account the following variables: type of substrate (sandy, calcareous, silty, etc.), the micro topography, the water turbulence, the distance from the spring, the relative humidity, and the presence of algae/cyanobacteria.

5. CONCLUSIONS

In this study, we highlighted the variables explaining the species richness and the distribution of moss assemblages of the 7220 habitat in the southern part of Belgium. The driving factors highlighted in this study can only be applied to the 7220 habitat and would probably not be valuable for studies at sites dominated by less calcareous waters in Wallonia and Brussels-Capital Region. Our study recommends that monitoring of water chemistry (eutrophication) should be applied to all 7220 habitats, and this information should be used to support practical management actions to maintain or achieve a good status of conservation.

ACKNOWLEDGMENTS

The authors would like to thank their colleagues Fabrice Etienne and Etienne Peiffer who collected the water samples and recoded the site variables. Other colleagues helped them when measuring the parameter, Daniel Galoux and Jonathan Devriese. Christine Keulen provided them with useful information about the thresholds of the different water courses in Wallonia. Mathias Engelbeen and Wim Van Den Eynden gave them access to the sites of the Brussels-Capital Region. They are also very grateful to David Zeleny who kindly accepted to check their multivariate analysis. They thank Gareth Farr (British Geological Survey) for undertaking a review of the English manuscript. Finally, they would like to thank Pierre Gerard who allowed them to carry out this study.

REFERENCES

1. J.-J. Symoens, P. Duvigneaud, and C. Vanden Berghen, "Aperçu sur la végétation des tufs calcaires de la Belgique," Bulletin de la Société Royale de Botanique de Belgique Tome, vol. 83, pp. 329–352, 1951.
2. J.-M. Couderc, "Les groupements muscinaux des tufs de Touraine," Documents Phytosociologiques N.S., vol. 1, pp. 37–50, 1977.
3. Pentecost, "The tufa deposits of the Malham district, north Yorkshire," Field Studies, vol. 5, no. 3, pp. 365–387, 1981.
4. P. De Zuttere, Aperçu bryosociologique des tufs calcaires actifs de moyenne et haute Belgique. Colloques phytosociologiques, X Les végétations aquatiques et amphibies (Lille, 1981), pp. 279–293, https://inpn.mnhn.fr/docs/cahab/habitats-declines/7220.pdf.
5. H. A. Viles and A. S. Goudie, "Tufas, travertines and allied carbonate deposits," Progress in Physical Geography, vol. 14, no. 1, pp. 19–41, 1990.
6. T. D. Ford and H. M. Pedley, "A review of tufa and travertine deposits of the world," Earth-Science Reviews, vol. 41, no. 3-4, pp. 117–175, 1996.

7. Janssen and R. Swennen, "Petrography and geochemistry of the travertine deposit at Treignes (S. Belgium)," Bulletin de la Société Belge de Géologie, vol. 106, pp. 97–115, 1997.
8. M. Merz-Preiß and R. Riding, "Cyanobacterial tufa calcification in two freshwater streams: ambient environment, chemical thresholds and biological processes," Sedimentary Geology, vol. 126, no. 1–4, pp. 103–124, 1999.
9. R. Boch, C. Spötl, J. M. Reitner, and J. Kramers, "A lateglacial travertine deposit in Eastern Tyrol (Austria)," Austrian Journal of Earth Sciences, vol. 98, pp. 78–91, 2005.
10. Franco, G. Houbrechts, J. Van Campehout, E. Hallot, and F. Petit, "Etude géomorphologique des barrages de travertin du Hoyoux," Bulletin de la Société Géographique de Liège, vol. 50, pp. 45–56, 2008.
11. G. Brusa and B. E. L. Cerabolini, "Ecological factors affecting plant species and travertine deposition in petrifying springs from an Italian 'Natura 2000' site," Botanica Helvetica, vol. 119, no. 2, pp. 113–123, 2009.
12. European Commission, "Interpretation Manual of European Union Habitats-Eur28. European Commission, DG Environment, Nature & Biodiversity," 2013, http://ec.europa.eu/environment/nature/legislation/habitatsdirective/docs/Int_Manual_EU28.pdf.
13. Pentecost, Travertine, Springer, Berlin, Germany, 2005.
14. M. O. Hill, N. Bell, M. A. Bruggeman-Nannenga et al., "An annotated checklist of the mosses of Europe and Macaronesia," Journal of Bryology, vol. 28, no. 3, pp. 198–267, 2006.
15. L.-M. Delescaille, L. Wibail, H. Claessens et al., Eds., Les Cahiers d'Habitats d'Intérêt Communautaire de Wallonie, Publication du Département de l'Étude du Milieu Naturel et Agricole (SPW-DGARNE), Faune—Flore—Habitats, Gembloux, Belgium, 2017.
16. P. Oosterlynck and W. Van Landuyt, "Kalktufbronnen in vlaanderen: mythe of werkelijkheid?" Muscillanea, vol. 32, pp. 36–52, 2012.
17. JNCC, Joint Nature Conservation Committee, Habitat account—7220 Petrifying springs with tufa formation (Cratoneurion), 2015, http://jncc.defra.gov.uk/ProtectedSites/SACselection/habitat.asp?FeatureIntCode=H7220.
18. K. van Dort, L. van Oirschot-Beerens, and H. Weinreich, "Bryophyte vegetation in petrifying springs with tufa in Limburg (The Netherlands)," Natuurhistorisch Maandblad, pp. 165–173, 2012.
19. Sotiaux and A. Vanderpoorten, Atlas des Bryophytes (Mousses, Hépatiques, Anthocérotes) de Wallonie (1980–2014), Série "Faune-Flore-Habitats" no. 9, Publication du Département de l'Etude du Milieu Naturel et Agricole (SPW-DGARNE), Gembloux, Belgium, 2015.
20. J.-J. Symoens, "Note sur les formations de tuf calcaire observées dans le Bois d'Hautmont (Wauthier-Braine)," Bulletin de la Société Royale de Botanique de Belgique, vol. 82, no. 1, pp. 81–95, 1949.
21. J. De Sloover and M. Goossens, "Les associations du Cratoneurion d'un travertin de Lorraine belge," Bulletin de la Société Royale Botanique de Belgique, vol. 117, pp. 37–50, 1984.
22. J. Duvigneaud and J. Saintenoy-Simon, "Les tufs calcaires des Fonds de Leffe à Dinant," Les Naturalistes Belges, vol. 72, no. 4, pp. 113–120, 1991.
23. J. Duvigneaud and J. Saintenoy-Simon, "Deux tufs calcaires sur Famennien à Hastière et à Blaimont (province de Namur)," Les Naturaliste Belges, vol. 73, no. 4, pp. 199–204, 1992.
24. M.-T. Romain, "Cron de neuville," Les Barbouillons, vol. 203, pp. 114–116, 2001.
25. H. Zechmeister and L. Mucina, "Vegetation of European springs: high-rank syntaxa of the Montio-Cardaminetea," Journal of Vegetation Science, vol. 5, no. 3, pp. 385–402, 1994.
26. J. Bardat and J.-C. Hauguel, "Synopsis bryosociologique pour la France," Cryptogamie, Bryologie, vol. 23, no. 4, pp. 279–343, 2002.
27. Van Gennip, J. A. M. Janssen, and E. J. Weeda, "De kalktufbron, kleinood met een grote status," Stratiotes, vol. 35, pp. 22–37, 2007.
28. G. Noulard-Weyembergh, "Une station de Carex lepidocarpa et de Palustriella commutata dans la vallée de l'Argentine (Overijse, district brabançon)," Adoxa, vol. 75, pp. 7–12, 2013.

29. M. Tomaselli, D. Spitale, and A. Petraglia, "Phytosociological and ecological study of springs in Trentino (south-eastern Alps, Italy)," Journal of Limnology, vol. 70, supplement 1, pp. 25–53, 2011.
30. G. Bailly, Identification des Habitats Aquatiques et des Formations Tufeuses de la Haute Seille, Conservatoire Botanique de Franche-Comté, Besançon, France, 2005.
31. M. L. H. Boyer and B. D. Wheeler, "Vegetation patterns in spring-fed calcareous fens: calcite precipitation and constraints on fertility," Journal of Ecology, vol. 77, no. 2, pp. 597–609, 1989.
32. Pentecost, "A note on the stable carbon isotope composition of bryophytes in calcareous aquatic habits and its relationship to carbon dioxide assimilation," Journal of Bryology, vol. 22, no. 1, pp. 13–15, 2000.
33. Pentecost and Z. Zhaohui, "Bryophytes from some travertine-depositing sites in France and the U.K.: relationships with climate and water chemistry," Journal of Bryology, vol. 24, no. 3, pp. 233–241, 2002.
34. J. Graham and G. Farr, "Petrifying springs in Wales. Report of the first systematic survey of 'tufa' springs in Wales," Field Bryology, vol. 112, pp. 19–29, 2014.
35. P. Oosterlynck and E. De Bie, "Kalktufbronnen in Vlaanderen. Bryologische en abiotische karakterisering van een Natura 2000 habitattype op de rand van zijn verspreiding," Rapporten van het Instituut voor Natuur- en Bosonderzoek, Brussels, Belgium.
36. R Core Team, R: A Language and Environment for Statistical Computing, R Foundation for Statistical Computing, Vienna, Austria, 2014, https://www.R-project.org/.
37. P. Legendre and E. D. Gallagher, "Ecologically meaningful transformations for ordination of species data," Oecologia, vol. 129, no. 2, pp. 271–280, 2001.
38. J. Oksanen, F. Guillaume Blanchet, R. Kindt et al., Vegan: Community Ecology Package. R package version 2.3-0, 2015, http://CRAN.R-project.org/package=vegan.
39. K. P. Burnham and D. R. Anderson, Model Selection and Multimodel Inference: A Practical Information-Theoretic Approach, Springer, New York, NY, USA, 2nd edition, 2002.
40. K. Sossey-Alaoui and F. Rosillon, "Macrophytic distribution and trophic state of some natural and impacted watercourses—Belgium Wallonia," International Journal of Water Sciences, vol. 2, pp. 1–11, 2013.
41. Galoux, F. Chérot, F. Rosillon, and K. Sossey-Alaoui, "Contribution to the macrophytic typology of belgian reference watercourses," Advances in Botany, vol. 2015, Article ID 651369, 13 pages, 2015.
42. A.-M. Geurts, "Formation des travertins de fond de vallée sous climat tempéré océanique," Comptes-Rendus de l'Académie des Sciences de Paris D, vol. 282, pp. 275–276, 1976.
43. G. Farr, J. Graham, and C. Stratford, "Survey characterisation and condition assessment of Palustriella dominated springs H7220 Petrifying springs with tufa formation (Cratoneurion)," Centre for Ecology and Hydrology and the British Geological Survey (NERC) Internal Report WL/NEC02832/13_14/T6, 2014.
44. Pentecost, "Moss growth and travertine deposition: the significance of photosynthesis, evaporation and degassing of carbon dioxide," Journal of Bryology, vol. 19, no. 2, pp. 229–234, 1996.
45. G. Arp, A. Bissett, N. Brinkmann et al., "Tufa-forming biofilms of German karstwater streams: microorganisms, exopolymers, hydrochemistry and calcification," Geological Society, London, Special Publications, vol. 336, pp. 83–118, 2010.
46. P. M. Dove and M. F. Hochella Jr., "Calcite precipitation mechanisms and inhibition by orthophosphate: in situ observations by Scanning Force Microscopy," Geochimica et Cosmochimica Acta, vol. 57, no. 3, pp. 705–714, 1993.
47. L. J. Plant and W. A. House, "Precipitation of calcite in the presence of inorganic phosphate," Colloids and Surfaces A: Physicochemical and Engineering Aspects, vol. 203, no. 1–3, pp. 143–153, 2002.
48. L. Denys and P. Oosterlynck, "Diatom assemblages of non-living substrates in petrifying Cratoneurion springs from lower Belgium," Fottea, vol. 15, no. 2, pp. 123–138, 2015.
49. Vandepoorten and B. Goffinet, Introduction to Bryophytes, Cambridge University Press, New York, NY, USA, 2009.

50. G. Bailly, J.-C. Vadam, and J.-P. Vergon, Guide Pratique d'Identification des Bryophytes Aquatiques, DIREN Franche-Comté, 2004.
51. L. Hedenäs and A. Kooijman, "Habitat differentiation within Palustriella," Lindbergia, vol. 29, no. 1, pp. 40–50, 2004.
52. H. Stieperaere, "Palustriella commutata en P. falcata in Vlaanderen," Muscillanea, vol. 26, pp. 32–36, 2007.
53. Vanderpoorten, A Bryological Survey of the Brussels Capital Region (Belgium), vol. 14, Scripta Botanica Belgica, 1997.

CHAPTER 12

Molecular Regulation of Flowering Time in Grasses

Fiorella D. B. Nuñez [1] and Toshihiko Yamada [2,*]

[1]*Graduate School of Environmental Science, Hokkaido University, Kita 10 Nishi 5, Sapporo, Hokkaido 060-0810, Japan*

[2]*Field Science Center for Northern Biosphere, Hokkaido University, Sapporo, Hokkaido 060-0810, Japan*

ABSTRACT

Flowering time is a key target trait for extending the vegetative phase to increase biomass in bioenergy crops such as perennial C_4 grasses. Molecular genetic studies allow the identification of genes involved in the control of flowering in different species. Some regulatory factors of the *Arabidopsis* pathway are conserved in other plant species such as grasses. However, differences in the function of particular genes confer specific responses to flowering. One of the major pathways is photoperiod regulation, based on the interaction of the circadian clock and environmental light signals. Depending on their requirements for day-length plants can be classified as long-day (LD), short-day (SD), and day-neutral. The *CONSTANS* (*CO*) and *Heading Date 1* (*Hd1*), orthologos genes, are central regulators in the flowering of *Arabidopsis* and rice, LD and SD plants, respectively. Additionally, *Early heading date 1* (*Ehd1*) induces the expression of *Heading date 3a* (*Hd3a*), conferring SD promotion and controls *Rice Flowering Locus T 1* (*RFT1*) in LD conditions, independently of *Hd1*. Nevertheless, the mechanisms promoting flowering in perennial bioenergy crops are poorly understood. Recent progress on the regulatory network of important gramineous crops and components involved in flowering control will be discussed.

AUTHOR CONTRIBUTION

F.D.B.N., written paper; T.Y., edited paper, mentored MS student (F.D.B.N.)

Keywords: flowering; *Arabidopsis*; grasses; photoperiod; circadian clock

1. INTRODUCTION

Global climate change and energy security issues have promoted interest in the production and increased availability of alternative energy sources. Lignocellulosic biomass is a promising feedstock source for biorefineries producing biofuel, which can mitigate greenhouse gas emissions [1,2,3] and reduce dependency on fossil oil [4,5]. Perennial C_4 bioenergy crops such as switchgrass (*Panicum virgatum* L.) and *Miscanthus* spp. provide good targets as non-edible plant species [6,7,8] having advantages with regard to land utilization and the avoidance of conflict with food security [9,10,11] to provide efficient production systems at low cost.

One of the most important traits in the plant life cycle is the timing of flowering, the floral transition between the vegetative and reproductive phases of plant development [12,13]. Consideration of flowering time is an important strategy in the cultivation of grain crops in northern latitudes. Early flowering is useful in regions where growing seasons are short to enhance grain yield stability by avoiding drought or adverse temperatures [14,15,16]. Flowering time is also a major determinant of biomass yield in perennial C_4 bioenergy crops, because delayed flowering time allows an extended period of vegetative growth and produces more biomass. Thus, earlier flowering will produce lower yields than late flowering in terms of feedstock production [17]. However, biomass potential also depends on environmental conditions. In switchgrass, lowland ecotypes that originate from southern areas flower later in high latitude areas, but the yield advantages of these southern switchgrasses are often not realized at northern latitudes due to high winter mortality [18].

Natural variation in flowering time is related to latitude in several plant species. Migration of plants into different latitudes often require the adoption of different signals to induce flowering and promote adaptive responses to diverse growing seasons [19]. Factors such as photoperiod and temperature that vary over large geographical scales are involved [20]. Plants possess an internal biological clock providing circadian rhythms that respond to fluctuations in day-length and thus anticipate upcoming seasonal changes [21] to regulate flowering. Depending on their requirements for day-length (light period in a 24-h cycle) to promote flowering, plants can be classified as long-day (LD) plants when photoperiod exceeds a critical day-length, short-day (SD) plants when photoperiod is shorter than a critical day-length and day-neutral plants when flowering occurs irrespective of day-length [22,23,24]. Winter annuals (e.g., wheat (*Triticum aestivum* L.)), biennials (e.g., sugar beet (*Beta vulgaris* L.)) and numerous perennials (e.g., orchardgrass (*Dactylis glomerata* L.)) are obligatory LD plants. These plants, however, flower only after vernalization during a cold period [23]. The molecular basis of flowering time regulation has been extensively studied using classical Quantitative Trait Loci (QTL) approaches in model plant species such as Arabidopsis thaliana [25,26,27], a LD plant, and rice (*Oryza sativa* L.), a SD plant [28]. Grasses have multiple pathways to control flowering time but only some of them are conserved in *Arabidopsis thaliana* (L.) Heynh. [29]. These studies have been crucial in establishing the multiple pathways that control flowering, of which the photoperiod pathway is of major importance [30]. The use of model species has played a major role in understanding the molecular mechanisms involved in flowering time to help in the genetic improvement of crop development. Nevertheless, little is known about the mechanisms promoting flowering in perennial C_4 bioenergy crops. In this review, we discuss recent progress concerning the regulatory network and components involved with flowering control in different species including C_4 grasses such as sorghum (*Sorghum bicolor* (L.) Moench), switchgrass and *Miscanthus* spp. To understand how plants initiate flowering is a crucial step in developing selection criteria in breeding programs of grasses used as bioenergy crops.

2. CONSERVATION AND DIVERGENCE IN FLOWERING PATHWAYS

2.1. Arabidopsis and Rice

In the last decades, studies on the model plant *Arabidopsis* have revealed that molecular mechanisms discovered in that species are evolutionally conserved in other species [15]. Genetic approaches in *Arabidopsis* have identified three genes that control flowering: *GIGANTEA (GI)—CONSTANS (CO)—FLOWERING LOCUS (FT)* [15,28]. Loss of function mutations in each gene for flowering control delay flowering under LD conditions but no effect is produced under SD [31]. *GI* is a key regulator of the photoperiodic pathway and in the evening promotes *CO* transcription under LD conditions [32]. The most extensively gene studied in *Arabidopsis* flowering is *CO* that confers LD responses. *CO* encodes a B-box zinc finger transcription factor and CCT domain genes that promote flowering under LD conditions and activate the expression of *FT* [33,34,35], a major component of the florigen that induces flower differentiation [29]. Its inactivation causes flowering delay, while its over-expression induces early flowering. *CO* and *FT* are expressed in the phloem and act there to promote flowering [31]. This signalling pathway is conserved in rice: *OsGI-Hd1-Hd3a* [28], mediated by LD responses. *OsGIGANTEA (OsGI)* acts an activator of *Heading date1 (Hd1)*, an ortholog of *CO*, and controls flowering time by modulating rhythmic flowering under SD [36]. *Hd1* encodes a zinc finger type transcriptional activator with the conserved CCT *(CO, CO-like, TIMING OF CAB EXPRESSION1* (*TOC1*)*)* domain [15,37]. In contrast to *Arabidopsis CO*, *Hd1* promotes *Heading date 3a (Hd3a)* expression in SD but expression is modified in LD conditions [38,39] where *Hd1* function is converted into a repressor. Rice involves at least two flowering pathways that control the expression of the florigens: *Hd1* that is conserved in rice and *Arabidopsis*, and *Early heading date 1*(*Ehd1*), without an ortholog in *Arabidopsis* [39,40] (Figure 1). Moreover, *Grain Number, Plant Height, and Heading Date7* (*Ghd7*) is unique in grasses [16]. *Ehd1* is a B-type response regulator that induces the expression of *Hd3a* in rice, conferring SD promotion of flowering in the absence of a functional allele of *Hd1*. It also controls *Rice Flowering Locus T (RFT)* gene in LD conditions independently of *Hd1* [40]. In LD conditions *Hd1* acts as a flowering repressor inhibiting *Hd3a* expression but promotes its expression and subsequent flowering in *SD* [41]. *Ghd7* is a small protein with a CCT-domain that represses *Ehd1* expression and downstream *Hd3a/RFT1* expression in LD conditions to delay flowering [42,43,44,45]. Recent studies demonstrated that the interaction between *Ghd7* and *Hd1* can play a critical role in repressing *Ehd1*. Under SD conditions *Hd1* activates the expression of *Ehd1* at night but not in the day while under LD conditions *Hd1* represses its expression in the morning. Indeed, *Hd1* repressor activity requires a proper *Ghd7* function under LD conditions to repress *Ehd1* in the morning. In contrast, *Ghd7* can repress the expression of *Ehd1, Hd3a* and *RFT1* by itself under all photoperiod conditions [15].

2.2. C_4 Grasses

Previous studies identified two floral activators in sorghum, a SD plant: *SbEhd1* and *SbCO*. *SbCO* is a homolog of the floral activator *CO* in *Arabidopsis* and an ortholog of *Hd1* in rice. It promotes early flowering in both LD and SD conditions, and increases the expression of *SbEhd1, SbCN8, SbCN12* and *SbCN15* [16]. Genetic analyses and expression studies in sorghum reveal that *SbCO* shares a conserved CCT-domain with *TOC1, PSEUDORESPONSE REGULATOR PROTEIN 37* (*PRR37*), *Ghd7* and *HEME*

ACTIVATOR PROTEINS (HAP). *SbCO* also increases expression of *SbEhd1*, a promoter of *Hd3a* in rice. In comparison with rice, *Ehd1* regulates positively the expression of *RFT1* to promote flowering; however, no ortholog of *RFT1* is present in the sorghum genome. *SbPRR37* (Ma_1) and *Ghd7* (Ma_6) inhibit flowering, reducing the expression of *SbEhd1* and *SbCN8/12* (florigens) under LD conditions, but not in SD (Figure 1). The ability of *SbPRR37* to inhibit their expression could be due to inhibition of *SbEhd1* or *SbCO*, activators of *SbCN8* and *SbCN12* expression [14,16]. The *PhyB* regulation of *SbCN15* expression may modify flowering time in a photoperiod-insensitive manner [46] (Figure 1). In switchgrass, the flowering time regulatory network is similarly to maize (*Zea mays* L.) and is regulated by both photoperiod-dependent and autonomous pathways. Some conserved flowering genes such as *FT-like gene* (*ZCN8* in maize) and *INDETERMINATE 1 (ID1)* have also been identified in the maize genome. The study of genes involved in flowering of switchgrass is relatively new. Hence, the functions of *FT-like* gene in switchgrass germplasm have not been clarified yet but may contribute to delayed flowering time as in maize [18]. Thus, the switchgrass *FT* homolog may have similar functions to the maize *FT* gene and is down-regulated by the expression of *LONG VEGETATIVE PHASE ONE (AtLOV1)* in switchgrass [18]. Overexpression of *AtLOV1* causes delayed flowering time in switchgrass but does not enhance cold tolerance as in *Arabidopsis* [18]. Sorghum is closely related to *Miscanthus* spp., a promising candidate C_4 bioenergy crop in temperate climates. The *CO/Hd1* sequence in *Miscanthus sinensis* Andersson was identified as *MsiHd1* with two types of diverged loci, *MsiHd1a* and *MsiHd1b*. The *MsiHd1* gene encodes two conserved B-box zinc finger domains and a CCT domain. Two to five different alleles of *MsiHd1* were found in *Miscanthus* accessions from mainland Asia and from Japan, suggesting that *MsiHd1* consists of at least three loci in the *Miscanthus* genome with small differences in the number of functional alleles [38]. From preliminary data we identified at least three alleles suggesting that *MsiEhd1* has two loci in the *Miscanthus* genome, *MsiEhd1*a and *MsiEhd1b* in comparison to sorghum, rice and maize, which have only one gene. We also detected two loci in *MsiGhd7*. The current diploid *M. sinensis* evolved from genome duplication of its progenitor that was very close to a sorghum ancestor [46]. Gene duplication is a key mechanism in evolution because it can provide genes with new functions.

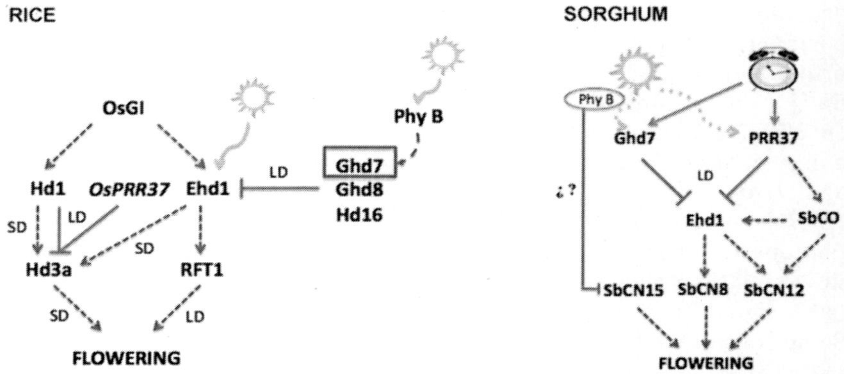

Figure 1. A simplified model of flowering time under short-day (SD) and long-day (LD) conditions in rice and sorghum. A dashed green arrow indicates transcriptional activation and a solid red line indicates transcriptional repression.

3. QTLS ANALYSIS

To understand the complex genetic network of flowering in perennial ryegrass (*Lolium perenne* L.), a C_3 forage grass, a number of genes have been identified through QTL mapping, using different plant material and genetic maps and by sequence homology with *Arabidopsis*, rice and maize [47]. The genomic and phenotypic variations associated with perennial ryegrass *LpFT3*, an ortholog of *FT*, were assessed in a diverse collection of nine European germplasm populations, identifying a total of 7 haplotypes. The results indicated a significant association between allelic variation in the *LpFT3* gene and flowering time. Haplotype C was associated with early flowering and the A and B haplotypes with late flowering. The variations were identified in the predicted sequence and in non-coding regions, mainly within the 5′ region of the coding sequence which is strongly conserved [48]. Comparative analysis established close proximity between genetic markers related to the *DGL1*, *Ph1* and *OsPIPK1* ortholoci and the corresponding perennial ryegrass QTLs. This suggests that *DGL1* and *Ph1* ortholoci may provide candidate genes for the herbage yield-related QTLs on linkage group 3 (LG3). The physical location of the *OsPIPK1* gene (a heading date locus) was located at the 28.2 Mb position of rice chromosome 3, close to the predicted *CDO795* ortholocus (23.1 Mb). Further studies have suggested that the *CDO795*-linked heading date QTL was equivalent to a rice heading date QTL, *dth3.3* (Gramene QTL Acc. ID AQFE011). As a consequence, the perennial ryegrass *OsPIPK1* ortholocus may be related to the heading date QTLs on LG4 [49]. In addition, the major QTL in the F_2/WSC and ILGI perennial ryegrass populations was identified on LG7, which is associated with the position of the genes *Hd3a* and *Hd1*, two heading date genes of rice on chromosome 6. However, analysis of the ILGI population grown in Japan identified a QTL on LG4, but not the QTL reported on LG7 [50,51]. This result emphasizes the importance of adaptation in plants to the broad range of agro-environmental conditions in which they grow. In sorghum, three significant QTL associated with flowering time, *PHYB (Ma3), PHYC (Ma5)* [52] and *SbGHD7 (Ma6)* [14,52] were identified, through analysis of flowering variation in LD using an F_2 population, which explained *~50% of the* phenotypic variance for flowering time [52]. Recessive *ma3R* alleles from 58 M populations associated with *Ma3* QTL produced early flowering time phenotypes; however, dominant alleles of Sb*Ghd7 (Ma6)* and Sb*PRR37* act in an additive manner to delay floral initiation for ~175 days until day-lengths decrease below 12.3 h [14,52]. Sorghum accessions exhibit significant variations in flowering time in response to day-length. One QTL controlling photoperiod sensitivity was detected on chromosome 1 under SD, and one QTL controlling photoperiod insensitivity expression was detected on chromosome 4 under 12 h and natural photoperiod conditions, from the SSR markers *Xtxp61* and *Xtxp51* respectively [53]. Under LD, *a cross between tropical and temperate sorghums* (Sorghum *propinquum* (Kunth) Hitchc.×*S. bicolor* (L.) Moench), revealed one QTL *FlrAvgD1* located in chromosome 6 in a 10 kb interval, which accounted for 85.7% of the variation in flowering time. This interval contains a single annotated gene, *Sb06g012260*, which is a member of the *FT* family of transcription factors. *Sb06g012260* is unique to panicoids and suppresses flowering, although it is quite distant evolutionarily from other *FT* family members that are floral suppressors [54]. In *M. sinensis* five putative flowering QTLs were detected using the Multiple QTL model (MQM) approach for plants grown in the years 2000 and 2001 [55]. Only QTL F12 was detected in both years on LG1, F11 and F13 were only detected in the first year while F14 and F15 were detected in the second year. So these QTLs may depend on interactions between genotype and environment. A genome-wide association study may be identified association with gene network in flowering time.

4. CIRCADIAN CLOCK AND PHOTOPERIOD RESPONSE

The circadian clock plays an important role in seasonal flowering time regulation of angiosperms; photoperiodic time measurement is based on the interaction between the endogenous circadian clock and environmental light signals in *Arabidopsis* [56,57]. The plant circadian system consists of biochemical timing mechanisms that temporarily modulate the function of several signalling pathways to measures changes in day-length and promote suitable timing of flowering to maximize reproductive success [2,3,6,7,9,10,13,14,15,21,24,26,28,29,31,36, 38,41, 42,46,48,49,56,57,58,59,60,61,62,63,64,65,66]. The photoperiod response on flowering time varies among grasses. Barley (*Hordeum vulgare* L.) and wheat are LD plants, while rice and sorghum are SD plants [52]. Flowering is regulated through the *CO* and *FT* genes [66]. *GI* plays an important role in regulating the circadian clock and flowering, promoting *CO* gene expression and light response. The rice ortholog of *GI, OsGI,* is a positive regulator of *Hd1* expression under both SDs and LDs [24]. Mutation in *OsGI* reduced photoperiod sensitivity in rice [36] and affected the expression of *LATE ELONGATED HYPOCOTYL (LHY)* and several *PSEUDO RESPONSE REGULATOR (PRR)* genes. However, *PRR37* expression was not affected in the *osgi* mutant, suggesting independent control of heading date by these factors [66]. *Hd1* is predominantly regulated by the circadian clock through *OsGIGANTEA (OsGI)* [64] and possesses two contrasting functions in the regulation of the rice ortholog of *Arabidopsis FT* gene, *Hd3a*. The bi-functionally mechanism of *Hd1* involves the action of the red-light photoreceptor *phytochrome B (phyB)*, a primary cause of long-day suppression of flowering in rice [39,60]. Over-expression of *Hd1* causes a delay in flowering under SD conditions and a single extension of day-length decreases *Hd3a* expression consistently with the duration of daylight [44,59]. The repression of flowering by *Hd1* under LD conditions is enhanced by the kinase activity of *Heading date 6 (Hd6),* a gene encoding the α subunit of protein kinase CK2 (CK2α) [58]. *Hd6* is a QTL involved in photoperiod response in rice. To induce delayed flowering under LD conditions, *Hd6* requires the presence of functional *Hd1* alleles and plays a critical role in *Hd1* activity. Despite this, *Hd6* regulation is not mediated by changes in the circadian clock [37,65]. The rice genome contains two important genes for photoperiodic regulation: *Ehd1* and *Ghd7*, specific to grass species such as rice, maize and sorghum but absent in the *Arabidopsis* genome [44]. The expression of *Hd3a* is also regulated by *Ehd1* conferring SD promotion of flowering and controlling *FT-like* gene expression independently of the *Hd1/CO* photoperiodic flowering pathway [30,40] (Figure 1). *Hd1* and *Ehd1* expression are controlled by the circadian clock, although *Ehd1* is also regulated by both blue and red light. In sorghum, *SbCO* expression is not altered significantly in response to day-length. However, *Ghd7*, a floral repressor regulated by the circadian clock and light, represses the expression of *SbEhd1* and *SbCN8* [14]. *Ehd1* expression in rice is strongly repressed by *Ghd7* in LD conditions but in SD conditions *Ghd7* rarely affects flowering time [36].

5. PHOTORECEPTORS INVOLVED IN FLOWERING TIME

Plants use the phytochrome system to regulate time of flowering and adjust growth based on the duration of dark and light periods (photoperiodism), while the spectrum of the light also affects flowering. Plants use many photoreceptors to detect the intensity and quality of light, including PHYTOCHROMES (PHY), which absorb the red and far-red region of the visible spectrum, and the CRYPTOCHROMES (CRY) [63]. *Arabidopsis* contains five PHYs (A-E), where accumulation of *CO* in LD is due to stabilization mediated by *phytochrome A* (*PHYA*), *cryptochromes* (*CRY1/2*) and *SUPPRESSOR OF PHYA-105 (SPA1)* [61]. However, it has been shown that PHYB signals delay flowering by destabilizing CO protein during the morning and have an inhibitory effect on *FT* expression [60]. The phytochromes PHYA, PHYB, CRY1 and CRY2 are directly clock regulated under specific light conditions. For example, over-expression of the photoreceptor PHYA under SD conditions promotes flowering, but *phyA* mutants delay flowering in LD conditions. In contrast to *Arabidopsis*, rice and sorghum encodes three phytochromes (PHYA, PHYB and PHYC) [12,52], where *phyA* mutants of rice do not produce significant alterations in flowering time [52,63]. This is despite the high similarity between the *PHYA* locus in *Arabidopsis* rice, sorghum and maize [62] suggesting that a similar response would be expected. However, PHYA mutations in combination with PHYB or PHYC cause early flowering in rice [61]. In addition, studies have shown that PHYC plays an essential role in the acceleration of wheat flowering under LD photoperiods. Moreover, it is stable and functionally active even in the absences of other phytochromes, compared with rice and *Arabidopsis* [67]. Blue and far-red lights promote flowering in *Arabidopsis* and rice, acting through the action of PHYA, CRY1 and CRY2 photoreceptors in *Arabidopsis*, while red light delays flowering [12,33,43,67]. On the other hand, PHYB modulates the expression of genes in response to red light and is the main component of the shade-avoidance mechanism in *Arabidopsis*. *PhyB* mutants revealed that PHYB inhibits flowering under both LD and SD photoperiods, but an over-expression of PHYB in LDs results in early flowering [22]. PHYB is responsible for delayed flowering and *Hd3a* suppression in the presence of a night-break (NB) treatment and activates the *Hd1* expression in rice. The NB treatment is a short exposure to light in the middle of night and was widely used to understand the role of the circadian clock and light on flowering. In recent studies, *photoperiodic sensitivity 5* (*SE5*) and PHYB also suppress *Ehd1* expression, by suppressing *Oryza sativa CO-like4 (OsCOL4)* [39]. Rice mutants deficient in PHYB have reduced sensitivity to red light and are early flowering [43,60]. In contrast, *phyB* null mutations in wheat are connected with delayed flowering [24]. In sorghum, under LD, PHYB (*Ma*$_3$) is required for elevated expression of *SbPRR37* and *SbGHD7* during the evening to inhibit flowering (Figure 1). This response results in repression of *SbEHD1*, *SbCN12*, *SbCN8* and floral initiation. *Ghd7* represses *Ehd1* expression in response to the red light signal in the morning mediated by phytochromes. In SD conditions, PHYB may have a limited effect on the expression of these genes as peak *SbPRR37* and *SbGHD7* expression is highest in the morning and lowest during the evening compared with expression in LD. The inactivation of PHYB results in early flowering in LD [43,52].

6. CONCLUSIONS AND PERSPECTIVES

Genetic analysis in *Arabidopsis* has allowed the identification of different pathways that promote flowering in response to environmental conditions and developmental regulation. The primary mechanism of the photoperiod pathway in plants is evolutionary conserved for flowering signalling. *CO* is the central regulator in promoting flowering and exhibits complex regulation. In addition day-length and the circadian clock control critical aspects of flowering. The effect of *GI* on flowering is associated with promoting expression of genes related to the circadian pathway. *Arabidopsis* is considered as a model plant to understand flower development while some grasses have defined their own responses and adaptation strategies. For example, *Ehd1* and *Ghd7* genes are unique in grasses in relation to the promotion and repression of flowering time respectively. Due to the lack of nucleotide information in *Miscanthus*, few genetic resources have been developed to clarify the relationship of the *Miscanthus* genome to its close relatives, sorghum and sugarcane (*Saccharum* spp.). Comparisons between the *Sorghum* genome and the genus *Miscanthus* reveal that whole genome duplication occurred in *Miscanthus* after its divergence from a common ancestor shared with sorghum. The base chromosome number of *Miscanthus* is approximately twice that of sorghum with nominally diploid and tetraploid species [46,68]. Analysis of natural variation in flowering in different ecotypes of grasses with economic value, such as *Miscanthus* spp. and switchgrass, is necessary to clarify the molecular network of flowering time control in these species. Through breeding programs, favorable alleles of QTLs then can then be efficiently introduced into elite cultivars to generate new varieties with high biomass productivity and beneficial adaptations to environmental changes.

ACKNOWLEDGMENTS

We gratefully acknowledge Mervyn O. Humphreys, for critical reading of the manuscript. F.D.B.N. acknowledgments the Ministry of Education, Culture, Sports, Science, and Technology (MEXT) scholarship for Master Course study at the Hokkaido University, Japan.

REFERENCES

1. Clifton-Brown, J.C.; Stampfl, P.F.; Jones, M.B. *Miscanthus* biomass production for energy in Europe and its potential contribution to decreasing fossil fuel carbon emission. *Glob. Chang. Biol.* **2004**, *10*, 509–518.
2. Lewandowski, I.; Kicherer, A.; Vonier, P. CO_2-balance for the cultivation and combustion of *Miscanthus*. *Biomass Bioenerg.* **1995**, *8*, 81–90.
3. Lewandowski, I.; Kicherer, A. Combustion quality of biomass: Practical relevance and experiments to modify the biomass quality of *Miscanthus × giganteus*. *Eur. J. Agr.* **1997**, *6*, 163–177.
4. Fargione, J.; Hill, J.; Tilman, D.; Polasky, S.; Hawthorne, P. Land clearing and the biofuel carbon debt. *Science* **2008**, *319*, 1235–1238.
5. Timothy, S.; Ralph, H.; Houghton, R.A.; Fengxia, D.; Amani, E.; Jacinto, F.; Simla, T.; Dermot, H.; Tun-Hsiang, Y. Use of US croplands for biofuels increases greenhouse gases through emissions from land-use change. *Science* **2008**, *319*, 1238–1240.
6. Oliver, R.J.; Finch, J.W.; Taylor, G. Second generation bioenergy crops and climate change: A review of the effects of elevated atmospheric CO_2 and drought on water use and the applications for yield. *GCB Bioenerg.* **2009**, *1*, 97–114.
7. Somerville, C. Biofuels. *Curr. Biol.* **2007**, *17*, 115–119.
8. Yuan, J.S.; Tiller, K.H.; Al-Ahmad, H.; Stewart, N.R.; Stewart, C.N. Plants to power: Bioenergy to fuel the future. *Trends Plant Sci.* **2008**, *13*, 421–429.
9. Heaton, E.A.; Frank, G.; Dohleman, F.G.; Long, S.P. Meeting biofuel goals with less land: The potential of *Miscanthus*. *Glob. Chang. Biol.* **2008**, *14*, 2000–2014.
10. Henry, R.J. Evaluation of plant biomass resources available for replacement of fossil oil. *Plant Biotechnol. J.* **2010**, *8*, 288–293.
11. Tilman, D.; Socolow, R.; Foley, J.A.; Hill, J.; Larson, E.; Lynd, L.; Pacala, S.; Reilly, J.; Searchinger, T.; Somerville, C.; et al. Beneficial biofuels-the food, energy, and environment trilemma. *Science* **2009**, *325*, 270–271.
12. Galvão, V.C.; Schmid, M. Regulation of flowering by endogenous signals. In *Advance in Botanical Research*; Fornara, F., Ed.; Elsevier: Amsterdam, The Netherlands, 2014; Volume 72, pp. 63–102.
13. Joshi, C.P.; DiFazio, S.P.; Chittaranjan, K. *Genetics, Genomics and Breeding of Poplar*; CRC Press: Boca Raton, FL, USA, 2011.
14. Murphy, R.L.; Morishige, D.T.; Brady, J.A.; Rooney, W.L.; Yang, S.; Klein, P.E.; Mullet, J.E. *Ghd7* (*Ma6*) represses sorghum flowering in long days: *Ghd7* alleles enhance biomass accumulation and grain production. *Plant Genome***2014**, *7*, 1–10.
15. Nemoto, Y.; Nonoue, Y.; Yano, M.; Izawa, T. *Hd1*, a *CONSTANS* ortholog in rice, functions as an *Ehd1* repressor through interaction with monocot-specific CCT-domain protein Ghd7. *Plant J.* **2016**, *86*, 221–233.
16. Yang, S.; Weers, B.D.; Morishige, D.T.; Mullet, J.E. *CONSTANS* is a photoperiod regulated activator of flowering in sorghum. *BMC Plant Biol.* **2014**, *14*, 148.

17. Grabowski, P.P.; Evans, J.; Daum, C.; Deshpande, S.; Barry, K.W.; Kennedy, M.; Casler, M.D. Genome-wide associations with flowering time in switchgrass using exome-capture sequencing data. *New Phytol.* **2017**, *213*, 154–169.
18. Xu, B.; Sathitsuksanoh, N.; Tang, Y.; Udvardi, M.K.; Zhang, J.; Shen, Z.; Balota, M.; Harich, K.; Zhang, P.; Zhao, B. Overexpression of *AtLOV1* in switchgrass alters plant architecture, lignin content, and flowering time. *PLoS ONE* **2012**, *7*, e47399.
19. Colasanti, J.; Coneva, V. Mechanisms of floral induction in grasses: Something borrowed, something new. *Plant Physiol.* **2009**, *149*, 56–62.
20. Brachi, B.; Faure, N.; Horton, M.; Flahauw, E.; Vazquez, A.; Nordborg, M.; Roux, F. Linkage and association mapping of *Arabidopsis thaliana* flowering time in nature. *PLoS Genet.* **2010**, *6*, e1000940.
21. Hayama, R.; Coupland, G. Shedding light on the circadian clock and the photoperiodic control of flowering. *Curr. Opin. Plant Biol.* **2003**, *6*, 13–19.
22. Blázquez, M.A.; Piñeiro, M.; Valverde, F. Bases moleculares de la floración. Prensa científica. Temas 61. *Investig. Cienc.* **2011**, *416*, 29–36.
23. Gardner, F.P.; Pearce, B.; Mitchell, R. *Physiology of Crop Plants*, 1st ed.; Iowa State University Press: Ames, IA, USA, 1985.
24. Pearce, S.; Kippes, N.; Chen, A.; Debernardi, J.M.; Dubcovsky, J. RNA-seq studies using wheat *PHYTOCHROME B* and *PHYTOCHROME C* mutants reveal shared and specific functions in the regulation of flowering and shade-avoidance pathways. *BMC Plant Biol.* **2016**, *16*, 141.
25. Coupland, G. Regulation of flowering time: Arabidopsis as a model system to study genes that promote or delay flowering. *Philos. Trans. R Soc. Lond. B Biol. Sci.* **1995**, *350*, 27–34.
26. Simpson, G.G.; Dean, C. Arabidopsis, the Rosetta stone of flowering time? *Science* **2002**, *296*, 285–289.
27. Valentin, F.L.; Van Mourik, S.; Posé, D.; Kim, M.C.; Schmid, M.; Van Ham, R.; Busscher, M.; Sanchez-Perez, G.F.; Molenaar, J.; Angenent, G.C.; et al. A quantitative and dynamic model of the Arabidopsis flowering time gene regulatory network. *PLoS ONE* **2015**, *10*, e0116973.
28. Peña, N. Caracterización de la Variación Natural de *Hd3a* y *RFT1* en Cultivares de Tipo Japónica de *Oryza sativa*. Máster's Thesis in Molecular and Cellular Biotechnology of Plants, The Technical University of Valencia, Valencia, Spain, 2013.
29. Mauro-Herrera, M.; Wang, X.; Barbier, H.; Brutnell, T.P.; Devos, K.M.; Doust, A.N. Genetic control and comparative genomic analysis of flowering time in *Setaria* (Poaceae). *G3 (Bethesda)* **2013**, *3*, 283–295.
30. Wada, M.; Shimazaki, K.; Lino, M. *Light Sensing in Plants*; Springer Science & Business Media: New York, NY, USA, 2005; pp. 333–346.
31. Mizoguchi, T.; Wright, L.; Fujiwara, S.; Cremer, F.; Lee, K.; Onouchi, H.; Mouradov, A.; Fowler, S.; Kamada, H.; Putterill, J.; et al. Distinct roles of *GIGANTEA* in promoting flowering and regulating circadian rhythms in *Arabidopsis*. *Plant Cell* **2005**, *17*, 2255–2270.
32. David, K.M.; Armbruster, U.; Tama, N.; Putterill, J. *Arabidopsis* GIGANTEA protein is post-transcriptionally regulated by light and dark. *FEBS Lett.* **2006**, *580*, 1193–1197.
33. Max Planck Institute for Plant Breeding Research. Control of Flowering Time. Available online: http://www.mpipz.mpg.de/coupland (accessed on 8 September 2016).
34. Valverde, F. *CONSTANS* and the evolutionary origin of photoperiodic timing of flowering. *J. Exp. Bot.* **2011**, *62*, 2453–2463.
35. Wang, C.C.; Chang, P.; Ng, K.; Chang, C.; Sheu, P.; Tsai, J. A model comparison study of the flowering time regulatory network in *Arabidopsis*. *BMC Syst. Biol.* **2014**, *8*, 15.
36. Lee, Y.S.; An, G. *OsGI* controls flowering time by modulating rhythmic flowering time regulators preferentially under short day in rice. *J. Plant Biol.* **2015**, *58*, 137–145.
37. Takahashi, Y.; Shomura, A.; Sasaki, T.; Yano, M. Hd6, a rice quantitative trait locus involved in photoperiod sensitivity, encodes the subunit of protein kinase CK2. *Proc. Natl. Acad. Sci. USA* **2001**, *98*, 7922–7927.
38. Nagano, H.; Clark, L.V.; Zhao, H.; Peng, J.; Yoo, J.H.; Heo, K.; Yu, C.Y.; Anzoua, K.G.; Matsuo, T.; Sacks, E.; et al. Contrasting allelic distribution of *CO/Hd1* homologues in *Miscanthus*

sinensis from the East Asian mainland and the Japanese archipelago. *J. Exp. Bot.* **2015**, *66*, 4227–4237.
39. Sun, C.; Chen, D.; Fang, J.; Wang, P.; Deng, X.; Chu, C. Understanding the genetic and epigenetic architecture in complex network of rice flowering pathways. *Protein Cell* **2014**, *5*, 889–898.
40. Doi, K.; Izawa, T.; Fuse, T.; Yamanouchi, U.; Kubo, T.; Shimatani, Z.; Yano, M.; Yoshimura, A. *Ehd1*, a B-type response regulator in rice, confers short-day promotion of flowering and controls *FT-like* gene expression independently of *Hd1*. *Genes Dev.* **2004**, *18*, 926–936.
41. Song, Y.H.; Ito, S.; Imaizumi, T. Similarities in the circadian clock and photoperiodism in plants. *Curr. Opin. Plant Biol.* **2010**, *13*, 594–603.
42. Itoh, H.; Nonoue, Y.; Yano, M.; Izawa, T. A pair of floral regulators sets critical day length for *Hd3a* florigen expression in rice. *Nat. Genet.* **2010**, *42*, 635–638.
43. Tsuji, H.; Taoka, K.; Shimamoto, K. Regulation of flowering in rice: Two florigen genes, a complex gene network, and natural variation. *Curr. Opin. Plant Biol.* **2011**, *14*, 45–52.
44. Tsuji, H.; Taoka, K. Florigen signaling. In *Signaling Pathways in Plants*; The Enzymes, Machida, Y., Lin, C., Tamanoi, F., Eds.; Elsevier: Amsterdam, The Netherlands, 2015; Volume 35, pp. 130–135.
45. Xue, W.; Xing, Y.; Weng, X.; Zhao, Y.; Tang, W.; Wang, L.; Zhou, H.; Yu, S.; Xu, C.; Li, X.; et al. Natural variation in *Ghd7* is an important regulator of heading date and yield potential in rice. *Nat. Genet.* **2008**, *40*, 761–767.
46. Ma, X.F.; Jensen, E.; Alexandrov, N.; Troukhan, M.; Zhang, L.; Thomas-Jones, S.; Farrar, K.; Clifton-Brown, J.; Donnison, I.; Swaller, T.; et al. High resolution genetic mapping by genome sequencing reveals genome duplication and tetraploid genetic structure of the diploid *Miscanthus sinensis*. *PLoS ONE* **2012**, *7*, e33821.
47. Fe, D.; Cericola, F.; Byrne, S.; Lenk, I.; Asraf, B.H.; Pedersen, M.G.; Roulund, N.; Asp, T.; Janss, L.; Jensen, C.S.; et al. Genomic dissection and prediction of heading date in perennial ryegrass. *BMC Genom.* **2015**, *16*, 921.
48. Skøt, L.; Sanderson, R.; Thomas, A.; Skøt, K.; Thorogood, D.; Latypova, G.; Asp, T.; Armstead, I. Allelic variation in the perennial ryegrass *FLOWERING LOCUS T* gene is associated with changes in flowering time across a range of populations. *Plant Physiol.* **2011**, 1013–1022.
49. Shinozuka, H.; Cogan, N.O.; Spangenberg, G.C.; Forster, J.W. Quantitative Trait Locus (QTL) meta-analysis and comparative genomics for candidate gene prediction in perennial ryegrass (*Lolium perenne* L.). *BMC Genet.* **2012**, *8*, 101.
50. Armstead, I.P.; Turner, L.B.; Marshall, A.H.; Humphreys, M.O.; King, I.P.; Thorogood, D. Identifying genetic components controlling fertility in the outcrossing grass species perennial ryegrass (*Lolium perenne*) by quantitative trait loci analysis and comparative genetics. *New Phytol.* **2008**, *178*, 559–571.
51. Yamada, T.; Jones, E.S.; Cogan, N.O.I.; Vecchies, A.C.; Nomura, T.; Hisano, H.; Shimamoto, Y.; Smith, K.F.; Hayward, M.D.; Forster, J.W. QTL analysis of morphological, developmental, and winter hardiness associated traits in perennial ryegrass. *Crop Sci.* **2004**, *44*, 925–935.
52. Yang, S.; Murphy, R.L.; Morishige, D.T.; Klein, P.E.; Rooney, W.L.; Mullet, J.E. Sorghum phytochrome B inhibits flowering in long days by activating expression of *SbPRR37* and *SbGHD7*, repressors of *SbEHD1*, *SbCN8* and *SbCN12*. *PLoS ONE* **2014**, *9*, e105352.
53. El Mannai, Y.; Shehzad, T.; Okuno, K. Variation in flowering time in sorghum core collection and mapping of QTLs controlling flowering time by association analysis. *Genet. Resour. Crop Evol.* **2011**, *58*, 983–989.
54. Cuevas, H.E.; Zhou, C.; Tang, H.; Khadke, P.P.; Das, S.; Lin, Y.; Ge, Z.; Clemente, T.; Upadhyaya, H.D.; Hash, T.C.; et al. The evolution of photoperiod-insensitive flowering in sorghum, a genomic model for panicoid grasses. *Mol. Biol. Evol.* **2016**.
55. Atienza, S.G.; Ramirez, M.C.; Martin, A. Mapping QTLs controlling flowering date in *Miscanthus sinensis* Anderss. *Cereal Res. Commun.* **2003**, *31*, 265–271.
56. Hajdu, A.; Ádám, É.; Sheerin, D.J.; Dobos, O.; Bernula, P.; Hiltbrunner, A.; Kozma-Bognar, L.; Nagy, F. High-level expression and phosphorylation of phytochrome B modulates flowering time in *Arabidopsis*. *Plant J.* **2015**, *83*, 794–805.

57. Izawa, T.; Mihara, M.; Suzuki, Y.; Gupta, M.; Itoh, H.; Nagano, A.J.; Motoyama, R.; Sawada, Y.; Yano, M.; Hirai, M.Y.; et al. *Os-GIGANTEA* confers robust diurnal rhythms on the global transcriptome of rice in the field. *Plant Cell* **2011**, *23*, 1741–1755.
58. Hori, K.; Nonoue, Y.; Ono, N.; Shibaya, T.; Ebana, K.; Matsubara, K.; Ogiso-Tanaka, E.; Tanabata, T.; Sugimoto, K.; Taguchi-Shiobara, F.; et al. Genetic architecture of variation in heading date among Asian rice accessions. *BMC Plant Biol.* **2015**, *15*, 115.
59. Ishikawa, R.; Aoki, M.; Kurotani, K.; Yokoi, S.; Shinomura, T.; Takano, M.; Shimamoto, K. Phytochrome B regulates *Heading date 1* (*Hd1*)-mediated expression of rice florigen *Hd3a* and critical day length in rice. *Mol. Genet. Genom.* **2011**, *285*, 461–470.
60. Ishikawa, R.; Tamaki, S.; Yokoi, S.; Inagaki, N.; Shinomura, T.; Takano, M.; Shimamoto, K. Suppression of the floral activator *Hd3a* is the principal cause of the night break effect in rice. *Plant Cell* **2015**, *17*, 3326–3336.
61. Jarillo, J.A.; Del Olmo, I.; Gómez-Zambrano, A.; Lázaro, A.; López-González, L.; Miguel, E.; Narro-Diego, L.; Sáez, D.; Piñeiro, M. Photoperiodic control of flowering time. *Span. J. Agric. Res.* **2008**, *6*, 221–244.
62. Morishige, D.T.; Childs, K.L.; Moore, L.D.; Mullet, J.E. Targeted analysis of orthologous *phytochrome A* regions of the sorghum, maize, and rice genomes using gene-island sequencing. *Plant Physiol.* **2002**, *130*, 1614–1625.
63. Mouradov, A.; Cremer, F.; Coupland, G. Control of flowering time: Interacting pathways as a basis for diversity. *Plant Cell* **2002**, S111–S130.
64. Naranjo, L.; Talon, M.; Domingo, C. Diversity of floral regulatory genes of japonica rice. *BMC Genom.* **2014**, *15*, 101.
65. Ogiso, E.; Takahashi, Y.; Sasaki, T.; Yano, M.; Izawa, T. The role of casein kinase II in flowering time regulation has diversified during evolution. *Plant Physiol.* **2010**, *152*, 808–820.
66. Shrestha, R.; Gómez-Ariza, J.; Brambilla, V.; Fornara, F. Molecular control of seasonal flowering in rice, *arabidopsis* and temperate cereals. *Ann. Bot.* **2014**, *114*, 1445–1458.
67. Chen, A.; Li, C.; Hu, W.; Lau, M.Y.; Lin, H.; Rockwell, N.C.; Martin, S.S.; Jernstedt, J.A.; Lagarias, J.C.; Dubcovsky, J. PHYTOCHROME C plays a major role in the acceleration of wheat flowering under long-day photoperiod. *Proc. Natl. Acad. Sci. USA* **2014**, *111*, 10037–10044.
68. Swaminathan, K.; Chae, W.B.; Mitros, T.; Varala, K.; Xie, L.; Barling, A.; Glowacka, K.; Hall, M.; Jezowski, S.; Ming, R.; et al. A framework genetic map for *Miscanthus sinensis* from RNAseq-based markers shows recent tetraploidy. *BMC Genom.* **2012**, *13*, 142.

CHAPTER 13

The Role of Specialized Photoreceptors in the Protection of Energy-Rich Tissues

Ian T. Major [1,†], Marcelo L. Campos [2,†] and Javier E. Moreno [3,*]

[1] *Michigan State University, 220 Trowbridge Rd, East Lansing, MI 48824, USA*

[2] *Departamento de Botânica, Instituto de Ciências Biológicas, Universidade de Brasília, Brasília, Distrito Federal 70910-900, Brazil*

[3] *Instituto de Agrobiotecnología del Litoral, UNL-Conicet, Facultad de Bioquímica y Ciencias Biológicas, Santa Fe 3000, Argentina*

ABSTRACT

The perception and absorption of light by plants is a driving force in plant evolutionary history, as plants have evolved multiple photoreceptors to perceive different light attributes including duration, intensity, direction and quality. Plant photoreceptors interpret these signals from the light environment and mold plant architecture to maximize foliar light capture. As active sites of the production and accumulation of energy-rich products, leaves are targets of pests and pathogens, which have driven the selection of physiological processes to protect these energy-rich tissues. In the last ten years, several research groups have accumulated evidence showing that plant photoreceptors control specific molecular programs that define plant growth and immune processes. Here, we discuss recent knowledge addressing these roles in Arabidopsis and show that (1) plant immune responses affect energy acquisition and partitioning; (2) plant photoreceptors interpret the light environment and control growth and immune processes; and finally; (3) defense and light signaling pathways can be genetically manipulated to obtain plants able to grow and defend at the same time. This basic knowledge from Arabidopsis plants should lead new lines of applied research in crops.

Keywords: light environment; phytochromes; defense; jasmonates; JAZ; photosynthesis; biomass; immune suppression; genetic rewiring

1. INTRODUCTION

As primary producers of the food chain, plants face constant attack from pests and pathogens that feed on their energy-rich tissues, while competing for sunlight with neighboring plants. Consequently, plants evolved strategies to protect their energy-rich tissues from attackers by initiating defense responses and to outperform competitors by reconfiguring plant architecture for rapid, extensile growth. However, both altered architecture and investment in defenses restricts the generation of new leaf tissues. Thus, plant strategies for competitive production and protection of biomass have future costs associated with reduced capacity for energy capture. It is therefore crucial for plants to perceive and integrate environmental cues and properly execute growth and defense programs when conditions warrant competitive growth or immunity from pests and pathogens.

Plant growth and health relies on the ability to perceive, absorb and transform solar radiation into photosynthates. Most of the absorbed light is funneled through chlorophyll and accessory pigments, but a reduced proportion of the light spectra is reflected and transmitted by green tissues. Plants perceive different attributes of the light environment using specialized photoreceptors, including phytochromes, cryptochromes, phototropins, and the ultraviolet-B photoreceptor. These photoreceptors, in turn, regulate morphological programs that shape plant architecture and defense responses to ensure efficient light absorption and tissue protection that maximize biomass production. In this review, we discuss the role of these specialized photoreceptors as key modulators of physiological responses that ultimately drive plant fitness through regulation of growth and defense strategies. We highlight recent evidence indicating that some photoreceptors are a fundamental component of a complex signaling hub that integrates growth and defense transcriptional modules to maximize fitness. In this sense, we present a model where light signaling cascades can be rewired to achieve robust growth concomitantly with strong defense.

2. SPECIALIZED PHOTORECEPTORS PERCEIVE THE LIGHT ENVIRONMENT

Light is the main source of external energy for plants and is also an important source of information. Using specialized photoreceptors, plants perceive the duration, quality, direction and intensity of sunlight, enabling more efficient "foraging for light". In this respect, neighboring plants are competitors for light and trigger an adaptive response to escape shade commonly referred to as the shade avoidance syndrome (SAS) [1]. This morphological response is characterized by elongation of hypocotyls, stems, and leaf petioles, as well as upward bending of leaves (hyponasty), reduced branching, and early flowering upon extended shading. SAS also includes reduced biomass, decreased leaf number, thinner leaves, lower leaf mass per area, and reduced chlorophyll and photosynthetic rates [2]. Overall, SAS expression reconfigures plant architecture to maximize future light absorption.

Chlorophyll-containing tissues strongly absorb photosynthetically active radiation (PAR) in the 400–700 nm range, preferentially in the blue (B, 480 nm) and red (R, 660 nm) bands. This causes a significant drop in the total amount of photons, as well as a change in light quality. Light reflected and transmitted by leaves is characterized by reduced PAR intensity, no ultraviolet-B (UV-B, 280–315 nm) photons and enrichment of far-red (FR, 730 nm). In crowded stands or under a plant canopy, these changes in light attributes elicit SAS. Interestingly, several of the photoreceptors involved with SAS expression are also associated with the protection of plant biomass through the modulation of defense responses. Here, we will focus on those photoreceptors that have been shown to modulate the plant immune system and consequently the protection of plant biomass: (1) UV-B absorbing UVR8 (UV RESISTANCE LOCUS 8); (2) B-absorbing cryptochromes (CRY); and (3) R- and FR-absorbing phytochromes (PHY).

Solar UV-B radiation is perceived by the photoreceptor UVR8 [3,4]. Upon UV-B perception, inactive dimers of UVR8 proteins undergo a conformational change, releasing monomers that accumulate in the nucleus. The nuclear-localized UVR8 monomers form a complex with and stabilize the E3 ubiquitin-ligase COP1 (CONSTITUTIVELY PHOTOMORPHOGENIC 1) [5], which induces expression of *HY5* (ELONGATED HYPOCOTYL 5), a master transcription factor for several UV-B responsive genes required for seedling UV-B acclimation. Attenuation of solar UV-B radiation can significantly promote leaf and stem elongation and reduce the load of DNA damage [5,6,7].

There is an important drop of B photons in crowded stands or under a plant canopy that elicits SAS [8,9,10], which is promoted by the cryptochromes

CRY1 and *CRY2*. In the case of low B, the transcription factors *PIF4* and *PIF5* (PHYTOCHROME INTERACTING FACTOR 4 and 5) are required for a full SAS response, since the single *pif4* and *pif5* mutants have a reduced expression of SAS while the *pif4 pif5* double mutant is unresponsive to B attenuation [8]. This photosensory pathway was further delineated by the recent finding that *CRY1/2* physically interacts with *PIF4/5* and that *CRY2* shares similar DNA-regulatory target regions as *PIF4/5* [11]. It thus appears that the *CRYs* modulate *PIF4/5* activity to promote SAS under low B [11].

The phytochrome photoreceptors oscillate between an active and inactive state depending on the R:FR ratio of the light environment [2]. Even subtle changes in the R:FR ratio by neighboring, non-shading plants changes the state of the phytochrome pool and triggers a SAS response, a feature that allows plants to detect the presence of future competitors. Phytochrome B (phyB) is the main photoreceptor of shading from neighboring competitors, as demonstrated by the strong and constitutive SAS phenotype of the single *phyB* mutant. phyB inactivation leads to enhanced stability of *PIF3* and *PIF4* [12], triggering cell elongation responses through the induction of genes involved in AUX [13,14] and GA synthesis [15], and of cell wall degrading enzymes [16,17]. The Arabidopsis genome encodes five phytochrome genes: PHYA, B, C, D and E. Recent studies are shedding light on the relative contribution of each phytochrome to the different developmental and physiological responses in Arabidopsis [18,19].

3. THE JASMONATE AND SALICYLATE PATHWAYS DRIVE INDUCIBLE DEFENSES

Plants have an important diversity of defense responses, many of which are activated only after stress. These inducible defenses minimize the costs of synthesizing defensive metabolites and limit detrimental effects on plant health. In general, plant inducible defenses against chewing insects and necrotrophic pathogens are controlled by the lipid-derived hormone jasmonic acid (JA), whereas the shikimate-derived hormone salicylic acid (SA) controls inducible defenses against sucking arthropods and biotrophic pathogens [20]. In both cases, hormone synthesis is initiated by perception of specific elicitors associated with attack. These "danger" signals include plant-derived signals produced by cellular damage (i.e., damaged self), so-called damage-associated molecular patterns (DAMPs), or specific elicitors produced by foreign organisms, known as herbivore-associated molecular patterns (HAMPs) from feeding arthropod herbivores, and as microbe-associated molecular patterns (MAMPs) from invading microbes [21]. In an example of co-evolution between host and attacker, herbivores and microbes secrete effectors that target host signaling to neutralize defenses, while plants deploy receptors that recognize specific effectors to elicit immunity.

Studies over the past decade have dramatically improved our understanding of the molecular controls of plant defense responses against pests and pathogens. Immune responses are maintained in a repressed "off" state in healthy plants by key transcriptional regulators, the JAZ (JASMONATE ZIM-DOMAIN) and NPR (NONEXPRESSOR OF PR GENES) proteins. These regulators perceive elevated JA or SA levels in plants challenged by pests or pathogens and activate appropriate responses. For JA signaling, JA promotes direct binding of JAZ proteins with the F-Box protein COI1 (CORONATINE INSENSITIVE 1), and this interaction targets the JAZs for degradation by the proteasome [22,23,24,25,26]. The *Arabidopsis thaliana* genome encodes 13 JAZ proteins with overlapping functions. Functional specificity of individual JAZ proteins is conferred by the diversity of their conserved domains, which is expanded by alternative splicing [27,28,29]. Differences in the conserved domains of JAZs influence their interaction with other proteins, including altered affinity for COI1 that determines JAZ degradation rate. Together, these differences fine-tune JA signaling and preference for target transcription factors that defines immune and developmental responses [28,30]. For SA signaling, NPR proteins may perceive SA as a multireceptor complex, where SA sensitivity would depend on the NPR protein in the receptor [31]. SA binding converts NPR1 into a transcriptional activator that initiates immune responses [32], while NPR3 and NPR4 proteins integrate SA levels by binding SA with different affinities to regulate turnover of NPR1 by the proteasome [33]. A high turnover of NPR1 is critical to plant immune responses [34] and also relies on other protein modifications including phosphorylation and SUMOylation [35].

Numerous laboratory and field experiments using different plant crops have shown a conserved antagonism between JA- and SA-defense responses, whereby the activation of JA-related responses is associated with suppression of SA-related responses and vice-versa [36]. For example, insect-damaged or JA-induced plants were more susceptible to infection by pathogens that trigger SA-dependent defense responses [37,38]. Similarly, Arabidopsis plants infected with the biotrophic pathogen *Pseudomonas syringae*, which elicits SA responses, were more susceptible to the necrotrophic pathogen *Alternaria brassicicola* by suppression of JA responses [39]. Interestingly, attackers have evolved to manipulate JA-SA antagonism; several biotrophic pathogens deploy effectors that hijack JA signaling to suppress SA-mediated immunity, while insects have strategies that hijack SA signaling to interfere with JA defenses [40]. Several points of crosstalk have been reported that contribute to JA-SA antagonism [41], including recent reports showing that NPR3 and NPR4 interact directly with several JAZ proteins to trigger JAZ degradation in a COI1-independent manner [42] and that JA activates a signaling cascade that inhibits SA accumulation [43]. It is becoming evident that SA-JA antagonism allows plants to tailor immune responses against specific attackers [36], which may mitigate the costs associated with immune responses.

4. PLANT IMMUNITY CONSTRAINS GROWTH AND DEVELOPMENT

Activation of defense processes can have negative consequences on plant growth and development [44], including reduced growth of shoots and roots, delayed development, and ultimately reduced biomass. The clearest examples of these trade-offs are constitutive defense mutants in which growth is severely hindered [45,46,47,48,49,50]. Photosynthesis is often reported to be decreased during defense responses, which would reduce growth and biomass if resources are limiting [51,52]. There is good evidence that wounding actively suppresses photosynthesis, as the decrease in photosynthesis is greater than the amount of tissue damaged and requires intact JA signaling [52,53]. Recently, the key defense regulator *MYC2* was shown to directly trigger expression of genes involved in chlorophyll degradation, suggesting a mechanism of JA-mediated suppression of photosynthesis [54,55]. However, other evidence questions whether photosynthesis is genuinely suppressed during defense responses. Chlorophyll degradation itself may be a defensive strategy, as degradation products have anti-insect activity [56]. Other studies have shown that photosynthetic rates are not perturbed when defense responses are activated in the absence of physical damage, suggesting that some reports of decreased photosynthesis could be a consequence of physical damage (e.g., localized water loss from disrupted cells) [48,49,57]. Importantly, active defense responses trigger reduced growth without a corresponding reduction in photosynthesis, demonstrating that decreased photosynthesis is not the main cause of reduced growth during defense responses [48,49].

Decreased biomass is also common during defense responses [58], and may be ascribed to tissue loss (i.e., damage or removal) and suppression of plant growth. However, the partitioning of resources to defense may play an important role, since the relationship between leaf area and leaf mass is determined by the partitioning of carbon among different physiological processes [59]. Interestingly, leaves treated with JA showed a delayed reduction in biomass relative to leaf area and these leaves had a higher ratio of leaf dry mass to area [60]. These data are consistent with a hypothesis that activation of defense responses prioritizes investment of resources in leaf defense and structural integrity, representing a strategy to protect energy-rich tissues. However, reduced investment in additional photosynthetic tissues is a lost opportunity, the consequence of which is compounding future losses in whole-plant photosynthesis and growth [60].

Several studies are now revealing that the processes to invest in growth and defense are determined by a complex signaling network, with cross-talk between hormone and light signaling pathways. Several pathogens perturb these pathways by manipulating growth-promoting levels of the hormone auxin (AUX) as a strategy to promote disease [61]. Correspondingly, plant immune responses stabilize the AUX/IAA repressor proteins to suppress auxin signaling [62]. Growth–defense trade-offs are also determined by competitive binding of JAZ proteins with DELLA proteins, which are repressors of growth responses promoted by the gibberellin (GA) hormone [63,64]. Growth cues promote GA-dependent degradation of DELLAs to allow transcription factors such as PIFs to activate growth traits, while stress cues trigger JA-mediated degradation of JAZs to allow transcription factors such as MYCs to activate defense responses. Interestingly, recent data suggests that the MYC transcription factors, which are associated primarily with defense processes, may also be repressors of growth themselves. *MYC2* negatively regulates root growth via the *PLT1* and *PLT2* transcription factors [65], while gain-of-function mutations of MYC2 (*myc2-322B*) and MYC3 (*atr2D*), which have reduced interaction with JAZ repressors, exhibit reduced shoot growth [66,67,68].

5. SPECIALIZED PHOTORECEPTORS INFORM GROWTH–DEFENSE TRADE-OFFS

Increasing evidence supports a positive correlation between higher crop densities and the occurrence of pests [69]. These observations in agricultural and ecological systems led to the proposed “dilemma of plants”; that is whether to invest limited resources for competitive growth with neighboring plants or for robust defense against insects and pathogens [70]. The high susceptibility of densely-cropped plants is partly related to changes in the light environment. Crowded stands combine lower PAR, less UV-B radiation, less B, and a lower R:FR ratio. In recent years, the use of Arabidopsis photoreceptor mutants and light-manipulation experiments has generated important evidence showing that these light components contribute to the immune-suppressed state of plants growing in crowded stands.

UV-B and inducible defenses: UV-B attenuation experiments in the field have clearly shown that natural herbivory is more severe on plants growing with attenuated levels of UV-B compared to those exposed to full sunlight

[7,71,72,73]. Most of the plants increase the accumulation of soluble phenolic compounds in green tissues in response to UV-B, where some of these compounds were shown to have a direct role on insect herbivore performance [51,74,75]. In addition, UV-B altered *Plutella xylostella* oviposition behavior in a JA-dependent manner, since adult moths laid more eggs on plants growing under attenuated UV-B [72]. Conversely, plants supplemented with UV-B experienced less herbivory and disease symptoms than control plants [51,76]. Resistance to *Botrytis cinerea* infection was significantly increased in plants grown with supplemental UV-B, an effect that required the UVR8 photoreceptor but not JA signaling [76]. This UV-B effect on *B. cinerea* resistance was mediated by the accumulation of sinapate in Arabidopsis [76].

Blue light and inducible defenses: B depletion does not appear to affect immunity. Arabidopsis plants expressing SAS, either due to B depletion or mutation of the *CRY1* photoreceptor, were not more susceptible to *B. cinerea* infection [77]. Moreover, *cry1* mutants showed similar resistance to infection by *P. syringae* [78,79]. However, the *cry1* mutant has also been reported more susceptible to *P. syringae* [80]. B may also indirectly influence defense responses since B controls stomatal openings, which are the main points of entry for pathogens [81,82]. B signals also synchronize the circadian clock, which regulates plant growth and defense processes. Immune responses against both herbivores and pathogens are 'gated' by the plant circadian clock such that defense is strongest in the morning when biotic attack is more likely [83,84,85]. Interestingly, *NPR1* also contributes to synchronizing the circadian clock, allowing SA to reinforce the circadian clock and gate immune responses [86]. Altogether, B signals contribute to circadian gating of immunity, which allows plants to reinforce morning defenses in 'anticipation' of challenge from herbivores and pathogens, while also minimizing the negative side effects on growth responses that take place at night.

R:FR ratio and inducible defenses: Plants expressing SAS, from either high-density planting or supplemental FR, displayed weak defense phenotypes against insect herbivores, necrotrophic fungus, and pathogenic bacteria [77,87,88]. The levels of defense metabolites, including phenolics, aliphatic glucosinolates, and camalexin, were lower in SAS-expressing plants upon *B. cinerea* infection [77,87,89]. In addition, growth in FR-rich environments suppressed both JA- and SA-dependent expression of defense genes [88]. Together, these results demonstrated that phytochrome inactivation by FR reduces JA and SA sensitivity. phyB is the principle phytochrome receptor responsible for FR-mediated perturbation of immune responses, as *phyB* mutants are less sensitive to JA and SA, and are more susceptible to insect herbivory and to infection by *B. cinerea* and *P. syringae* [77,87,88,90]. The molecular mechanisms behind these growth–defense trade-offs include multiple layers of regulation by FR: (1) JA-mediated degradation of JAZ10 is significantly reduced in seedlings exposed to FR radiation and in the *phyB* mutant background [91]; (2) FR destabilizes MYC2, MYC3 and MYC4 proteins, key transcription factors of JA responses, in a phyB-dependent manner [92]; (3) FR triggers turnover of DELLA proteins [91], which releases repression of PIFs to activate growth processes, as well as the release of JAZ proteins from DELLA–JAZ interactions to enhance repression of JA responses (see Section 3) [62,63]; and (4) FR interferes with NPR1 phosphorylation, contributing to reduced NPR1-mediated immune responses [88].

6. GROWTH–DEFENSE DECISIONS ARE CONTROLLED BY THE PHYTOCHROME B SIGNALING PATHWAY

Several studies are now revealing that the classic and simple view that growth and defense processes are limited by resource constraints is too simplistic; instead, growth and defense are determined by signaling programs. For example, studies have shown that JA-mediated growth repression involves reduced cell division and expansion by perturbation of the cell cycle [58,93,94]. Moreover, Arabidopsis mutants that no longer produced glucosinolates (the main defense compound in brassica plants) still exhibited growth suppression, demonstrating that growth and defense are not strictly coupled genetically [50,95]. Similarly, the "decision" to prioritize growth over defense during SAS appears to be a specific physiological response mediated by phyB, rather than the limitation of plant energy constraints [87,91]. As such, several studies are revealing that SAS-mediated growth promotion and defense suppression can be processes that are uncoupled in various mutants. For example, the *cpr1* and *cev1* mutants, which have constitutively active SA- and JA-dependent defense responses, respectively, both express partial SAS in response to FR, suggesting that these mutants can direct resources to growth and defense [88]. In the *sav3-2* mutant, FR exposure does not trigger expression of SAS but does suppress immune responses, demonstrating that the morphological investment of energy during SAS expression cannot per se explain the increased susceptibility to insects or *B. cinerea* [77,87]. Defense suppression is also uncoupled from SAS in the *jaz10 phyB* double mutant, which maintains constitutive SAS but with wild type JA sensitivity [91]. These points of cross-talk among growth and defense signaling pathways are establishing how complex networks of regulators govern these processes to properly respond to external conditions [49,95,96]. An emerging question is whether we can use the knowledge of these networks to manipulate growth–defense trade-offs (see Section 7 below).

7. MANIPULATING REPRESSORS OF DEFENSE AND LIGHT SIGNALING FOR IMPROVED PROTECTION AND GROWTH OF ENERGY-RICH TISSUES

The molecular basis of how the JA signaling pathway integrates growth and defense remained obscured by the genetic redundancy among the JAZ transcriptional repressors. As a consequence, the majority of single *jaz* mutants showed no clear phenotype [25]. In an effort to overcome this situation, Campos et al. (2016) developed a higher-order mutant that lacks five of the thirteen JAZ repressors present in the Arabidopsis genome. Removal of multiple JAZ genes in this *jaz quintuple* (*jazQ*) mutant led to constitutive activation of JA responses, causing hypersensitivity to exogenous JA treatment, upregulation of defense-related genes, increased production of secondary metabolites and higher resistance to insect herbivory attack [49]. On the other hand, rosette expansion was restricted in this higher-order mutant, as *jazQ* exhibited stunted rosettes and overall delayed development when compared to wild type plants (Figure 1). In addition, *jazQ* was ~50% lighter than wild type in terms of rosette dry weight, indicating strong down-regulation of plant biomass accumulation. Taken together, these results support the premise that JA modulates a transcriptional route whose activation (through removal of multiple JAZ transcriptional repressors) culminates in strong promotion of defense along with a reduction in growth-related processes (see Section 3).

The evident shift observed in *jazQ* growth–defense parameters can be explained in light of its genetic lesions. The five JAZ impaired in *jazQ* (JAZ1, JAZ3, JAZ4, JAZ9 and JAZ10) were shown to physically interact with DELLA proteins [64]. The concept that plant hormones integrate a regulatory network that determines growth–defense outputs was further confirmed by Campos et al. (2016), who genetically rewired crosstalk among defense and growth signaling pathways to obtain an Arabidopsis genotype where growth and defense are transcriptionally and physiologically uncoupled. The authors showed that the combination of the "more defenses, less growth" *jazQ* with the "more growth, less defense" *phyB* genotypes produced a higher-order mutant (named *jazQ phyB*) where growth and defense processes were simultaneously activated (Figure 1). These results are consistent with, and greatly expand upon, previous work showing that suppression of defenses is uncoupled from shade avoidance traits in the *jaz10 phyB* double mutant [91,97]. The *jazQ phyB* plants showed increased

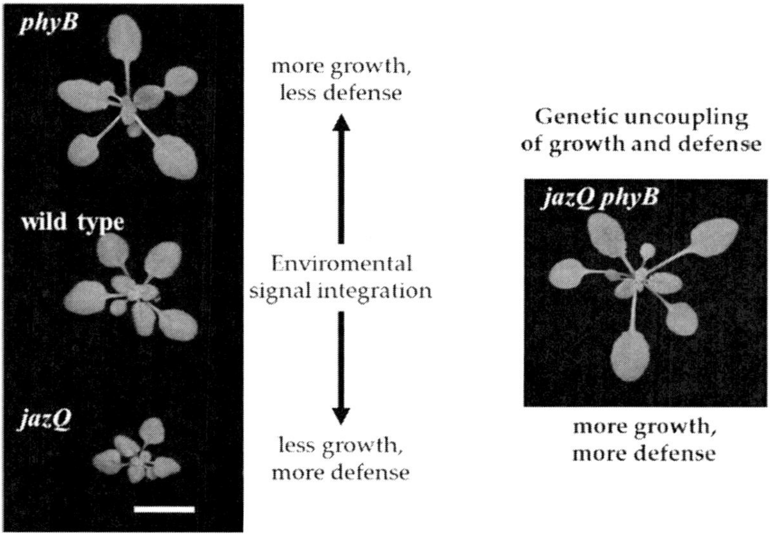

Figure 1. Genetic uncoupling of growth and defense in Arabidopsis. The *jazQ phyB* mutant combines the large rosette of *phyB* with elevated defenses of *jazQ*. Photograph of 3 weeks old wild type, *jazQ phyB*, *phyB* and *jazQ* rosettes. Plants were grown in soil and maintained at 20 °C (±1 °C) under 16 h at a light intensity of 120 μmol $m^{-2} \cdot s^{-1}$ and 8 h dark. Bar = 1 cm.

expression of defense-related genes, higher production of secondary metabolites and enhanced resistance against insect herbivory, as well as growth promotion and faster development, even when compared with wild type Arabidopsis plants [49]. Full transcriptome sequencing indicated that the *jazQ phyB* phenotype is the result of an additive combination of the immune program activated in *jazQ* and the growth program activated in *phyB*. Those results indicate that depletion of the JAZ and DELLA repressors caused by *jazQ* and *phyB* mutations respectively leads to concomitant de-repression of growth and defense transcriptional cascades in the *jazQ phyB* mutant [49,91]. Specific traits found in *jazQ phyB* were not observed in the *jazQ* or *phyB* mutants. For instance, the biomass of both *jazQ* and *phyB* was about 50% of that observed for wild type plants, yet the combination of these mutations restored plant biomass to similar levels as wild type plants. Projected leaf area and rosette diameter were also larger in *jazQ phyB* compared to the parental mutants, as well as wild type plants. At the transcriptional level, more than a thousand genes were differentially expressed in *jazQ phyB* but not in its parental mutants. Taken together, these results indicate, at one side, a strong modular effect of defense and growth transcriptional programs mainly dependent on MYC and DELLA transcriptional regulators and, at the other side, demonstrate that growth and photomorphogenic signaling pathways show a synergistic control of transcriptional changes to achieve a balanced physiological response to the dilemma of plants.

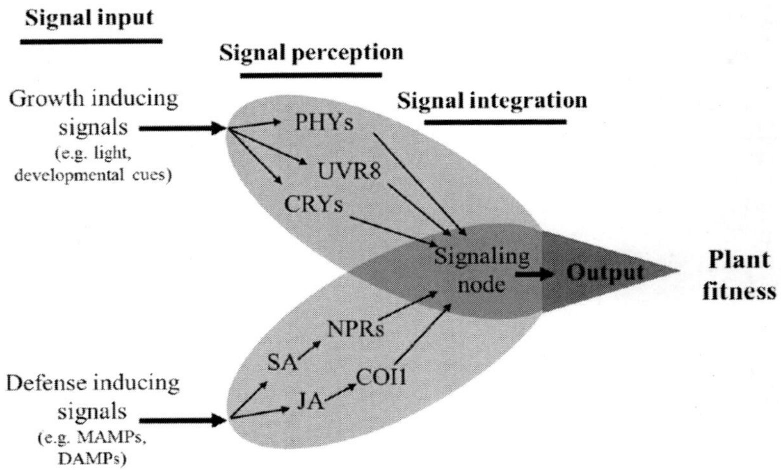

Figure 2. Environmental cues are perceived and integrated through light and defense signaling to appropriately drive plant growth and defense processes. Proposed model of the integration of growth and defense signals, whereby a signaling node integrates multiple signals perceived from the environment to define the plant phenotype (output), which will ultimately determine plant fitness. The nature of the signal elicits different receptors (signal perception) to trigger signaling cascades that are integrated by molecular players acting as nodes of integration. The final plant phenotype is determined by the degree of activation and repression of specific molecular pathways.

8. CONCLUSIONS

Recently, a wealth of studies have begun to unveil the molecular basis of the trade-offs between growth and defense, indicating the presence of a complex signaling hub where hormone and light signaling pathways converge to optimize fitness. In light of this recent evidence, we propose a model showing that growth and defense trade-offs are not just the result of metabolic constraints (such as resource limitation), but rather a consequence of a transcriptional network that evolved to maximize plant fitness (Figure 2) [49,95]. As such, growth and defense environmental cues are initially perceived by receptors, which in turn, activate transcriptional cascades responsible for the activation of specialized responses. A signaling node integrates these transcriptional programs to precisely control the elicitation of growth and defense traits in order to optimize plant performance. The proposed interaction of regulatory modules is prone to evolve in natural settings, where plants are constantly facing a blend of environmental signals that must be integrated and properly translated into growth and defense traits to maximize fitness. A relevant application of this premise is the use of genetic rewiring to achieve concomitant activation of growth and defense [49,91]. This strategy could lead, in a relative short time, to the development of new crop cultivars where high biomass production is combined with lower demand for pesticides.

ACKNOWLEDGMENTS

This work was supported by grants to J.E.M. Consejo Nacional de Investigaciones Científicas y Técnicas (PIP2014-2016 112201301 00267CO), Universidad Nacional del Litoral (50220140100011LI), and the Agencia Nacional de Promoción Científica y Tecnológica (PICT2013-3285). MLC is supported by a fellowship from the Conselho Nacional de Desenvolvimento Cientifico e Tecnológico (CNPq-Brazil).

REFERENCES

1. Ballaré, C.L.; Scopel, A.L.; Sanchez, R.A. Far-red radiation reflected from adjacent leaves: An early signal of competition in plant canopies. *Science* **1990**, *247*, 329–332.
2. Casal, J.J. Photoreceptor signaling networks in plant responses to shade. *Ann. Rev. Plant Biol.* **2013**, *64*, 403–427.
3. Rizzini, L.; Favory, J.J.; Cloix, C.; Faggionato, D.; O'Hara, A.; Kaiserli, E.; Baumeister, R.; Schafer, E.; Nagy, F.; Jenkins, G.I.; et al. Perception of UV-B by the Arabidopsis UVR8 protein. *Science* **2011**, *332*, 103–106.
4. Heijde, M.; Ulm, R. UV-B photoreceptor-mediated signalling in plants. *Trends Plant Sci.* **2012**, *17*, 230–237.
5. Favory, J.J.; Stec, A.; Gruber, H.; Rizzini, L.; Oravecz, A.; Funk, M.; Albert, A.; Cloix, C.; Jenkins, G.I.; Oakeley, E.J.; et al. Interaction of COP1 and UVR8 regulates UV-B-induced photomorphogenesis and stress acclimation in Arabidopsis. *EMBO J.* **2009**, *28*, 591–601.
6. Mazza, C.A.; Ballaré, C.L. Photoreceptors UVR8 and phytochrome B cooperate to optimize plant growth and defense in patchy canopies. *New Phytol.* **2015**, *207*, 4–9.
7. Mazza, C.A.; Gimenez, P.I.; Kantolic, A.G.; Ballaré, C.L. Beneficial effects of solar UV-B radiation on soybean yield mediated by reduced insect herbivory under field conditions. *Physiol. Plant.* **2013**, *147*, 307–315.
8. Keller, M.M.; Jaillais, Y.; Pedmale, U.V.; Moreno, J.E.; Chory, J.; Ballaré, C.L. Cryptochrome 1 and phytochrome B control shade-avoidance responses in Arabidopsis via partially independent hormonal cascades. *Plant J.* **2011**, *67*, 195–207.

9. Keuskamp, D.H.; Keller, M.M.; Ballaré, C.L.; Pierik, R. Blue light regulated shade avoidance. *Plant. Signal. Behav.***2012**, *7*, 514–517.
10. Keuskamp, D.H.; Sasidharan, R.; Vos, I.; Peeters, A.J.; Voesenek, L.A.; Pierik, R. Blue-light-mediated shade avoidance requires combined auxin and brassinosteroid action in Arabidopsis seedlings. *Plant J.* **2011**, *67*, 208–217.
11. Pedmale, U.V.; Huang, S.S.; Zander, M.; Cole, B.J.; Hetzel, J.; Ljung, K.; Reis, P.A.; Sridevi, P.; Nito, K.; Nery, J.R.; et al. Cryptochromes interact directly with PIFs to control plant growth in limiting blue light. *Cell* **2016**, *164*, 233–245.
12. De Lucas, M.; Daviere, J.M.; Rodriguez-Falcon, M.; Pontin, M.; Iglesias-Pedraz, J.M.; Lorrain, S.; Fankhauser, C.; Blazquez, M.A.; Titarenko, E.; Prat, S. A molecular framework for light and gibberellin control of cell elongation. *Nature* **2008**, *451*, 480–484.
13. Hornitschek, P.; Kohnen, M.V.; Lorrain, S.; Rougemont, J.; Ljung, K.; Lopez-Vidriero, I.; Franco-Zorrilla, J.M.; Solano, R.; Trevisan, M.; Pradervand, S.; et al. Phytochrome interacting factors 4 and 5 control seedling growth in changing light conditions by directly controlling auxin signaling. *Plant J.* **2012**, *71*, 699–711.
14. Tao, Y.; Ferrer, J.L.; Ljung, K.; Pojer, F.; Hong, F.; Long, J.A.; Li, L.; Moreno, J.E.; Bowman, M.E.; Ivans, L.J.; et al. Rapid synthesis of auxin via a new tryptophan-dependent pathway is required for shade avoidance in plants. *Cell***2008**, *133*, 164–176.
15. Feng, S.; Martinez, C.; Gusmaroli, G.; Wang, Y.; Zhou, J.; Wang, F.; Chen, L.; Yu, L.; Iglesias-Pedraz, J.M.; Kircher, S.; et al. Coordinated regulation of *Arabidopsis thaliana* development by light and gibberellins. *Nature* **2008**, *451*, 475–479.
16. Sasidharan, R.; Chinnappa, C.C.; Staal, M.; Elzenga, J.T.; Yokoyama, R.; Nishitani, K.; Voesenek, L.A.; Pierik, R. Light quality-mediated petiole elongation in Arabidopsis during shade avoidance involves cell wall modification by xyloglucan endotransglucosylase/hydrolases. *Plant Physiol.* **2010**, *154*, 978–990.
17. Li, L.; Ljung, K.; Breton, G.; Schmitz, R.J.; Pruneda-Paz, J.; Cowing-Zitron, C.; Cole, B.J.; Ivans, L.J.; Pedmale, U.V.; Jung, H.S.; et al. Linking photoreceptor excitation to changes in plant architecture. *Genes Dev.* **2012**, *26*, 785–790.
18. Sanchez-Lamas, M.; Lorenzo, C.D.; Cerdán, P.D. Bottom-up assembly of the phytochrome network. *PLoS Genet.* **2016**, *12*, e1006413.
19. Strasser, B.; Sanchez-Lamas, M.; Yanovsky, M.J.; Casal, J.J.; Cerdán, P.D. *Arabidopsis thaliana* life without phytochromes. *Proc. Natl. Acad. Sci. USA* **2010**, *107*, 4776–4781.
20. Pieterse, C.M.; van der Does, D.; Zamioudis, C.; Leon-Reyes, A.; Van Wees, S.C. Hormonal modulation of plant immunity. *Annu. Rev. Cell Dev. Biol.* **2012**, *28*, 489–521.
21. Campos, M.L.; Kang, J.H.; Howe, G.A. Jasmonate-triggered plant immunity. *J. Chem. Ecol.* **2014**, *40*, 657–675.
22. Katsir, L.; Schilmiller, A.L.; Staswick, P.E.; He, S.Y.; Howe, G.A. COI1 is a critical component of a receptor for jasmonate and the bacterial virulence factor coronatine. *Proc. Natl. Acad. Sci. USA* **2008**, *105*, 7100–7105.
23. Sheard, L.B.; Tan, X.; Mao, H.; Withers, J.; Ben-Nissan, G.; Hinds, T.R.; Kobayashi, Y.; Hsu, F.F.; Sharon, M.; Browse, J.; et al. Jasmonate perception by inositol-phosphate-potentiated COI1-JAZ co-receptor. *Nature* **2010**, *468*, 400–405.
24. Chini, A.; Fonseca, S.; Fernandez, G.; Adie, B.; Chico, J.M.; Lorenzo, O.; Garcia-Casado, G.; Lopez-Vidriero, I.; Lozano, F.M.; Ponce, M.R.; et al. The JAZ family of repressors is the missing link in jasmonate signalling. *Nature***2007**, *448*, 666–671.
25. Thines, B.; Katsir, L.; Melotto, M.; Niu, Y.; Mandaokar, A.; Liu, G.; Nomura, K.; He, S.Y.; Howe, G.A.; Browse, J. JAZ repressor proteins are targets of the SCF^{coi1} complex during jasmonate signalling. *Nature* **2007**, *448*, 661–665.
26. Yan, Y.; Stolz, S.; Chetelat, A.; Reymond, P.; Pagni, M.; Dubugnon, L.; Farmer, E.E. A downstream mediator in the growth repression limb of the jasmonate pathway. *Plant Cell* **2007**, *19*, 2470–2483.
27. Chung, H.S.; Cooke, T.F.; Depew, C.L.; Patel, L.C.; Ogawa, N.; Kobayashi, Y.; Howe, G.A. Alternative splicing expands the repertoire of dominant JAZ repressors of jasmonate signaling. *Plant J.* **2010**, *63*, 613–622.

28. Moreno, J.E.; Shyu, C.; Campos, M.L.; Patel, L.C.; Chung, H.S.; Yao, J.; He, S.Y.; Howe, G.A. Negative feedback control of jasmonate signaling by an alternative splice variant of JAZ10. *Plant Physiol.* **2013**, *162*, 1006–1017.
29. Chung, H.S.; Howe, G.A. A critical role for the TIFY motif in repression of jasmonate signaling by a stabilized splice variant of the JASMONATE ZIM-domain protein JAZ10 in Arabidopsis. *Plant Cell* **2009**, *21*, 131–145.
30. Chini, A.; Gimenez-Ibanez, S.; Goossens, A.; Solano, R. Redundancy and specificity in jasmonate signalling. *Curr. Opin. Plant Biol.* **2016**, *33*, 147–156.
31. Pajerowska-Mukhtar, K.M.; Emerine, D.K.; Mukhtar, M.S. Tell me more: Roles of NPRs in plant immunity. *Trends Plant Sci.* **2013**, *18*, 402–411.
32. Wu, Y.; Zhang, D.; Chu, J.Y.; Boyle, P.; Wang, Y.; Brindle, I.D.; de Luca, V.; Despres, C. The Arabidopsis NPR1 protein is a receptor for the plant defense hormone salicylic acid. *Cell Rep.* **2012**, *1*, 639–647.
33. Fu, Z.Q.; Yan, S.; Saleh, A.; Wang, W.; Ruble, J.; Oka, N.; Mohan, R.; Spoel, S.H.; Tada, Y.; Zheng, N.; et al. NPR3 and NPR4 are receptors for the immune signal salicylic acid in plants. *Nature* **2012**, *486*, 228–232.
34. Spoel, S.H.; Mou, Z.; Tada, Y.; Spivey, N.W.; Genschik, P.; Dong, X. Proteasome-mediated turnover of the transcription coactivator *NPR1* plays dual roles in regulating plant immunity. *Cell* **2009**, *137*, 860–872.
35. Saleh, A.; Withers, J.; Mohan, R.; Marques, J.; Gu, Y.; Yan, S.; Zavaliev, R.; Nomoto, M.; Tada, Y.; Dong, X. Posttranslational modifications of the master transcriptional regulator NPR1 enable dynamic but tight control of plant immune responses. *Cell Host Microbe* **2015**, *18*, 169–182.
36. Thaler, J.S.; Humphrey, P.T.; Whiteman, N.K. Evolution of jasmonate and salicylate signal crosstalk. *Trends Plant Sci.***2012**, *17*, 260–270.
37. Thaler, J.S.; Fidantsef, A.L.; Bostock, R.M. Antagonism between jasmonate- and salicylate-mediated induced plant resistance: Effects of concentration and timing of elicitation on defense-related proteins, herbivore, and pathogen performance in tomato. *J. Chem. Ecol.* **2002**, *28*, 1131–1159.
38. Cipollini, D. Costs and benefits of induced resistance to herbivores and pathogens in plants. *CAB Rev. Perspect. Agric. Vet. Sci. Nutr. Nat. Res.* **2010**, *5*.
39. Spoel, S.H.; Johnson, J.S.; Dong, X. Regulation of tradeoffs between plant defenses against pathogens with different lifestyles. *Proc. Natl. Acad. Sci. USA* **2007**, *104*, 18842–18847.
40. Zhang, L.; Zhang, F.; Melotto, M.; Yao, J.; He, S.Y. Jasmonate signaling and manipulation by pathogens and insects. *J. Exp. Bot.* **2017**.
41. Gimenez-Ibanez, S.; Solano, R. Nuclear jasmonate and salicylate signaling and crosstalk in defense against pathogens. *Front. Plant Sci.* **2013**, *4*, 72.
42. Liu, L.; Sonbol, F.M.; Huot, B.; Gu, Y.; Withers, J.; Mwimba, M.; Yao, J.; He, S.Y.; Dong, X. Salicylic acid receptors activate jasmonic acid signalling through a non-canonical pathway to promote effector-triggered immunity. *Nat. Commun.* **2016**, *7*, 13099.
43. Zheng, X.Y.; Spivey, N.W.; Zeng, W.; Liu, P.P.; Fu, Z.Q.; Klessig, D.F.; He, S.Y.; Dong, X. Coronatine promotes *Pseudomonas syringae* virulence in plants by activating a signaling cascade that inhibits salicylic acid accumulation. *Cell Host Microbe* **2012**, *11*, 587–596.
44. Agrawal, A.A.; Hastings, A.P.; Johnson, M.T.; Maron, J.L.; Salminen, J.P. Insect herbivores drive real-time ecological and evolutionary change in plant populations. *Science* **2012**, *338*, 113–116.
45. Gomez-Gomez, L.; Felix, G.; Boller, T. A single locus determines sensitivity to bacterial flagellin in *Arabidopsis thaliana*. *Plant J.* **1999**, *18*, 277–284.
46. Clarke, J.D.; Volko, S.M.; Ledford, H.; Ausubel, F.M.; Dong, X. Roles of salicylic acid, jasmonic acid, and ethylene in *cpr*-induced resistance in Arabidopsis. *Plant Cell* **2000**, *12*, 2175–2190.
47. Zhang, Y.; Goritschnig, S.; Dong, X.; Li, X. A gain-of-function mutation in a plant disease resistance gene leads to constitutive activation of downstream signal transduction pathways in suppressor of *npr1-1, constitutive 1*. *Plant Cell***2003**, *15*, 2636–2646.

48. Attaran, E.; Major, I.T.; Cruz, J.A.; Rosa, B.A.; Koo, A.J.; Chen, J.; Kramer, D.M.; He, S.Y.; Howe, G.A. Temporal dynamics of growth and photosynthesis suppression in response to jasmonate signaling. *Plant Physiol.* **2014**, *165*, 1302–1314.
49. Campos, M.L.; Yoshida, Y.; Major, I.T.; de Oliveira Ferreira, D.; Weraduwage, S.M.; Froehlich, J.E.; Johnson, B.F.; Kramer, D.M.; Jander, G.; Sharkey, T.D.; et al. Rewiring of jasmonate and phytochrome B signalling uncouples plant growth-defense tradeoffs. *Nat. Commun.* **2016**, *7*, 12570.
50. Zust, T.; Joseph, B.; Shimizu, K.K.; Kliebenstein, D.J.; Turnbull, L.A. Using knockout mutants to reveal the growth costs of defensive traits. *Proc. Biol. Sci.* **2011**, *278*, 2598–2603.
51. Izaguirre, M.M.; Scopel, A.L.; Baldwin, I.T.; Ballaré, C.L. Convergent responses to stress. Solar ultraviolet-B radiation and *Manduca sexta* herbivory elicit overlapping transcriptional responses in field-grown plants of *Nicotiana longiflora*. *Plant Physiol.* **2003**, *132*, 1755–1767.
52. Nabity, P.D.; Zavala, J.A.; DeLucia, E.H. Herbivore induction of jasmonic acid and chemical defences reduce photosynthesis in *Nicotiana attenuata*. *J. Exp. Bot.* **2013**, *64*, 685–694.
53. Zangerl, A.R.; Hamilton, J.G.; Miller, T.J.; Crofts, A.R.; Oxborough, K.; Berenbaum, M.R.; de Lucia, E.H. Impact of folivory on photosynthesis is greater than the sum of its holes. *Proc. Natl. Acad. Sci. USA* **2002**, *99*, 1088–1091.
54. Qi, T.; Song, S.; Ren, Q.; Wu, D.; Huang, H.; Chen, Y.; Fan, M.; Peng, W.; Ren, C.; Xie, D. The Jasmonate-ZIM-domain proteins interact with the WD-repeat/bHLH/MYB complexes to regulate jasmonate-mediated anthocyanin accumulation and trichome initiation in *Arabidopsis thaliana*. *Plant Cell* **2011**, *23*, 1795–1814.
55. Zhu, X.; Chen, J.; Xie, Z.; Gao, J.; Ren, G.; Gao, S.; Zhou, X.; Kuai, B. Jasmonic acid promotes degreening via MYC2/3/4- and ANAC019/055/072-mediated regulation of major chlorophyll catabolic genes. *Plant J.* **2015**, *84*, 597–610.
56. Hu, X.; Makita, S.; Schelbert, S.; Sano, S.; Ochiai, M.; Tsuchiya, T.; Hasegawa, S.F.; Hortensteiner, S.; Tanaka, A.; Tanaka, R. Reexamination of chlorophyllase function implies its involvement in defense against chewing herbivores. *Plant Physiol.* **2015**, *167*, 660–670.
57. Bilgin, D.D.; Zavala, J.A.; Zhu, J.; Clough, S.J.; Ort, D.R.; DeLucia, E.H. Biotic stress globally downregulates photosynthesis genes. *Plant Cell Environ.* **2010**, *33*, 1597–1613.
58. Zhang, Y.; Turner, J.G. Wound-induced endogenous jasmonates stunt plant growth by inhibiting mitosis. *PLoS ONE***2008**, *3*, e3699.
59. Weraduwage, S.M.; Chen, J.; Anozie, F.C.; Morales, A.; Weise, S.E.; Sharkey, T.D. The relationship between leaf area growth and biomass accumulation in *Arabidopsis thaliana*. *Front. Plant Sci.* **2015**, *6*, 167.
60. Havko, N.E.; Major, I.T.; Jewell, J.B.; Attaran, E.; Browse, J.; Howe, G.A. Control of carbon assimilation and partitioning by jasmonate: An accounting of growth-defense tradeoffs. *Plants* **2016**, *5*, 7.
61. Huot, B.; Yao, J.; Montgomery, B.L.; He, S.Y. Growth-defense tradeoffs in plants: A balancing act to optimize fitness. *Mol. Plant* **2014**, *7*, 1267–1287.
62. Navarro, L.; Bari, R.; Achard, P.; Lison, P.; Nemri, A.; Harberd, N.P.; Jones, J.D. DELLAs control plant immune responses by modulating the balance of jasmonic acid and salicylic acid signaling. *Curr. Biol.* **2008**, *18*, 650–655.
63. Hou, X.; Lee, L.Y.; Xia, K.; Yan, Y.; Yu, H. DELLAs modulate jasmonate signaling via competitive binding to JAZs. *Dev. Cell* **2010**, *19*, 884–894.
64. Yang, D.L.; Yao, J.; Mei, C.S.; Tong, X.H.; Zeng, L.J.; Li, Q.; Xiao, L.T.; Sun, T.P.; Li, J.; Deng, X.W.; et al. Plant hormone jasmonate prioritizes defense over growth by interfering with gibberellin signaling cascade. *Proc. Natl. Acad. Sci. USA* **2012**, *109*, E1192–E1200.
65. Chen, Q.; Sun, J.; Zhai, Q.; Zhou, W.; Qi, L.; Xu, L.; Wang, B.; Chen, R.; Jiang, H.; Qi, J.; et al. The basic helix-loop-helix transcription factor MYC2 directly represses *PLETHORA* expression during jasmonate-mediated modulation of the root stem cell niche in Arabidopsis. *Plant Cell* **2011**, *23*, 3335–3352.
66. Smolen, G.A.; Pawlowski, L.; Wilensky, S.E.; Bender, J. Dominant alleles of the basic helix-loop-helix transcription factor ATR2 activate stress-responsive genes in Arabidopsis. *Genetics* **2002**, *161*, 1235–1246.

67. Gasperini, D.; Chetelat, A.; Acosta, I.F.; Goossens, J.; Pauwels, L.; Goossens, A.; Dreos, R.; Alfonso, E.; Farmer, E.E. Multilayered organization of jasmonate signalling in the regulation of root growth. *PLoS Genet.* **2015**, *11*, e1005300.
68. Goossens, J.; Swinnen, G.; Vanden Bossche, R.; Pauwels, L.; Goossens, A. Change of a conserved amino acid in the MYC2 and MYC3 transcription factors leads to release of JAZ repression and increased activity. *New Phytol.* **2015**, *206*, 1229–1237.
69. Moreno, J.E.; Ballaré, C.L. Phytochrome regulation of plant immunity in vegetation canopies. *J. Chem. Ecol.* **2014**, *40*, 848–857.
70. Herms, D.A.; Mattson, W.J. The dilemma of plants: To grow or defend. *Q. Rev. Biol.* **1992**, *67*, 283–335.
71. Mazza, C.A.; Zavala, J.; Scopel, A.L.; Ballaré, C.L. Perception of solar UVB radiation by phytophagous insects: Behavioral responses and ecosystem implications. *Proc. Natl. Acad. Sci. USA* **1999**, *96*, 980–985.
72. Caputo, C.; Rutitzky, M.; Ballaré, C.L. Solar ultraviolet-B radiation alters the attractiveness of Arabidopsis plants to diamondback moths (*Plutella xylostella* L.): Impacts on oviposition and involvement of the jasmonic acid pathway. *Oecologia* **2006**, *149*, 81–90.
73. Zavala, J.A.; Mazza, C.A.; Dillon, F.M.; Chludil, H.D.; Ballaré, C.L. Soybean resistance to stink bugs (*Nezara viridula* and *Piezodorus guildinii*) increases with exposure to solar UV-B radiation and correlates with isoflavonoid content in pods under field conditions. *Plant Cell Environ.* **2015**, *38*, 920–928.
74. Mazza, C.A.; Boccalandro, H.E.; Giordano, C.V.; Battista, D.; Scopel, A.L.; Ballaré, C.L. Functional significance and induction by solar radiation of ultraviolet-absorbing sunscreens in field-grown soybean crops. *Plant Physiol.* **2000**, *122*, 117–126.
75. Izaguirre, M.M.; Mazza, C.A.; Svatos, A.; Baldwin, I.T.; Ballaré, C.L. Solar ultraviolet-B radiation and insect herbivory trigger partially overlapping phenolic responses in *Nicotiana attenuata* and *Nicotiana longiflora*. *Ann. Bot.* **2007**, *99*, 103–109.
76. Demkura, P.V.; Ballaré, C.L. UVR8 mediates UV-B-induced Arabidopsis defense responses against *Botrytis cinerea* by controlling sinapate accumulation. *Mol. Plant* **2012**, *5*, 642–652.
77. Cerrudo, I.; Keller, M.M.; Cargnel, M.D.; Demkura, P.V.; de Wit, M.; Patitucci, M.S.; Pierik, R.; Pieterse, C.M.; Ballaré, C.L. Low red/far-red ratios reduce Arabidopsis resistance to *Botrytis cinerea* and jasmonate responses via a COI1-JAZ10-dependent, salicylic acid-independent mechanism. *Plant Physiol.* **2012**, *158*, 2042–2052.
78. Griebel, T.; Zeier, J. Light regulation and daytime dependency of inducible plant defenses in Arabidopsis: Phytochrome signaling controls systemic acquired resistance rather than local defense. *Plant Physiol.* **2008**, *147*, 790–801.
79. Jeong, R.D.; Chandra-Shekara, A.C.; Barman, S.R.; Navarre, D.; Klessig, D.F.; Kachroo, A.; Kachroo, P. Cryptochrome 2 and phototropin 2 regulate resistance protein-mediated viral defense by negatively regulating an E3 ubiquitin ligase. *Proc. Natl. Acad. Sci. USA* **2010**, *107*, 13538–13543.
80. Wu, L.; Yang, H.Q. Cryptochrome 1 is implicated in promoting R protein-mediated plant resistance to *Pseudomonas syringae* in Arabidopsis. *Mol. Plant* **2010**, *3*, 539–548.
81. Zeiger, E.; Iino, M.; Shimazaki, K.I.; Ogawa, T. The blue-light response of stomata: Mechanism and function. In *Stomatal Function*; Zeiger, E., Farquhar, G.D., Cowan, I.R., Eds.; Stanford University Press: Stanford, CA, USA, 1981; pp. 209–227.
82. Zhang, C.; Xie, Q.; Anderson, R.G.; Ng, G.; Seitz, N.C.; Peterson, T.; McClung, C.R.; McDowell, J.M.; Kong, D.; Kwak, J.M.; et al. Crosstalk between the circadian clock and innate immunity in Arabidopsis. *PLoS Pathog.* **2013**, *9*, e1003370.
83. Wang, W.; Barnaby, J.Y.; Tada, Y.; Li, H.; Tor, M.; Caldelari, D.; Lee, D.U.; Fu, X.D.; Dong, X. Timing of plant immune responses by a central circadian regulator. *Nature* **2011**, *470*, 110–114.
84. Goodspeed, D.; Chehab, E.W.; Covington, M.F.; Braam, J. Circadian control of jasmonates and salicylates: The clock role in plant defense. *Plant. Signal. Behav.* **2013**, *8*, e23123.
85. Goodspeed, D.; Chehab, E.W.; Min-Venditti, A.; Braam, J.; Covington, M.F. Arabidopsis synchronizes jasmonate-mediated defense with insect circadian behavior. *Proc. Natl. Acad. Sci. USA* **2012**, *109*, 4674–4677.

86. Zhou, M.; Wang, W.; Karapetyan, S.; Mwimba, M.; Marques, J.; Buchler, N.E.; Dong, X. Redox rhythm reinforces the circadian clock to gate immune response. *Nature* **2015**, *523*, 472–476.
87. Moreno, J.E.; Tao, Y.; Chory, J.; Ballaré, C.L. Ecological modulation of plant defense via phytochrome control of jasmonate sensitivity. *Proc. Natl. Acad. Sci. USA* **2009**, *106*, 4935–4940.
88. de Wit, M.; Spoel, S.H.; Sanchez-Perez, G.F.; Gommers, C.M.; Pieterse, C.M.; Voesenek, L.A.; Pierik, R. Perception of low red:far-red ratio compromises both salicylic acid- and jasmonic acid-dependent pathogen defences in Arabidopsis. *Plant J.* **2013**, *75*, 90–103.
89. Cargnel, M.D.; Demkura, P.V.; Ballaré, C.L. Linking phytochrome to plant immunity: Low red : Far-red ratios increase Arabidopsis susceptibility to *Botrytis cinerea* by reducing the biosynthesis of indolic glucosinolates and camalexin. *New Phytol.* **2014**, *204*, 342–354.
90. Gangappa, S.N.; Berriri, S.; Kumar, S.V. PIF4 coordinates thermosensory growth and immunity in Arabidopsis. *Curr. Biol. CB* **2017**, *27*, 243–249.
91. Leone, M.; Keller, M.M.; Cerrudo, I.; Ballaré, C.L. To grow or defend? Low red : Far-red ratios reduce jasmonate sensitivity in Arabidopsis seedlings by promoting DELLA degradation and increasing JAZ10 stability. *New Phytol.***2014**, *204*, 355–367.
92. Chico, J.M.; Fernandez-Barbero, G.; Chini, A.; Fernandez-Calvo, P.; Diez-Diaz, M.; Solano, R. Repression of jasmonate-dependent defenses by shade involves differential regulation of protein stability of MYC transcription factors and their JAZ repressors in Arabidopsis. *Plant Cell* **2014**, *26*, 1967–1980.
93. Noir, S.; Bomer, M.; Takahashi, N.; Ishida, T.; Tsui, T.L.; Balbi, V.; Shanahan, H.; Sugimoto, K.; Devoto, A. Jasmonate controls leaf growth by repressing cell proliferation and the onset of endoreduplication while maintaining a potential stand-by mode. *Plant Physiol.* **2013**, *161*, 1930–1951.
94. Pauwels, L.; Morreel, K.; De Witte, E.; Lammertyn, F.; van Montagu, M.; Boerjan, W.; Inze, D.; Goossens, A. Mapping methyl jasmonate-mediated transcriptional reprogramming of metabolism and cell cycle progression in cultured Arabidopsis cells. *Proc. Natl. Acad. Sci. USA* **2008**, *105*, 1380–1385.
95. Kliebenstein, D.J. False idolatry of the mythical growth versus immunity tradeoff in molecular systems plant pathology. *Physiol. Mol. Plant Pathol.* **2016**, *95*, 55–59.
96. Ballaré, C.L. Light regulation of plant defense. *Ann. Rev. Plant Biol.* **2014**, *65*, 335–363.
97. Cerrudo, I.; Caliri-Ortiz, M.E.; Keller, M.M.; Degano, M.E.; Demkura, P.V.; Ballaré, C.L. Exploring growth-defense tradeoffs in Arabidopsis. Phytochrome B inactivation requires JAZ10 to suppress plant immunity but not to trigger shade avoidance responses. *Plant Cell Environ.* **2016**.

CHAPTER 14

Effect of Long-Term Continuous Fumigation on Soil Microbial Communities

Sadikshya R. Dangi [1,2,*], Rebecca Tirado-Corbalá [1,3], James Gerik [1] and Bradley D. Hanson [3]

[1] *United States Department of Agriculture (USDA), Agricultural Research Service, San Joaquin Valley Agricultural Sciences Center, 9611 S. Riverbend Ave., Parlier, CA 93648, USA*

[2] *Department of Plant Sciences, University of California, Davis, One Shields Ave., Davis, CA 95616, USA*

[3] *Agro-Environmental Sciences Department, University of Puerto Rico-Mayagüez, Box 9000, Mayagüez, PR 00681, USA*

ABSTRACT

High value crop producers in California rely heavily on soil fumigation to control a wide array of soil borne pests including nematodes, pathogens and weeds. Fumigants with broad biocidal activity can affect soil microbial communities that contribute to nutrient cycling and plant nutrient uptake which can impact soil health. It is often thought that soil microbial communities make a relatively rapid recovery following fumigation. However, recently it has been found that repeated application of fumigants over time can have greater and longer lasting impacts on soil microorganisms than single fumigation events. Therefore, the main objective of this study was to determine the effect of long-term repeated application of fumigants on soil microbial communities and compare them with non-fumigated and organic sites. Soil samples were collected from fields in Watsonville, CA. Chronosequence sites were defined by number of years of annual fumigation (yaf) with methyl bromide (15, 26, 33, 39 yaf) at the time of sampling, and representative non-fumigated sites were also included for comparison. Phospholipid fatty acid (PLFA) analysis was used to analyze the samples. The canonical variate analysis showed that microbial communities in sites with a longer history of fumigation (33 and 39 yaf) were similar to one another; however, they differed significantly from 15 yaf site and further analysis concluded that non-fumigated sites were significantly different than fumigated sites. This study showed that the proportion of arbuscular mycorrhizal fungi (AMF) was lower in all fumigated (15, 33 and 39 yaf) sites as compared to their non-fumigated counterparts, which could be a threat to sustainability since AMF plays a major role in soil health and quality.

AUTHOR CONTRIBUTIONS

Sadikshya R. Dangi and James Gerik conceived and designed experiments; Bradley D. Hanson contributed in critical revision of the manuscript; Rebecca Tirado Corbala performed soil chemical analysis and analyzed data; Sadikshya R. Dangi performed the experiments, analyzed data and wrote the paper.

Keywords: fumigation; soil microbial communities; non-fumigated; organic sites

1. INTRODUCTION

For over 50 years, California has been the US's top agricultural state generating approximately $54 billion a year in farm-gate value [1]. California is one of the largest users of methyl bromide (MeBr) and other soil fumigants in the United States (US) [2]. Due to its deleterious effects on stratospheric ozone, the import and manufacture of MeBr has largely been phased out under the provisions of the U.S. Clean Air Act and the Montreal Protocol [3,4]. California strawberry producers have continued using MeBr under a critical use exemption subject to annual review by the parties of the Montreal Protocol [5]. In the US, 1,3-dichloropropene (1,3-D), chloropicrin, dimethyl disulfide (DMDS), and methyl isothiocyanate (MITC) have been proposed as the most likely chemical alternatives to MeBr [3,5,6]. Of these, DMDS is not currently registered in California.

To reduce the use of methyl bromide fumigation and alleviate its detrimental effects to the soil, rotation with cover crops has been recommended [7]. Cover crops grown in rotation with cash crops can alleviate the accumulation of weeds and pathogens in the soil, increase soil organic matter and improve soil physical, chemical, and biological properties [8]. Therefore, strawberry fields in California are rotated with cover crops such as rye and barley, or other cash crops; however, in areas where land and water costs are high, cover crops are not economically feasible.

Several short-term studies on the recovery of soil microbial communities in fumigated fields concluded that soil microbial communities make a relatively rapid recovery (less than a year) after the application of fumigants [9,10,11,12]. However, there are limited studies on the effect of long-term, repeated application of fumigants on soil microbial communities. Reeve et al. [13] proposed that long-term repeated application of fumigants can have significantly different and longer lasting impacts on soil microorganisms compared to single fumigation events.

It is important to study the effect of fumigants on soil microbial communities in regularly-fumigated fields, as soil microorganisms are crucial in sustaining long-term health of agricultural soil [9]. Hence, understanding the impact of fumigants on beneficial soil microbial communities is essential to help further preserve the health and sustainability of these fumigated lands. Therefore, the main objective of this study was to determine the effect of long-term annual application of fumigants (15, 26, 33 and 39 years of annual fumigation) on soil microbial communities and compare them with non-fumigated sites.

2. MATERIALS AND METHODS

2.1. Study Area and Soil Sampling

The study was conducted in the Watsonville, California area (Table 1) which accounts for almost half of the state's strawberry acreage. In California, fumigation with MeBr and chloropicrin became an integral part of strawberry cultivation around 1960 and, until recently, nearly all strawberry acres were fumigated with a mixture of both fumigants prior to planting [5]. The chronosequence selected for this study contained four sites defined by number of years of annual fumigation (yaf) with

MeBr and chloropicrin. The sites were a strawberry field with 15 years of annual fumigation (15 yaf; 36°56′26.8″ N, 121°43′11.4″ W) with no crop rotation, a 26 yaf site where beans, peas or barley were rotated annually with strawberries, a 33 years of annual fumigation site where either barley, peas and beans were rotated with strawberries (33 yaf; 36°52′35.2″ N, 121°40′42.3″ W), two 39 yaf sites where strawberries were rotated with barley (39a; 36°54′15.01″ N, 121°49′30.4″ W and 39b yaf; 36°54′11.3″ N, 121°49′25.2″ W) every other year, and three representative non-fumigated (NF) sites. Two non-fumigated sites, NF-15 (36°56′26.8″ N, 121°43′11.4″ W) and NF-33 (36°52′24.06″ N, 121°41′31.05″ W) were vegetable gardens that had never been fumigated adjacent to the 15 yaf and 33 yaf sites. An organic strawberry farm established in 2006 adjacent to the 39a and 39b yaf sites which had been previously fumigated until 2002 and fallowed for 3 years to get organic certification (NF-39; 36°54′12.9″ N, 121°49′23.08″ W) was chosen as a non-fumigated control for 39 yaf sites. The soil texture at the 15 yaf, NF-15 and 26 yaf sites are clay loam while the 33 yaf, NF-33, 39a yaf, 39b yaf and NF-39 sites are sandy loam soils. Methyl bromide-fumigated and non-fumigated fields were selected across the Watsonville area on the basis of data provided by representatives from the fumigation industry and grower interviews.

Table 1. Description of the five long-term fumigated study sites in Watsonville, CA, USA.

Site	Age * (Yaf)	Crop	Fumigation Starting Year	Fumigation before Sampling	Fumigants Used
1	15	Strawberry	1998	May 2013	Yearly with MeBr: chloropicrin
2	26	Strawberry rotated with either barley, peas and beans	1990	May 2013	Every other year with MeBr: chloropicrin
3	33	Strawberry rotated either with barley, peas and beans	1980	May 2013	Every other year with MeBr
4	39a	Strawberry and barley	1974	May 2013	Every other year with MeBr: chloropicrin
5	39b	Strawberry and barley	1974	May 2013	Every other year with MeBr: chloropicrin

* Yaf = Years of annual fumigation.

Soil samples were collected in May 2013 from 0–5, 5–15 and 15–30 cm depths. Three 100 m transects were randomly placed at each site and soil samples were collected from four equally spaced points along the transects. Due to the difficulty in finding fumigated sites of the same age and the non-fumigated treatments, each transect was considered a replicate and the four sampling points were not composited. Samples for phospholipid fatty acids (PLFA) analysis were placed in sealed plastic bags, stored on dry ice immediately after collection, and then returned to the laboratory where they were placed in a −20 °C freezer until analyzed.

2.2. Soil Physicochemical Analysis

Soil pH was determined by using 1:1 soil to deionized water (DI) ratio. Dissolved organic carbon (DOC) was determined after saturating the soil with DI water (1:1 soil: water) for 24 h, shaken for 1 h on a reciprocal shaker, and filtered through a Whatman, no. 42 filter. Carbon recovered in the water extract was determined by using Fusion Total Organic Carbon Analyzer from Teledyne Tekmar. Total nitrogen (TN) and total carbon (TC) contents were determined by dry combustion with a Flash 2000 N & C Soil Analyzer (Thermo Scientific®, Pittsburgh, PA, USA). Macronutrients such as Ca, Mg, S, K and Na were determined using ICP-OES (Varian, Palo Alto, CA, USA).

2.3. Soil PLFA Analysis

Phospholipid fatty acids (PLFA) were extracted from 5 g soil samples using a modified Bligh-Dyer methodology [14]. Fatty acids were directly extracted from soil samples using a mixture of chloroform: methanol: phosphate buffer (1:2:0.8). Phospholipid fatty acids were separated from neutral and glycolipid fatty acids in a solid phase extraction column. After mild alkaline methanolysis, PLFA samples were qualitatively and quantitatively analyzed using an Agilent 6890 gas chromatograph (Agilent Technologies, Palo Alto, CA, USA) and fatty acids were identified using the MIDI PLFAD1 calibration mix and naming table [15], (MIDI Inc., Newark, NJ, USA).

Individual PLFA signatures were used to quantify the abundances of specific microbial groups in soil samples [15]. Gram + bacteria were identified and quantified by the presence of iso- and anteiso-branched fatty acids, Gram − bacteria with monounsaturated fatty acids and cyclopropyl 17:0 and 19:0, and eubacteria with 15:0, 17:0 cyclo, 19:0 cyclo, 15:1 iso, and 17:1 iso and 17:1 anteiso. Fungi were identified and quantified with 18:2 ω6c, arbuscular mycorrhizal fungi (AMF) with 16:1 ω5c, and actinomycetes with 10-methyl fatty acids [16]. As indicators of physiological or nutritional stress in bacterial communities, the ratio of saturated (SAT) and monounsaturated PLFAs were used in conjunction with the ratios of the sum of cyclopropyl PLFAs to the sum of their monoenoic precursors (cy 17:0 + cy 19:0)/(16:1ω7 + 18:1ω7; abbreviated as Cy/pre) [17].

2.4. Data Analysis

Two-way analysis of variance was used to analyze fumigation effects on soil chemical properties using SAS 9.4 [18]. Fatty acids were summed into biomarker groups according to [15]. The ANOVA was conducted on total PLFA and on each biomarker group to determine fumigation effects on the total microbial biomass and biomass of each group of organisms. Additionally, ANOVA (generalized linear model) using proc GLIMMIX and a link-logit function were conducted on the proportion of each biomarker. A multivariate method (canonical analysis) was used to compare soil microbial communities in fumigated and non-fumigated sites. In this analysis, MANOVA on the relative area of each biomarker was used to identify the linear combination of variables (referred to as canonical variates) that best separated soil microbial community structure at different treatments. The canonical variates were graphed to summarize group differences [19,20]. All statistical analyses were performed at the $p < 0.05$ significance level.

3. RESULTS

3.1. Response of Carbon, Nitrogen and Macronutrients to Fumigation and Non-Fumigation

In order to understand the long-term repeated application of fumigants on soil properties, macronutrients, TC, TN, and DOC were determined (Table 2). Soil pH was lower in the fumigated 39b site as compared to other fumigated and non-fumigated sites (Table 2). Concentration of Ca was significantly higher in 39 yaf site compared to 33 and lowest in 15 and 26 yaf sites. Concentration of K, Mg and Na were significantly higher in NF-15 site and concentration of S was significantly higher in 39a and 39b yaf sites compared to other fumigated and non-fumigated sites where it remained similar to each other (Table 2). The DOC, TN and TC varied between fumigated and non-fumigated sites (Table 2). The DOC was higher in NF-15 and 15 yaf sites than fumigated 26, 33, 39a, 39b sites and non-fumigated 33 and 39 sites. Total N was higher in 15 yaf site than NF-15 site. Total C was significantly higher in NF-33 and 39b yaf sites.

Table 2. pH, macronutrients, dissolved organic carbon (DOC), total nitrogen (TN), and total carbon in fumigated and non-fumigated sites at the 0–5 cm depth.

Site	pH	Ca	K	Mg	Na	S	DOC	TN	TC
				mg kg^{-1}			ppm	%	
15 yaf	7.72 ^{a}	40.0 ^{c}	13.4 ^{b}	18.1 ^{b}	51.1 ^{b}	39 ^{b}	494 ^{a}	0.24 ^{a}	1.38 ^{b}
NF-15	7.94 ^{a}	433 ^{a}	50.9 ^{a}	162 ^{a}	145 ^{a}	310 ^{b}	512 ^{a}	0.08 ^{b}	0.40 ^{b}
26 yaf	7.89 ^{a}	45.7 ^{c}	2.93 ^{c}	12.2 ^{b}	8.34 ^{c}	8.16 ^{c}	18.8 ^{b}	0.11 ^{a,b}	0.86 ^{b}
33 yaf	7.20 ^{a}	13.8 ^{d}	7.13 ^{b}	6.9 ^{b}	14.6 ^{c}	7.6 ^{c}	58.7 ^{b}	0.10 ^{a,b}	0.89 ^{b}
NF-33	7.27 ^{a}	70.8 ^{b}	14.8 ^{b}	26.5 ^{b}	21.9 ^{b,c}	37 ^{b}	68 ^{b}	0.16 ^{a,b}	2.35 ^{a}
39a yaf	6.90 ^{a}	197 ^{b}	9.57 ^{b}	46.1 ^{b}	34.2 ^{b,c}	1002^{a}	26 ^{b}	0.15 ^{a,b}	0.87 ^{b}
39b yaf	5.92 ^{b}	189 ^{b}	8.61 ^{b}	44.4 ^{b}	47.4 ^{b,c}	1338 ^{a}	22.2 ^{b}	0.76 ^{a,b}	4.13 ^{a}
NF-39	6.93 ^{a}	92.6 ^{b}	13.9 ^{b}	26.5 ^{b}	20.6 ^{b,c}	329 ^{b}	24.3 ^{b}	0.17 ^{a,b}	0.85 ^{b}
Pr > F	0.002 *	0.001 *	0.003 *	<0.0001 *	<0.0001 *	<0.0001 *	0.044 *	0.042 *	0.040 *
SE	0.12	54.5	6.72	9.4	10.0	40.5	130	0.05	0.37

* Means followed by different letters in a column are significantly different by Tukey's test at $p < 0.05$. Ca, Calcium; K, potassium; Mg, magnesium; Na, sodium; S, sulfur; dissolved organic carbon, DOC; TN, total nitrogen; TC, total carbon.

3.2. Response of Soil Microbial Communities to Long-Term Fumigation and Non-Fumigation

Soil microbial community biomass represented by total PLFA was significantly ($p < 0.05$) higher in NF-33 site as compared to the fumigated 33 yaf site at the first 5 cm depth (Figure 1A). At the 5–15 cm depth, total PLFA in NF-33 site was higher than in fumigated 26, 39a, 39b and non-fumigated 39 sites; whereas at 15–30 cm depth, total microbial biomass in non-fumigated sites (33 and 15) was higher in comparison to 26 yaf site (Figure 1B,C).

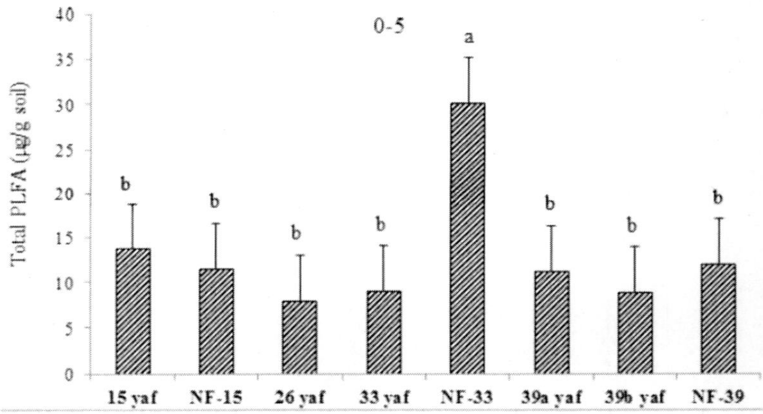

(A)

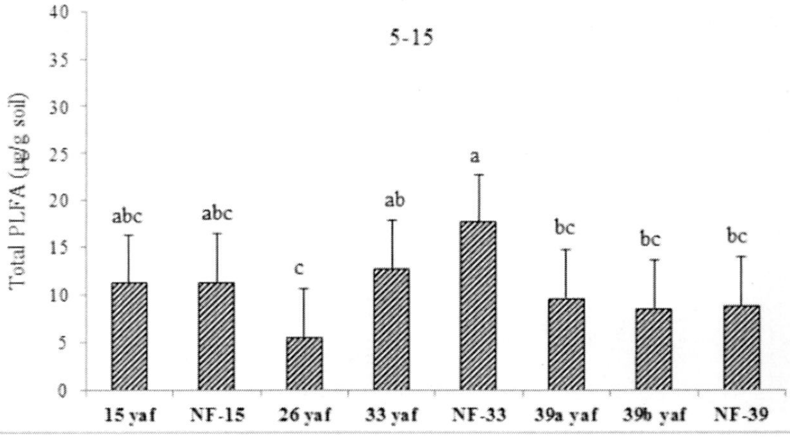

(B)

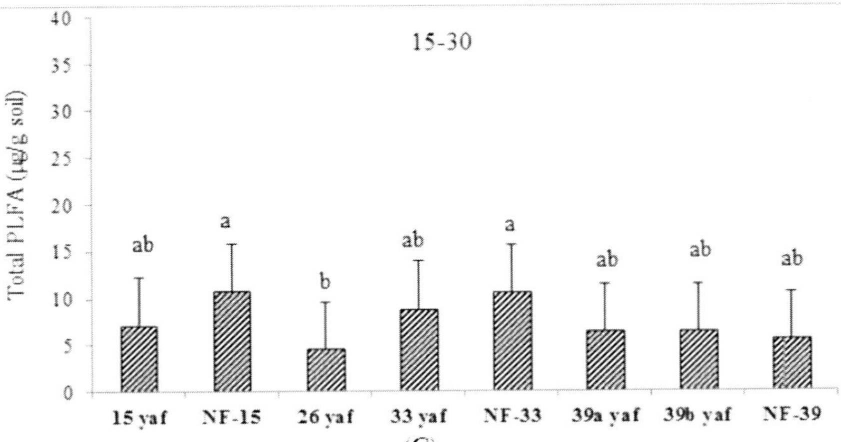

(C)

Figure 1. Total phospholipid fatty acid (PLFA) for fumigated and non-fumigated sites at (**A**) 0–5 cm; (**B**) 5–15 cm; and (**C**) 15–30 cm depths. Different letters indicate significant differences ($p < 0.05$). Error bars indicate standard error. Soil samples were collected from 15, 26, 33, 39a, 39b yaf sites as well as non-fumigated (NF-15, NF-33 and NF-39) sites.

The proportion of fungal PLFA was much higher in the 15 and 26 yaf sites than NF-15 site at 0–5 cm depth (Table 3). The AMF PLFA proportion was higher in all non-fumigated (NF-15, NF-33 and NF-39) sites as compared to their fumigated (15, 33 and 39 yaf) counterparts. Gram positive bacterial proportion was lower in NF-33 than 33 yaf, whereas Gram positive proportion was lower in the NF-15 site compared to 15 and 26 yaf sites. In addition, proportional levels of Gram negative and actinomycetes PLFA were higher in the NF-33 and the NF-15 sites compared to the 33 yaf and the 26 yaf sites. In the 5–15 cm depth, the 15 yaf site had significantly higher proportions of fungal PLFA than the NF-15 site (Table 4). Moreover, the amount of AMF was higher in the NF-15 site than in the 15 and 26 yaf sites. Proportional levels of Gram + and actinomycetes PLFAs were significantly higher in the NF-15 compared to the 15 yaf site, whereas eukaryotic PLFA was higher at the 15 yaf site compared to the 26 yaf site. At the 15–30 cm depth, significantly higher proportions of AMF in the NF-33 and the NF-15 sites than in the 33 and the 15 yaf sites were observed (Table 5). Also, actinomycetes proportions were higher in the 26 yaf site compared to the 15 yaf site.

Table 3. PLFA biomarker proportions at the 0–5 cm depth.

Site	Fungi	AMF	Gram Positive	Gram Negative	Actinomycetes	Eukaryote
Sandy loam						
33 yaf	0.020 (0.004) b	0.010 (0.002) c	0.365 (0.023) a	0.374 (0.010) b	0.099 (0.015) b	0.092 (0.031) a
NF-33	0.229 (0.001) b	0.025 (0.003) b	0.282 (0.024) b	0.438 (0.023) a	0.149 (0.006) a	0.064 (0.013) a
39a yaf	0.051 (0.009) a	0.021 (0.004) b,c	0.318 (0.008) a,b	0.396 (0.010) a,b	0.094 (0.009) b	0.102 (0.020) a
39b yaf	0.039 (0.004) a,b	0.025 (0.0005) b	0.326 (0.008) a,b	0.392 (0.012) a,b	0.105 (0.005) a,b	0.074 (0.015) a
NF-39	0.035 (0.002) a,b	0.046 (0.002) a	0.343 (0.002) a,b	0.412 (0.003) a,b	0.117 (0.001) a,b	0.028 (0.003) a
Clay loam						
15 yaf	0.108 (0.008) a	0.025(0.001) b	0.265 (0.011) b	0.415 (0.005) a,b	0.090 (0.003) b	0.083 (0.007) a
NF-15	0.026 (0.002) b	0.033 (0.002) a	0.346 (0.014) a	0.423 (0.009) a	0.130 (0.004) a	0.021 (0.002) b
26 yaf	0.109 (0.008) a	0.027(0.0009) b	0.277 (0.013) b	0.397 (0.004) b	0.094 (0.003) b	0.072 (0.010) a

Values are least square means. Values for each effect within a column with the same letter were not significantly different ($p \leq 0.05$). Standard errors are in parentheses.

Table 4. PLFA biomarker proportions at the 5–15 cm depth.

Site	Fungi	AMF	Gram + Bacteria	Gram − Bacteria	Actinomycetes	Eukaryote
			Sandy loam			
33 yaf	0.034 (0.006) ^{a,b}	0.013 (0.002) ^{b}	0.346 (0.043) ^{a}	0.421 (0.050) ^{a}	0.107 (0.015) ^{a}	0.040 (0.010) ^{a}
NF-33	0.214 (0.003) ^{b}	0.029 (0.002) ^{a,b}	0.332 (0.014) ^{a}	0.404 (0.007) ^{a}	0.134 (0.005) ^{a}	0.049 (0.014) ^{a}
39a yaf	0.039 (0.008) ^{a,b}	0.029 (0.004) ^{a,b}	0.313 (0.021) ^{a}	0.390 (0.009) ^{a}	0.104 (0.007) ^{a}	0.071 (0.013) ^{a}
39b yaf	0.065 (0.010) ^{a}	0.028 (0.001) ^{a,b}	0.314 (0.014) ^{a}	0.389 (0.008) ^{a}	0.099 (0.003) ^{a}	0.074 (0.013) ^{a}
NF-39	0.032 (0.005) ^{a,b}	0.041 (0.007) ^{a}	0.324 (0.014) ^{a}	0.419 (0.007) ^{a}	0.103 (0.005) ^{a}	0.057 (0.019) ^{a}
			Clay loam			
15 yaf	0.106 (0.014) ^{a}	0.023 (0.001) ^{c}	0.305 (0.012) ^{b}	0.388 (0.008) ^{a}	0.092 (0.004) ^{b}	0.067 (0.009) ^{a}
NF-15	0.023 (0.0005) ^{b}	0.040 (0.0008) ^{a}	0.360 (0.007) ^{a}	0.399 (0.009) ^{a}	0.128 (0.005) ^{a}	0.030 (0.006) ^{a,b}
26 yaf	0.057 (0.005) ^{a,b}	0.033 (0.001) ^{b}	0.335 (0.003) ^{a,b}	0.405 (0.006) ^{a}	0.120 (0.002) ^{a}	0.018 (0.008) ^{b}

Values are least square means. Values for each effect within a column with the same letter were not significantly different ($p \leq 0.05$). Standard errors are in parentheses.

Table 5. PLFA biomarker proportions at the 15–30 cm depth.

Site	Fungi	AMF	Gram + Bacteria	Gram − Bacteria	Actinomycetes	Eukaryote
Sandy loam						
33 yaf	0.018 (0.004) [a]	0.008 (0.003) [c]	0.328 (0.047) [a]	0.437 (0.056) [a]	0.072 (0.019) [a]	0.064 (0.014) [a]
NF-33	0.013 (0.002) [a]	0.023 (0.002) [a,b]	0.308 (0.012) [a]	0.379 (0.035) [a]	0.118 (0.013) [a]	0.137 (0.047) [a]
39a yaf	0.028 (0.006) [a]	0.024 (0.004) [a,b]	0.355 (0.006) [a]	0.380 (0.009) [a]	0.096 (0.013) [a]	0.089 (0.038) [a]
39b yaf	0.027 (0.005) [a]	0.027 (0.001) [a]	0.329 (0.011) [a]	0.372 (0.010) [a]	0.104 (0.004) [a]	0.100 (0.024) [a]
NF-39	0.018 (0.002) [a]	0.034 (0.004) [a]	0.345 (0.013) [a]	0.396 (0.007) [a]	0.108 (0.013) [a]	0.065 (0.017) [a]
Clay loam						
15 yaf	0.041 (0.009) [a]	0.022(0.003) [b]	0.341 (0.016) [a]	0.385 (0.001) [a]	0.113 (0.004) [b]	0.059 (0.007) [a]
NF-15	0.021 (0.001) [a]	0.035 (0.001) [a]	0.354 (0.0003) [a]	0.399 (0.008) [a]	0.127 (0.001) [a,b]	0.042 (0.010) [a]
26 yaf	0.037 (0.003) [a]	0.027(0.001) [b]	0.389 (0.006) [a]	0.360 (0.008) [a]	0.135 (0.003) [a]	0.013 (0.004) [b]

Values are least square means. Values for each effect within a column with the same letter were not significantly different ($p \leq 0.05$). Standard errors are in parentheses.

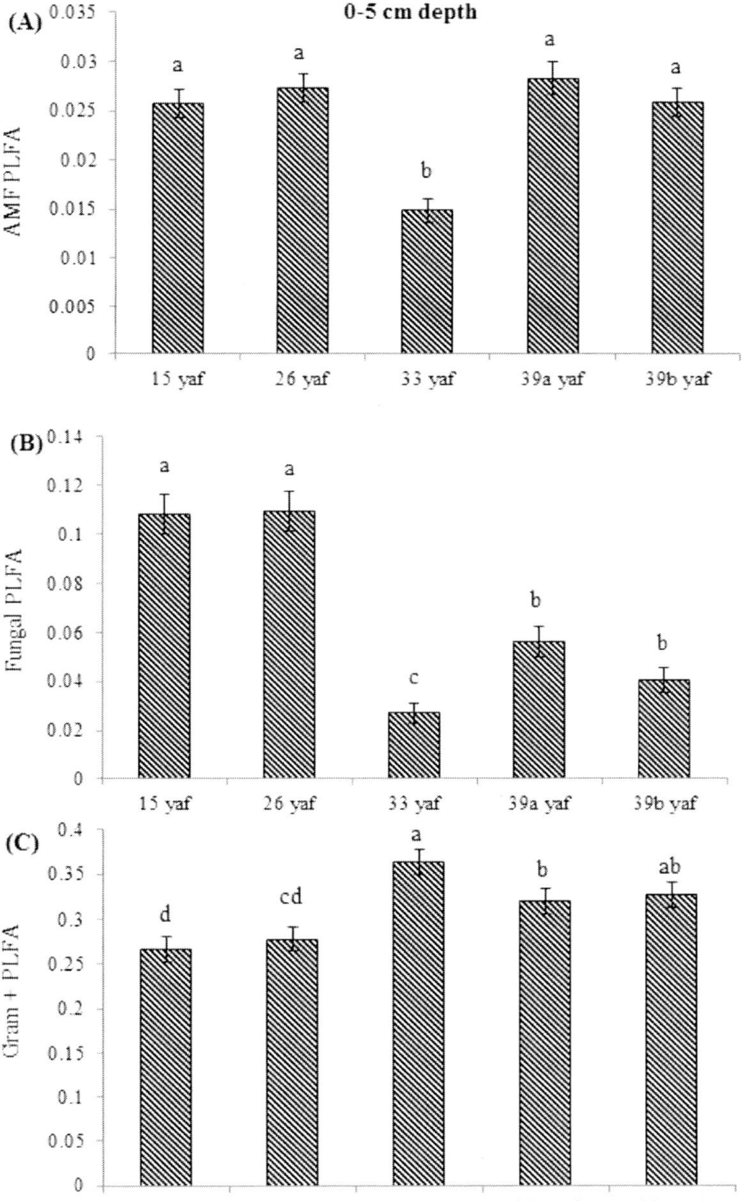
(A)
0-5 cm depth
AMF PLFA
0.035
0.03
0.025
0.02
0.015
0.01
0.005
0
a
a
b
a
a
15 yaf
26 yaf
33 yaf
39a yaf
39b yaf
(B)
Fungal PLFA
0.14
0.12
0.1
0.08
0.06
0.04
0.02
0
a
a
c
b
b
15 yaf
26 yaf
33 yaf
39a yaf
39b yaf
(C)
Gram + PLFA
0.4
0.35
0.3
0.25
0.2
0.15
0.1
0.05
0
d
cd
a
b
ab
15 yaf
26 yaf
33 yaf
39a yaf
39b yaf

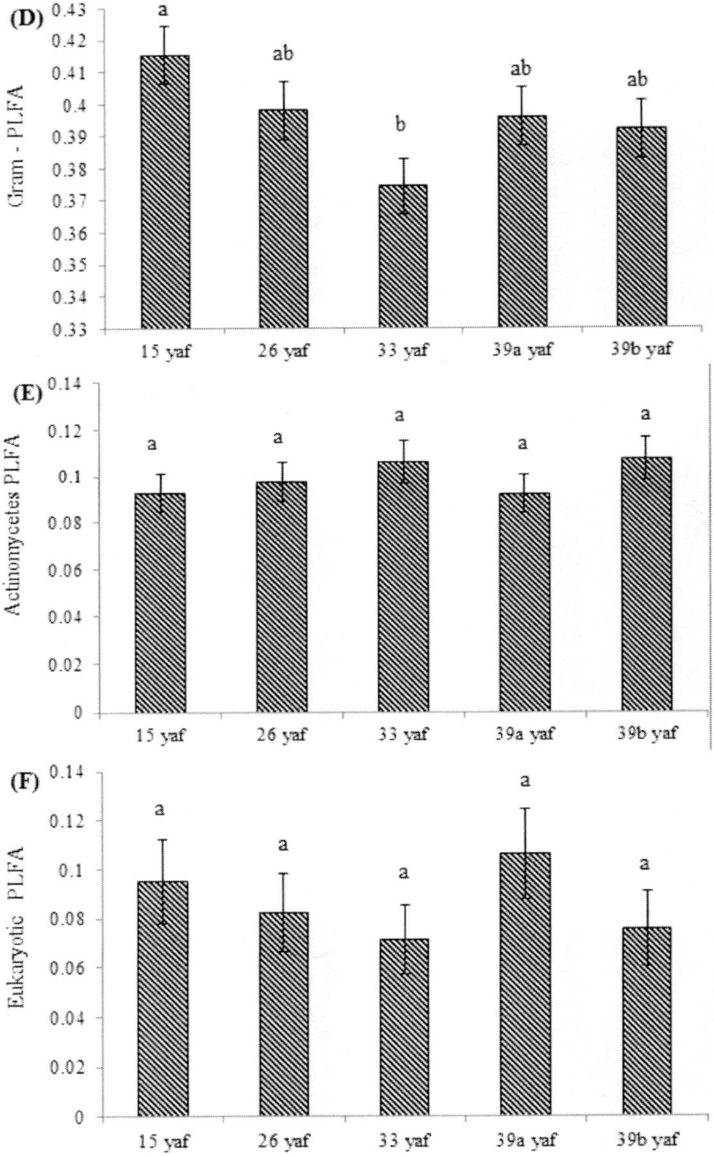

Figure 2. (**A**) AMF, (**B**) fungal, (**C**) gram positive, (**D**) gram negative, (**E**) actinomycetes, and (**F**) eukaryotic PLFA proportions at the 0–5 cm depth. Different letters indicate significant differences ($p < 0.05$). Error bars indicate standard error. Samples were collected from 15, 26, 33, 39a and 39b yaf sites.

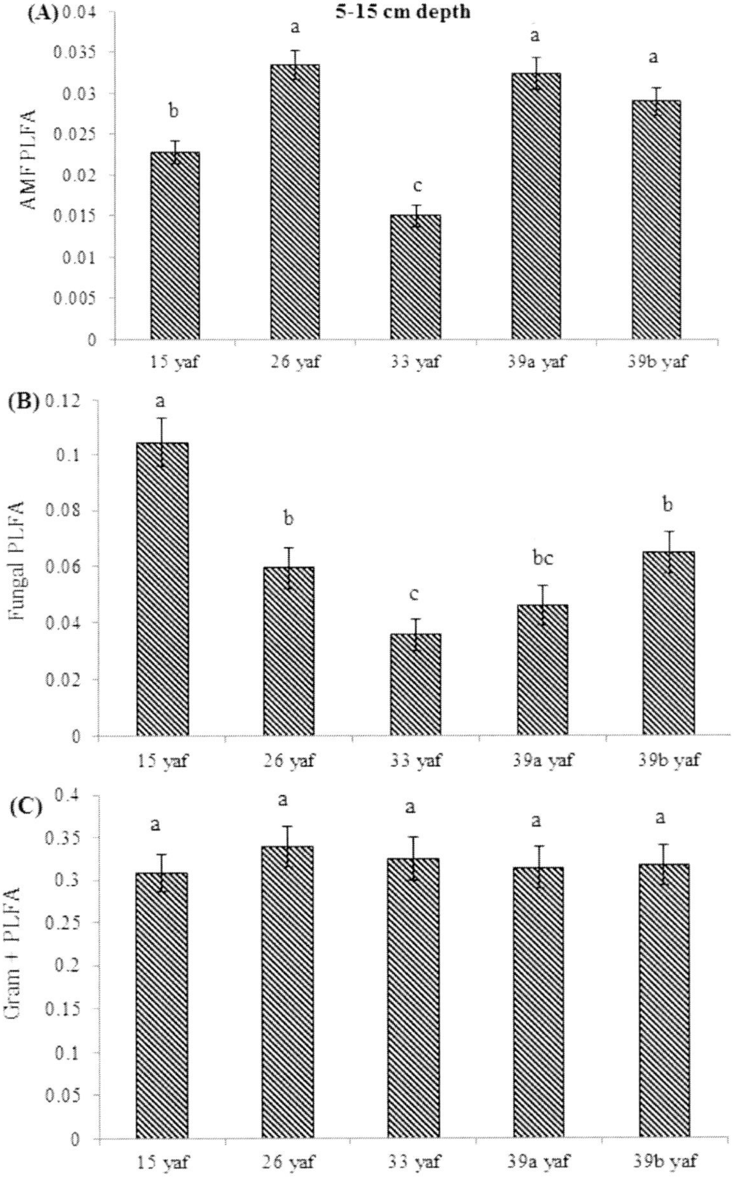
(A)
5-15 cm depth
AMF PLFA
0.04
0.035
0.03
0.025
0.02
0.015
0.01
0.005
0
b
a
c
a
a
15 yaf
26 yaf
33 yaf
39a yaf
39b yaf
(B)
Fungal PLFA
0.12
0.1
0.08
0.06
0.04
0.02
0
a
b
c
bc
b
15 yaf
26 yaf
33 yaf
39a yaf
39b yaf
(C)
Gram + PLFA
0.4
0.35
0.3
0.25
0.2
0.15
0.1
0.05
0
a
a
a
a
a
15 yaf
26 yaf
33 yaf
39a yaf
39b yaf

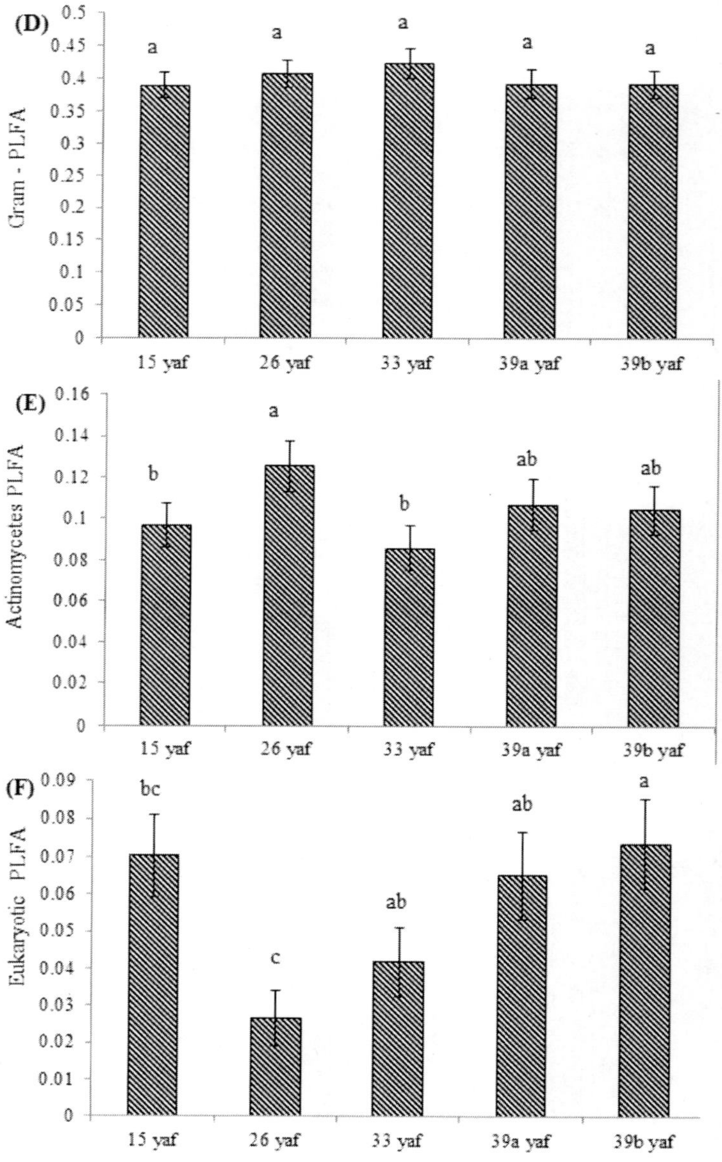

Figure 3. (**A**) AMF, (**B**) fungal, (**C**) gram positive, (**D**) gram negative, (**E**) actinomycetes, and (**F**) eukaryotic PLFA proportions at the 5–15 cm depth. Different letters indicate significant differences ($p < 0.05$). Error bars indicate standard error. Samples were collected from 15, 26, 33, 39a and 39b yaf sites.

At the 0–5 cm depth, the proportion of AMF PLFA was significantly lower in 33 yaf site as compared to fumigated 15, 26, 39a and 39b sites (Figure 2). Fungal PLFA in 15 and 26 yaf sites were significantly higher in proportion than 33 and 39 yaf sites. Additionally, the fungal PLFA proportion was higher in fumigated 39 yaf sites compared to the 33 yaf site. Gram + bacterial PLFA at the 33 yaf site was significantly higher than the 15, 26 and 39 yaf sites and Gram – bacterial PLFA was lower in 33 yaf site compared to 15 yaf site. No differences among locations were observed in the proportion of actinomycetes and eukaryotic biomarkers. At the 5–15 cm depth, the proportion of AMF PLFA was much higher in 26, 39a and 39b yaf sites than in 15 and 33 yaf sites (Figure 3). However, fungal biomarkers were significantly higher in the 15 yaf site than in 26, 33 and 39 yaf sites. No significant differences in proportion of Gram + and Gram – bacteria were detected. The 26 yaf field had higher proportional levels of actinomycetes biomarkers than 15 and 33 yaf fields. Moreover, lower eukaryotic biomarker proportions were detected in the 26 yaf site than 33, 39a and 39b sites. At the 15–30 cm depth, proportions of AMF, fungal, actinomycetes and eukaryotic PLFAs were lower in the 33 yaf site compared to other fumigated fields (Figure 4). Gram + bacterial proportion was significantly higher in the 26 yaf site compared to the 33 yaf site.

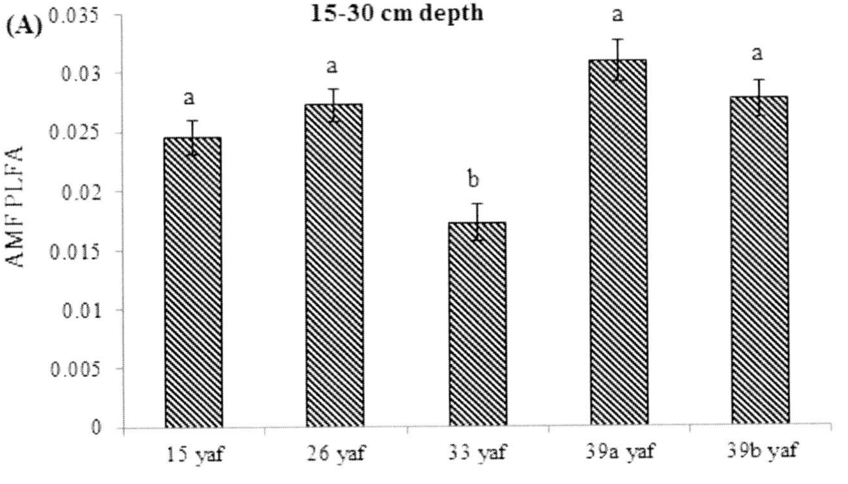

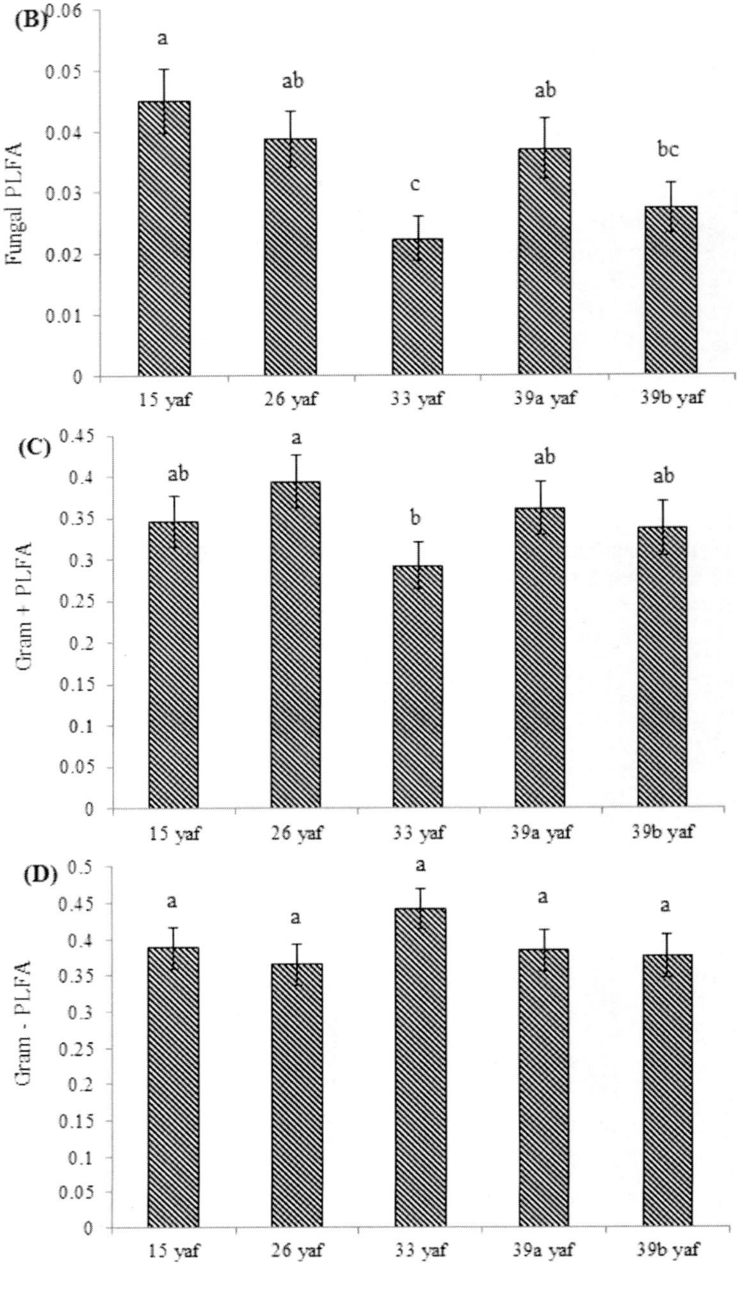
(B)
Fungal PLFA
0.06
0.05
0.04
0.03
0.02
0.01
0
a
ab
c
ab
bc
15 yaf
26 yaf
33 yaf
39a yaf
39b yaf
(C)
Gram + PLFA
0.45
0.4
0.35
0.3
0.25
0.2
0.15
0.1
0.05
0
ab
a
b
ab
ab
15 yaf
26 yaf
33 yaf
39a yaf
39b yaf
(D)
Gram - PLFA
0.5
0.45
0.4
0.35
0.3
0.25
0.2
0.15
0.1
0.05
0
a
a
a
a
a
15 yaf
26 yaf
33 yaf
39a yaf
39b yaf

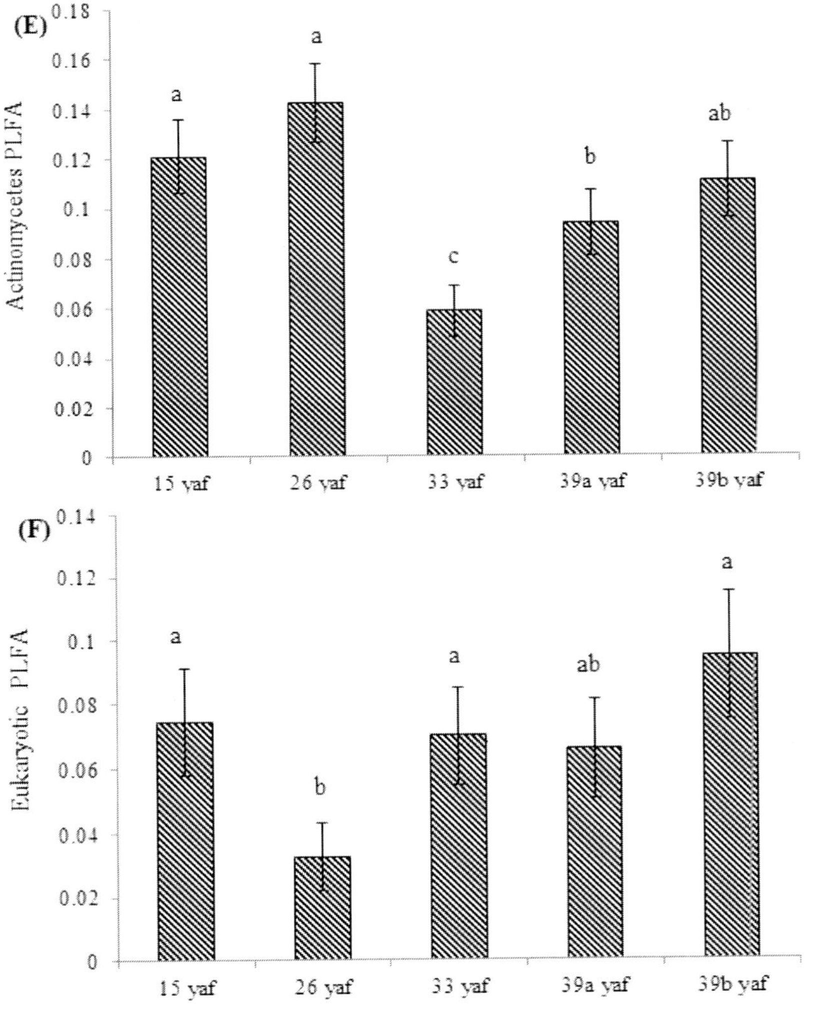

Figure 4. (**A**) AMF, (**B**) fungal, (**C**) gram positive, (**D**) gram negative, (**E**) actinomycetes, and (**F**) eukaryotic PLFA proportions at the 15–30 cm depth. Different letters indicate significant differences ($p < 0.05$). Error bars indicate standard error. Samples were collected from 15, 26, 33, 39a and 39b yaf sites.

3.3. Microbial Community Structure

Canonical multivariate analysis of variance suggests that differences in the microbial communities in fumigated and non-fumigated sites are significantly ($p < 0.05$) different. However, similarities were observed among clay loam 15 and 26 yaf sites; while the older fumigated sandy loam 33, 39a and 39b yaf sites were also similar to each other. Similarities were observed among NF-15 and NF-39 sites. Moreover, the NF-33 and 15 yaf sites are distinct from one another as well as the other fumigated sites at 0–5 cm depth (Figure 5). At the 5–15 cm depth, differences between fumigated and non-fumigated sites remain significant; all the fumigated sites except 33 yaf site remained in a close proximity as compared to non-fumigated sites where differences were greater (Figure 3). At the deeper depth, the 33 yaf and NF-33 sites were distinct from all other fumigated and non-fumigated sites (Figure 4).

Discriminant analysis indicated that microbial community structure in the fumigated site was completely different than non-fumigated sites. Canonical variates one and two discriminated 33 yaf site vs. NF-33, 15 yaf vs. NF-15 and NF-39 vs. 39 yaf site at the three soil depths (Table 6). At the 0–5 cm depth, the 33 yaf site contained higher actinomycetes.

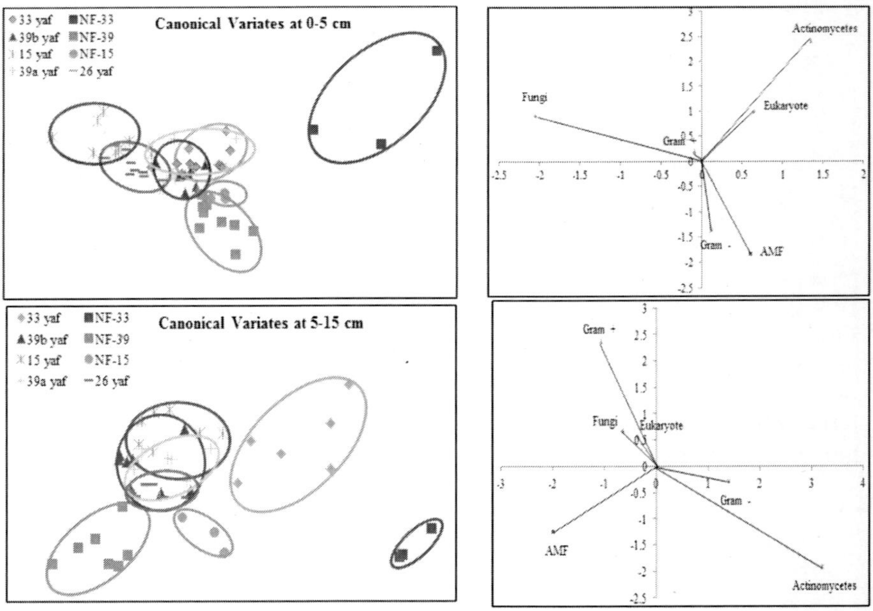

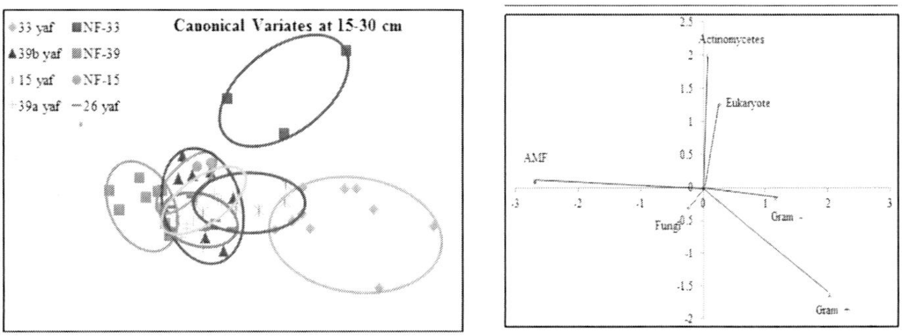

Figure 5. Canonical multivariate analysis of variance of phospholipid fatty acid biomarkers for 0–5 cm, 5–15 and 15–30 cm depths. Vectors represent standardized canonical coefficients and indicate the relative contribution of each biomarker group to each canonical variate. Soil samples were collected from 15, 26, 33, 39a, 39b yaf sites as well as non-fumigated (NF-15, NF-33 and NF-39) sites.

Table 6. Structure matrix (pooled with canonical structure) and function at group centroid for 0–5, 5–15, and 15–30 cm soil depths

Structure Loadings						
	0–5 cm		5–15 cm		15–30 cm	
Parameter	CV† 1	CV 2	CV 1	CV 2	CV 1	CV 2
Fungi	0.32	0.38	−0.05	0.40	0.02	−0.06
AMF	0.28	−0.13	0.0006	−0.44	−0.16	0.29
Gram +	0.307	0.16	0.31	−0.05	0.13	0.19
Gram −	0.34	0.33	0.28	0.02	0.25	0.17
Actino	0.51	0.34	0.39	−0.22	0.04	0.30
Eukaryote	0.09	0.56	0.10	0.17	0.20	0.70
Group Centroids						
15 yaf	−2.71	1.35	−0.84	1.71	0.17	−0.19
NF-15	1.36	−1.51	0.57	−1.69	−0.86	0.83
26 yaf	−1.98	0.24	−0.80	−0.51	−0.75	−0.78
33 yaf	0.50	0.55	3.26	1.23	3.20	−0.62
NF-33	7.31	3.31	4.90	−2.07	1.81	4.11
39a yaf	−0.09	0.39	−0.45	0.01	−0.46	−0.47
39b yaf	0.07	−0.42	−1.00	0.56	−0.61	0.19
NF-39	1.25	−2.77	−1.91	−1.41	−2.04	0.30

† CV, Canonical variate.

Physiological stress was determined using the ratios of saturated to monounsaturated (S/M) and the ratios of cyclopropyl PLFAs to their monoenoic precursors (cy/pre; Table 7). The S/M ratio was significantly greater for 15 yaf compared to NF-39 at 5–15 cm depth but not between other fumigated and non-fumigated sites at 0–5 and 5–15 cm depths. However, Cy/P ratio was significantly higher in the 15 yaf site compared to other fumigated and non-fumigated sites at 5–15 cm depth. Elevated levels of these indicate greater levels of microbial community stress in 15 years old fumigated sites at 5–15 cm depth.

Table 7. Changes in the ratio of saturated/monounsaturated PLFAs and cyclopropyl PLFAs to their precursors for fumigated and non-fumigated sites at 0–5, 5–15, and 15–30 cm depths. Different letters indicate significant differences ($p < 0.05$). Soil samples were collected from 15, 26, 33, 39a, 39b yaf sites as well as non-fumigated (NF-15, NF-33 and NF-39) sites.

Site	0–5 cm		5–15 cm		15–30 cm	
	S/M	Cy/P	S/M	Cy/P	S/M	Cy/P
15 yaf	1.19	0.64	1.29 [a]	1.31 [a]	1.19	0.82
NF-15	0.89	0.46	0.99 [a,b]	0.56 [b]	0.95	0.59
26 yaf	1.16	0.58	1 [a,b]	0.65 [b]	1.13	0.75
33 yaf	0.93	0.7	1.07 [a,b]	0.81 [b]	0.88	0.79
NF-33	0.92	0.56	0.97 [a,b]	0.64 [b]	0.58	0.73
39a yaf	1.08	0.41	0.99 [a,b]	0.62 [b]	1.04	0.53
39b yaf	1.19	0.53	1.19 [a,b]	0.69 [b]	1.1	0.76
NF-39	0.92	0.49	0.81 [b]	0.52 [b]	0.98	0.56

4. DISCUSSION

California is one of the largest users of MeBr and other soil fumigants in the US. To ensure better and more consistent crop yields and provide greater benefits to growers, soil fumigation has become an important agricultural practice. Therefore, producers of strawberries, tomatoes and other high value cash crops have been applying soil fumigants to control a wide array of target soil borne pests including nematodes, pathogens and weeds [21]. For example, California alone used 3550 tons of MeBr in 2004 before it was banned [22]. With the phase-out of MeBr because of its role in depleting stratospheric ozone, several alternatives have been used or in registration. However, the broad biocidal activity of fumigants and the continuous application can affect both target or harmful as well as non-target or beneficial soil organisms. Beneficial microorganisms are crucial in sustaining the health of agricultural soil by contributing to nutrient cycling, plant nutrient uptake, and agricultural productivity. This study helps to understand the effect of long-term repeated MeBr fumigation and early importance in studying the impact of alternative fumigants which will help to protect and create awareness among farmers on the importance of beneficial microorganisms.

Several short-term studies on the effect of soil microbial communities after fumigation have concluded that fumigation does not harm non-target soil microorganisms. Other studies have suggested there may be differences in deleterious and beneficial rhizosphere colonizers following fumigation in strawberry rhizosphere fumigated and non-fumigated soils [23]. A study by Ibekwe et al. [24] concluded that the lack of negative effects after fumigation may be due to the reduction in populations of plant pathogens and weeds, thus enhancing bacteria to rebound after the initial decline in population and use the available carbon sources for growth. Moreover, Ibekwe et al. [25] confirmed that MeBr has the greatest impact on soil microbial communities and that 1,3-D has the least impact. Therefore, whether it is a short-term or a long-term study, the residual effects of different fumigants on beneficial microorganisms are still unclear. Almost all of the studies conducted to investigate the effect of fumigants on soil microbial communities have focused on the short-term effect. In the present study, we have focused on the field sites where MeBr was applied annually for 15 years or every other year for 26, 33 and 39 years.

Soil microbial biomass (total PLFA) was affected by long-term repeated fumigation in the 33 yaf site. Soil microbial biomass in the non-fumigated 33 year old site was higher than in the 33 yaf site (Figure 1); however, the different years (15, 33 and 39) did not affect soil microbial biomass. Nevertheless, canonical multivariate analysis showed that microbial communities in fumigated and non-fumigated sites are completely different and similarities were also observed between 15 and 26 yaf clay loam fields and 33, 39a, and 39b yaf sandy loam fields at the soil surface. Also, the shift in microbial community occurred in the 15 yaf continuously fumigated site at 5–15 cm depth, as evidenced by elevated levels of the stress rations S/M and Cy/Pre. These stress markers reflect the degree to which microorganisms are affected by temperature, nutrient, pesticide application and tillage [17,26]. However, other fumigated and non-fumigated sites in 0–5 and 15–30 cm depths did not show differences in S/M and Cy/Pre ratios.

Reeve et al. [13] proposed that the long-term repeated use of fumigants may have more significant and lasting effects on soil microbial populations than have been measured in a single fumigation events study. To test the long-term effect on microbial community structure, we collected soil samples from different years of continuous fumigation. Our study shows that the proportion of fungal biomarkers was significantly higher in 15 and 26 yaf as compared to sites with 33 and 39 years of annual fumigation. However, AMF proportion was higher in 39a and 39b yaf sites compared to 33 yaf site. Lower fungal biomass is typical of intensively cultivated agricultural soils and it has been attributed to different factors such as physical disturbance, and altered amount and complexity of nutrient inputs and decrease in soil organic matter as compared to undisturbed soil [27]. Gram + bacterial proportion was lower in 15 and 26 yaf sites than older fumigated sites, i.e., 33, 39a and 39b yaf sites. A study by Ibekwe et al. [25] and Zelles et al. [28] also found that Gram + bacteria were less injured by fumigation and concluded that this is due to protection by the cell wall structure of the bacteria by formation of spores and ability to adapt to fumigant vapor more quickly. Other studies on the effect of fumigation on soil microbial composition determined that application of high doses of metam sodium (MS) fumigant increased actinomycetes and Gram + bacterial population following fumigation; moreover, fungal and arbuscular mycorrhizal fungal populations were sensitive to MS [29].

The soil textural effect (sandy loam and clay loam) of long-term annual fumigation on soil microbial communities was also studied. Ibekwe et al. [30] suggested clayey soils might protect microorganisms from the effects of fumigation, and clay loam soil in 15 and 26 yaf sites might have protected fungal communities somewhat from the effects of fumigation as compared to the sandy loam at the 33, 39a and 39b yaf sites. However, the proportional level of AMF PLFA was significantly lower in fumigated 33 and 39 yaf sandy loam soils and 15 and 26 yaf clay loam soils compared to their non-fumigated counterparts. Similarly, a study on effects of soil type and farm management on soil ecological functional genes and microbial activities conducted by Reeve et al. [13] concluded that over the long term under field conditions some genes were more affected by management on silty clay loam soil compared to sandy loam.

Cover crops can improve soil quality [31] by reducing chemical inputs such as methyl bromide and other fumigations in strawberry production. Continuous strawberry production on the same site can decrease strawberry yield by causing proliferation of weeds and accumulation of pathogens in the soil [32]. Limited information is available on the effect on microbial community in strawberry after rotation with cover crops in comparison with continuous strawberry production. However, continuous strawberry production at the 15 yaf site was associated with significantly higher proportions of fungal biomarkers as compared to rotation sites. Similar to our results, a study by Zhang et al. [33] on short-term effects of crop rotation on microbial communities found that soils under corn-soybean rotation had lower fungal biomass compared to soils under monoculture corn. In the present study, cover crop such as barley in 39 yaf site and cash crop such as peas or beans at the 33 yaf site possibly increased Gram + bacterial proportion. Long-term repeated fumigation without any crop rotation at the 15 yaf site was associated with a lower proportion of Gram + bacteria and actinomycetes.

In the current study, to present comparative data from fumigated and non-fumigated sites, we collected soil samples from differently aged fumigated sites as well as from never-fumigated sites, and organic sites that have not been fumigated for at least 10 years. Arbuscular mycorrhizal fungi play a crucial role in nutrient acquisition and soil fertility. Our results show that proportions of AMF were significantly lower at the 15, 33 and 39 yaf sites compared to their non-fumigated counterparts. Other studies have also found MeBr to be clearly detrimental to AMF [34], which causes stunting or poor production in several cropping systems [35]. Also, [36] showed that AMF spore abundance and species diversity was significantly higher in the organic than in the conventional systems.

Microbial biomass is often correlated with soil organic C [37,38] and may be more highly correlated with active C than total C [39]. In the present study, the non-fumigated site compared to the 33 yaf site had significantly greater total C as well as greater microbial biomass (total PLFA) at the surface soil. Total carbon (organic carbon) can have a beneficial impact on soil quality, enhancing soil structure and fertility, and increasing water infiltration and storage [40]. However, no notable difference in DOC between fumigated and non-fumigated sites was found. Also, TN was significantly higher at the 15 yaf site compared to its representative non-fumigated site. Levels of macronutrients were significantly lower in 15 yaf surface soil compared to its non-fumigated counterpart, similar to the 39 years old yaf surface soil in comparison to organic surface soil, excluding Ca being notably higher in NF-33 soil.

5. CONCLUSIONS

Strawberry fields had lower AMF populations as compared to their non-fumigated sites in all fumigated fields, which is a risk to fumigated fields as AMF plays a major role in soil health and fertility. Crop rotations and cover crops likely helped bacterial proportions; however, proportions of Gram + bacteria and actinomycetes were affected without any crop rotations in the 15 yaf site. This study stresses a need for further research on the effects of crop rotation as well as the long-term impact of currently used MeBr alternative soil treatments on beneficial soil microorganisms to select soil disinfestation methods that have low negative environmental impacts, maintain soil health, and enhance sustainable crop production in California.

ACKNOWLEDGMENTS

The authors would like to acknowledge Allie Marquez, Margaret Axiak and David Correia for assistance with field and lab work; Ernie Levya from TriCal, Inc. and collaborating farmers; Rod Koda, Adrian Espinoza, Luis Martin, Miguel Ramos and Darrel Uyeda who participated in this project by helping find fumigated and non-fumigated sites. Financial support for this research was provided by California Department of Food and Agriculture (CDFA) Specialty Crop Block Grant Program (Grant No. 12043). Mention of trade names or commercial products in this publication is solely for the purpose of providing specific information and does not imply recommendation or endorsement by the U.S. Department of Agriculture. USDA is an equal opportunity provider and employer.

REFERENCES

1. CDFA, California Department of Food and Agriculture. California Agricultural Production Statistics. Sacramento, CA, USA, 2014. Available online: https://www.cdfa. ca. gov/statistics/ (accessed on 12 January 2017).
2. Carpenter, J.; Lynch, L.; Trout, T. Township limits on 1,3-D will impact adjustment to methyl bromide phase-out. *Calif. Agric.* **2001**, *55*, 8–12.
3. Hanson, B.D.; Gao, S.; Gerik, J.S.; Qin, R.; Cabrera, J.A.; Jhala, A.J.; Abit, M.J.M.; Cox, D.; Correiar, B.; Wang, D.; et al. Preplant 1,3-D treatments test well for perennial crop nurseries, but challenges remain. *Calif. Agric.* **2013**, *67*, 181–189.
4. McCarthy, J.E.; Copeland, C.; Parker, L.; Schierow, L.J. *Clean Air Act: A Summary of the Act and Its Major Requirements*; Congressional Research Service (CRS): Washington, DC, USA, 2011.
5. Fennimore, S.A.; Serohijos, R.; Samtani, J.B.; Ajwa, H.A.; Subbarao, K.V.; Martin, F.N.; Daugovish, O.; Legard, D.; Browne, G.T.; Muramoto, J.; et al. TIF film, substrates and nonfumigant soil disinfestation maintain fruit yields. *Calif. Agric.* **2013**, *67*, 139–146.
6. Fennimore, S.A.; Martin, F.N.; Miller, T.C.; Broome, J.C.; Dorn, N.; Greene, I. Evaluation of a mobile steam applicator for soil disinfection in California strawberry. *HortScience* **2014**, *49*, 1542–1549.
7. Reicosky, D.; Forcella, F. Cover crop and soil quality interactions in agroecosystems. *J. Soil Water Conserv.* **1998**, *53*, 224–229.
8. Seigies, A.; Pritts, M. Cover crop rotations alter soil microbiology and reduce replant disorders in strawberry. *HortScience* **2006**, *41*, 1303–1308.
9. Dangi, S.R.; Tirado-Corbalá, R.; Cabrera, J.A.; Wang, D.; Gerik, J. Soil biotic and abiotic responses to dimethyl disulfide spot drip fumigation in established grape vines. *Soil Sci. Soc. Am. J.* **2014**, *78*, 520–530.
10. Dangi, S.R.; Gerik, J.; Tirado-Corbalá, R.; Ajwa, H. Soil microbial community structure and target organisms under different fumigation treatments. *Appl. Environ. Soil Sci.* **2015**.
11. Dungan, R.S.; Ibekwe, A.M.; Yates, S.R. Effect of propargyl bromide and 1, 3 dichloropropene on microbial communities in an organically amended soil. *FEMS Microbiol. Ecol.* **2003**, *43*, 75–87.
12. Klose, S.; Ajwa, H.A. Enzyme activities in agricultural soils fumigated with methyl bromide alternatives. *Soil Biol. Biochem.* **2004**, *36*, 1625–1635.

13. Reeve, J.R.; Schadt, C.W.; Carpenter-Boggs, L.; Kang, S.; Zhou, J.; Reganold, J.P. Effects of soil type and farm management on soil ecological functional genes and microbial activities. *ISME J.* **2010**, *4*, 1099–1107.
14. Buyer, J.S.; Teasdale, J.R.; Roberts, D.P.; Zasada, I.A.; Maul, J.E. Factors affecting soil microbial community structure in tomato cropping systems. *Soil Biol. Biochem.* **2010**, *42*, 831–841.
15. Buyer, J.S.; Sasser, M. High throughput phospholipid fatty acid analysis of soils. *Appl. Soil Ecol.* **2012**, *61*, 127–130.
16. Blackwood, C.B.; Buyer, J.S. Soil microbial communities grown with Bt and non—Bt corn in three soils. *J. Environ. Qual.* **2004**, *33*, 832–836.
17. Moore-Kucera, J.; Dick, R.P. PLFA profiling of microbial community structure and seasonal shifts in soils of a Douglas-fir chronosequence. *Microb. Ecol.* **2008**, *55*, 500–511.
18. SAS Institute. *SAS/STAT User's Guide*; Version 9.4; SAS Institute: Cary, NC, USA, 2003.
19. Buyer, J.S.; Roberts, D.P.; Russek-Cohen, E. Microbial community structure and function in the spermosphere as affected by soil and seed type. *Can. J. Microbiol.* **1999**, *45*, 138–144.
20. Buyer, J.S.; Roberts, D.P.; Russek-Cohen, E. Soil and plant effects on microbial community structure. *Can. J. Microbiol.***2002**, *48*, 955–964.
21. Gerik, J.S.; Hanson, B.D. Drip application of methyl bromide alternative chemicals for control of soilborne pathogens and weeds. *Pest Manag. Sci.* **2011**, *67*, 1129–1133.
22. Meadows, R. News overview: Researchers develop alternatives to methyl bromide fumigation. *Calif. Agric.* **2013**, *67*, 125–127.
23. Martin, F.N. Development of alternative strategies for management of soil borne pathogens currently controlled with methyl bromide. *Annu. Rev. Phytopathol.* **2003**, *41*, 325–350.
24. Ibekwe, A.M.; Papiernik, S.K.; Yang, C.H. Influence of soil fumigation by methyl bromide and methyl iodide on rhizosphere and phyllosphere microbial community structure. *J. Environ. Sci. Health Part B* **2010**, *45*, 427–436.
25. Ibekwe, A.M.; Papiernik, S.K.; Gan, J.; Yates, S.R.; Yang, C.H.; Crowley, D.E. Impact of fumigants on soil microbial communities. *Appl. Environ. Microbiol.* **2001**, *67*, 3245–3257.
26. Kaur, A.; Chaudhary, A.; Kaur, A.; Choudhary, R.; Kaushik, A. Phospholipid fatty acid- A bioindicator of environment monitoring and assessment in soil ecosystem. *Curr. Sci.* **2005**, *89*, 1103–1112.
27. Lazcano, C.; Gomez-Brandon, M.; Revilla, P.; Dominguez, J. Short-term effects of organic and inorganic fertilizers on soil microbial community structure and function. *Biol. Fertil. Soils* **2012**, *49*, 723–733.
28. Zelles, L.; Palojarvi, A.; Kandeler, E.; Von lutzow, M.; Winter, K.; Bai, Q.Y. Changes in soil microbial properties and phospholipid fatty acid fractions after chloroform fumigation. *Soil Biol. Biochem.* **1997**, *29*, 1325–1336.
29. Macalady, J.L.; Fuller, M.E.; Scow, K.M. Effects of metam sodium fumigation on soil microbial activity and community structure. *J. Environ. Qual.* **1998**, *27*, 54–63.
30. Ibekwe, A.M. Effects of fumigants on non-target organisms in soil. *Adv. Agron.* **2004**, *83*, 1–35.
31. Nonnecke, G.; Pritts, M. Sustainable strawberry production and management. In *Our Strawberries*; Khanizadeh, S., DeEll, J., Eds.; Public Works and Government Services Canada, Publishing and Depository Services: Ottawa, ON, Canada, 2005; pp. 84–96.
32. LaMondia, J.; Elmer, W.; Mervosh, T.; Cowles, R. Integrated management of strawberry pests by rotation and intercropping. *Crop Protect.* **2002**, *21*, 837–846.
33. Zhang, B.; Li, Y.; Ren, T.; Tian, Z.; Wang, G.; He, X. Short-term effect of tillage and crop rotation on microbial community structure and enzyme activities of a clay loam soil. *Biol. Fertil. Soils* **2014**, *50*, 1077–1085.
34. Menge, J.A. Effect of soil fumigants and fungicides on vesicular-arbuscular fungi. *Phytopathology* **1982**, *72*, 1125–1132.
35. Hass, J.H.; Bar-Yosef, B.; Krikun, J.; Barak, R.; Markovitz, T.; Kramer, S. Vesicular-arbuscular mycorrhizal-fungus infestation and phosphorus fertigation to overcome pepper stunting after methyl bromide fumigation. *Agron. J.* **1987**, *79*, 905–910.

36. Oehl, F.; Sieverding, E.; Mader, P.; Dubois, D.; Ineichen, K.; Boller, T.; Wiemken, A. Impact of long-term conventional and organic farming on the diversity of arbuscular mycorrhizal fungi. *Oecologia* **2004**, *138*, 574–583.
37. Franzluebbers, A.J.; Haney, R.L.; Hons, F.M.; Zuberer, D.A. Active fraction of organic matter in soils with different texture. *Soil Biol. Biochem.* **1996**, *28*, 1367–1372.
38. Hassink, J. Effect of soil texture on the size of the microbial biomass and on the amount of C and N mineralized per unit of microbial biomass in Dutch grassland soils. *Soil Biol. Biochem.* **1994**, *33*, 1713–1722.
39. Weil, R.R.; Islam, K.R.; Stine, M.A.; Gruver, J.B.; Samson-Liebig, S.E. Estimating active carbon for soil quality assessment: A simplified method for laboratory and field use. *Am. J. Altern. Agric.* **2003**, *18*, 3–17.
40. Weil, R.R.; Magdoff, F. Significance of soil organic matter to soil quality and health. In *Soil Organic Matter in Sustainable Agriculture*; Magdoff, F., Weil, R.R., Eds.; CRC Press: Boca Raton, FL, USA, 2004; pp. 1–43.

CHAPTER 15

Yield gaps in oil palm: A quantitative review of contributing factors

Lotte S. Woittiez[a,*], Mark T. van Wijk[b], Maja Slingerland[a], Meine van Noordwijk[a, c], Ken E. Giller[a]

[a] *Plant Production Systems, Wageningen University, P.O. Box 430, 6700 AK Wageningen, The Netherlands*

[b] *International Livestock Research Institute (ILRI), Old Naivasha Road, PO Box 30709, Nairobi 00100, Kenya*

[c] *World Agroforestry Centre (ICRAF) Southeast Asia Regional Office, Jl. Cifor, Situ Gede, Sindang Barang, Bogor 16115, Indonesia*

ABSTRACT

Oil palm, currently the world's main vegetable oil crop, is characterised by a large productivity and a long life span (≥25 years). Peak oil yields of 12 t ha^{-1} yr^{-1} have been achieved in small plantations, and maximum theoretical yields as calculated with simulation models are 18.5 t oil ha^{-1} yr^{-1}, yet average productivity worldwide has stagnated around 3 t oil ha^{-1} yr^{-1}. Considering the threat of expansion into valuable rainforests, it is important that the factors underlying these existing yield gaps are understood and, where feasible, addressed. In this review, we present an overview of the available data on yield-determining, yield-limiting, and yield-reducing factors in oil palm; the effects of these factors on yield, as measured in case studies or calculated using computer models; and the underlying plant-physiological mechanisms. We distinguish four production levels: the potential, water-limited, nutrient-limited, and the actual yield. The potential yield over a plantation lifetime is determined by incoming photosynthetically active radiation (PAR), temperature, atmospheric CO_2 concentration and planting material, assuming optimum plantation establishment, planting density (120–150 palms per hectares), canopy management (30–60 leaves depending on palm age), pollination, and harvesting. Water-limited yields in environments with water deficits >400 mm year^{-1}can be less than one-third of the potential yield, depending on additional factors such as temperature, wind speed, soil texture, and soil depth. Nutrient-limited yields of less than 50% of the potential yield have been recorded when nitrogen or potassium were not applied. Actual yields are influenced by yield-reducing factors such as unsuitable ground vegetation, pests, and diseases, and may be close to zero in case of severe infestations. Smallholders face particular constraints such as the use of counterfeit seed and insufficient fertiliser application. Closing yield gaps in existing plantations could increase global production by 15–20 Mt oil yr^{-1}, which would limit the drive for further area expansion at a global scale. To increase yields in existing and future plantations in a sustainable way, all production factors mentioned need to be understood and addressed.

Keywords: Palm oil; Perennial; Yield; Intensification; Physiology; Management

HIGHLIGHTS

- Potential oil palm yields are >8 t oil ha^{-1} year^{-1} in most regions but average actual yields are about 3.3 t oil ha^{-1} year^{-1}.
- Basic physiological processes underlying bunch production are not well understood.
- Water-limited yields are less than half of potential yields in drier growing regions of Thailand, Africa and the Americas.
- Average nutrient application rates in Southeast Asia are 40–90% of recommended rates.
- Pest pressure is limited in most regions, but actual yields can be strongly reduced because of disease pressure (*Ganoderma* and bud rot).
- There is much scope for closing the yield gap and simultaneously improving plantation sustainability.

1. INTRODUCTION

African oil palm (*Elaeis guineensis* Jacq.) has its centre of origin in the humid lowland tropics of West Africa. Wild oil palms are uncommon in primary forests but rather grow in disturbed and very wet locations, such as swamps and river banks, where sunlight is abundant and water available throughout the year (Zeven, 1967). The oil palm is a member of the monocotyledonous palm family (*Arecaceae*). The woody stem carries a single terminal growing point, from which leaves appear at regular intervals in a double spiral (Rees, 1964). Each leaf supports a single inflorescence, which can be either male or female. The harvested product is a fruit bunch comprising 1500–2000 fruitlets. Crude palm oil (CPO) is extracted from the orange-yellow mesocarp, and palm kernel oil (PKO) from the white kernel.

Over the last 100 years, oil palm has changed from a smallholder agroforestry crop and ornamental palm into the world's most important vegetable oil crop. Current worldwide production is estimated at 63 Mt crude palm oil per year, or 36% of the total world vegetable oil production (USDA, 2014). Expansion of oil palm plantations has been suggested as a key cause of deforestation in both Indonesia (Carlson *et al.*, 2012 ; Stibig *et al.*, 2014) and Malaysia (Miettinen *et al.*, 2011 ; Stibig *et al.*, 2014), although other drivers such as logging also play a major role (Laurance, 2007 ; Lambin *et al.*, 2001). The increasing demand for palm oil over the coming decades will probably be met both through expansion of the area planted and increased productivity (Carter *et al.*, 2007 ; Corley, 2009).

Since oil palm expansion may lead to the displacement of biodiverse rainforests (Gaveau et al., 2014), increased productivity, combined with targeted expansion into degraded areas (Fairhurst and McLaughlin, 2009), are the preferred strategies to meet the growing demand for palm oil. Increasing productivity does not, *per se*, lead to reduction in deforestation unless supporting policies are in place and are properly enforced (Angelsen, 2010), but is a necessary step towards reducing pressure on land. A thorough understanding and quantification of the contribution of different production factors to oil palm yield is urgently needed to estimate the scope to increase productivity in existing stands, and in ongoing (re)planting programs.

Yield gap analysis has been commonly used as a tool to explore the possibilities for improving land productivity (Lobell *et al.*, 2009 ; van Ittersum *et al.*, 2013; see also www.yieldgap.org). The 'yield gap' is defined as the difference between potential and actual yield (van Ittersum and Rabbinge, 1997), with the upper limit of productivity per hectare being the 'potential yield'. This potential yield is defined as the theoretical yield at a given temperature, ambient atmospheric CO_2 concentration, and incoming photosynthetically active radiation (PAR), with optimum agronomic management and without water, nutrient, pest and disease limitations (van Ittersum and Rabbinge, 1997). It refers to current germplasm or to the best currently available material.

Yield gap analysis has been carried out for a range of annual crops such as wheat (Aggarwal and Kalra, 1994; Bell *et al.*, 1995 ; Anderson, 2010), cassava (Fermont et al., 2009), rice (Yang *et al.*, 2008 ; Laborte *et al.*, 2012), and cereals in general (Neumann et al., 2010). A limited number of perennial cropping systems has been subjected to yield gap analysis, including coffee (Wairegi and Asten, 2012), highland banana (Wairegi et al., 2010), and cocoa (Zuidema et al., 2005). Perennial crops such as oil palm are structurally different from annual crops in several ways. In annual crops, growers can take advantage of new seeds with each growing season. By contrast, the yield potential for perennial crops, with a lifespan of up to several decades, is fixed for each planting cycle. Events early in the plantation lifetime, especially in the nursery and at planting, may have strong effects on yield in later years, which complicates the interpretation of yield data (Breure and Menendez, 1990). In addition, oil palm fruit bunches take several years to develop, and there is a time lag of 20–30 months between the onset of stress factors and their impact on yield. This makes it difficult to separate and quantify the effects of individual factors (Adam et al., 2011).

Quantitative data on yield responses of oil palm to different production factors, particularly planting density, irrigation, and fertiliser use, are available from trials carried out by companies or research stations. Results of many such trials are reported only in the grey literature and can be difficult to access, but Corley and Tinker (2016) provide a very complete overview. Recently, Fairhurst and Griffiths (2014) performed a yield gap analysis in oil palm from a practical planters' perspective, with a step-by-step guidance on the identification and resolution of yield constraints in the field. However, an assessment of the underlying causes of yield gaps in oil palm production systems worldwide is lacking. In this review, we explore existing knowledge on oil palm productivity from a plant physiological perspective, to provide a coherent picture of factors contributing to yield gaps in oil palm. We start with a discussion on plantation life cycle, vegetative growth, and leaf area development in Section 2. In Section 3 we provide a detailed assessment of bunch production, focusing on bunch number and bunch weight, the two main determinants of yield. In Section 4 we review the yield gap concept and the different production levels (i.e. potential, water-limited, nutrient-limited, and actual yield), and discuss the different factors that affect generative productivity in oil palm, including climatic factors, nutrition, and the main pests and diseases. In Section 5 we consider the most important constraints to yield in the oil palm producing regions around the world, with focus on both large-scale commercial and smallholder systems. Finally in Section 6 we identify the existing knowledge gaps and propose directions for future action and research.

2. PLANTATION LIFE CYCLE AND VEGETATIVE GROWTH

In this section we discuss the oil palm production system, the different yield profiles during the plantation life time and the vegetative growth of the oil palm, with a focus on leaf area development.

2.1. Plantation life cycle

Oil palms are commercially grown in plantation systems, with a density of 120–150 palms per hectare. Pre-germinated seeds are raised in polybags in a nursery for 6–12 months (Rankine and Fairhurst, 1999a), after which the seedlings are planted in the field at final density with limited options for replacing plants that do not survive or prove to have less-desirable properties beyond the first 12 months. Plantations have an average lifetime of 25 years, of which 21–23 are productive. Four yield phases have been described (Fig. 1): 1) the immature or 'yield building phase', up to 2–3 years after planting (YAP), before harvestable production begins and when the canopy is not yet closed; 2) the young mature phase or 'steep ascent yield phase', 4–7 YAP, when leaf area and yield increase linearly; 3) the mature or 'plateau yield phase', 8–14 YAP, when yield and leaf area are stable; and 4) a phase of yield decline, 15–25 YAP (Ng, 1983; Goh *et al.*, 1994 ; Fairhurst and Griffiths, 2014).

The first year of harvest typically yields 10–15 t fruit bunches ha^{-1} (with and oil to bunch ratio of 10–15%) under favourable circumstances; initial yields of >20 t fruit bunches ha^{-1}have been achieved in commercial plantings (Rao et al., 2008). Under favourable conditions, bunch production peaks 6–7 YAP, with typical peak yields of 35 t fruit bunches ha^{-1} (Ng, 1983 ; Donough *et al.*, 2009). Maximum yields of 60 t fruit bunches ha^{-1} have been obtained with selected clonal planting materials (Ng et al., 2003). During the mature phase, bunch production stabilises somewhat below the peak achieved at six YAP, with typical commercial yields of 25–30 t fruit bunches ha^{-1} in well-managed plantations (Ng, 1983 ; Donough *et al.*, 2010). In the phase of yield decline, leaf production rate and bunch numbers decrease, but increased bunch size partly compensates for the reduction in bunch number (Hardon *et al.*, 1969; Goh *et al.*, 1994 ; Jacquemard and Baudouin, 1998: 21). Oil palms continue to produce fruit bunches until death, but replanting is required at 20–25 YAP when palms become too tall for economic harvesting or when yields decline due to the loss of palms to pests and diseases.

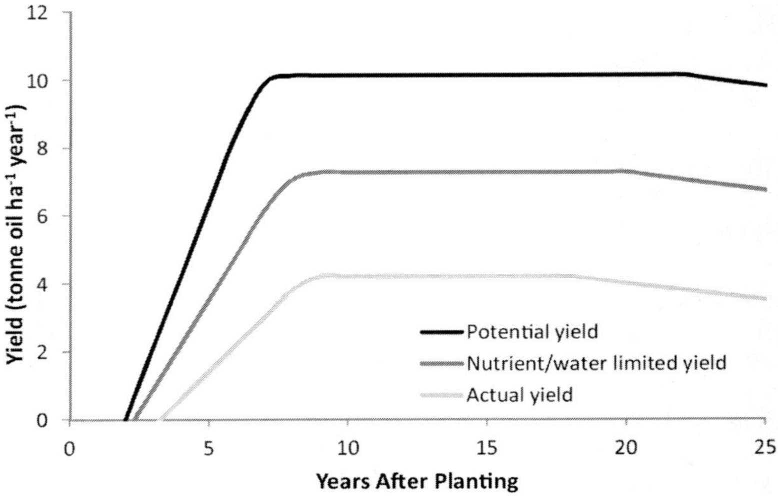

Figure. 1. Development of oil palm yield over time in three hypothetical plantations (after Ng, 1983; Goh *et al.*, 1994 ; Fairhurst and Griffiths, 2014). The light grey (bottom), dark grey (middle) and black grey (top) lines show the yield progress at different productivity levels: actual yield (average 3.5 t oil ha^{-1} $year^{-1}$), nutrient/water limited yield (average 6.1 t oil ha^{-1} $year^{-1}$) and potential yield (average 8.9 t oil ha^{-1} $year^{-1}$), respectively, with a large gap between the three levels. The yield building (no yield), young mature (increasing yield), mature (plateau) and yield decline phase can be discerned.

2.2. Vegetative growth

The average yearly above-ground dry matter production per hectare for mature palms (> 10 YAP) planted with triangular spacing at planting densities of 120–150 palms ha^{-1} ranges from 19 t DM ha^{-1} yr^{-1} in Nigeria (Rees and Tinker, 1963) to 32 t DM ha^{-1} yr^{-1} in Malaysia (Corley et al., 1971a). Dry matter production can be described by the following equation:

$$DMP = PAR \times f \times RUE \text{ (1)}$$

where DMP = dry matter production (kg m^2 yr^{-1}), *PAR* = yearly photosynthetically active radiation (MJ m^2 yr^{-1}; 50% of total incoming solar radiation (Monteith, 1972)), *f* = fraction of radiation intercepted by the canopy, and RUE = radiation use efficiency (kg DM MJ^{-1} PAR) (Monteith, 1977 ; Corley, 2006). Estimated values for RUE are 0.6–1.3 g MJ^{-1} PAR (Rees and Tinker, 1963; Squire, 1986 ; Squire and Corley, 1987). RUE does not change with age in oil palm (Squire and Corley, 1987) but is decreased in dry climates and on poor soils and enhanced by fertiliser use (15–30% increase in response to the application of N-P-K) (Squire, 1986). Radiation interception (*f*) depends mainly on the leaf area index (LAI), i.e. the area of leaves per surface area (m^2 m^{-2}), although leaf orientation with respect to light angle can modify effective interception. The LAI increases linearly from planting until 5–6 YAP and peaks around 10 YAP, when the leaves reach their maximum size (Gerritsma and Soebagyo, 1999). The maximum LAI typically varies between 4 and 6 depending on genotype (Gerritsma and Soebagyo, 1999 ; Breure, 2010), environment (Corley et al., 1973), planting density (Corley *et al.*, 1973 ; Gerritsma and Soebagyo, 1999), pruning (Squire and Corley, 1987), fertiliser use (Breure, 1985 ; Corley and Mok, 1972), and general agronomic management. In plantations where old leaves are not removed, LAI may exceed 10 (Squire and Corley, 1987). At an LAI of 4.5 interception of PAR is at least 80%, increasing up to 90–95% at an LAI of 6–7 (Gerritsma, 1988 ; Breure, 1988). Yields are reduced when LAI exceeds a value of 6 due to competition among palms (Breure, 2010).

In older plantations, most of the standing biomass is contained in the trunk (Rees and Tinker, 1963). Of an estimated gross primary production of 160 t DM ha^{-1} yr^{-1} in 10-year-old palms in Malaysia, around 70 t ha^{-1} yr^{-1} was allocated to trunk, root, and rachis respiration, and 55 t ha^{-1} yr^{-1} allocated to leaflet respiration, leaving 30–35 t ha^{-1} yr^{-1} of dry matter production (Corley, 1976b). Estimates of standing root biomass at 15 YAP from different experiments were listed by Henson and Chai (1997), ranging from 9 t DM ha^{-1} (Corley et al., 1971a) to 20 t DM ha^{-1} (Teoh and Chew, 1988). Under conditions without water limitation, about 10–12% of assimilates are allocated to the roots (Henson and Chai, 1997), but under water limited conditions, assimilate allocation to roots may be up to 35% (Dufrène *et al.*, 1990 ; van Noordwijk *et al.*, 2015).

In productive palms planted at standard densities, about 45–50% of the aboveground dry matter production is allocated to generative growth (male inflorescences and female inflorescences and bunches; Corley et al., 1971b). It has been proposed that allocation of assimilates to inflorescences and bunches will not occur until demands for vegetative production are met (the "overflow" model; Corley et al., 1971b). Yet later research has shown that both vegetative and generative growth are source-limited and that competition occurs between the different sinks, although priority is given to vegetative growth (Corley and Tinker, 2016: 103).

3. FRUIT DEVELOPMENT

A number of key stages can be distinguished during inflorescence and fruit bunch development (Fig. 2; for a detailed review, see Adam et al., 2005). Oil yield depends on the number of harvested bunches, the bunch weight, and the oil content of the fruit (Breure et al., 1990). These factors are discussed in detail below.

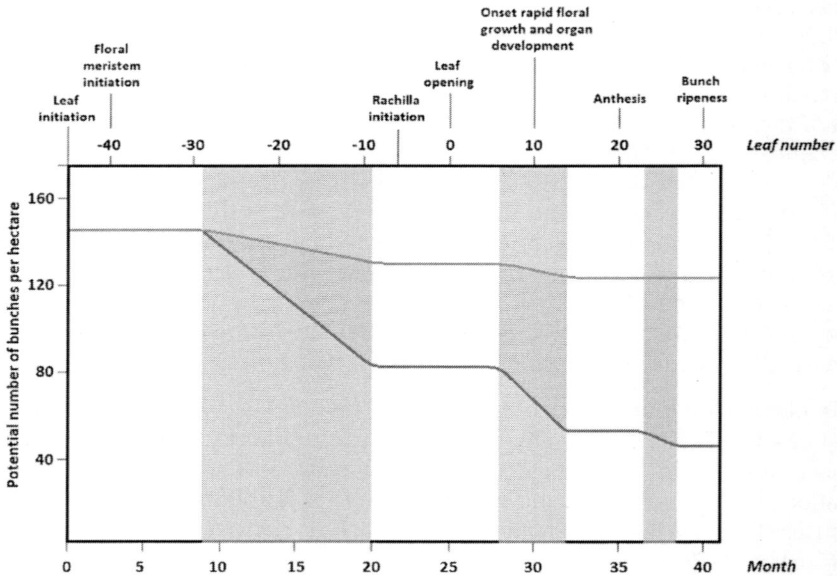

Figure. 2. Schematic representation of inflorescence and bunch development, showing key developmental stages and the effects of stress on potential bunch number (after Uexküll and Fairhurst, 1991; Corley *et al.*, 1995 ; Adam *et al.*, 2005). Time starts at leaf initiation (point zero) and progresses until bunch ripeness, and is indicated in months since leaf initiation (bottom x-axis) and leaf number (upper x-axis, assuming an average phyllochron length of 1.9 month^{-1}). The y-axis shows the number of potential bunches per hectare. The two lines show the progress of two hypothetical batches of potential bunches, starting at one per palm in a plantation with a planting density of 142 palms per hectare. Over time the number of potential bunches decreases as the batches pass through several critical phases. Severe stress (bottom line) leads to larger reductions in bunch number than mild stress (top line). The bars represent the stress-sensitive periods: sex determination (left), inflorescence abortion (middle) and bunch failure (right).

3.1. Bunch number

The number of ripe bunches available for harvest is determined by 1) the number of inflorescences initiated (which in turn depends on the rate of leaf production; Gerritsma and Soebagyo, 1999); 2) sex ratio (Corley *et al.*, 1995; Adam *et al.*, 2011 ; Heel *et al.*, 1987); 3) abortion of female inflorescences before anthesis (Pallas et al., 2013); and 4) failure of developing bunches between anthesis and bunch ripeness (Combres et al., 2013).

3.1.1. Number of developing inflorescences

Leaf initiation rate determines directly the potential number of inflorescences, as a single inflorescence is initiated in the axil of each leaf. An average oil palm carries 45–50 unopened leaves in varying stages of development and 32–48 opened leaves (Breure, 1994). The youngest fully opened leaf is denoted as Leaf 1, with unopened leaves being numbered negatively (Fig. 2). Leaf initiation rate is determined primarily by palm age (Broekmans, 1957), with opening rates declining rapidly in the first 10 YAP (Gerritsma and Soebagyo, 1999). Typically 40–45 leaves palm^{-1} year^{-1} are produced at two YAP, 25–35 leaves year^{-1} at six YAP, 20–25 leaves year^{-1} at 12–14 years YAP (Broekmans, 1957 ; Gerritsma and Soebagyo, 1999) and 17–20 leaves year^{-1} at 21 YAP (Broekmans, 1957 ; Rafii *et al.*, 2013). Leaf initiation rate may vary between different planting materials by ±1 leaf palm^{-1} year^{-1} (Gerritsma and Soebagyo, 1999) or three days per phyllochron (the time elapsed between the appearance of two consecutive leaves) (Lamade et al., 1998). Leaf initiation rates of individual palms respond positively to light availability: initiation rates increased by 19% two years after thinning of palms 11–15 YAP at high density (186 palms ha^{-1}; Breure, 1994). Sink limitation in 13 year old palms, resulting from complete removal of developing fruits, reduced phyllochron length from 17 days to 15 days (Legros et al., 2009b), possibly because of increased carbohydrate availability to young leaves. This suggests that oil palm is able to respond to abundant carbohydrate supply by increasing its rate of inflorescence initiation (Pallas et al., 2013). The rate of leaf opening is reduced rapidly in response to drought (Chang et al., 1988), resulting in the accumulation of unopened leaves in the centre of the palm crown (Broekmans, 1957 ; Nouy *et al.*, 1999). Drought may also reduce leaf initiation rates (Chang *et al.*, 1988 ; Breure, 1994).

3.1.2. Sex determination, inflorescence abortion, and sex ratio

In contrast to other palms, such as coconut, that carry male and female flowers in the same inflorescence, sex is determined at inflorescence level in oil palm. The earliest morphological difference between male and female inflorescences is the increased number of bracts initiated on male rachillae (Leaf -6; Corley, 1976a; Heel *et al.*, 1987 ; Adam *et al.*, 2005). The timing of sex determination varies among experiments, research sites and planting material, ranging from 29 to 30 months before harvest (Broekmans, 1957) to 20 months before harvest (Breure and Menendez, 1990; Fig. 2). Corley et al. (1995) found that the timing of sex determination varies among clones: either at bract initiation, Leaf -29, or just before first rachilla initiation, Leaf -10, or both. This led Corley and Tinker (2016: 121) to speculate that sex differentiation occurs at Leaf -29 but is reversible up to Leaf -10 (Cros et al., 2013). The physiological mechanisms underlying sex determination and the role of carbohydrate balance and plant hormones remain poorly understood (Corley, 1976a; Corley and Tinker, 2016: 120; for a review on the effects of environmental factors on sex determination see Adam et al., 2011).

Sex ratio (i.e. the ratio of female inflorescence number to total inflorescence number) is affected by both sex determination and the preferential abortion of female or male inflorescences; the two effects are difficult to separate (Corley, 1976a). In the absence of severe stress, the average sex ratio is 0.9–1.0 in the first four YAP (Henson and Dolmat, 2004), 0.6–0.9 until 12 YAP (Jones, 1997 ; Henson and Dolmat, 2004), and then steadily declines (Corley and Gray, 1976). Severe water deficit, such as occurs in the dry season in West Africa, can reduce the sex ratio to 0.1–0.2 (Broekmans, 1957; Bredas and Scuvie, 1960 ; Corley, 1976a). Sex ratio, particularly inflorescence abortion, is affected by fruiting activity (Corley and Breure, 1992). The combined effects of environmental and internal signals result in annual oscillations in sex ratio and yield (Cros et al., 2013). Developing inflorescences are most sensitive to abortion 4–6 months before anthesis, which coincides with the onset of floral organ development and elongation (Broekmans, 1957). Whereas several authors reported a preferential abortion of female inflorescences during (part of) the sensitive period of inflorescence development (Bredas and Scuvie, 1960; Breure and Menendez, 1990 ; Pallas *et al.*, 2013), others observed preferential abortion of female inflorescences only in specific lines (Corley et al., 1995), preferential abortion of male inflorescences (Legros et al., 2009b), or equal abortion rates for inflorescences of both sexes (Henry, 1960). Inflorescence abortion rates of 25–40% were measured in young mature palms that experienced prolonged dry seasons in Nigeria, decreasing to 5–10% in palms >15 YAP (Broekmans, 1957). Much smaller abortion rates of 2–13% were measured in palms of 4–17 YAP planted on deep peat soils with a high water table in Malaysia, and no clear age trend was observed (Henson and Dolmat, 2004). A reduction in source availability through defoliation down to 16 leaves increased inflorescence abortion rates in Leaves +2 to +12 from 10% to 40%, on average, in clonal palms of 9 YAP in Malaysia (Corley et al., 1995). While the sex ratio at the moment of peak abortion did not change significantly in all clones but one, the average percentage of leaf axils with male inflorescences increased from 50% in the control to 60% in the pruned palms, in the period 11–25 months after defoliation. Conversely, a decrease in sink activity induced by fruit pruning in palms of 14 YAP in Sumatra increased the fraction of female inflorescences in the trough and the peak season from 0.15–0.6 in the control to 0.25–0.8 in the pruned palms. Simultaneously, the aborted fractions decreased from 0.2–0.6 to 0.1–0.2, and the number of male inflorescences in the trough season increased from 0.1 to 0.5 (Legros et al., 2009b). Thresholds of specific assimilate availability that trigger sex determination and floral abortion responses remain to be identified, due to the large variation in response among planting materials, research sites, and experiments (Breure, 1987; Corley and Breure, 1992; Corley *et al.*, 1995 ; Cros *et al.*, 2013).

3.1.3. Bunch failure

Bunch failure, the abortion of a bunch before full ripening, occurs 2–4 months after anthesis (Sparnaaij, 1960). Bunch failure may be caused by poor pollination or acute and severe assimilate shortage, usually caused by lack of water or radiation (Corley and Tinker, 2016: 125; Combres et al., 2013). Bunch failure rates between 1.5% (Corley, 1973b) and >25% (Sparnaaij, 1960; Corley and Tinker, 2016: 124–125) have been observed, but the available data is scarce, and the phenomenon remains poorly described and understood (Corley and Tinker, 2016: 124–125).

3.2. Bunch weight and oil content

Bunch weight and oil content are less responsive to stress than bunch number, but have a major impact on yield. We briefly describe inflorescence and bunch development, and then discuss the regulation of the various components of bunch weight and oil content.

3.2.1. Inflorescence and bunch development

Both male and female inflorescences consist of a peduncle, carrying spikelets on which the flowers are set, each subtended by a single bract. The male peduncle and spikelets are 40 and 10–30 cm in length, respectively, and each of the 100–300 spikelets carries 400–1500 male flowers 3–4 mm in length. The female peduncle is shorter (20–30 cm) and thicker and carries around 150 spikelets, each 6–15 cm in length. A spikelet carries 5–30 flowers that are subtended by a bract in the shape of a sharp spine (Jacquemard and Baudouin, 1998). The number of spikelets and the number of flowers per spikelet increase with palm age but reach a plateau at 10–12 YAP (Corley and Gray, 1976). The number of female flowers that develops into fruitlets ranges from 30–60% (Corley and Tinker, 2016: 49) to 80% (Harun and Noor, 2002) when insect pollinators are present. In palms 10–15 YAP, bunches contain 1500–2000 fruitlets. The bunch maturation time (from anthesis to bunch ripeness) varies from 140 to 180 days, depending on both genetic and environmental factors (Lamade *et al.*, 1998 ; Henson, 2005). Fruit maturation starts two weeks after anthesis and occurs in several distinct phases (Oo et al., 1986). Oil starts to accumulate in the endosperm of fruitlets about 12 weeks after anthesis, and four weeks later the endocarp and endosperm (which together form the kernel) have hardened (Oo *et al.*, 1986 ; Sambanthamurthi *et al.*, 2000). Oil deposition in the mesocarp begins around 15 weeks after anthesis and continues until fruit ripeness, 5–6 months after anthesis (Oo et al., 1986), when fruitlet mesocarp oil content is about 60% and water content has decreased from more than 80% to less than 40% (Bafor and Osagie, 1986 ; Bille Ngalle *et al.*, 2013).

3.2.2. Regulating mechanisms of bunch weight and oil content

The main components that determine bunch weight are the number of spikelets, number of flowers per spikelet, fruit set, weight per fruitlet, and weight of non-fruit bunch components (Broekmans, 1957). Bunch fresh weight (with 53% dry matter, on average; Corley et al., 1971b) increases with palm age, starting at 3–5 kg at 24 MAP and increasing to over 30 kg by 25 YAP (Lim and Chan, 1998, cited by Corley and Tinker, 2003: 113; Sutarta and Rahutomo, 2016). All components of bunch weight respond positively to increased assimilate availability (Breure and Menendez, 1990; Corley and Breure, 1992 ; Pallas *et al.*, 2013). Removal of 75% of the inflorescences in palms of 4–7 YAP increased total bunch weight to 12.7 kg from 7.6 kg in control palms, resulting from an increase in all components mentioned above (Corley and Breure, 1992 ; Breure and Corley, 1992). Fruit set is determined mainly by pollination efficiency (4.2.8).

Oil content is primarily affected by planting material (4.2.4). A single gene determines kernel shell thickness, which in turn affects the thickness of the mesocarp and therefore fruit bunch oil content (Beirnaert and Vanderweyen, 1941). Wildtype oil palm (*dura*) has a thick shell and a typical oil extraction rate of 16–18%, whereas the *tenera* hybrid, a cross between *dura* and the shell-less *pisifera* mutant, has an intermediate shell thickness and oil extraction rates of 22–30% (Jalani *et al.*, 2002 ; Rajanaidu and Kushairi, 2006). Fertiliser use affects bunch oil content (Ochs and Ollagnier, 1977), with increased tissue chloride concentrations leading to an increase in kernel-to-fruit from 7.8 to 9.3%, and a reduction in mesocarp-to-fruit from 81.7 to 79.2% in palms of 8 YAP in Papua New Guinea (Breure, 1982). Oil content is negatively correlated with rainfall, and positively correlated with available radiation; high rainfall in Malaysia in 1996 resulted in a 0.8–1.5% decrease of oil extraction rate (OER) compared with 1993 (Hoong and Donough, 1998). It is positively related with the concentration of Mg in leaf tissue (Ochs and Ollagnier, 1977) but sometimes negatively correlated with the application of potassium chloride (Ochs and Ollagnier, 1977 ; Zin *et al.*, 1993), probably as a consequence of increased Cl concentrations in the plant tissue resulting in increased kernel-to-fruit ratio (Breure, 1982).

4. MAGNITUDE, CAUSES, AND MANAGEMENT OF YIELD GAPS

Oil palm is grown in large-scale monoculture plantations or as a smallholder crop, with fruit bunches as the primary output and crude palm oil (CPO) and palm kernel oil (PKO) as the final products. Productivity is best measured as oil yield (t ha^{-1}), calculated from the yield of fruit bunches (t ha^{-1}) and the extraction rate (%). In this review yields are expressed either in t ha^{-1} fruit bunches (with 53% DM) or in t ha^{-1} oil. PKO is not considered, as it is a by-product which is extracted and traded by a limited number of mills. Kernel extraction rate is usually about 5% (Carter et al., 2007).

4.1. The different yield gaps in oil palm

In production ecology, three production levels are commonly distinguished: the potential yield (Yp) determined by yield-defining factors (PAR, temperature, ambient CO_2concentration, and crop genetic characteristics); the water-limited (Yw) and nutrient-limited yield (Yn) determined by yield-limiting factors (water and nutrition); and the actual yield (Ya) determined by yield-reducing factors (weeds, pest, diseases) (van Ittersum and Rabbinge, 1997). Yield gap analysis is the analysis of the difference between Yp (assuming genotype and management are optimal) and Ya in a particular physical environment (van Ittersum and Rabbinge, 1997; for recent reviews on yield gap analysis see also Lobell *et al.*, 2009 ; van Ittersum *et al.*, 2013).

We define the potential yield as *the yield of a cultivar, when grown in environments to which it is adapted; with nutrients and water non-limiting; and with pests, diseases, weeds, lodging and other stresses effectively controlled* (Evans, 1993). The theoretical limit to genetic gain in crop yield can be calculated using simulation models (Lobell et al., 2009). This number is sometimes also referred to as the 'potential yield' in oil palm literature (Breure, 2003 ; Corley, 2006), and can be used to set a target for breeders and to explore future scenarios, such as for land use. Oil palm management literature refers to the "site yield potential" (Tinker, 1984 ; Goh *et al.*, 2000), defined as *the yield obtained on a specified site, with natural water supply, nutrients supplied at optimum rates, and agronomic and disease control measures implemented to a high standard* (Corley and Tinker, 2016: 322). This is similar to what we call the water-limited yield, but includes management decisions taken at planting, specifically planting material and density. For thorough reviews on the approach to yield gap analysis from the oil palm management perspective, see Goh et al. (1994); Griffiths et al. (2002), and Fairhurst and Griffiths (2014), among others.

Accurate analysis of yield gaps depends on the correct assessment of the various production levels (Fig. 3). The yield-determining, yield-limiting and yield-reducing factors relevant in oil palm and their quantitative effects on productivity are discussed in detail below.

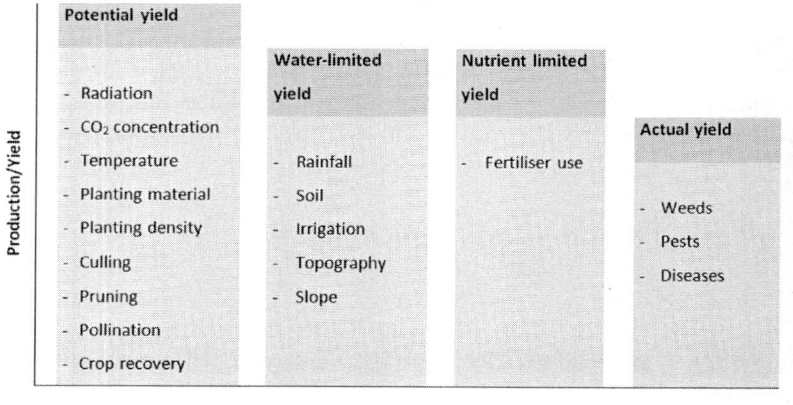

Figure. 3. Different oil palm production levels and the contributing factors.

4.2. Potential yield and yield-determining factors

The potential oil yield, as defined by fruit bunch yield and oil content, is determined by PAR, temperature, ambient CO_2 concentration, and crop genetic characteristics, under perfect crop management (van Ittersum and Rabbinge, 1997; Table 1). We discuss the different factors that determine the potential yield in further detail below.

4.2.1. Available radiation and PAR

As a perennial with a permanent leaf canopy, oil palm is able to intercept radiation throughout the year, which is one of the main reasons why its productivity is so large compared with other vegetable oil crops. In the tropics, available radiation is mostly limited by cloudiness. The range of total daily incoming short-wave radiation and sunshine hours per day in oil palm growing regions are shown in Table 1.

A minimum of 15 MJ m^{-2} day^{-1} total solar radiation (equivalent to ~7.5 MJ m^{-2} day^{-1}PAR) or 5.5 h day^{-1} of sunshine is optimal for oil palm growth, indicating a lesser yield potential in parts of Africa and the Americas (Paramananthan, 2003). Modelling work by Kraalingen et al. (1989) indicated that each hour per day of bright sunshine results in 15–20 kg bunch dry matter production $palm^{-1}$ $year^{-1}$ in excess of the bunch dry matter produced under cloudy circumstances, assuming a planting density of 110 palms ha^{-1}. Thus potential yields in regions with eight sunshine hours per day would be >60% larger than in regions with three sunshine hours per day (Kraalingen et al., 1989). Light saturation in oil palm leaves typically occurs at a photosynthetic photon flux density (PPFD) of >1100–1200 μmol m^{-2} s^{-1}, roughly equivalent to 250 W m^{-2} PAR (Dufrène et al., 1990). A light-saturated net assimilation rate of about 20 μmol CO_2 m^{-2} s^{-1} was measured at 1100 μmol m^{-2} s^{-1} PPFD in leaf 8 and 9 of palms planted in Ivory Coast (Dufrène and Saugier, 1993), which is similar to the average rate of 17.8 μmol $CO_2 m^{-2}$ s^{-1} found in palms 12–13 YAP in Malaysia (Henson, 1991b). Reduction of available PAR due to haze, caused by forest burning, is a common issue in Indonesia. Forest burning occurs mostly during the dry season when available radiation is at its peak, and is likely to reduce yields significantly (Table 1). In Africa, dust from the Harmattan and smog cause periodic reductions in radiation.

4.2.2. CO_2 concentration

Under current circumstances the rate of photosynthesis in C3 crops such as oil palm is limited by the availability of CO_2. Yield increases of 10–30% in response to doubling atmospheric CO_2 concentrations have been observed in other C3 crops such as wheat (Kimball *et al.*, 1993 ; Fuhrer, 2003), and may be expected in future in as oil palm is well adapted to high temperature-environments (Dufrène and Saugier, 1993). Increases in photosynthetic rates in oil palm seedlings from 5 to 12 μmol m^{-2} s^{-1} have been observed in response to changes in atmospheric CO_2 concentrations from 400 to 800 ppm (Ibrahim et al., 2010). Whether increased rates of photosynthesis are translated into improved yields depends on multiple factors, particularly the source/sink balance (e.g. Paul and Foyer, 2001) and the air temperature (4.2.3). Mature palms are usually source-limited (Breure, 2003) making an actual yield response to rising CO_2 concentrations likely, if the temperature remains stable. No research has been carried out to date on the actual effect of available CO_2 on oil palm yield in mature plantations. The expected effects of climate change on worldwide palm oil production are reviewed by Corley and Tinker (2016, Section 17.3).

Table 1. Yield-determining factors in oil palm systems: potential yield (Yp).

Yield-determining factors	Range in oil-palm growing areas	Yield effects measured in case studies	Selected references
Radiation: solar radiation	• All regions: average 15 to 23 MJ total radiation m^{-2} day^{-1} • Africa and parts of the Americas: < 10 MJ m^{-2} day^{-1} during the wet season	• Modelled increases of 1.7–2.1 t fruit bunches ha^{-1} yr^{-1} per additional MJ m^{-2} day^{-1} • Modelled 15–20% annual yield loss after two months reduction from 15 to 12 MJ total radiation m^{-2} day^{-1} due to haze	(Paramananthan et al., 2000) (Henson, 2000) (Goh, 2000) (Caliman et al., 1998)
Radiation: sunshine hours day^{-1}	• Asia: 5.3–6.9 • Americas: 2.2–7.7 • Africa: 3.6–6.3	• Productivity constraints if <5.5 h day^{-1} • One additional hr day^{-1} yields an additional 15–20 kg bunch DM $palm^{-1}$ yr^{-1} compared with productivity under cloudy conditions	(Hartley, 1988: 100-101) (Kraalingen et al., 1989) (Paramananthan, 2003)
CO_2 concentration	• 1960: 317 ppm • 1980: 339 ppm • 2000: 370 ppm • 2015: 399 ppm	• Modelled bunch DM production (t ha^{-1} yr^{-1}) in site without water deficit: Bunch DM / CO_2 (ppm) / Temperature (°C) 11 / 350 / +0 30 / 550 / +0 18 / 550 / +2 10 / 550 / +4	(Ibrahim et al., 2010) (Henson, 2006) (Tans and Keeling, 2015)
Temperature	Lowest monthly minimum: 17.7 °C (Bahia, Brazil) Highest monthly maximum: 34.6 °C (Aracataca, Colombia)	• Undefined strong yield reductions at minimum monthly average temperatures of less than 18–19 °C • Seedling growth inhibited at 15 °C, seven times slower at 17.5 °C and three times slower at 20 °C than at 25 °C • Immature period in cold conditions up to 1 year longer	(Hartley, 1988: 102-103; 110) (Henry, 1958) (Olivin, 1986)

Planting material	• *Tenera* clones • *Tenera* semi-clones • DxP *tenera* seed • *Dura* seed • Seed of unknown origin	• *Tenera* clones: 15.7 t oil ha^{-1} yr^{-1} at 7 YAP • *Tenera* semi-clones: 11.1 t oil ha^{-1} yr^{-1} at 5 YAP • DxP *tenera* seed: 8.9 t oil ha^{-1} yr^{-1} • *Dura* seed: ~ 35–50% reduced bunch oil content • Seed of unknown origin: reductions potentially very large depending on percentage *pisifera* in population (zero yield from *pisifera* palms) and potential of parent materials	(Simon et al., 1998) (Ng et al., 2003) (Rajanaidu et al., 2005) (Sharma, 2007)
Planting density	• 110–156 palms ha^{-1} in favourable environments • 160–170 palms ha^{-1} in unfavourable soils	• Optimum fixed planting density: 140–160 palms ha^{-1}; optimum LAI: 5.5–6.0 • 1-2% reduction in cumulative plantation yield when density ± 10 palms from optimum • On deep peat: higher optimum densities (> 160 palms ha^{-1}) • Yield increase of 4 t fruit bunches ha^{-1} yr^{-1} from 9 to 16 YAP in response to thinning from 160 to 120 palms ha^{-1} at 8 YAP compared with no thinning or a fixed density of 143 palms ha^{-1} in Thailand	(Corley and Tinker, 2016: 282) (Breure, 2010) (Corley, 1973a) (Breure, 1977) (Gurmit et al., 1986) (Goh et al., 1994) (Uexküll et al., 2003)
Culling	• Good: 20–30% of seedlings removed • Poor: incorrect or insufficient culling	• No culling: 20–30% abnormal seedlings producing 40–100% less yield than normal seedlings	(Tam, 1973) (Gillbanks, 2003)
Pruning	• 50–60 leaves at 0–3 YAP • 40–50 leaves at 4–10 YAP • 32–40 leaves at >10 YAP	• Over-pruning palms 8–12 YAP planted at 138 palms ha^{-1} in Malaysia: <2, 12, 19, 24, and 25 t fruit bunches ha^{-1} with 8, 16, 24, 32, and 40 leaves $palm^{-1}$, respectively • Under-pruning: direct but unquantified yield loss due to reduced harvesting efficiency	(Hartley, 1988: 441–442) (Henson, 2002) (Corley and Hew, 1976)

Fruit set and pollination	• Pollinating weevil present in all regions • Average fruit set 70–80%	• Quadratic asymptotic relation between fruit set and bunch weight with an average bunch weight of 24, 20 and 14 kg at 90, 50 and 20% fruit set, respectively • Quadratic relation between fruit set and oil to bunch ratio with an average O/B of 25, 20 and 13% at a fruit set of 75, 40 and 20%, respectively	(Harun and Noor, 2002) (Syed et al., 1982) (Rao and Law, 1998) (Henson, 2001)
Harvesting frequency	• Plantations: 7-day, 10-day or 14-day harvesting interval • Smallholders: usually 14 or 15-day harvesting interval, sometimes up to 30 days	• Yield increase of 5–20% when reducing length of harvesting round from 14 to 10 days	(Donough et al., 2013) (Lee et al., 2013) (Corley, 2001) (Donough, 2003)
Crop recovery in the field	• Varying from near complete recovery to less than 70% of fruit	• Reported yield losses of up to 5 t fruit bunches ha^{-1} due to poor crop recovery • Yearly losses under strict harvesting regime at 7-day interval: 200 kg fruit bunches ha^{-1} unharvested bunches and 65 kg ha^{-1} uncollected loose fruits • Incomplete collection of loose fruit: on average >5% yield loss • ~30% less oil yield from unripe bunches	(Fairhurst and Griffiths, 2014: Chapter 6) (Donough et al., 2013) (Corley, 2001) (Wood, 1985)

4.2.3. Temperature

The temperature range in the oil palm growing regions is shown in Table 1. The upper temperature limit for efficient photosynthesis in oil palm leaves is >38 °C, provided that vapour pressure deficit is small (Dufrène *et al.*, 1990; Dufrène and Saugier, 1993 ; Paramananthan, 2003). Temperature and maintenance respiration in plants are strongly positively related, with an average factor two increase in maintenance respiration at every 10 °C temperature rise (Amthor, 1984 ; Ryan, 1991). Whether this estimate holds for oil palm remains unclear, and yield responses to increasing temperatures have not been quantified (Henson, 2004 ; Henson, 2006). Oil palm is sensitive to cold (Table 1). In cooler regions such as in Bahia (Brazil) and Tela (Honduras), strong reductions in yield occur during the second half of the cold season and the beginning of the warmer season, and in Sumatra low temperatures at higher elevations were found to extend the immature period by at least one year (Hartley, 1988: 110).

4.2.4. Planting material

Estimates of theoretical ceiling oil yields (with future planting materials under the best possible environmental and management condition) range from 10.6 (Breure, 2003) and 14.0 (Henson, 1992) to 18.5 t oil ha^{-1} yr^{-1} (Corley, 1998 ; Corley, 2006) on average over the plantation lifetime. While the larger estimates may be based on some unrealistic assumptions (Breure, 2003), best yields achieved in small plantations or experimental fields already fall within the estimated range (Table 1). Non-clonal planting materials, raised from seed, consist of a population of offspring from a *dura* mother and a *pisifera* father (DxP), and individuals vary in terms of potential for vegetative growth and productivity (Okwuagwu et al., 2008). Potential yields of DxP planting materials have increased by an estimated 1.5% per year through breeding with specific male/female parent combinations that show an early track record of performance: this trend in yield increase is expected to continue (Soh, 2004 ; Corley, 2006). Breeding has particularly improved photosynthetic conversion efficiency (Corley and Lee, 1992) and bunch oil content (Corley and Lee, 1992; Prasetyo *et al.*, 2014 ; Soh, 2015). Varieties with improved tolerance for cold (Chapman et al., 2003) and drought (Rao et al., 2008) are being further developed.

Clones from carefully selected ortets can outyield conventional seed material by 20–30%, due to a combination of better uniformity, increased fruit bunch yield and greater oil to bunch ratio (Khaw and Ng, 1998; Simon *et al.*, 1998 ; Kushairi *et al.*, 2010; Soh, 2012; Table 1). Although field experiments have confirmed the superior yields of selected clones under circumstances of rigorous culling, key issues with multiplication of embryos and somaclonal variation limit the current planting of clones at commercial scale (Soh, 2004 ; Soh *et al.*, 2011). The recent finding of the epigenetic factor underlying the *mantling* phenotype (a floral malformation that results in failure to form fruitlets or reduced fruitlet oil content) is likely to boost the planting and performance of clonal oil palm (Ong-Abdullah et al., 2015).

4.2.5. Planting density

Planting density is an important determinant of potential yield (Corley, 1973a; Breure, 1977; Breure, 1982 ; Uexküll *et al.*, 2003). An optimum planting density (Table 1) balances the requirement for rapid canopy closure in the immature phase with a large number of palms (i.e. bunches) in the young mature phase and limited inter-palm competition for light in the mature phase. On deep peat, vegetative growth is reduced and denser planting has been recommended (Table 1; Gurmit et al., 1986). High-density planting followed by selective thinning at 8–9 YAP is an effective strategy for yield maximisation (Uexküll et al., 2003; Palat et al., 2012; Table 1).

4.2.6. Culling

The quality and uniformity of field palms depends on the planted material and on the selection of individuals during the nursery phase, termed 'culling' (Tam, 1973). Due to genetic diversity and stresses during the nursery and field planting phase, large differences in productivity between palms have been observed even when rigorous culling has been carried out (Okwuagwu et al., 2008), with the most productive individuals yielding two to three times more than average, and the least productive individuals yielding no bunches (Yeow et al., 1982; Hartley, 1988: 222). Normally the prevalence of stunted or abnormal seedlings is 20–30%. Abnormal seedlings, identified by phenotypic selection in the nursery phase, give strongly reduced yields when planted out (Tam, 1973; Table 1). All abnormal seedlings should be removed during the nursery phase or replaced within 12 months after planting (Gillbanks, 2003; Jacquemard and Baudouin, 1998: 56).

4.2.7. Pruning

Pruning, the removal of selected leaves, is a management practice specific for perennial crops. Pruning aims to optimise source availability while minimising loss of assimilates due to respiration in senescing leaves. Newly-opened leaves in oil palm show a stable or slightly increasing photosynthetic activity until 4–10 months after opening (in palms of 3 and 10–12 YAP, respectively), after which activity decreases until the leaves senesce and die (Corley, 1983 ; Corley, 1976b). Leaves at the bottom of the canopy remain photosynthetically active and are net sources until senescence (Henson, 1991a), and retaining all living leaves but removing senescing leaves is the best way to maximise assimilate availability irrespective of plantation age (Hartley, 1988: 441; Henson, 2002). Pruning in immature and young mature palms is usually limited to the removal of senescing or dead leaves, as reductions in leaf area have a strong negative effect on light interception and total assimilate availability during this phase (Gerritsma, 1988 ; Breure, 2003). Yield penalties when pruning from >48 down to 32–40 leaves per palm in mature plantations are not significant (Corley and Hew, 1976) and sufficient pruning of tall palms to facilitate complete and correct harvesting and quick recycling of nutrients is recommended (Fairhurst and Griffiths, 2014).

4.2.8. Pollination

A quadratic function describes the relationship between fruit set and bunch weight, with a maximum bunch weight at 90%, and a maximum oil to bunch ratio at 75% fruit set (Harun and Noor, 2002; Table 1). Seasonal episodes of poor (10–20%) fruit set have been observed in Malaysia, caused by strong reductions of pollinating weevil populations due to excessive rain, absence of sufficient male flowers and infection with parasitic nematodes (Rao and Law, 1998). As a consequence oil extraction rate (OER) fell from 21.2 to 18.8%, and kernel extraction rate from 4.7 to 3.5% in Malaysia between 1993 and 1996. A minimum of two male palms per hectare in plantations with a high sex ratio is thought to supply sufficient pollen and maintain weevil populations (Rao and Law, 1998).

4.2.9. Crop recovery

The goal of harvesting, or crop recovery, is to collect all fruit bunches at the moment of optimum ripeness (i.e. maximum oil content with a minimum concentration of free fatty acids in the extracted oil; PORLA, 1995). Infrequent, incomplete or incorrect harvesting practices (i.e. harvesting unripe or overripe bunches) directly reduce both the quantity of fruit and the oil quality (Donough et al., 2010; Table 1). The harvesting interval (i.e. the number of days between two harvesting rounds) should be adapted to the speed at which loose fruits detach from the ripe bunch, to minimise losses from uncollected loose fruit and overripe bunches (Gan, 1998). An optimal harvesting interval of 10 days has been proposed (Gan, 1998; Rankine and Fairhurst, 1999b ; Donough *et al.*, 2010). Harvesting of unripe bunches is likely to affect the source/sink balance as bunch sink requirements increase strongly towards the last phase of ripening (Henson, 2007), but this has not been quantified.

Table 2. Yield-limiting factors in oil palm systems: water-limited yield (Yw).

Yield-limiting factors	Range in oil-palm growing areas	Yield effects measured in case studies	Selected references
Total rainfall and distribution	Rainfall (mm year^{-1}) • Malaysia and Indonesia: 1700–4000 • Africa: 1200–3500 • Americas: 1600–3500 Dry months (less than 100 mm rain month^{-1}) • Malaysia and Indonesia: 0–3 • Africa: 3–6 • Americas: 0–5	• Yield reduced if rainfall <2000 mm yr^{-1} or >3500 mm yr^{-1} and/or <100 mm month^{-1} • Yield reductions in relation to water deficit: ○ None if water deficit is less than threshold of 50–200 mm yr^{-1}, depending on local conditions; ○ 10–20% yield loss per 100 mm deficit after the threshold; ○ Exponential decline down to <10 t fruit bunches ha^{-1} yr^{-1} at water deficits of >500 mm • See also: Table 3	(Dufrêne et al., 1990) (Hartley, 1988: 98–99) (Paramananthan, 2003) (Goh, 2000) (Olivin, 1986)
Irrigation	• Most areas are rain-fed • Plantations in Thailand, parts of Africa, and parts of the Americas use irrigation • Some smallholders in the Americas and Thailand use irrigation; rare in other areas	• Estimated response according to IRHO method for calculating soil water deficit: 20–30 kg ha^{-1} year^{-1} fruit bunches per mm irrigation water in areas where the potential soil water deficit is 200–600 mm year^{-1} • Approximately linear relationship between water volume (mm water dry day^{-1}) and yield response (t fruit bunches ha^{-1} year^{-1}) in irrigation trial in Thailand (soil water deficit 235 mm year^{-1} over 3–4 months): ○ 18 t fruit bunches at 0 mm ○ 24 t fruit bunches at 3.2 mm ○ 28 t fruit bunches at 6.4 mm	(Palat et al., 2008) (Ochs and Daniel, 1976) (Carr, 2011)

Soil type	Most common soil types (according to the USDA soil taxonomy) • SE Asia: ultisols, oxisols and histosols • Africa: oxisols, ultisols and mullisols • Americas: oxisols and ultisols	Most soil types are not constraining apart from: • Shallow soils (Malacca series and Baiayo family): <30 t fruit bunches ha^{-1} yr^{-1} • Coarse textured soils (psamments): yield 'poor' but not quantified • Biochemically constrained soils (saline soils, peat soils, acid sulphate soils): 20–30 t fruit bunches ha^{-1} yr^{-1} when managed well	(Goh et al., 1994) (Gurmit et al., 1986) (Paramananthan, 2013) (Mutert et al., 1999) (Paramananthan, 2000)
Soil texture	• Sandy soils to heavy clay	• Large but unquantified yield losses in very sandy soils and in heavy clay soils	(Paramananthan, 2003) (Paramananthan, 2013)
Topography and slope	• Flat to hilly	• Slopes >20° are considered unsuitable; slopes 10–20° require soil conservation measures • Estimated 10–30% yield reduction on slopes of 2–7° without conservation measures • Measured 20–30% yield increase (mature plantation; 3-month dry season) after implementation of soil conservation measures on slope of 2–5°	(Paramananthan, 2003) (Paramananthan, 2013) (Balasundram et al., 2006) (Kee and Soh, 2002) (Murtilaksono et al., 2011)
Waterlogging	• Common in low-lying areas in all regions • Localised flooding or water logging for several days to months per year	• Mortality of up to 75% of immature palms in regions of frequent inundation • Yield losses of 20–30% in poorly drained mature plantations • Yield increases of >5 t ha^{-1} fruit bunches after drainage of frequently flooded fields	(Lee and Ong, 2006) (Abram et al., 2014) (Henson et al., 2008) (Chuah and Lim, 1992; cited by Lim et al., 1994)

4.3. Water-limited yield and yield-limiting factors

The water-limited yield (Yw; Table 2) is an important benchmark as most oil palm cropping systems are rain-fed (Ludwig et al., 2011). Water availability depends on rainfall and soil characteristics and is strongly site-specific (Lobell *et al.*, 2009 ; van Ittersum *et al.*, 2013). Yw can be approximated by crop simulation models using plausible physiological and agronomic assumptions (Evans and Fischer, 1999), by field experiments, estimates of best farmers' yields, or growers' contests (van Ittersum et al., 2013).

4.3.1. Rainfall

Oil palm transpires about 6 mm water day^{-1} under non-limiting conditions, and requires sufficient rainfall throughout the year (Table 2). Average actual transpiration rates in oil palm plantations are 4.0–6.5 mm day^{-1} in the rainy season and 1.0–2.5 mm day^{-1} on dry days (Carr, 2011). Moderate to severe water stress strongly suppresses yield (Table 3). Oil palm leaves do not wilt, but the opening of new leaves is delayed in response to water stress, and stomatal opening is strongly affected by air vapour pressure deficit (VPD) and soil water availability (Smith, 1989 ; Caliman, 1992). Henson and Harun (2005) measured potential evapotranspiration rates of 1.3 mm day^{-1} at 1.9 kPa VPD and 75% available soil water content, in palms of 3 YAP planted at a site with a regular dry season in Malaysia. In another site, an increased VPD from 0.4 to 2.0 kPa resulted in a decline in photosynthetic rate from 18–19 to 10–12 μmol CO_2 m^{-1} s^{-1} in palms of 1–2 YAP, even under conditions of sufficient soil water availability (Henson and Chang, 1990).

Table 3. Water-limited yield in Africa related to soil class and water deficit, assuming all other production factors are sufficient (after Olivin, 1968).

Soil class		Water deficit (mm)				
		0	100	200	300	400
		Yield (t fruit bunches ha^{-1} yr^{-1})				
I	Very suitable	≥27	24	18	14	12
IIa	Suitable	25	20	16	13	10
IIb	Moderately suitable	25	20	16	11	8
III	Somewhat suitable	22	16	13	9	6
IV	Unsuitable	16	13	9	6	4

A linear relationship between applied water volume and yield has been found in irrigation trials in drier environments (Corley, 1996 ; Palat *et al.*, 2008; Carr, 2011; Table 2). Although yield responses to irrigation have been observed in areas with occasional dry spells in Malaysia, irrigation is not always economically feasible (Corley and Hong, 1982 ; Henson and Chang, 1990). Critical water deficit thresholds at different stages of palm development and optimum volumes of water to be applied remain to be defined (Carr, 2011).

4.3.2. Soil

Soil water availability depends on the influx of water (rainfall, irrigation, and groundwater), the loss of water (evapotranspiration, drainage, and surface water run-off), and the previous soil water reserve. A simplified calculation was proposed by Surre (1968) to allow for a quick assessment of the suitability of soil-climate combinations for oil palm development. This calculation is based on the following equation:

$$B = Res + R - Etp \quad (2)$$

where B is the water balance at the end of a period, Res is the soil water reserve at the beginning of a period, R is rainfall and Etp is the potential evapotranspiration (Surre, 1968). Using this equation, Olivin (1968) estimated water-limited yields in Africa for five scenarios of water deficit on five soil classes ranging from I (excellent, such as young alluvial soils) to IV (unsuitable, such as very sandy or gravelly soils; Table 3). In Malaysia, yields of >30 t fruit bunches ha^{-1} have been reported on most soil types apart from shallow soils, which cause problems such as reduced root proliferation, increased susceptibility to drought and waterlogging, and risk of palms falling over (Goh *et al.*, 1994 ; Fairhurst and McLaughlin, 2009; Paramananthan, 2013; Table 2). On peat soils, yields of 30 t fruit bunches ha^{-1} have been reported (Gurmit et al., 1986) but yields are generally less than on mineral soils because of palms leaning or falling over, waterlogging, and soil drying (Paramananthan, 2013).

4.3.3. Topography and slope

Cultivation on slopes increases surface run-off which reduces the amount of water available for the crop. A maximum slope of 10° without soil conservation, or 20° with terraces, has been proposed to maintain economic yields (Paramananthan, 2003), but yield responses to soil conservation on slopes of 2–10° have been reported (Table 2). Water losses by run-off varies from zero to >30%, with erosion and fertiliser loss occurring mostly from weeded circles and harvesting paths where soils are bare and become compacted (Comte *et al.*, 2012; Bah *et al.*, 2014 ; Banabas *et al.*, 2008). Water and fertilisers flow from summits and side slopes to valleys, creating heterogeneity in soil fertility and yield (Balasundram et al., 2006), as well as environmental problems (Comte et al., 2012).

4.3.4. Waterlogging

Oil palm is tolerant of temporary flooding, which may be partly due to the ability of the roots to form pneumatodes (Purvis, 1956 ; Jourdan and Rey, 1997). However, submerged roots are unable to respire normally, leading to impaired water and nutrient uptake, delayed frond opening, and reduced carbohydrate availability (Corley and Tinker, 2016: 109). Henson et al. (2008) demonstrated that photosynthetic activity and transpiration rates are 3–4 times less in oil palms under waterlogged conditions compared with palms in well-drained soils. Waterlogging is a common problem in plantations in Southeast Asia (Paramananthan, 2003 ; Lee and Ong, 2006) and severe, but poorly quantified, reductions in yield have been observed (Abram et al., 2014; Carr, 2011; Table 2). Data on effects of drainage on productivity in waterlogged fields is scarce (Table 2). In peat soils and acid sulphate soils, sufficient drainage while maintaining the water table at 40–50 cm below ground level or above the acid sulphate layer is critical to prevent soil degradation, reduce greenhouse gas emissions, and obtain high yields (Toh and Poon, 1981 ; Othman *et al.*, 2011).

4.4. Nutrient-limited yield and yield-limiting factors

The nutrient-limited yield (Yn; Table 4) is location dependent, mostly due to the effects of soil properties on nutrient availability.

The nutrient needs of oil palm are well-researched and reviewed (Uexküll and Fairhurst, 1991; Goh *et al.*, 2003; Ng, 1977 ; Breure, 1982). Oil palm requires particularly large quantities of potassium, as well as nitrogen, phosphorus, magnesium, and boron (Table 5). Fertilisation with copper and zinc is required on peat soils. In case of severe deficiencies, foliar symptoms become visible (Broeshart et al., 1957; Table 5). Critical tissue nutrient concentrations that indicate nutrient deficiencies have been established (Uexküll and Fairhurst, 1991), but are site and soil specific (Foster and Chang, 1977 ; Foster, 2003). The availability, uptake, and allocation of the different nutrients are strongly interdependent (Tohiruddin *et al.*, 2010 ; Foster and Prabowo, 2002). In order to provide correct fertiliser recommendations, accurate measurements of the concentrations of N, P, K and Mg in both the leaflet and the rachis tissues are required (Foster and Prabowo, 2006 ; Prabowo *et al.*, 2011).

Table 4. Yield-limiting factors in oil palm systems: nutrient-limited yield (Yn).

Yield-limiting factors	Range in oil-palm growing areas	Yield effect	Selected references
Fertilisation	In kg ha^{-1} year^{-1}, assuming 140 palms ha^{-1} • N: ≤260 kg • P: ≤130 kg • K: ≤350 (up to 430 on peat soils) • Mg: ≤70 • B: ≤20 • Cu (on peat): ≤10 • Zn (on peat): ≤10 • Mn, Cl, Ca, Fe, S: occasionally applied	• For N, P, K, and Mg: see Table 5 • B: yield reductions of >35% in palms with severe B deficiency symptoms • Cu (on peat): 10–25% yield increase • Zn (on peat): 10–80% yield increase	(Rankine and Fairhurst, 1999b) (Rajaratnam, 1973) (Cheong and Ng, 1977) (Gurmit, 1988) (Osman and Kueh, 1996) (Ng, 2002)

Table 5. Role of key nutrients in oil palm physiology.

Element	Physiological role	Effect of deficiency on oil palm growth and yield	Visual deficiency symptoms	Selected references
Nitrogen	Formation of chlorophyll, amino acids, DNA, and ATP	Suppressed net assimilation rate; decreased vegetative dry matter production; increased phyllochron time; decreased bunch weight and number	Chlorosis in younger leaves; stunting	(Corley and Mok, 1972) (Bah Lias, 2011)
Phosphorus	Formation of DNA, RNA, and ATP	Yield decrease on some soils; reduced yield response to N and K fertiliser	Conical trunk shape	(Kraip and Nake, 2006) (Bah Lias, 2011) (Ng, 1986)
Potassium	Transport of photosynthates; control of stomatal opening	Decreased vegetative dry matter production; strongly decreased bunch weight and number	Yellow spotting in older leaves	(Corley and Mok, 1972) (Bah Lias, 2011) (Braconnier and d'Auzac, 1985) (Zakaria et al., 1990)
Magnesium	Chlorophyll formation; ribosome aggregation; enzyme functioning	Yield decrease on some soils; reduced yield response to N and K fertiliser; reduced oil/bunch ratio	Yellow/orange colour in leaflets of older leaves exposed to sunlight	(Dubos et al., 1999) (Härdter, 1999) (Shaul, 2002)
Boron	RNA formation; pollen formation; flavonoid synthesis; seed and cell wall formation	Decreased LAI (occurrence of 'little leaf'); decreased bunch number and yield when leaf deficiency symptoms are present	Crinkling of leaflets in older leaves; stunting of young leaves ('little leaf')	(Rajaratnam, 1973) (Rajaratnam and Lowry, 1974)
Copper, zinc	Electron transport; photosynthesis	Reduced photosynthesis (Zn); decreased vegetative dry matter production (Zn); reduced bunch number and size (Zn, Cu)	Yellowing and necrosis in older leaves starting at the leaflet tip	(Cheong and Ng, 1977) (Gurmit, 1988) (Osman and Kueh, 1996)

Table 6. Effects of N, P, K and Mg on yield in three different fertiliser experiments. Significant responses are printed in bold.

Source	Corley and Mok (1972)				Kraip and Nake (2006)				Bah Lias (2011)			
Location	South Johore, Malaysia				Milne Bay, PNG				South Sumatra			
Soil type	Sandy clay loam granite-derived red-yellow oxisol (Rengam series)				Recent alluvial sandy clay loam (fluvent)				Low-pH loamy kaolinitic inceptisol (typic dystrudept)			
Palm age	10–20 YAP				5 YAP				14 YAP			
Duration of trial	10 years				7 years				> 14 years			
Palms ha^{-1}	114				127				143			
Treatments (kg $palm^{-1}$ yr^{-1})		0	1	2		0	1	2		0	1	2
	N	0	0.8	1.5	N	0	0.4	0.7	N	0	0.9	1.8
	P	0	0.9	1.7	P	0	0.2	0.2	P	–	0.2	0.5
	K	0	1.8	3.7	K	0.4	1.2	2.2	K	0	1.0	2.0
	Mg	0	0.4	0.8					Mg	0	0.2	–
Yield (converted to t fruit bunches ha^{-1})		0	1	2		0	1	2		0	1	2
	N	25	**29**	**30**	N	31	31	29	N	15	**20**	**21**
	P	27	29	28	P	30	30	–	P	–	19	18
	K	26	**29**	**29**	K	28	**31**	**32**	K	10	**23**	**22**
	Mg	27	28	**29**					Mg	19	18	–
Remarks					Significant K effect 4 and 5 YAP, but not 6 and 7 YAP				Significant yield increases by application of P and Mg at the highest levels of N and K			

In oil palm plantation systems, nutrients are removed through harvesting of fruit bunches, leaching, run-off, and immobilisation in the trunk; recycled through pruned fronds and male inflorescences; and supplied by rainfall, soil nutrient stocks, mill waste products, and fertilisers (Ng et al., 1999). Chemical fertilisers are usually required to maintain the balance between nutrient removal and supply. Yield responses to chemical fertiliser application are location dependent and vary widely, and numerous randomised factorial N-P-K(-Mg) fertiliser experiments are described in literature (see Tohiruddin et al., 2006 for a good overview of results from Sumatra). Three experiments are summarised in Table 6 to highlight the type of yield responses observed. The range of nutrient use efficiencies (NUE) at different levels of fertiliser use was 0–45, 0–20, and 15–90 kg fruit bunches per kg nutrient ha^{-1} yr^{-1} for N, P, and K, respectively, when comparing no fertiliser with average applications (Treatment 0–1; Table 6). When quantities of nutrients applied were increased from average to large quantities (Treatment 1–2; Table 6), NUE became negative in some cases, and maximum NUE were 13, 0, and 8 kg fruit bunches per kg nutrient ha^{-1} yr^{-1} for N, P, and K, respectively.

In none of the experiments in Table 6 a clear yield response to phosphorus application was observed, but yield increases of 50–100% in response to P fertilisers have been observed elsewhere (Vossen, 1970; Ng, 1986 ; Sidhu *et al.*, 2001). Yield increases of up to 45% in response to magnesium application as kieserite were observed on yellow podzolic sandy loams in north Sumatra (Akbar et al., 1976). Because of the variability in NUE, site-specific factorial fertiliser experiments are required to optimise fertiliser applications (Tohiruddin *et al.*, 2010 ; Webb, 2009).

4.5. Actual yield and yield-reducing factors

The actual yield (Ya) is the water and nutrient limited yield, reduced by weeds, pests, and diseases (van Ittersum and Rabbinge, 1997). The cumulative yield over the plantation lifetime is the most important productivity indicator, which takes into account the duration of the unproductive yield-building phase. This has similarities to milk production from cows, but while individuals in a dairy herd can be replaced at any time (van der Linden et al., 2015), abnormal palms can only be replaced during the nursery phase and the first 12 months after planting. The yield-reducing effects of pests and diseases in oil palm unfold over a period of at least three years (Corley and Gray, 1976; Corley, 1976b; Legros *et al.*, 2009a ; Adam *et al.*, 2011). This time lag, combined with seasonal variations in fruit production, complicates the interpretation of oil palm yield data (Legros et al., 2009a). The calculation of 'rolling yields' over a 12-month period is useful to filter out seasonal variability when analysing yield trends (Uexküll and Fairhurst, 1991). Pest and disease damage early in the plantation lifetime often have a large effect on total yield, especially when they lead to palm death. The different yield-reducing factors are summarised in Table 7 and are further discussed in the sections below.

Table 7. Yield-reducing factors in oil palm systems: actual yield (Ya).

Yield-reducing factors	Range in oil-palm growing areas	Yield effects measured in case studies	Selected references
Ground cover management	• Good practice: closed legume cover plant canopy in the first six YAP; afterwards a closed canopy of soft weeds without noxious or woody weeds • Common practices: clear-weeding (companies, smallholders) or no weeding (smallholders)	• Uncontrolled weed growth: 50–60% yield reduction in young plantations at first harvest; no data for mature plantations • Clear weeding: up to 50% yield reduction in plantations 4–6 YAP • Planting of legume cover crops: yield increase of 10–20% in first productive years compared with non-leguminous weeds	(Ojuederie et al., 1983) (Samedani et al., 2014a) (Wood, 1977)
Pests: Leaf-eating insects	• Common in all regions • In case of severe infestation complete defoliation of palm clusters can occur	• Yield loss in case of complete defoliation: ~50%, 25% and 15% in year 1, 2 and 3 after defoliation, respectively	(Wood, 1977) (Wood et al., 1973)
Pests: *Oryctes*	• Common in immature plantations in all regions	• Yield reductions of 50% in first year and 20% in second year of production following severe attacks in young plantations • Rarely: death of severely damaged immature palms • In mature stands: yield reductions when LAI reduced below 5 (rare)	(Wood, 1977) (Wood et al., 1973) (Cahyasiwi et al., 2010) (Sushil and Mukhtar, 2008)
Pests: Rats	• Common in all regions • In case of severe infestation populations reach >300 individuals per hectare	• Estimated 5% loss of oil (130–240 kg oil ha^{-1} $year^{-1}$) in mature plantations with rat populations at 'saturation' level • Death of immature palms leading to incomplete stand or extended immature period	(Wood and Liau, 1984) (Wood and Chung, 2003) (Puan et al., 2011)

Diseases: *Ganoderma*	• Common in all regions, especially Southeast Asia • Potentially severe in Malaysia and Sumatra with up to 80% mortality at > 15 YAP	• Palm losses of up to 30–40% at 12 YAP and > 50% at 25 YAP in affected areas • When > 10% of stand lost: yield reduction of 0.16 t fruit bunches ha^{-1} per additional palm death • Around 35% yield loss at 50% palm mortality • One-year fallowing before replanting: 4% reduction in cumulative yield due to one-year increase of unproductive period; infection rate down from 30% to 3–6% at 9 YAP	(Flood et al., 2000) (Idris et al., 2004) (Ariffin et al., 2000; Cooper et al., 2011) (Cooper et al., 2011) (Flood et al., 2002) (Virdiana et al., 2010)
Diseases: bud rot	• Common in South America with up to 100% mortality in severe outbreaks	• Disease progress: linear phase (several years, ~1% of palms lost/year, limited or no yield effects), exponential phase (destruction of up to 100% of palms, complete loss of yield) • When > 10% of stand lost: yield reduction of 0.16 t fruit bunches ha^{-1} per additional palm death	(Uexküll et al., 2003) (De Franqueville, 2003) (Cooper et al., 2011) (Lopez, 2010)

4.5.1. Weeds

Weeds protect the soil against erosion and provide a habitat for natural pest enemies, while interacting with the water and nutrient cycles (Table 7). If weeds are allowed to grow uncontrolled, physical access to the plantation is reduced resulting in incomplete and inefficient harvesting. Certain weeds, especially woody and creeping weeds and speargrass (*Imperata cylindrica*), compete strongly with oil palms for water and nutrients (Wood, 1977 ; Surre, 1968). Weeds in immature plantations can reduce yields by overgrowing the young palms and thus limit available radiation (Table 7). In mature plantations the actual yield response to various weed populations remains unknown. Shade-tolerant legumes as well as certain soft grasses and *Nephrolepis* spp. are preferred as undergrowth in mature plantations because they are easy to manage, able to out-compete woody weeds once established, and not very competitive (Samedani et al., 2014b). Legumes provide an additional advantage because of their capacity to fix nitrogen. A well-established legume cover crop can provide over 150 kg N ha^{-1} $year^{-1}$ in the first 4 years after field planting (Agamuthu and Broughton, 1985; Giller, 2001 ; Giller and Fairhurst, 2003). Livestock integration, practiced regularly by smallholders, improves weed management and plantation productivity (Devendra and Thomas, 2002 ; Devendra, 2009).

4.5.2. Pests

Leaf-eating insects are present in all oil palm producing regions and large-scale outbreaks periodically occur, especially of bagworm (*Psychidae* spp.) and nettle caterpillar (*Lamicodidae* spp.) in Southeast Asia (Wood, 1968) and leaf miner (*Coelaenomenodera* spp.) in West-Africa (Mariau, 1976 ; Mariau and Lecoustre, 2000). Effects of mild infestations are small but yields can be strongly affected when severe defoliation reduces the LAI to less than 5 (Wood, 1977; Table 7). Rhinoceros beetle (*Oryctes rhinoceros*) is a pest in both immature and mature oil palm plantations (Bedford, 1980). Whereas the effects are usually limited in mature plantations, rhinoceros beetle is a problem in young plantings as it is capable of reducing growth by damaging the growing point, and on rare occasions killing the immature palms (Table 7). The effects are a delay in time to maturity or an incomplete stand of productive palms and hence a reduction in yield during the beginning of the productive phase (Wood et al., 1973). Rats (*Rattus* spp.) are common in all oil palm producing regions in the world, with unchecked populations reaching over 300 individuals per hectare in mature plantations. Rats eat the developing fruitlets and cause direct losses in oil yield (Wood and Liau, 1984; Wood and Chung, 2003; Table 7). In the immature phase, rats can eat through the bole of seedlings and destroy the growing point, causing palm death.

4.5.3. Diseases

Two diseases cause significant yield losses in oil palm plantations: basal stem rot in Southeast Asia and Africa, and bud rot in Latin America. Basal stem rot, caused by the pathogenic fungi *Ganoderma boninense*, can devastate old plantations (Flood *et al.*, 2000; Flood and Hasan, 2004 ; Idris *et al.*, 2004; for a review on previous research see Paterson, 2007). The onset of infection happens earlier at each replanting if no sanitation measures are taken, and can occur as soon as 1–2 years after planting when oil palm is planted after oil palm or coconut (Ariffin et al., 2000). The implementation of a one-year fallow can significantly reduce infection rates, but increases the immature/fallow to mature ratio from 0.12 to 0.15 (Virdiana et al., 2010; Table 7). Sanitation, the removal of diseased material, has been recommended as a management strategy in mature plantations (Chung, 2011 ; Hushiarian *et al.*, 2013) but there is no experimental evidence that shows conclusively that it reduces disease incidence (Idris *et al.*, 2004 ; Hoong, 2007). Breeding for resistant planting material is an important strategy to prevent future yield losses (Durand-Gasselin *et al.*, 2005 ; Ho and Tan, 2015). Bud rot is a fatal disease in the Americas, with incidental outbreaks having caused the destruction of complete stands across thousands of hectares since the 1960s (De Franqueville, 2003). The causal agent of bud rot in Colombia may be the oomycete *Phytophtora palmivora* (Martínez et al., 2010), but other pathogens such as the fungus *Fusarium* and the bacterium *Erwinia* spp. have also been associated with the occurrence of bud rot symptoms, as have the pest *Rhynchophorus palmarum* and a variety of abiotic factors (Benítez and García, 2014). Remediation and prevention measures are available but expensive and labour-consuming (Fontanilla et al., 2014).

4.6. Interactions between stress factors

While each production factor has certain quantifiable effects on yield, in reality multiple factors interact. For example, good ground cover management increases water retention in the soil, prevents the establishment of more competitive weeds, increases the population of natural enemies to pests, and reduces *Oryctes rhinoceros* breeding, each of which may affect yield. In order to close yield gaps it is necessary to take these interactions into consideration and to address multiple stresses simultaneously. Examples of such efforts are the Maximum Exploitation of Genetic Yield Potentials (MEGYP) approach (Henson and Chang, 1990) and the Best Management Practices (BMP) approach (Griffiths and Fairhurst, 2003 ; Witt *et al.*, 2005). The accurate recording of yields, input use and climatic and environmental factors is an essential component of all yield improvement strategies in oil palm (Griffiths et al., 2002).

Table 8. Potential yields over the plantation lifetime in six selected countries from Southeast Asia, Africa, and the Americas.

Country	Potential yield (t ha^{-1} yr^{-1})		Remark	Source
	Fruit bunches[a]	Oil[b]		
Indonesia	32–40	8–10	Low-lying areas, modelled	(Hoffmann et al., 2014)
	16–32	4–8	Higher elevations, modelled	(Hoffmann et al., 2014)
Malaysia	38	9.5	Progeny trial	(Rajanaidu and Kushairi, 2006)
	24–32	6–8	Low-lying coastal areas, modelled	(Hoffmann et al., 2014)
	8–24	2–6	Inland, modelled	(Hoffmann et al., 2014)
Thailand	36	9	Progeny trial	(Univanich, 2011; Rao et al., 2008)
Ghana	30–36	7.5–9	Modelled	(Hoffmann et al., 2015)
Ecuador	28	7	At research station	(Mite et al., 1999b; Pulver and Guerrero, 2014)
Costa Rica	36	9	Progeny trial	(ASD de Costa Rica, 2014)
Guatemala	32	8	Progeny trial	(ASD de Costa Rica, 2014)

A Peak yields in single years were converted to 25-year averages by assuming that yield over plantation lifetime = 0.8 × yield from peak year (adapted from Goh et al., 1994).

B Assumed oil extraction rate: 25%.

5. CURRENT CAUSES OF YIELD GAPS AND FUTURE OUTLOOK

In this section, the main factors contributing to the worldwide yield gaps are discussed with special attention to smallholders, who face a number of unique constraints. Smallholders, with a plantation area of <50 ha, produce about 40% of the total CPO volume worldwide (RSPO, 2015).

Potential palm oil yields in the main palm oil producing countries are shown in Table 8. Specific estimates have been made for Indonesia, Malaysia and Ghana using the PALMSIM model (Hoffmann *et al.*, 2014 ; Rhebergen *et al.*, 2014). For the other countries no potential yield profiles are available but data from best-yielding trials or plantations can provide a benchmark. Large variations in potential yields may exist within countries, depending mostly on radiation (cloudiness) and elevation (temperature).

The actual yields achieved in the 16 largest oil-palm producing countries in the world in 2013 are shown in Table 9. Worldwide average yields have been rising steadily and are currently around 15 t fruit bunches or 3.0 t oil ha^{-1}, but yield increases are slow compared with other crops (Fry, 2009 ; Murphy, 2009). When comparing the numbers in Table 9 with the potential yields as estimated in Table 8, it is clear that the yield gaps in most countries are large. In Southeast Asia the average oil yield from the top producing plantation companies is 5.5 t oil (23 t fruit bunches) ha^{-1} yr^{-1} (Fairhurst and Griffiths, 2014). The estimated average production from smallholder plantations in Indonesia is only 13 t fruit bunches ha^{-1} yr^{-1} (Molenaar *et al.*, 2013 ; FAO, 2013), but positive exceptions exist, such as the Ophir scheme smallholders in West Sumatra who consistently achieved yields of 22–29 t fruit bunches ha^{-1} yr^{-1} (Jelsma et al., 2009). In Africa average actual yields are less than 8 t fruit bunches ha^{-1} yr^{-1} (Table 9).

The water-limited yields in currently planted areas are around 3.5 t oil ha^{-1} yr^{-1} in Africa (e.g. Rhebergen et al., 2014), 4.5 t ha^{-1} yr^{-1} in the Americas (e.g. Melling and Henson, 2011) and Thailand (e.g. Palat et al., 2008), and 5.5 t ha^{-1} yr^{-1} in Indonesia and Malaysia (e.g. Corley, 2009). Current yield gaps range from 2 to 4 t oil ha^{-1} yr^{-1} in smallholder systems and from 1 to 3 t oil ha^{-1} yr^{-1} in large plantations. Closing these yield gaps to only 80% of Yw could realistically increase global production by 15–20 Mt oil yr^{-1}–the equivalent to clearing 4–6 Mha of new land.

Table 9. Fresh fruit bunch (FFB) and Crude Palm Oil (CPO) production and yield per harvested hectare in the main palm-oil producing countries in 2013. Sources: FAO (2013); USDA-FAS (2016). Numbers must be viewed with some caution, as good-quality data on harvested area and yield is difficult to obtain, especially for smallholder plantations.

Country	Area harvested[1] (Mha)	Annual production (Mt)		Yield (t ha^{-1} yr^{-1})		OER[2] (%)	Data source
		FFB	CPO	FFB	CPO[3]		
Indonesia	7.1	120	26.9	17	3.8	22.4	FAO, unofficial figure
	8.1		30.5		3.8		USDA
Malaysia	4.6	95.7	19.2	21	4.2	20.0	FAO, unofficial figure
	4.5		20.2		4.5		USDA
Nigeria	3.0	8.0	1.0	2.7	0.32	12.0	FAO, estimate
	2.5		1.0		0.39		USDA
Thailand	0.63	12.8	2.0	20.5	3.1	15.1	FAO, official data
	0.66		2.0		3.0		USDA
Colombia	0.45	5	1.0	20	3.5	17.5	FAO, official data
	0.34		1.0		3.1		USDA
Ghana	0.36	2.1	0.12	5.8	0.30	5.2	FAO, estimate
	0.37		0.49		1.3		USDA
Guinea	0.31	0.8	0.05	2.7	0.20	7.4	FAO, estimate
	0.31		0.05		0.16		USDA
DRC (Congo)	0.28	1.8	0.30	6.6	1.1	16.7	FAO estimate
	0.18		0.22		1.2		USDA
Côte d'Ivoire	0.27	1.7	0.42	6.5	1.5	23.1	FAO, unofficial figure
	0.27		0.42		1.5		USDA
Ecuador	0.22	2.3	0.33	10.6	1.5	14.2	FAO, official data
	0.22		0.57		2.6		USDA
Papua New Guinea	0.15	2.1	0.50	14	3.3	23.6	FAO, unofficial figure
	0.15		0.50		3.4		USDA
Cameroon	0.14	2.5	0.23	18.2	1.7	9.3	FAO, unofficial figure
	0.13		0.29		2.2		USDA
Honduras	0.13	2	0.43	16	3.4	21.3	FAO, unofficial figure
	0.13		0.46		3.7		USDA
Brazil	0.11	1.3	0.34	11.5	3.1	27.0	FAO, official data
	0.12		0.34		2.8		USDA
Guatemala	0.07	1.5	0.40	22.8	6.2	27.2	FAO, unofficial figure
	0.10		0.43		4.3		USDA
Costa Rica	0.07	1.3	0.30	17.5	4.0	22.9	FAO, estimate
	0.06		0.21		3.5		USDA
World	18.1	266.5	54.4	14.8	3.0	20.3	FAO, aggregate
	18.6		59.4		3.2		USDA

1 Area harvested excludes immature area.

2 Oil extraction rate (OER) was calculated from the yield data (tonCPO/tonFFB×100).

3 CPO yield was calculated by dividing production over harvested area (MtonCPO/mHaharvestedarea).

Traditional village plantations in Africa are usually planted with 100% *dura*, which partly explains the poor oil extraction rates found in most African countries (Table 9). In Indonesia, *dura* presence in smallholder plantations is likely to be common, with an estimated 50% of independent smallholders in some areas having planted non-hybrid materials (Molenaar et al., 2013). Early replanting (i.e. replanting before the 25 year cycle has been completed) with new, high-yielding varieties is an important strategy to improve productivity. In Malaysia, slow replanting has led to aging of oil palm plantations and resulting declines in yield (Wahid and Simeh, 2010 ; USDA-FAS, 2012). The production in 25–30 year old palms is estimated to be 60–90% of peak productivity (Goh et al., 1994). For smallholder farmers, delayed replanting due to lack of financial means is a serious threat to current and future productivity (Government of Malaysia, 2011 ; Molenaar *et al.*, 2013). On the other hand, the ratio of immature to mature plantations is high worldwide due to area expansion. In Indonesia 22% of the planted area in 2014 was immature (USDA-FAS, 2015), while in a static area that is replanted every 25 years, 12% of the area is immature.

Drought is a key constraining factor to yield in Africa, parts of Latin and Central America, and parts of Southeast Asia. To allow for expansion into drier areas or for further yield improvements, irrigation has been used successfully in Ecuador (Mite et al., 1999a), Thailand (Palat *et al.*, 2008 ; Univanich, 2011) and India (Prasad et al., 2010), but is uncommon (and uneconomic) in most plantations. As a consequence of global warming, irrigation is likely to become increasingly relevant due to projected increases in frequency of droughts, especially in Africa and Latin America (Fischer *et al.*, 2007; Paeth *et al.*, 2009 ; Marengo *et al.*, 2009). The costs and benefits of different irrigation regimes under a range of environmental conditions need urgent further investigation. Waterlogging and flooding are largely un-quantified yield limiting factors which are likely to suppress yields especially in Malaysia and Indonesia (Lee and Ong, 2006 ; Malay Mail Online, 2015). Whether these are serious issues in other oil palm growing regions is unclear, and research efforts on the effects of flooding and waterlogging on yield in the different phases of the plantation life cycle are needed. Due to scarcity of suitable land, 2.1 Mha of peatlands in Southeast Asia were cleared for oil palm planting by 2010 (Miettinen *et al.*, 2012 ; Koh *et al.*, 2011). Proper water management in peat soils requires the establishment of drainage canals, dams and flood gates over a larger area (Othman *et al.*, 2011 ; Lim *et al.*, 2012). Smallholders cannot implement such practices at field scale and are therefore likely to obtain poor yields, especially in deep peat areas. Due to subsidence, drainage, and fire, cultivated peat soils progressively degrade, which threatens the future livelihoods of farmers established on peat areas (Könönen et al., 2015) and causes serious environmental problems including estimated greenhouse gas emissions of 60 Mg CO_2 ha^{-1} yr^{-1} in the first 25 years after forest clearing (Murdiyarso et al., 2010). Carbon stocks in mineral soils planted with oil palm remain stable, provided that pruned fronds are recycled within the plantation (Khasanah et al., 2015) and trunks are left in the field at replanting (Khalid et al., 2000).

Total N, P and K fertiliser use in oil palm in 2010 as estimated by the International Fertilizer Industry Association (Heffer, 2013) for Indonesia, Malaysia and Thailand are presented in Table 10. Applications are far less than optimal, ranging from 40 to 90% of recommended rates (Rankine and Fairhurst, 1999b). Almost double the amount of P and K are applied on plantations in Malaysia than in Indonesia, and three times more K than in Thailand, yet the average K application is still insufficient to replace the nutrients removed with a yield of 30 t fruit bunches ha^{-1} (Corley and Tinker, 2016: 365). Data for other countries are not available.

In smallholder plantations in Indonesia (Molenaar *et al.*, 2013; Lee *et al.*, 2013 ; Woittiez *et al.*, 2015) and Africa (Rafflegeau *et al.*, 2010; Kim *et al.*, 2013 ; Nkongho *et al.*, 2014) limited amounts of mineral fertilisers are applied, with potassium application rates being especially small (Rafflegeau *et al.*, 2010 ; Woittiez *et al.*, 2015). Site-specific recommendations are usually not available because tissue analysis and on-site fertiliser experiments can only be implemented when fields are managed and sampled collectively (Jelsma et al., 2009). Organic fertilisers from mill waste streams may not be accessible for smallholders due to competition or lack of infrastructure. Alternatively, smallholders sometimes integrate livestock within their oil palm systems and therefore have access to manure.

Pest problems in oil palm are relatively mild, apart from leaf miner in West Africa (Chung, 2015). In Malaysia, the estimated incidence of *Ganoderma* in 2009 was around 3.7% of the mature area, with a yearly increase in incidence rate of >10%, corresponding to an estimated 270,000 ha of affected palms in 2015 (Roslan and Idris, 2012). In Indonesia, *Ganoderma* is most prevalent in Sumatra, and losses of 40–50% of the palms at the time of replanting are reported to be common in North Sumatra (Cooper et al., 2011). In Latin America bud rot disease remains an important cause of yield loss (Benítez Sastoque, 2011). The disease currently affects an estimated 15% of the oil palm area in Colombia (Fontanilla et al., 2014) and similar areas in other Latin American countries (Tapia and Velasco, 2015 ; Gálvez Intriago, 2014).

Table 10. Fertilizer use (N, P and K) on oil palm in Indonesia, Malaysia, and Thailand in 2010/11 (Heffer, 2013). Data for other oil palm producing countries were not available.

Nutrient	Application, total (1000 t year^{-1})			Application per hectare (kg ha^{-1} year^{-1})[a]			Nutrient removal[b] (kg ha^{-1} year^{-1})
	Indonesia	Malaysia	Thailand	Indonesia	Malaysia	Thailand	
Nitrogen (N)	548	374	41	95	91	72	146
Phosphate (P)	61	78	9	11	19	16	19
Potassium (K)	643	821	39	111	199	69	248

[a] The application per hectare was calculated by diving the total application over the oil palm area in 2010 (FAO, 2013).
[b] The final right column shows the nutrient removal, assuming a yield of 30 t fruit bunches ha^{-1} (Corley and Tinker, 2003: 358).

A The application per hectare was calculated by diving the total application over the oil palm area in 2010 (FAO, 2013).

B The final right column shows the nutrient removal, assuming a yield of 30 t fruit bunches ha^{-1} (Corley and Tinker, 2003: 358).

Lack of labour, especially for harvesting, is a key issue in Malaysia, and to a lesser extent in Indonesia, leading to longer harvesting rounds, which result in reduced oil extraction rates, loss of loose fruits and unharvested bunches (Murphy, 2014). Plantations in Malaysia report manpower shortages of 20–30% and consequent yield losses of 15% (Murphy, 2014). In South and Latin America, labour is more expensive leading to a competitive disadvantage. Mechanisation options for spreading fertilisers, spraying pesticides, and harvesting are being developed but have not yet been sufficiently successful to resolve labour shortages (Carter *et al.*, 2007; Yahya *et al.*, 2013 ; Khalid and Shuib, 2014).

6. CONCLUSIONS

Yield gaps in oil palm plantations are large, and there is considerable scope for improving yields and environmental performance. Yield responses to waterlogging, drainage, micronutrient fertilisers, and biotic stresses in mature plantations are poorly understood. A number of basic processes underlying bunch production need further investigation, especially sex determination and bunch failure. Also, the signalling pathway leading to drought stress responses needs to be unravelled, so that breeding and irrigation strategies can be further developed. Considering that smallholders produce 40% of the world palm oil supply, but often lag behind in terms of yield, particular effort should be put into understanding all the factors that limit yield in smallholder plantations, and to identify effective ways in which large numbers of smallholders can be supported to improve the sustainability and yield in their plantations. Increasing global yields to 80% of Yw could substitute the clearing of 4–6 Mha of new land. Improving yields in existing plantations in ways that are environmentally sound, while targeting expansion of oil palm cultivation into degraded lands only, appears to be the most responsible way forward for producing sufficient palm oil to meet future demands while preventing further loss of tropical rainforests.

ACKNOWLEDGEMENTS

The writing of this paper was funded by Wageningen University, The Netherlands. We thank Tiemen Rhebergen and an anonymous specialist for their useful comments on earlier drafts of the paper. We remain responsible for any errors or omissions.

REFERENCES

1. Abram, N.K., Xofis, P., Tzanopoulos, J., MacMillan, D.C., Ancrenaz, M., Chung, R.,Peter, L., Ong, R., Lackman, I., Goossens, B., Ambu, L., Knight, A.T., 2014.Synergies for improving oil palm production and forest conservation infloodplain landscapes. PLoS One 9 (6), 1–12, http://dx.doi.org/10.1371/journal.pone.0095388.
2. Adam, H., Jouannic, S., Escoute, J., Duval, Y., Verdeil, J.L., Tregear, J.W., 2005.Reproductive developmental complexity in the African oil palm (Elaeisguineensis, Arecaceae). Am. J. Bot. 92 (11), 1836–1852, http://dx.doi.org/10.3732/ajb.92.11.1836.
3. Adam, H., Collin, M., Richaud, F., Beule, T., Cros, D., Omore, A., Nodichao, L., Nouy,B., Tregear, J.W., 2011. Environmental regulation of sex determination in oilpalm: current knowledge and insights from other species. Ann. Bot. 108 (8),1529–1537, http://dx.doi.org/10.1093/aob/mcr151.
4. Agamuthu, P., Broughton, W.J., 1985. Nutrient cycling within the developing oilpalm-legume ecosystem. Agric. Ecosyst. Environ. 13 (2), 111–123, http://dx.doi.org/10.1016/0167-8809(85)90054-4.
5. Aggarwal, P.K., Kalra, N., 1994. Analyzing the limitations set by climatic factors,genotype, and water and nitrogen availability on productivity of wheat. 2.Climatically potential yields and management strategies. Field Crops Res. 38(2), 93–103, http://dx.doi.org/10.1016/0378-4290(94)90003-5.
6. Akbar, U., Tampubolon, F., Amiruddin, D., Ollagnier, M., 1976. Fertilizerexperimentation on oil palm in North Sumatra. Oléagineux 31 (7), 305–316.
7. Amthor, J.S., 1984. The role of maintenance respiration in plant growth. Plant CellEnviron. 7 (8), 561–569, http://dx.doi.org/10.1111/j.1365-3040.1984.tb01856.x.
8. Anderson, W.K., 2010. Closing the gap between actual and potential yield ofrainfed wheat. The impacts of environment, management and cultivar. FieldCrops Res. 116 (1-2), 14–22, http://dx.doi.org/10.1016/j.fcr.2009.11.016.
9. Angelsen, A., 2010. Policies for reduced deforestation and their impact onagricultural production. Proc. Natl. Acad. Sci. U. S. A. 107 (46), 19639–19644,http://dx.doi.org/10.1073/pnas.0912014107.
10. Ariffin, D., Idris, A.S., Gurmit Singh, 2000. Status of Ganoderma in oil palm. In:Flood, J., Bridge, P.D., Holderness, M. (Eds.), Ganoderma Diseases of PerennialCrops. CABI Publishing, Wallingford, pp. 49–68.
11. ASD de Costa Rica, 2014. Guía de Variedades y Clones. ASD de Costa Rica, San José.
12. Bafor, M.E., Osagie, A.U., 1986. Changes in lipid class and fatty acid compositionduring maturation of mesocarp of oil palm (Elaeis guineensis) variety dura. J.Sci. Food Agric. 37 (9), 825–832, http://dx.doi.org/10.1002/jsfa.2740370902.
13. Bah Lias, 2011. Oil palm NPKMg fertiliser trial (Sei Lakitan) EXPT EG00.1412A. In:Bah Lias Annual Report Volume 1. Bah Lias Research Station, PTPP LondonSumatra Indonesia, Medan, pp. 89–92.
14. Bah, A., Husni, M.H.A., Teh, C.B.S., Rafii, M.Y., Syed Omar, S.R., Ahmed, O.H., 2014.Reducing runoff loss of applied nutrients in oil palm cultivation usingcontrolled-release fertilizers. Adv. Agric. 2014, http://dx.doi.org/10.1155/2014/285387.
15. Balasundram, S.K., Robert, P.C., Mulla, D.J., Allan, D.L., 2006. Relationship betweenoil palm yield and soil fertility as affected by topography in an Indonesianplantation. Commun. Soil Sci. Plant Anal. 37 (9-10), 1321–1337, http://dx.doi.org/10. 1080/00103620600626817.
16. Banabas, M., Turner, M.A., Scotter, D.R., Nelson, P.N., 2008. Losses of nitrogenfertiliser under oil palm in Papua New Guinea: 1. Water balance, and nitrogenin soil solution and runoff. Aust. J. Soil Res. 46 (4), 332–339, http://dx.doi.org/10.1071/sr07171.
17. Bedford, G.O., 1980. Biology. ecology and control of palm rhinoceros beetles. Annu.Rev. Entomol. 25, 309–339, http://dx.doi.org/10.1146/annurev.en.25.010180.001521.

18. Beirnaert, A., Vanderweyen, R., 1941. Contribution a l'étude génetique etbiométrique des varietés d'Elaeis guineensis Jacquin. Institut National pourl'Etude Agronomique de Congo Belge, Yangambi.
19. Bell, M.A., Fischer, R.A., Byerlee, D., Sayre, K., 1995. Genetic and agronomiccontributions to yield gains: a case study for wheat. Field Crops Res. 44 (2–3),55–65, http://dx.doi.org/10.1016/0378-4290(95)00049-6.
20. Benítez Sastoque, E.R., 2011. Epidemiología de la pudrición del cogollo de la palmade aceite. PhD Thesis. Universidad Nacional de Colombia, Bogotá.
21. Benítez, E., García, C., 2014. The history of research on oil palm bud rot (Elaeisguineensis Jacq.) in Colombia. Agronomía Colombiana 32 (3), 390–398.
22. Bille Ngalle, H., Bell, J.M., Ngando Ebongue, G.F., Nyobe, L., Ngangnou, F.C.,Ntsomboh, G.N., 2013. Morphogenesis of oil palm fruit (Elaeis guineensis Jacq.)in mesocarp and endocarp development. J. Life Sci. 7 (2), 153–158.
23. Braconnier, S., d'Auzac, J., 1985. Anatomical study and cytological demonstrationof potassium and chlorine flux associated with oil palm and coconut stomatalopening. Oléagineux 40, 547–551.
24. Bredas, J., Scuvie, L., 1960. Aperc, u des influences climatiques sur les cycles deproduction du palmier à huile. Oléagineux 15 (4), 211–222.
25. Breure, C.J., 1977. Preliminary results from an oil palm density × fertiliserexperiment on young volcanic soils in West New Britain. In: Earp, D.A., Newall,W. (Eds.), International Developments in Oil Palm: Proceedings of theMalaysian International Agricultural Oil Palm Conference, 14–17 June 1976.The Incorporated Society of Planters, Kuala Lumpur, pp. 192–207.
26. Breure, C.J., 1982. Factors affecting yield and growth of oil palm tenera in WestNew Britain. Oléagineux 37 (5), 213–223.
27. Breure, C.J., 1985. Relevant factors associated with crown expansion in oil palm(Elaeis guineensis Jacq.). Euphytica 34 (1), 161–175, http://dx.doi.org/10.1007/BF00022876.
28. Breure, C.J., 1987. Factors associated with the allocation of carbohydrates to bunchdry matter production in oil palm (Elaeis guineensis Jacq.) PhD Thesis.Wageningen University, Wageningen.
29. Breure, C.J., 1988. The effect of planting densities on yield trends in oil palm. Exp.Agric. 24 (1), 37–52, http://dx.doi.org/10.1017/s0014479700015684.
30. Breure, C.J., 1994. Development of leaves in oil palm (Elaeis guineensis) anddetermination of leaf opening rate. Exp. Agric. 30 (4),467472http:// dx. doi.org/10.1017/5001447900024728.
31. Breure, C.J., 2003. The search for yield in oil palm: basic principles. In: Fairhurst, T.,Härdter, R. (Eds.), The Oil Palm—Management for Large and Sustainable Yields.Potash & Phosphate Institute of Canada, Potash & Phosphate Institute,International Potash Institute, Singapore, pp. 59–98.
32. Breure, C.J., 2010. Rate of leaf expansion: a criterion for identifying oil palm (Elaeisguineensis Jacq.) types suitable for planting at high densities. NJAS Wagening. J.Life Sci. 57 (2), 141–147, http://dx.doi.org/10.1016/j.njas.2010.03.001.
33. Breure, C.J., Corley, R.H.V., 1992. Fruiting activity, growth and yield of oil palm 2.Observations in untreated populations. Exp. Agric. 28 (1), 111–121, http://dx.doi.org/10.1017/s001447970002305x.
34. Breure, C.J., Menendez, T., 1990. The determination of bunch yield components inthe development of inflorescences in oil palm (Elaeis guineensis). Exp. Agric. 26(1), 99–115, http://dx.doi.org/10.1017/s0014479700015441.
35. Breure, C.J., Menendez, T., Powell, M.S., 1990. The effect of planting density on theyield components of oil palm (Elaeis guineensis). Exp. Agric. 26 (1), 117–124,http:// dx.doi.org/10.1017/S0014479700015453.
36. Broekmans, A.F.M., 1957. Growth, flowering and yield of the oil palm in Nigeria. J.West Afr. Inst. Oil Palm Res. 2, 187–220.
37. Broeshart, H., Ferwerda, J.D., Kovachich, W.G., 1957. Mineral deficiency symptomsof the oil palm. Plant Soil 8 (4), 289–300, http://dx.doi.org/10.1007/Bf01666319.

38. Cahyasiwi, L., Wood, B., Lubis, F., Caudwell, R., 2010. The economics of Oryctesattack in oil palm replants. In: Proceedings of the International Oil PalmConference, 1–3 June 2010, Yogyakarta.
39. Caliman, J.P., 1992. Oil palm and water deficit, production, adapted croppingtechniques. Oléagineux 47 (5), 205–216.
40. Caliman, J.P., Southworth, A., Jatmika, A., Bangun, D., Asmono, D., Sutarta, E.S.,Pamin, K., Guritno, P., Prawirosukarto, S., Wahyono, T., Herawan, T., Hutomo,T., Darmosarkoro, W., Adiwiganda, Y.T., Poeloengan, Z., 1998. Effect of droughtand haze on the performance of oil palm. In: Proceedings of the InternationalOil Palm Conference 'Commodity of the Past, Today, and the Future', 23–25September 1998, International oil Palm Research Institude, Medan, pp250–274.
41. Carlson, K.M., Curran, L.M., Ratnasari, D., Pittman, A.M., Soares, B.S., Asner, G.P.,Trigg, S.N., Gaveau, D.A., Lawrence, D., Rodrigues, H.O., 2012. Committedcarbon emissions, deforestation, and community land conversion from oilpalm plantation expansion in West Kalimantan, Indonesia. Proc. Natl. Acad. Sci.U. S. A. 109 (19), 7559–7564, http://dx.doi.org/10.1073/pnas.1200452109.
42. Carr, M.K.V., 2011. The water relations and irrigation requirement of oil palm(Elaeis guineensis): a review. Exp. Agric. 47 (4), 629–652, http://dx.doi.org/10.1017/s0014479711000494.
43. Carter, C., Finley, W., Fry, J., Jackson, D., Willis, L., 2007. Palm oil markets and futuresupply. Eur. J. Lipid Sci. Technol. 109 (4), 307–314, http://dx.doi.org/10.1002/ejlt.200600256.
44. Chang, K.C., Foster, H.L., Abas, Z., 1988. Monthly frond production of oil palm inMalaysia. Oléagineux 43 (12), 439–443.
45. Chapman, K.R., Escobar, R., Griffee, P., 2003. Cold tolerant or altitude adapted oilpalm hybrid development initiatives in the Asia/Pacific region. Assump. Univ. J.Technol. 6 (3), 134–138.
46. Cheong, S.P., Ng, S.K., 1977. Copper deficiency of oil palms on peat. In: Earp, D.A.,Newall, W. (Eds.), International Developments in Oil Palm: Proceedings of theMalaysian International Agricultural Oil Palm Conference, 14–17 June 1976.The Incorporated Society of Planters, Kuala Lumpur, pp. 362–370.
47. Chuah, J.H., Lim, K.H., 1992. Water management and irrigation of oil palm oncoastal soils – Sime Darby's experience. Planter 65 (761), 334–344.
48. Chung, G.F., 2011. Management of Ganoderma diseases in oil palm plantations.Planter 87 (1022), 325–339.
49. Chung, G.F., 2015. Effects of pests and diseases on oil palm yield. In: Lai, O.M., Tan,C.P., Akoh, C.C. (Eds.), Palm Oil: Production Processing Characterization, andUses. Elsevier Amsterdam, pp. 163–210.
50. Combres, J.C., Pallas, B., Rouan, L., Mialet-Serra, I., Caliman, J.P., Braconnier, S.,Soulie, J.C., Dingkuhn, M., 2013. Simulation of inflorescence dynamics in oilpalm and estimation of environment-sensitive phenological phases: a modelbased analysis. Funct. Plant Biol. 40 (3), 263–279, http://dx.doi.org/10.1071/fp12133.
51. Comte, I., Colin, F., Whalen, J.K., Grunberger, O., Caliman, J.P., 2012. Agriculturalpractices in oil palm plantations and their impact on hydrological changes,nutrient fluxes and water quality in Indonesia: a review. Adv. Agron. 116,71–124, http://dx.doi.org/10.1016/b978-0-12-394277-7.00003-8.
52. Cooper, R.M., Flood, J., Rees, R.W., 2011. Ganoderma boninense in oil palmplantations: current thinking on epidemiology, resistance and pathology.Planter 87 (1024), 515–526.
53. Corley, R.H.V., Ng, M., Donough, C.R., 1995. Effects of defoliation on sexdifferentiation in oil palm clones. Exp. Agric. 31 (2), 177–190, http://dx.doi.org/10.1017/S0014479700025266.
54. Corley, R.H.V., 1973a. Effects of plant density on growth and yield of oil palm. Exp.Agric. 9, 169–180, http://dx.doi.org/10.1017/S0014479700005639.
55. Corley, R.H.V., 1973b. Oil palm physiology: a review. In: Wastie, R.L., Easp, D.A.(Eds.), Advances in Oil Palm Cultivation: Proceedings of the International OilPalm Conference, 16–10 November 1972. Incorporated Society of Planters,Kuala Lumpur, pp. 37–51.

56. Corley, R.H.V., 1976a. Inflorescence abortion and sex differentiation. In: Corley,R.H.V., Hardon, J.J., Wood, B.J. (Eds.), Oil Palm Research. Elsevier Amsterdam,pp. 37–54.
57. Corley, R.H.V., 1976b. Photosynthesis and productivity. In: Corley, R.H.V., Hardon,J.J., Wood, B.J. (Eds.), Oil Palm Research. Elsevier Amsterdam, pp. 55–76.
58. Corley, R.H.V., 1983. Photosynthesis and age of oil palm leaves. Photosynthetica 17(1), 97–100.
59. Corley, R.H.V., 1996. Irrigation of oil palms—a review. J. Plant. Crops 24, 45–52.
60. Corley, R.H.V., 1998. What Is the Upper Limit to Oil Extraction Ratio? In: Rajanaidu,N., Henson, I.E., Jalani, B.S. (Eds.), Proceedings of the 1996 InternationalConference on Oil and Kernel Production in Oil Palm—A Global Perspective,27–28 September 1996. Palm Oil Research Institute Malaysia, Kuala Lumpur,pp. 256–269.
61. Corley, R.H.V., 2001. Ripening, harvesting and oil extraction. Planter 77 (906),507–524.
62. Corley, R.H.V., 2006. Potential Yield of Oil Palm—An Update. Symposium on YieldPotential in the Oil Palm. International Society of Oil Plam Breeders, Phuket.
63. Corley, R.H.V., 2009. How much palm oil do we need? Environ. Sci. Policy 12 (2),134–139, http://dx.doi.org/10.1016/j.envsci.2008.10.011.
64. Corley, R.H.V., Breure, C.J., 1992. Fruiting activity, growth and yield of oil palm. 1.Effects of fruit removal. Exp. Agric. 28 (1), 99–109, http://dx.doi.org/10.1017/ s0014 479700023048.
65. Corley, R.H.V., Gray, B.S., 1976. Yield and yield components. In: Corley, R.H.V.,Hardon, J.J., Wood, B.J. (Eds.), Oil Palm Research. Elsevier Amsterdam, pp.77–86.
66. Corley, R.H.V., Hew, C.K., 1976. Pruning. In: Corley, R.H.V., Hardon, J.J., Wood, B.J.(Eds.), Oil Palm Research. Elsevier Amsterdam, pp. 307–313.
67. Corley, R.H.V., Hong, T.K., 1982. Irrigation of oil palms in Malaysia. In: Pushparajah,E., Chew, P.S. (Eds.), The Oil Palm in Agriculture in the Eighties. IncorporatedSociety of Planters, Kuala Lumpur, pp. 343–356.
68. Corley, R.H.V., Lee, C.H., 1992. The physiological basis for genetic improvement ofoil palm in Malaysia. Euphytica 60 (3), 179–184, http://dx.doi.org/10.1007/bf00039396.
69. Corley, R.H.V., Mok, C.K., 1972. Effects of nitrogen, phosphorus, potassium andmagnesium on growth of the oil palm. Exp. Agric. 8 (4), 347–353, http://dx.doi.org/10.1017/ s0014479700005470.
70. Corley, R.H.V., Tinker, P.B., 2003. The Oil Palm. Blackwell Science, Oxford, http://dx.doi.org/10.1007/9780470750971.
71. Corley, R.H.V., Tinker, P.B., 2016. The Oil Palm. John Wiley & Sons, New York,http://dx.doi.org/10.1002/9781118953297.
72. Corley, R.H.V., Gray, B.S., Kee, N.S., 1971a. Productivity of the oil palm (Elaeisguineensis Jacq.) in Malaysia. Exp. Agric. 7 (02), 129–136, http://dx.doi.org/10. 1017/ S0014479700004907.
73. Corley, R.H.V., Hardon, J.J., Tan, G.Y., 1971b. Analysis of growth of the oil palm(Elaeis guineensis Jacq.) I. Estimation of growth parameters and application inbreeding. Euphytica 20 (2), 307–315, http://dx.doi.org/10.1007/bf00056093.
74. Corley, R.H.V., Hew, C.K., Tam, T.K., Lo, K.K., 1973. Optimal spacing for oil palms. In:Wastie, R.L., Earp, D.A. (Eds.), Advances in Oil Palm Cultivation: Proceedings ofthe International Oil Palm Conference, 16–18 November 1972. IncorporatedSociety of Planters, Kuala Lumpur, pp. 52–69.
75. Cros, D., Flori, A., Nodichao, L., Omore, A., Nouy, B., 2013. Differential response towater balance and bunch load generates diversity of bunch production profilesamong oil palm crosses (Elaeis guineensis). Trop. Plant Biol. 6 (1), 26–36, http://dx.doi.org/ 10.1007/512042-013-9116-2.
76. De Franqueville, H., 2003. Oil palm bud rot in Latin America. Exp. Agric. 39 (03),225–240, http://dx.doi.org/10.1017/50014479703001315.
77. Devendra, C., 2009. Intensification of integrated oil palm-ruminant systems.Enhancing increased productivity and sustainability in South-East Asia.Outlook Agric. 38 (1), 71–81, http://dx.doi.org/10.5367/000000009787762734.
78. Devendra, C., Thomas, D., 2002. Crop–animal interactions in mixed farmingsystems in Asia. Agric. Syst. 71 (1–2), 27–40 http://dx.doi.org/10.1016/S0308-521X(01)00034-8.

79. Donough, C., 2003. Harvesting study at IJM plantations QTR centre. Presentationfor the Seminar on Managing Soils of Miri-Bintulu Area, 8–11 December 2003,Sarawak.
80. Donough, C., Witt, C., Fairhurst, T., 2009. Yield intensification in oil palmplantations through best management practice. Better Crops Plant Food 93,12–14.
81. Donough, C., Witt, C., Fairhurst, T., 2010. Yield intensification in oil palm using BMPas a management tool. In: Proceedings of the International Oil PalmConference, 1–3 June 2010, Yogyakarta, Indonesian Oil Palm Research Institute.
82. Donough, C.R., Cock, J., Oberthür, T., Indrasuara, K., Gatot, A.R., Dolong, T., 2013.Estimating oil content of commercially harvested oil palm fresh fruitbunches—A step towards increasing palm oil yields. Poster Presented at thePIPOC 2013 International Palm Oil Congress and Exhibition, 19–21 November2013, Kuala Lumpur.
83. Dubos, B., Caliman, J.P., Corrado, F., Quencez, P., Siswo, S., Tailez, B., 1999.Importance of magnesium nutrition in oil palm. Plant. Recherche Dév. 6 (5),25–39.
84. Dufrène, E., Saugier, B., 1993. Gas exchange of oil palm in relation to light,vapor-pressure deficit, temperature and leaf age. Oléagineux 48 (8–9),347–356.
85. Dufrène, E., Ochs, R., Saugier, B., 1990. Oil palm photosynthesis and productivitylinked to climatic factors. Oléagineux 45 (8–9), 345–353.
86. Durand-Gasselin, T., Asmady, H., Flori, A., Jacquemard, J., Hayun, Z., Breton, F., DeFranqueville, H., 2005. Possible sources of genetic resistance in oil palm (Elaeisguineensis Jacq.) to basal stem rot caused by Ganoderma boninense–prospectsfor future breeding. Mycopathologia 159 (1), 93–100, http://dx.doi.org/10.1007/511046-004-4429-1.
87. Evans, L.T., Fischer, R.A., 1999. Yield potential: its definition, measurement, andsignificance. Crop Sci. 39 (6), 1544–1551, http://dx.doi.org/ 10.2135/ crop sci1999.396.3961544x.
88. Evans, L.T., 1993. Crop Evolution, Adaptation and Yield. Cambridge UniversityPress, Cambridge.
89. FAO, 2013. FAOSTAT Database Collections. Food and Agriculture Organization ofthe United Nations, Rome, URL: faostal3.fao.org/home/E, accessed on 12.12.15.
90. Fairhurst, T., Griffiths, W., 2014. Oil Palm: Best Management Practices for YieldIntensification. International Plant Nutrition Institute (IPNI), Singapore.
91. Fairhurst, T., McLaughlin, D., 2009. Sustainable Oil Palm Development on DegradedLand in Kalimantan. World Wildlife Fund.
92. Fermont, A.M., Asten, P. J. A. v., Tittonell, P., Wijk, M. T. v., Giller, K.E., 2009. Closingthe cassava yield gap: an analysis from smallholder farms in East Africa. FieldCrops Res. 112 (1), 24–36, http://dx.doi.org/10.1016/j.fcr.2009.01.009.
93. Fischer, G., Tubiello, F.N., Van Velthuizen, H., Wiberg, D.A., 2007. Climate changeimpacts on irrigation water requirements: effects of mitigation, 1990–2080.Technol. Forecast. Soc. Change 74 (7), 1083–1107, http://dx.doi.org/10.1016/j.techfore.2006.05.021.
94. Flood, J., Hasan, Y., 2004. Basal stem rot – taxonomy, biology, epidemiology,economic status and control in Southeast Asia and Pacific islands. In:Proceedings of the International Conference on Pests and Diseases ofImportance to the Oil Palm Industry, 18–19 May 2004, Kuala Lumpur.
95. Flood, J., Bridge, P.D., Holderness, M., 2000. Ganoderma Diseases of Perennial Crops.CABI Publishing, Wallingford, http://dx.doi.org/10.1079/9780851993881.0000.
96. Flood, J., Hasan, Y., Foster, H., 2002. Ganoderma diseases of oil palm—aninterpretation from Bah Lias Research Station. Planter 78 (921), 689–710.
97. Fontanilla, C.A., Montoya, M.M., Ruiz, E., Sánchez, A.C., Arias, N., Guerreo, J.M.,Castro, W., Penagos, Y., 2014. Estimación de costos de manejo de la Pudriciόndel cogollo (PC) de la palma de aceite. Rev. Palmas 35 (2), 23–37.
98. Foster, H., 2003. Assessment of oil palm fertilizer. In: Fairhurst, T., Härdter, R.(Eds.), Oil Palm: Management for Large and Sustainable Yields. Potash &Phosphate Institute of Canada, Potash & Phosphate Institute, InternationalPotash Institute, Singapore, pp. 231–257.
99. Foster, H.L., Chang, K.C., 1977. The diagnosis of the nutrient status of oil palms inWest Malaysia. In: Earp, D.A., Newall, W. (Eds.), International Developments inOil Palm—

Malaysian International Agricultural Oil Palm Conference, 14–17June 1976. Incorporated Society of Planters, Kuala Lumpur, pp. 290–312.
100. Foster, H.L., Prabowo, N.E., 2002. Overcoming the limitations of foliar diagnosis inoil palm. In: International Oil Palm Conference and Exhibition, 8–12 July 2002,Bali, Indonesian Oil Plam Research Institute.
101. Foster, H.L., Prabowo, N.E., 2006. Partition and transfer of nutrients in the reservetissues and leaves of oil palm. In: Workshop on Nutrient Needs in Oil Palm,17–18 October 2006, Singapore.
102. Fry, J., 2009. The Challenges Facing Palm Oil in the 21st Century—Presentation tothe SCI. LMC International, Oxford.
103. Fuhrer, J., 2003. Agroecosystem responses to combinations of elevated CO2, ozone,and global climate change. Agric. Ecosyst. Environ. 97 (1–3), 1–20, http://dx. doi. org/10.1016/s0167-8809(03)00125-7.
104. Gálvez Intriago, D.A., 2014. Análisis de mercado para la exportación de aceite depalma a Venezuela. Universidad Católica de Santiago de Guayaquil, Guayaquil.
105. Gan, L.T., 1998. Critical operational challenges for maximising oil extractions in oilpalm. Planter 74 (870), 487–499.
106. Gaveau, D.L.A., Sloan, S., Molidena, E., Yaen, H., Sheil, D., Abram, N.K., Ancrenaz, M.,Nasi, R., Quinones, M., Wielaard, N., Meijaard, E., 2014. Four decades of forestpersistence, clearance and logging on Borneo. PLoS One 9 (7), e101654, http://dx.doi.org/ 10.1371/journal.pone.0101654, 101611 pages.
107. Gerritsma, W., 1988. Light Interception, Leaf Photosynthesis and Sink-SourceRelations in Oil Palm. PhD Thesis. Wageningen University, Wageningen.
108. Gerritsma, W., Soebagyo, F.X., 1999. An analysis of the growth of leaf area of oilpalms in Indonesia. Exp. Agric. 35 (3), 293–308, http://dx.doi.org/10.1017/S0014479799003038.
109. Gillbanks, R.A., 2003. Standard agronomic procedures and practices. In: Fairhurst,T., Härdter, R. (Eds.), The Oil Palm—Management for Large and SustainableYields. Potash & Phosphate Institute of Canada, Potash & Phosphate Institute,International Potash Institute, Singapore, pp. 115–150.
110. Giller, K.E., 2001. Nitrogen Fixation in Tropical Cropping Systems. CABI Publishing,Wallingford.
111. Giller, K.E., Fairhurst, T.H., 2003. Legume cover plants. In: Fairhurst, T., Härdter, R.(Eds.), The Oil Palm—Management for Large and Sustainable Yields. Potash &Phosphate Institute of Canada, Potash & Phosphate Institute, InternationalPotash Institute, Singapore, pp. 151–162.
112. Goh, K.J., 2000. Climatic requirements of the oil palm for high yields. In: Goh, K.(Ed.), Managing Oil Palm for High Yields: Agronomic Principles. MalaysianSociety of Soil Science and Param Agricultural Surveys, Kuala Lumpur, pp. 1–17.
113. Goh, K.J., Chew, P.S., Teo, C.B., 1994. Maximising and maintaining oil palm yields oncommercial scale in Malaysia. In: Chee, K.H. (Ed.), Proceedings of the 1994Internation Planters Conference on Management for Enhanced Profitability inPlantations. Incorporated Society of Planters, Kuala Lumpur, pp. 121–141.
114. Goh, K.J., Härdter, R., Fairhurst, T., 2003. Fertilizing for maximum return. In:Fairhurst, T., Härdter, R. (Eds.), The Oil Palm—Management for Large andSustainable Yields. Potash & Phosphate Institute of Canada, Potash &Phosphate Institute, International Potash Institute, Singapore, pp. 279–306.
115. Goh, K.J., Kee, K.K., Chew, P.S., Gan, H.H., Heng, Y.C., Ng, H.C.P., 2000. Concept of siteyield potential and its applications in oil palm plantations. In: OFIC 2000, 4September, Kuala Lumpur.
116. Government of Malaysia, 2011. Economic Transformation Programme:Accelerating Replanting of Oil Palm. Performance Management & Delivery Unit(PEMANDU), Kuala Lumpur, http://etp.pemandu.gov.my/8 March 2011-@-Accelerating Replanting of Oil Palm.aspx. (accessed on 01.07.15).
117. Griffiths, W., Fairhurst, T., 2003. Implementation of best management practices inan oil palm rehabilitation project. Better Crops Int. 17, 16–19.

118. Griffiths, W., Fairhurst, T., Rankine, I., Kerstan, A., Taylor, C., 2002. Identificationand elimination of yield gaps in oil palm. Use of OMP7 and GIS. In: Proceedingsof the 2002 International Oil Palm Conference and Exhibition, 8–12 July 2002,Bali.
119. Gurmit, S., 1988. Zinc nutrition of oil palms on peat soils. In: Proceedings of theInternational Oil Palm/Palm Oil Conferences: Progress andProspects—Agriculture, 23–26 June 1987, Palm Oil Research Institute ofMalaysia, Kuala Lumpur, pp. 321–328.
120. Gurmit, S., Tan, Y.P., Rajah Padman, C.V., Lee, F.W., 1986. Experiences on thecultivation and management of oil palms on deep peat in United PlantationsBerhad. In: Second International Soil Management Workshop: ClassificationCharacterization and Utilization of Peat Land, 7–18 April 1986, Thailand &Malaysia.
121. Härdter, R., 1999. A review of magnesium nutrition in oil palm. In: Branch Seminaron Reforming the Mindset of Plantation Management into the NextMillennium, 29 July 1999, Taipeng, Perak, Malaysia.
122. Hardon, J.J., Williams, C.N., Watson, I., 1969. Leaf area and yield in the oil palm inMalaya. Exp. Agric. 5 (1), 25–32, http://dx.doi.org/10.1017/S0014479700009935.
123. Hartley, C.W.S., 1988. The Oil Palm (Elaeis Guineensis Jacq.). Longman GroupLimited, London.
124. Harun, M.H., Noor, M.R.M.D., 2002. Fruit set and oil palm bunch components. J. OilPalm Res. 14 (2), 24–33.
125. Heel, W. A. v., Breure, C.J., Menendez, T., 1987. The early development ofinflorescences and flowers of the oil palm (Elaeis guineensis Jacq.) seen throughthe scanning electron microscope. Blumea 32, 67–78.
126. Heffer, P., 2013. Assessment of Fertilizer Use by Crop at the Global Level2010-2010/11. International Fertilizer Industry Association (IFA), Paris.
127. Henry, P., 1958. Croissance et développement chez Elaeis guineensis Jacq. de lagermination a la première floraison. Revue Générale de Botanique 66, 5–34.
128. Henry, P., 1960. Recherches cytologiques sur l'appareil floral et la graine chez Elaeisguineensis et Cocos nucifera. I. La formation de l'appareil floral. Revue Généralede Botanique 68, 111–133.
129. Henson, I.E., 1991a. Adaptation to light environment by leaves of oil palm (Elaeisguineensis). PORIM Bull. 22, 1–8.
130. Henson, I.E., 1991b. Age-related changes in stomatal and photosyntheticcharacteristics of leaves of oil palm (Elaeis guineensis Jacq.). Elaeis 3 (2),336–348.
131. Henson, I.E., 1992. Estimating potential productivity of oil palm. In: Rao, V.,Henson, I.E., Rajanaidu, N. (Eds.), Proceedings of the 1990 ISOPB InternationalWorkshop on Yield Potential in the Oil Palm in Phuket, Thailand, on 29–30October. Palm Oil Research Institute of Malaysia, Kuala Lumpur, pp. 98–108.
132. Henson, I.E., 2000. Modelling the effects of 'haze' on oil palm productivity andyield. J. Oil Palm Res. 12 (1), 123–134.
133. Henson, I.E., 2001. Marco analítico para identificar los factores que determinan lastasas de extracción de aceite. PALMAS 22 (3).
134. Henson, I.E., 2002. Oil palm pruning and relationships between leaf area and yield- A review of previous experiments. Planter 78, 351–362.
135. Henson, I.E., 2004. Estimating maintenance respiration of oil palm. Oil Palm Bull.48, 1–10.
136. Henson, I.E., 2005. Modelling seasonal variation in oil palm bunch productionusing a spreadsheet programme. J. Oil Palm Res. 17 (1), 27–40.
137. Henson, I.E., 2006. Modelling the impact of climatic and climate-related factors onoil palm growth and productivity. MPOB Technol. 28, 1–43.
138. Henson, I.E., 2007. Modelling oil palm yield based on source and sink. Oil Palm Bull.54, 27–51.
139. Henson, I.E., Chang, K.C., 1990. Evidence for water as a factor limiting performanceon field palms in West Malaysia. In: Proceedings of the PORIM InternationalPalm Oil Development Conference: Module II—Agriculture, 5–9 September1989, Palm Oil Researh Institute of Malaysia, Kuala Lumpur, pp. 487–498.
140. Henson, I.E., Chai, S.H., 1997. Analysis of oil palm productivity. II. Biomassdistribution, productivity and turnover of the root system. Elaeis 9 (2), 78–92.

141. Henson, I.E., Dolmat, M.T., 2004. Seasonal variation in yield and developmentalprocesses in an oil palm density trial on a peat soil: 2. Bunch weightcomponents. J. Oil Palm Res. 16 (2), 106–120.
142. Henson, I.E., Harun, M.H., 2005. The influence of climatic conditions on gas andenergy exchanges above a young oil palm stand in North Kedah, Malaysia. J. OilPalm Res. 17, 73–91.
143. Henson, I.E., Harun, M.H., Chang, K.C., 2008. Some observations on the effects ofhigh water tables and flooding on oil palm, and a preliminary model of oilpalm water balance and use in the presence of a high water table. Oil PalmBull. 56, 14–22.
144. Ho, C.L., Tan, Y.C., 2015. Molecular defense response of oil palm to Ganodermainfection. Phytochemistry 114, 168–177, http://dx.doi.org/10.1016/j.phytochem.2014.10.016.
145. Hoffmann, M.P., Castaneda Vera, A., Wijk, M.T.v., Giller, K.E., Oberthür, T., Donough,C., Whitbread, A.M., 2014. Simulating potential growth and yield of oil palm(Elaeis guineensis) with PALMSIM: model description, evaluation andapplication. Agric. Syst. 131, 1–10, http://dx.doi.org/10.1016/j.agsy.2014.07.006.
146. Hoffmann, M.P., Donough, C., Oberthür, T., Castaneda Vera, A., v, Wijk M.T., Lim,C.H., Asmono, D., Samosir, Y., Lubis, A.P., Moses, D.S., Whitbread, A.M., 2015.Benchmarking yield for sustainable intensification of oil palm production inIndonesia using PALMSIM. Planter 91 (1067), 81–96.
147. Hoong, H.W., Donough, C.R., 1998. Recent trends in oil extraction rate (OER) andkernel extraction rate (KER) in Sabah. Planter 74 (865), 181–202.
148. Hoong, H., 2007. Ganoderma disease of oil palm in Sabah. Planter 83 (974),299–313.
149. Hushiarian, R., Yusof, N.A., Dutse, S.W., 2013. Detection and control of Ganodermaboninense: strategies and perspectives. SpringerPlus 2 (555), 12 p., http://dx.doi.org/10.1186/2193-1801-2-555.
150. Ibrahim, M.H., Jaafar, H.Z.E., Harun, M.H., Yusop, M.R., 2010. Changes in growth andphotosynthetic patterns of oil palm (Elaeis guineensis Jacq.) seedlings exposedL.S. Woittiez et al. / Europ. J. Agronomy 83 (2017) 57–77 75 to short-term CO2enrichment in a closed top chamber. Acta Physiol. Plant. 32(2), 305–313, http://dx.doi.org/10.1007/s11738-009-0408-y.
151. Idris, A.S., Ismail, S., Ariffin, D., 2004. Innovative technique of sanitation forcontrolling Ganoderma at replanting. MPOB Information Series 213, 4 p.
152. Jacquemard, J.C., Baudouin, L., 1998. Oil Palm. MacMillan, London.
153. Jalani, B.S., Basiron, Y., Arrifin, D., Chan, K.W., Rajanaidu, D., 2002. Prospects ofelevating national oil palm productivity: a Malaysian perspective. Oil Palm Ind.Econ. J. 2 (2), 1–9.
154. Jelsma, I., Giller, K., Fairhurst, T., 2009. Smallholder Oil Palm Production Systems inIndonesia: Lessons from the NESP Ophir Project. Wageningen University andResearch centre, Wageningen.
155. Jones, L.H., 1997. The effects of leaf pruning and other stresses on sexdetermination in the oil palm and their representation by a computersimulation. J. Theor. Biol. 187 (2), 241–260, http://dx.doi.org/10.1006/jtbi.1997.0433.
156. Jourdan, C., Rey, H., 1997. Architecture and development of the oil-palm (Elaeisguineensis Jacq.) root system. Plant Soil 189 (1), 33–48, http://dx.doi.org/10.1023/A:1004290024473.
157. Kee, K.K., Soh, A.C., 2002. Management of oil palms on slopelands in Malaysia. In:International Oil Palm Conference, 8–12 July 2002, Bali, Indonesian Oil PalmResearch Institute.
158. Khalid, M.R., Shuib, A.R., 2014. Field evaluation of harvesting machines for tall oilpalms. J. Oil Palm Res. 26 (1), 125–132.
159. Khalid, H., Zin, Z.Z., Anderson, J.M., 2000. Soil nutrient dynamics and palm growthperformance in relation to residue management practices following replantingof oil palm plantations. J. Oil Palm Res. 12 (1), 25–45.
160. Khasanah, N.m., van Noordwijk, M., Ningsih, H., Rahayu, S., 2015. Carbon neutral?No change in mineral soil carbon stock under oil palm plantations derivedfrom forest or non-forest in Indonesia. Agric. Ecosyst. Environ. 211, 195–206,http://dx.doi.org/10.1016/j.agee.2015.06.009.

161. Khaw, C.H., Ng, S.K., 1998. Performance of commercial scale clonal oil palm (Elaeisguineensis Jacq.) plantings in Malaysia. Acta Hortic. 461, 251–258, http://dx.doi.org/10.17660/ActaHortic.1998.461.27.
162. Kim, H., Kim, H., Madhavan, M., Suarez, A., 2013. Measuring EnvironmentalExternalities to Agriculture in Africa. Case Study: Ghana Palm Oil Sector. TheGeorge Washington University and the FAO, Rome.
163. Kimball, B.A., Mauney, J.R., Nakayama, F.S., Idso, S.B., 1993. Effects of increasingatmospheric CO2on vegetation. Vegetatio 104, 65–75, http://dx. doi. org/10.1007/bf00048145.
164. Koh, L.P., Miettinen, J., Liew, S.C., Ghazoul, J., 2011. Remotely sensed evidence oftropical peatland conversion to oil palm. Proc. Natl. Acad. Sci. U. S. A. 108 (12),5127–5132, http://dx.doi.org/10.1073/pnas.1018776108.
165. Könönen, M., Jauhiainen, J., Laiho, R., Kusin, K., Vasander, H., 2015. Physical andchemical properties of tropical peat under stabilised land uses. Mires Peat 16,1–13.
166. Kraalingen, D.W.G.v., Breure, C.J., Spitters, C.J.T., 1989. Simulation of oil palmgrowth and yield. Agric. Forest Meteorol. 46 (3), 227–244.
167. Kraip, J., Nake, S., 2006. Response of oil palm (Elaeis guineensis Jacq.) to additions ofnitrogen, phosphorus and potassium from planting to 7 years after planting:PNG OPRA Trial 501. Oil Palm Research Association, Papua New Guinea.
168. Kushairi, A., Tarmizi, A.H., Zamzuri, I., Ong-Abdullah, M., Samsul Kamal, R., Ooi, S.E.,Rajanaidu, N., 2010. Production, performance and advances in oil palm tissueculture. In: International Seminar on Advances in Oil Palm Tissue Culture, 29May 2010, Yogyakarta.
169. Laborte, A.G., Bie, K.d., Smaling, E.M.A., Moya, P.F., Boling, A.A., Van Ittersum, M.K.,2012. Rice yields and yield gaps in Southeast Asia: past trends and futureoutlook. Eur. J. Agron 36 (1), 9–20, http://dx.doi.org/10.1016/j.eja.2011.08.005.
170. Lamade, E., Bonnot, F., Pamin, K., Setyo, I.E., 1998. Quantitative approach of oilpalm phenology in different environments for La Me Deli and Yangambi Delimaterials investigations in the inflorescence cycles process. In: Proceedings ofthe International Oil Palm Conference 'Commodity of the Past, Today, and theFuture', 23–25 September 1998, Indonesian Oil Palm Research Institute,Medan, pp. 287–301.
171. Lambin, E.F., Turner, B.L., Geist, H.J., Agbola, S.B., Angelsen, A., Bruce, J.W., Coomes,O.T., Dirzo, R., Fischer, G., Folke, C., George, P.S., Homewood, K., Imbernon, J.,Leemans, R., Li, X., Moran, E.F., Mortimore, M., Ramakrishnan, P.S., Richards,J.F., Skånes, H., Steffen, W., Stone, G.D., Svedin, U., Veldkamp, T.A., Vogel, C., Xu,J., 2001. The causes of land-use and land-cover change: moving beyond themyths. Global Environ. Change 11 (4), 261–269, http://dx.doi.org/10.1016/S0959-3780(01)00007-3.
172. Laurance, W.F., 2007. Forest destruction in tropical Asia. Curr. Sci. 93 (11),1544–1550.
173. Lee, W.K., Ong, B.K., 2006. The unseen flood: waterlogging in large oil palmplantations. JURUTERA (January), 28–31.
174. Lee, J.S.H., Ghazoul, J., Obidzinski, K., Koh, L.P., 2013. Oil palm smallholder yieldsand incomes constrained by harvesting practices and type of smallholdermanagement in Indonesia. Agron. Sustain. Dev. 34 (2), 501–513, http://dx.doi.org/10.1007/s13593-013-0159-4.
175. Legros, S., Mialet-Serra, I., Caliman, J.P., Siregar, F.A., Clement-Vidal, A., Dingkuhn,M., 2009a. Phenology and growth adjustments of oil palm (Elaeis guineensis) tophotoperiod and climate variability. Ann. Bot. 104 (6), 1171–1182, http://dx.doi. org/10.1093/aob/mcp214.
176. Legros, S., Mialet-Serra, I., Caliman, J.P., Siregar, F.A., Clement-Vidal, A., Fabre, D.,Dingkuhn, M., 2009b. Phenology, growth and physiological adjustments of oilpalm (Elaeis guineensis) to sink limitation induced by fruit pruning. Ann. Bot.104 (6), 1183–1194, http://dx.doi.org/10.1093/aob/mcp216.
177. Lim, K.C., Chan, K.W., 1998. Bunch component studies over the past two decades.In: Rajanaidu, N., Henson, I.E., Jalani, B.S. (Eds.), Proceedings of the 1996International Conference on Oil and Kernel Production in Oil Palm—A GlobalPerspective, 27–28 September 1996. Palm Oil Research Institute Malaysia,Kuala Lumpur, pp. 133–150.

178. Lim, K.H., Lim, S.S., Parish, F., Suharto, R., 2012. RSPO Manual on Best ManagementPractices (BMPs) for Existing Oil Palm Cultivation on Peat. RSPO, Kuala Lumpur.
179. Lobell, D.B., Cassman, K.G., Field, C.B., 2009. Crop yield gaps: their importance,magnitudes, and causes. Annu. Rev. Environ. Resour. 34, 179–204, http://dx.doi.org/10.1146/annurev.environ.041008.093740.
180. Lopez, G., 2010. Bud rot, sudden wilt, red ring and lethal wilt of oil palm in theAmericas. PALMAS 31 (1), 43–53.
181. Ludwig, F., Biemans, H., Jacobs, C., Supit, I., Diepen, K.V., Fawell, J., 2011. Water Useof Oil Crops: Current Water Use and Future Outlooks. International LifeSciences Institute (ILSI) Europe, Brussels.
182. Malay Mail Online, 2015. Recent Malaysian Flood to Dent Palm Oil Supply in Q1.Kuala Lumpur, (accessed on 27.07.15) http://www.themalaymailonline.com/money/article/recent-malaysian-flood-to-dent-palm-oil-supply-in-q1.
183. Marengo, J.A., Jones, R., Alves, L.M., Valverde, M.C., 2009. Future change oftemperature and precipitation extremes in South America as derived from thePRECIS regional climate modeling system. Int. J. Climatol. 29 (15), 2241–2255,http://dx.doi.org/ 10.1002/joc.1863.
184. Mariau, D., 1976. Insect pests in West Africa. In: Corley, R.H.V., Hardon, J.J., Wood,B.J. (Eds.), Oil Palm Research. Elsevier, Amsterdam, pp. 369–383.
185. Mariau, D., Lecoustre, R., 2000. Rôle de facteurs eco-climatiques et edaphiques surla fécondité au champ de Coelaenomenodera lameensis, mineur des feuilles dupalmier à huile en Afrique de l'Ouest. Int. J. Trop. Insect Sci 20 (01), 7–21,http:// dx.doi.org/10.1017/S1742758400017781.
186. Martínez, G., Sarria, G.A., Torres, G.A., Varón, F., 2010. Phytophthora palmivora es elagente causal de la pudrición del cogollo de la palma de aceite. PALMAS 31,334–344 (Especial, Tomo I).
187. Melling, L., Henson, I.E., 2011. Greenhouse gas exchange of tropical peatlands – areview. J. Oil Palm Res. 23, 1087–1095.
188. Miettinen, J., Shi, C.H., Liew, S.C., 2011. Deforestation rates in insular Southeast Asiabetween 2000 and 2010. Global Change Biol. 17 (7), 2261–2270, http://dx.doi.org/10.1111/j.1365-2486.2011.02398.x.
189. Miettinen, J., Hooijer, A., Shi, C., Tollenaar, D., Vernimmen, R., Liew, S.C., Malins, C.,Page, S.E., 2012. Extent of industrial plantations on Southeast Asian peatlandsin 2010 with analysis of historical expansion and future projections. GCBBioenergy 4 (6), 908–918, http://dx.doi.org/10.1111/j.1757-1707.2012.01172.x.
190. Mite, F., Carrillo, M., Espinosa, J., 1999a. Fertilizer use efficiency in oil palm isincreased under irrigation in Ecuador. Better Crops Int. 13 (1), 30–32.
191. Mite, F., Carrillo, M., Espinosa, J., 1999b. Influencia de la fertilizacion y el riegosobre el desarrollo, nutricion y rendimiento de la palma Africana en Ecuador.Informaciones Agronómicas 36, 1–5.
192. Molenaar, J.W., Persch-Ort, M., Lord, S., Taylor, C., Harms, J., 2013. Oil PalmSmallholders—Developing a Better Understanding of Their Performance andPotential. International Finance Corporation, Jakarta.
193. Monteith, J.L., 1972. Solar-radiation and productivity in tropical ecosystems. J.Appl. Ecol. 9 (3), 747–766, http://dx.doi.org/10.2307/2401901.
194. Monteith, J.L., 1977. Climate and efficiency of crop production in Britain. Philos.Trans. R. Soc. Lond. Ser. B: Biol. Sci. 281 (980), 277–294, http://dx.doi.org/10.1098/ rstb.1977.0140.
195. Murdiyarso, D., Hergoualc'h, K., Verchot, L.V., 2010. Opportunities for reducinggreenhouse gas emissions in tropical peatlands. Proc. Natl. Acad. Sci. U. S. A.107 (46), 19655–19660, http://dx.doi.org/10.1073/pnas.0911966107.
196. Murphy, D.J., 2009. Oil palm: future prospects for yield and quality improvements.Lipid Technol. 21 (11/12), 257–260, http://dx.doi.org/10.1002/lite.200900067.Murphy, D.J.,

2014. The future of oil palm as a major global crop: opportunities andchallenges. J. Oil Palm Res. 26 (1), 1–24.

197. Murtilaksono, K., Darmosarkoro, W., Sutarta, E.S., Siregar, H.H., Hidayat, Y., Yusuf,M.A., 2011. Feasibility of soil and water conservation techniques on oil palmplantation. AGRIVITA 33 (1), 63–69.
198. Mutert, E., Fairhurst, T.H., von Uexküll, H.R., 1999. Agronomic management of oilpalms on deep peat. Better Crops Int. 13 (1), 22–27.
199. Neumann, K., Verburg, P.H., Stehfest, E., Müller, C., 2010. The yield gap of globalgrain production: a spatial analysis. Agric. Syst. 103 (5), 316–326, http://dx.doi.org/10.1016/j.agsy.2010.02.004.
200. Ng, S.K., 1977. Review of oil palm nutrition and manuring—scope for greatereconomy in fertilizer usage. In: Earp, D.A., Newall, W. (Eds.), InternationalDevelopments in Oil Palm: Proceedings of the Malaysian InternationalAgricultural Oil Palm Conference, 14–17 June 1976. Incorporated Society ofPlanters, Kuala Lumpur, pp. 209–233.
201. Ng, S.K., 1983. Advances in oil palm nutrition: agronomy and productivity inMalaysia. PORIM Occas. Paper 12, 1–20.
202. Ng, S.K., 1986. Phosphorus nutrition and fertilization of oil palms. Oléagineux 41,307–313.
203. Ng, S.K., 2002. Nutrition and nutrient management of oil palm—New thrust for thefuture perspective. In: Pasricha, N.S., Bansal, S.K. (Eds.), Potassium forSustainable Crop Production: International Symposium on the Role ofPotassium. Potash Research Institute of India and International PotashInstitute, New Delhi, pp. 415–429.
204. Ng, H.C.P., Chew, P.S., Goh, K.J., Kee, K.K., 1999. Nutrient requirements andsustainability in mature oil palms – an assessment. Planter 75 (880), 331–345.
205. Ng, S.K., Cheong, T.K., Haw, K.C., Ooi, H.S.H., Yee, L.K., Kayaroganam, P., Uexküll,H.R.v., Härdter, R., 2003. Clonal oil palm production, yield performance, andnutritional requirements. In: Fairhurst, T., Härdter, R. (Eds.), The OilPalm—Management for Large and Sustainable Yields. Potash & PhosphateInstitute of Canada, Potash & Phosphate Institute, International PotashInstitute, Singapore, pp. 99–114.
206. Nkongho, R.N., Feintrenie, L., Levang, P., 2014. Strengths and weaknesses of thesmallholder oil palm sector in Cameroon. Oilseeds and fats, Crops and Lipids21 (2), D208.
207. Nouy, B., Baudouin, L., Djegui, N., Omore, A., 1999. Le palmier à huile en conditionshydriques limitantes. Plant. Recherche Dév. 6, 31–45.
208. Ochs, R., Daniel, C., 1976. Research on techniques adapted to dry regions. In:Corley, R.H.V., Hardon, J.J., Wood, B.J. (Eds.), Oil Palm Research. Elsevier,Amsterdam, pp. 315–330.
209. Ochs, R., Ollagnier, M., 1977. Effect of fertilizers on composition of lipids producedby perennial tropical oil plants and on their yield. Oléagineux 32 (10), 409–426.
210. Ojuederie, B.M., Iremiren, G.O., Utulu, S.N., 1983. Effects of various inter-rowslashing regimes and size of weeded rings on the early growth, flowering andbunch yield of the oil palm. J. Nig. Inst. Oil Palm Res. 6 (24), 322–334.
211. Okwuagwu, C.O., Okoye, M.N., Okolo, E.C., Ataga, C.D., Uguru, M.I., 2008. Geneticvariability of fresh fruit bunch yield in Deli/dura × tenera breeding populationsof oil palm (Elaeis guineensis Jacq.) in Nigeria. J. Trop. Agric. 46 (1–2), 52–57.
212. Olivin, J., 1968. Étude pour la localisation d'un bloc industriel de palmiers à huile.Oléagineux 23 (8–9), 499–504.
213. Olivin, J., 1986. Study for the siting of a commercial oil palm plantation. Oléagineux41 (3), 113–118.
214. Ong-Abdullah, M., Ordway, J.M., Jiang, N., Ooi, S.-E., Kok, S.-Y., Sarpan, N., Azimi, N.,Hashim, A.T., Ishak, Z., Rosli, S.K., Malike, F.A., Bakar, N.A.A., Marjuni, M.,Abdullah, N., Yaakub, Z., Amiruddin, M.D., Nookiah, R., Singh, R., Low, E.-T.L.,Chan, K.-L., Azizi, N., Smith, S.W., Bacher, B., Budiman, M.A., Van Brunt, A.,Wischmeyer, C., Beil, M., Hogan, M., Lakey, N., Lim, C.-C., Arulandoo, X., Wong,C.-K., Choo, C.-N., Wong, W.-C., Kwan, Y.-Y., Alwee, S.S.R.S., Sambanthamurthi,R., Martienssen, R.A., 2015. Loss of Karma transposon methylation underliesthe mantled somaclonal variant of oil palm. Nature 525 (7570), 533–

537,http://dx.doi.org/10.1038/nature15365, http://www.nature.com/nature/journal/v525/n7570/ abs/nature15365.html#supplementary-information.

215. Oo, K.C., Lee, K.B., Ong, A.S.H., 1986. Changes in fatty acid composition of the lipidclasses in developing oil palm mesocarp. Phytochemistry 25 (2), 405–407.
216. Osman, J., Kueh, H.S., 1996. Evaluation of micronutrients omission on theperformance of oil palm grown on deep peat. In: Proceedings of the SoilScience Conference of Malaysia 1996: Managing Soil Resources withEnvironmental Consideration, 15–17 April 1996, Kuching, Sarawak.
217. Othman, H., Mohammed, A.T., Darus, F.M., Harun, M.H., Zambri, M.P., 2011. Bestmanagement practices for oil palm cultivation on peat: ground water-tablemaintenance in relation to peat subsidence and estimation of CO2emissions atSessang, Sarawak. J. Oil Palm Res. 23, 1078–1086.
218. PORLA, 1995. Fresh Fruit Bunch Grading Manual. Palm Oil Registration andLicensing Authority, Kuala Lumpur.
219. Paeth, H., Born, K., Girmes, R., Podzun, R., Jacob, D., 2009. Regional climate changein tropical and northern Africa due to greenhouse forcing and land usechanges. J. Clim. 22 (1), 114–132, http://dx.doi.org/10.1175/2008jcli2390.1.
220. Palat, T., Nakharin, C., Clendon, J.H., Corley, R.H.V., 2008. A review of 15 years of oilpalm irrigation research in southern Thailand. In: Indian National Conferenceon Oil Palm, Andhra Pradesh, India, pp. w537–w545.
221. Palat, T., Chayawat, N., Corley, R.H.V., 2012. Maximising oil palm yield by highdensity planting and thinning. Planter 88, 241–256.
222. Pallas, B., Mialet-Serra, I., Rouan, L., Clement-Vidal, A., Caliman, J.P., Dingkuhn, M.,2013. Effect of source/sink ratios on yield components, growth dynamics andstructural characteristics of oil palm (Elaeis guineensis) bunches. Tree Physiol.33 (4), 409–424, http://dx.doi.org/10.1093/treephys/tpt015.
223. Paramananthan, S., 2000. Soil requirements of oil palm for high yields. In: Goh, K.J.(Ed.), Managing Oil Palm for High Yields: Agronomic Principles. MalaysianSociety of Soil Science and Param Agricultural Surveys, Kuala Lumpur, pp.18–38.
224. Paramananthan, S., 2003. Land selection for oil palm. In: Fairhurst, T., Härdter, R.(Eds.), The Oil Palm—Management for Large and Sustainable Yields. Potash &Phosphate Institute of Canada, Potash & Phosphate Institute, InternationalPotash Institute, Singapore, pp. 27–58.
225. Paramananthan, S., 2013. Managing marginal soils for sustainable growth of oilpalms in the tropics. J. Oil Palm Environ. 4, 1–16.
226. Paramananthan, S., Chew, P., Goh, K., 2000. Towards a practical framework for landevaluation for oil palm in the 21st Century. In: Proceedings of the InternationalPlanters Conference 'Plantation Tree Crops in the New Millennium: the WayAhead', 17–20 May 2000, Kuala Lumpur, pp. 869–885.
227. Paterson, R.R.M., 2007. Ganoderma disease of oil palm—a white rot perspectivenecessary for integrated control. Crop Prot. 26 (9), 1369–1376, http://dx.doi.org/10.1016/j.cropro.2006.11.009.
228. Paul, M.J., Foyer, C.H., 2001. Sink regulation of photosynthesis. J. Exp. Bot. 52 (360),1383–1400, http://dx.doi.org/10.1093/jexbot/52.360.1383.
229. Prabowo, N.E., Tohiruddin, L., Tandiono, J., Foster, H.L., 2011. Prediction of ExpectedYields of Oil Palm from the Weights of Leaf Nutrients. In: Proc. Int. Palm OilConf. Agriculture, biotechnology and sustainability, Vol. II. Malaysian Palm OilBoard, Kuala Lumpur, pp. 463–470.
230. Prasad, M.V., Sarkar, A., Jameema, J., 2010. Performance of oil palm productiontechnologies. Indian J. Ext. Educ. 10 (3), 10–15.
231. Prasetyo, J.H.H., Sitepu, B., Djuhjana, J., Nelson, S.P.C., 2014. Ekona × Deli duramaterials: promising high oil-to-bunch dura materials. Oil Palm Bull. 68, 1–4.Puan, C.L., Goldizen, A.W., Zakaria, M., Hafidzi, M.N., Baxter, G.S., 2011.
232. Relationships among rat numbers, abundance of oil palm fruit and damagelevels to fruit in an oil palm plantation. Integr. Zool. 6 (2), 130–139, http://dx.doi.org/10.1111/j.1749-4877.2010.00231.x.

233. Pulver, E., Guerrero, J.M., 2014. Improving the Competitiveness of Small Scale OilPalm Farmers and Production in Latin America and the Caribbean: Bridging theYield Gap. Fondo Latinoamericano de Innovación en Palma de Aceite (FLIPA),Cali.
234. Purvis, C., 1956. The root system of the oil palm: its distribution, morphology, andanatomy. J. West Afr. Inst. Oil Palm Res. 4, 60–82.
235. RSPO, 2015. Smallholders. Roundtable on Sustainable Palm Oil (RSPO), KualaLumpur. URL: www.rspo.org/smallholders, (accessed on 24.07.15).
236. Rafflegeau, S., Michel-Dounias, I., Tailliez, B., Ndigui, B., Papy, F., 2010. UnexpectedN and K nutrition diagnosis in oil palm smallholdings using references ofhigh-yielding industrial plantations. Agron. Sustain. Dev. 30 (4), 777–787,http:// dx.doi.org/ 10.1051/agro/2010019.
237. Rafii, M.Y., Isac, Z.A., Kushairi, A., Saleh, G.B., Latif, M.A., 2013. Variation in yieldcomponents and vegetative traits in Malaysian oil palm (Elaeis guineensis jacq.)dura × pisifera hybrids under various planting densities. Ind. Crops Prod. 46,147–157, http://dx.doi.org/10.1016/j.indcrop.2012.12.054.
238. Rajanaidu, N., Kushairi, A., 2006. Oil palm planting materials and their yieldpotential. Presentation for the International Seminar on Yield Potential in OilPalm II, 27–28 November 2006, Phuket, Thailand.
239. Rajanaidu, N., Kushairi, A., Din, A., Isa, Z.A., Maizura, I., Noh, A., 2005. Oil palmplanting materials: current developments and competetiveness with other oilbearing crops. In: Proceedings of the 2005 International Palm Oil Congress:Technological Breakthroughs and Commercialization—The Way Forward,25–29 September 2005, Malaysian Palm Oil Board, Selangor, pp. 645–659.
240. Rajaratnam, J.A., 1973. Effect of boron deficiency on the production of oil palm inMalaysia. Oléagineux 28 (7), 329–332.
241. Rajaratnam, J.A., Lowry, J.B., 1974. The role of boron in the oil palm (Elaeisguineensis). Ann. Bot. 38, 193–200.
242. Rankine, I.R., Fairhurst, T.H., 1999a. Field Handbook: Oil Palm Series, Volume1—Nursery. Potash & Phosphate Institute (PPI), Singapore.
243. Rankine, I.R., Fairhurst, T.H., 1999b. Field Handbook: Oil Palm Series, Volume3—Mature. Potash & Phosphate Institute (PPI), Singapore.Rao, V., Law, I.H., 1998. The problem of poor fruit set in parts of East Malaysia.Planter 74 (870), 21.
244. Rao, V., Palat, T., Chayawat, N., Corley, R.H.V., 2008. The Univanich oil palmbreeding programme and progeny trial results from Thailand. Planter 84 (989),519–531.
245. Rees, A.R., 1964. The apical organization and phyllotaxis of the oil palm. Ann. Bot.28 (1), 57–69.
246. Rees, A.R., Tinker, P.B., 1963. Dry-matter production and nutrient content ofplantation oil palms in Nigeria. I. Growth and dry-matter production. Plant Soil19 (1), 19–32, http://dx.doi.org/10.1007/BF01347859.
247. Rhebergen, T., Hoffmann, M., Zingore, S., Oberthür, T., Acheampong, K., Dwumfour,G., Zutah, V., Adu-Frimpong, C., Ohipeni, F., Fairhurst, T., 2014. The effects ofclimate, soil and oil palm management practices on yield in Ghana. In:International Oil Palm Conference (IOPC), 17–19 June 2014, Bali.
248. Roslan, A., Idris, A.S., 2012. Economic impact of Ganoderma incidence on Malaysianoil palm plantation—a case study in Johor. Oil Palm Ind. Econ. J. 12 (1), 24–30.
249. Ryan, M.G., 1991. Effects of climate change on plant respiration. Ecol. Appl. 1 (2),157–167, http://dx.doi.org/10.2307/1941808.
250. Sambanthamurthi, R., Sundram, K., Tan, Y.A., 2000. Chemistry and biochemistry ofpalm oil. Prog. Lipid Res. 39 (6), 507–558, http://dx.doi.org/10.1016/S0163-7827(00)00015-1.
251. Samedani, B., Juraimi, A.S., Abdullah, S.A.S., Rafii, M.Y., Rahim, A.A., Anwar, M.P.,2014a. Effect of cover crops on weed community and oil palm yield. Int. J.Agric. Biol. 16, 23–31.
252. Samedani, B., Juraimi, A.S., Rafii, M.Y., Awadz, S.A.S., Anwar, M.P., Anuar, A.R.,2014b. Effect of cover crops on weed suppression in oil palm plantation. Int. J.Agric. Biol. 17 (2), 251–260.

253. Sharma, M., 2007. Challenges facing the Malaysian palm oil industry -multi-pronged strategies for raising oil yield, productivity and profitability.Planter 83 (981), 797–833.
254. Shaul, O., 2002. Magnesium transport and function in plants: the tip of the iceberg.Biometals 15 (3), 309–323, http://dx.doi.org/10.1023/A:1016091118585.
255. Sidhu, M., Kim, M., Sinuraya, Z., 2001. Yield responses of oil palm to NPK fertiliserapplication on an alluvial soil in Indonesia. Planter 77 (908), 635–651, http://dx.doi.org/10.1093/jxb/40.6.647.
256. Simon, S., Hendry, T., Chang, S.W., Kiaw, C.W., 1998. Early yield performance ofclonal oil palm (Elaeis guineensis Jacq.) plantings in PPB Oil Palms Bhd Sabah—acase study. Planter 74 (866), 257–269.
257. Smith, B.G., 1989. The effects of soil water and atmospheric vapour pressure deficiton stomatal behaviour and photosynthesis in the oil palm. J. Exp. Bot. 40 (215),647–651, http://dx.doi.org/10.1093/jxb/40.6.647.
258. Soh, A.C., 2004. Selecting the ideal oil palm: what you see is not necessarily whatyou get. J. Oil Palm Res. 16 (2), 121–128.
259. Soh, A.C., 2012. The future of oil palm clones in Malaysia. J. Oil Palm Environ. 3,93–97.
260. Soh, A.C., 2015. Breeding and genetics of the oil palm. In: Lai, O.M., Tan, C.P., Akoh,C.C. (Eds.), Palm Oil: Production, Processing, Characterization, and Uses. AOCSPress, Urbana, pp. 31–58.
261. Soh, A.C., Wong, G., Tan, C.C., Chew, P.S., Chong, S.P., Ho, Y.W., Wong, C.K., Choo,C.N., Nor Azura, H., Kumar, K., 2011. Commercial-scale propagation andplanting of elite oil palm clones: research and development towardsrealization. J. Oil Palm Res. 23 (1), 935–952.
262. Sparnaaij, L.D., 1960. The analysis of bunch production in the oil palm. J. West Afr.Inst. Oil Palm Res. 3, 109–180.
263. Squire, G.R., 1986. A physiological analysis for oil palm trials. PORIM Bull. 12,12–31.
264. Squire, G.R., Corley, R.H.V., 1987. Oil palm. In: Sethuraj, M.R., Raghavendra, A.S.(Eds.), Tree Crop Physiology. Elsevier Science Publishers, Amsterdam, pp.141–167.
265. Stibig, H.J., Achard, F., Carboni, S., Raˇsi, R., Miettinen, J., 2014. Change in tropicalforest cover of Southeast Asia from 1990 to 2010. Biogeosciences 11, 247–258,http://dx.doi.org/10.5194/bg-11-247-2014.
266. Surre, C., 1968. Les besoins en eau du palmier à huile. Oléagineux 3, 165–167.
267. Sushil, K., Mukhtar, A., 2008. Impact of various methods of management ofrhinoceros beetle, Oryctes rhinoceros Linn. in oil palm. J. Appl. Zool. Res. 19 (2),157–162.
268. Sutarta, E.S., Rahutomo, S., 2016. New Standard for FFB Yield of IOPRI's PlantingMaterials Based on Land Suitability Class. Indonesian Oil Palm ResearchInstitute, Medan.
269. Syed, R.A., Law, I.H., Corley, R.H.V., 1982. Insect pollination of oil palm:introduction, establishment and pollinating efficiency of Elaeidobiuskamerunicus in Malaysia. Planter 58 (681), 547–561.
270. Tam, T.K., 1973. Culling and selection in oil palm nurseries. In: Wastie, R.L., Earp,D.A. (Eds.), Advances in Oil Palm Cultivation: Proceedings of the InternationalPalm Conference, 16–18 November 1972. Incorporated Society of Planters,Kuala Lumpur, pp. 104–113.
271. Tans, P., Keeling, R., 2015. Trends in Atmospheric Carbon Dioxide. Mauna Loa,Hawaii. Earth System Research Laboratory, Boulder, USA (accessed on14.08.15) http://www.esrl.noaa.gov/gmd/ccgg/trends/.
272. Tapia, E., Velasco, B., 2015. Una enfermedad merma la bonanza de la palmaafricana. El Comercio, Quito (accessed on 02.12.15) http://www.elcomercio.com/actualidad/enfermedad-bonanza-palma-africana.html.
273. Teoh, K.C., Chew, P.S., 1988. Potassium in the oil palm eco-system and someimplications to manuring practice. In: Hassan, A.H., Soon, C.P., Wood, B.J.,Pushparajah, E. (Eds.), International Oil Palm/Palm Oil Conferences: Progressand Prospects—Agriculture, 23–26 June 1987. Palm Oil Research Institute ofMalaysia & Incorporated Society of Planters, Kuala Lumpur, pp. 277–286.
274. Tinker, P.B., 1984. Site-specific yield potentials in relation to fertilizer use. In:Peter, A.v. (Ed.), Nutrient Balances and Fertilizer Needs in TemperateAgriculture. International Potash Institute, Bern, Switzerland, pp. 193–208.

275. Toh, P.Y., Poon, Y.C., 1981. Effect of water management on field performance of oilpalms on acid sulphate soils in Peninsular Malaysia. In: Symposium on AcidSulphate Soils, Bangkok.
276. Tohiruddin, L., Prabowo, N.E., Foster, H.L., 2006. Comparison of the response of oilpalm to fertilisers at different locations in North and South Sumatra. In:International Oil Palm Conference (IOPC), 19–23 June 2006, Nusa Dua, Bali,Indonesian Oil Palm Research Institute.
277. Tohiruddin, L., Tandiono, J., Silalahi, A.J., Prabowo, N.E., Foster, H.L., 2010. Effects ofN, P and K fertilizers on leaf trace element levels of oil palm in Sumatra. J. OilPalm Res. 22, 869–877.
278. Uexküll, H.R.v., Fairhurst, T., 1991. Fertilizing for high yield and quality—the oilpalm. IPI Bull. 12.
279. Uexküll, H.v., Henson, I.E., Fairhurst, T., 2003. Canopy management to optimizeyield. In: Fairhurst, T., Härdter, R. (Eds.), Oil Palm Management for Large andSustainable Yields. Potash & Phosphate Institute of Canada, Potash &Phosphate Institute, International Potash Institute, Singapore, pp. 163–180.
280. Univanich, 2011. Some Best Practices in Thailand's Oil Palm Industry. UnivanichPalm Oil PCL, Krabi, Thailand.
281. USDA, 2014. Oilseeds: World Markets and Trade. USDA Foreign AgriculturalService, Washington, D.C.
282. USDA-FAS, 2012. MALAYSIA: Stagnating Palm Oil Yields Impede Growth.Commodity Intelligence Report, USDA Foreign Agricultural Service.Washington, D.C. (accessed on 24.06.15) URL: www.pecad.fas.usda.gov/highlights/2012/12/Malaysia/
283. .USDA-FAS, 2015. Production, Supply and Distribution online database, oil seeds,Table 11: Palm Oil: World Supply and Distribution. USDA Foreign Agriculturalservice, Washington, D.C. (accessed on 01.12.15) URL: http://apps.fas.usda.gov/ psdonline/ psdHome.aspx.
284. USDA-FAS, 2016. Production, Supply and Distribution Online: Oilseeds. USDAForeign Agricultural Service, Washington, D.C (accessed on 09.03.16) http://apps.fas.usda.gov/psdonline/psdDownload.aspx.
285. van Ittersum, M.K., Rabbinge, R., 1997. Concepts in production ecology for analysisand quantification of agricultural input-output combinations. Field Crops Res.52 (3), 197–208, http://dx.doi.org/10.1016/S0378-4290(97)00037-3.
286. van Ittersum, M.K., Cassman, K.G., Grassini, P., Wolf, J., Tittonell, P., Hochman, Z.,2013. Yield gap analysis with local to global relevance—a review. Field CropsRes. 143, 4–17, http://dx.doi.org/10.1016/j.fcr.2012.09.009.
287. van Noordwijk, M., Lawson, G., Hairiah, K., Wilson, J., 2015. Root distribution oftrees and crops: competition and/or complementarity. In: Ong, C.K., Black, C.R.,Julia, W. (Eds.), Tree-Crop Interactions: Agroforestry in a Changing Climate.CABI, Wallingford, pp. 221–257.
288. van der Linden, A., Oosting, S.J., van de Ven, G.W.J., de Boer, I.J.M., van Ittersum,M.K., 2015. A framework for quantitative analysis of livestock systems usingtheoretical concepts of production ecology. Agric. Syst. 139, 100–109, http://dx.doi.org/10.1016/ j.agsy.2015.06.007.
289. Virdiana, I., Hasan, Y., Aditya, R., Flood, J., 2010. Testing the effects of oil palmreplanting practices (windrowing, fallowing and poisoning) on incidence ofGanoderma. In: Proceedings of the International Oil Palm Conference, 1–3 June2010, Yogyakarta, Indonesian Oil Palm Research Institute.
290. Vossen, H.A.M.v.d., 1970. Nutrient status and fertilizer responses of oil palms inGhana. Ghana J. Agric. Sci. 3, 109–129.
291. Wahid, M.B., Simeh, M.A., 2010. Accelerated oil palm replanting: the way forwardfor a sustainable and competetive industry. Oil Palm Ind. Econ. J. 10 (2), 29–38.
292. Wairegi, L.W.I., Asten, P.J.A.v., 2012. Norms for multivariate diagnosis of nutrientimbalance in arabica and robusta coffee in the East African Highlands. Exp.Agric. 48 (3), 448–460, http://dx.doi.org/10.1017/50014479712000142.

293. Wairegi, L.W.I., Asten, P.J.A.v., Tenywa, M.M., Bekunda, M.A., 2010. Abioticconstraints override biotic constraints in East African highland bananasystems. Field Crops Res. 117 (1), 146–153.
294. Webb, M., 2009. A conceptual framework for determining economically optimalfertiliser use in oil palm plantations with factorial fertiliser trials. Nutr. Cycl.Agroecosyst. 83 (2), 163–178, http://dx.doi.org/10.1007/s10705-008-9207-x.
295. Witt, C., Fairhurst, T., Griffiths, W., 2005. Key principles of crop and nutrientmanagement in oil palm. Better Crops Plant Food 89, 27–31.
296. Woittiez, L.S., Slingerland, M., Giller, K.E., 2015. Yield gaps in Indonesiansmallholder plantations: causes and solutions. In: PIPOC 2015 InternationalPalm Oil Congress and Exhibition, 6–8 October, Kuala Lumpur.
297. Wood, B., 1968. Pests of Oil Palms in Malaysia and Their Control. IncorporatedSociety of Planters, Kuala Lumpur.
298. Wood, B.J., 1977. The economics of crop protection in oil palms. PANS 23 (3),253–267.
299. Wood, B.J., 1985. Some consequences of weevil pollination of the oil palm in SouthEast Asia. In: Oil Palm Symposium, 27–28 March 1985, Medan.
300. Wood, B.J., Chung, G.F., 2003. A critical review of the development of rat control inMalaysian agriculture since the 1960. Crop Prot. 22 (3), 445–461, http://dx.doi.org/10.1016/S0261-2194(02)00207-7.
301. Wood, B.J., Corley, R.H.V., Goh, K.H., 1973. Studies on the effect of pest damage onoil palm yield. In: Wastie, R.L., Earp, D.A. (Eds.), Advances in Oil PalmCultivation: Proceedings of the International Oil Palm Conference, 16–18November 1972. Incorporated Society of Planters, Kuala Lumpur, pp. 360–377.
302. Wood, B.J., Liau, S.S., 1984. A long-term study of Rattus tiomanicus populations in anoil palm plantation in Johore, Malaysia: II. Recovery from control and economicaspects. J. Appl. Ecol. 21 (2), 465–472, http://dx.doi.org/10.2307/2403422.
303. Yahya, A., Su, V.H., El Pebrian, D., Ishola, T.A., 2013. Performance and economiccomparisons of manual and mechinised fertiliser applications for mature oilpalm. Planter 89 (1046), 323–340.
304. Yang, W., Peng, S., Laza, R.C., Visperas, R.M., Dionisio-Sese, M.L., 2008. Yield gapanalysis between dry and wet season rice crop grown under high-yieldingmanagement conditions. Agron. J. 100 (5), 1390–1395, http://dx.doi.org/10.2134/agronj2007.0356.
305. Yeow, K.H., Tam, K.T., Poon, Y.C., Toh, P.Y., 1982. Recent innovation in agronomicpractice: palm replacement during immaturity. In: Pushparajah, E., Chew, P.S.(Eds.), The Oil Palm in Agriculture in the Eighties. Incorporated Society ofPlanters, Kuala Lumpur, pp. 357–365.
306. Zakaria, Z.Z., Foster, H.L., Mohammed, A.T., Dolmat, M.T., Bachik, A.T., Bidin, A.,1990. Yield responses to phosphate fertilizer in oil palm. In: Proceedings of theWorkshop on Phosphate Sources for Acid Soils in the Humid Tropics of Asia,6–7 November 1990, Malaysian Society of Soil Science, Kuala Lumpur, pp.104–109.
307. Zeven, A., 1967. The Semi-Wild Oil Palm and Its Industry in Africa. Centrum voorlandbouwpublikaties en landbouwdocumentatie, Wageningen.
308. Zin, Z.Z., Hamdan, A.B., Gurmit Singh, 1993. Palm oil extraction rates in peninsularMalaysia: the effect of fertilizers. In: National Seminar on Palm Oil ExtractionRate: Problems and Issues, Kuala Lumpur.
309. Zuidema, P.A., Leffelaar, P.A., Gerritsma, W., Mommer, L., Anten, N.P.R., 2005. Aphysiological production model for cocoa (Theobroma cacao): modelpresentation, validation and application. Agric. Syst. 84 (2), 195–225, http://dx doi.org/10. 1016/j.agsy.2004.06.015.

INDEX

O

P

R

S

W